EDA精品智汇馆

AMD FPGA 设计优化宝典

面向Vivado/VHDL

高亚军 编著

电子工业出版社
Publishing House of Electronics Industry
北京·BEIJING

内 容 简 介

本书以 Xilinx 公司（目前已被 AMD 公司收购）的 7 系列 FPGA、UltraScale/UltraScale+ FPGA 和 Versal ACAP 内部架构为基础，介绍与之匹配的 RTL 代码的风格（采用 VHDL 语言）和基于 Vivado 的设计分析方法。全书共 10 章，包括时钟网络、组合逻辑、触发器、移位寄存器、存储器、乘加运算单元和状态机的代码风格和优化方法，也包含扇出和布线拥塞的优化方法。

本书可供电子工程领域的本科生和研究生学习参考，也可供 FPGA 工程师和自学者参考使用。

未经许可，不得以任何方式复制或抄袭本书之部分或全部内容。
版权所有，侵权必究。

图书在版编目（CIP）数据

AMD FPGA 设计优化宝典：面向 Vivado/VHDL / 高亚军编著. —北京：电子工业出版社，2023.4（2025.9 重印）
（EDA 精品智汇馆）
ISBN 978-7-121-45098-3

Ⅰ. ①A… Ⅱ. ①高… Ⅲ. ①可编程序逻辑器件－系统设计 Ⅳ. ①TP332.1

中国国家版本馆 CIP 数据核字（2023）第 030059 号

责任编辑：张　楠　　　　特约编辑：田学清
印　　刷：北京七彩京通数码快印有限公司
装　　订：北京七彩京通数码快印有限公司
出版发行：电子工业出版社
　　　　　北京市海淀区万寿路 173 信箱　　邮编：100036
开　　本：787×1092　1/16　　印张：27.75　　字数：710.4 千字
版　　次：2023 年 4 月第 1 版
印　　次：2025 年 9 月第 6 次印刷
定　　价：125.00 元

凡所购买电子工业出版社图书有缺损问题，请向购买书店调换。若书店售缺，请与本社发行部联系，联系及邮购电话：(010) 88254888，88258888。
质量投诉请发邮件至 zlts@phei.com.cn，盗版侵权举报请发邮件至 dbqq@phei.com.cn。
本书咨询联系方式：(010) 88254579。

◇ 作者简介 ◇

高亚军，FPGA 技术分享者，设计优化和时序收敛专家，Vivado 工具使用专家，数字信号处理专家，现任 Xilinx 资深战略应用工程师，多年来使用 Xilinx FPGA 实现数字信号处理算法，对 Xilinx FPGA 器件架构、开发工具 Vivado/Vitis HLS/Model Composer 及其设计理念有深厚的理论和实战经验。

主要著作：

- 2011 年出版《基于 FPGA 的数字信号处理》
- 2012 年发布网络视频课程《Vivado 入门与提高》
- 2015 年出版《基于 FPGA 的数字信号处理（第 2 版）》
- 2016 年出版《Vivado 从此开始（基础篇）》
- 2016 年发布网络视频课程《跟 Xilinx SAE 学 HLS》
- 2020 年出版《Vivado 从此开始（进阶篇）》
- 2021 年出版《Vivado/Tcl 零基础入门与案例实战》
- 2018 年创建 FPGA 技术分享公众号：FPGA 技术驿站。每周更新一篇原创文章，累计发表原创文章 370 余篇，获得大量粉丝的认可和赞誉。

前 言

设计收敛是 FPGA 工程师面临的一个重要课题：既要保证功耗收敛，又要保证时序收敛。两者均与设计自身有很大关系。笔者在多年的工程实践和技术支持过程中发现，很多设计未能收敛都与代码风格或对 Vivado 工具的理解有很大关系。

就代码风格而言，目前已有越来越多的工程师意识到其重要性。往往"良好的代码风格"能起到事半功倍的效果。"良好的代码风格"的一个重要指标就是代码风格与 FPGA 内部结构相匹配，保证综合工具能够完美地推断出期望结果，而这一点被很多工程师忽略。例如，7 系列 FPGA 内部的 DSP48E1 是不支持异步复位的，如果乘法器使用了异步复位，那么相应的触发器是无法被吸收到 DSP48E1 内部的，不仅消耗了额外的触发器（SLICE 内部的触发器），还会导致时序恶化。鉴于此，本书以 FPGA 内部结构为基础，以 VHDL 语言为描述方式，结合大量实际案例，力求帮助读者深入理解两者之间的对应关系。Vivado 从 2015.3 版本开始支持 VHDL-2008，随着版本的不断升级，对 VHDL-2008 的支持力度也不断增大，并引入 VHDL-2008 的更多特性。相比于 VHDL-93 版本和 VHDL-87 版本，VHDL-2008 版本更灵活、更易用。因此，本书采用 VHDL-2008 版本，同时列出所用到的 VHDL-2008 新特性。

就工具而言，Vivado 的功能越来越强大，自身越来越智能，分析手段越来越多。这就需要工程师深入理解工具在各个阶段所提供的选项含义，能够在工程实践中正确、合理地使用这些选项对应的功能，同时，面对未能收敛的设计，能够找到其中的根本原因。为此，本书也介绍了什么是有缺陷的设计、如何借助 Tcl 脚本找到这些缺陷，以及如何解决这些缺陷。

全书共 10 章。第 1 章从 FPGA 发展历程的角度对 FPGA 技术进行了分析。第 2 章以时钟网络架构为基础阐述了优化时钟网络的经典方法，涉及改善时钟偏移和降低时钟抖动。第 3 章～第 7 章分别介绍了优化组合逻辑、优化触发器、优化移位寄存器、优化存储器和优化乘加运算单元的经典方法。每章第 1 节均重点介绍优化对象的基本结构，包含目前主流的 FPGA（7 系列 FPGA、UltraScale/UltraScale+ FPGA 和 Versal ACAP）。每章最后 1 节均重点介绍如何发现缺陷单元及如何处理这些缺陷。第 8 章介绍了优化状态机的经典方法，包括状态机的"两段式"和"三段式"描述方法、状态机编码方式、Vivado 提供的状态机编码选项等。第 9 章阐述了优化扇出的经典方法。第 10 章阐述了优化布线拥塞的经典方法。这两章以 Vivado 工具为核心，给出了相应的 Tcl 脚本。

全书既阐述了 FPGA 内部结构，又阐述了 RTL 代码风格（采用 VHDL 语言）；既介绍了以 Vivado 图形界面为主的分析手段，又介绍了以 Tcl 脚本为主的分析手段。力求帮助读者从设计输入和设计分析两个维度理解设计。

为便于读者阅读和理解书中内容，本书给出了 469 张图片、85 个表格、189 个 VHDL 代码片段、56 个 Tcl 脚本片段、65 条设计规则和 29 个应用案例。同时，为加深印象，本书在每章结束之后还列出一些常见问题留给读者思考，共有 101 个问题。

  FPGA 设计收敛不是一蹴而就的，既需要依靠大脑智慧，又需要工具协同，两者缺一不可。希望您阅读本书之后对 FPGA 架构和 RTL 代码风格的理解能够更上一层楼。

  您在阅读本书的过程中，如果发现书中内容有任何不当之处，或对本书内容有任何建议或意见，都可发送邮件到 laurengao@126.com，不胜感激。

  如果需要获取代码示例，可关注编著者微信公众号"FPGA 技术驿站"，回复关键字"设计优化 VHDL"即可。

高亚军

2022/9/12

目 录

第1章 FPGA 技术分析 ································· 1
1.1 芯片架构的演变 ································· 1
1.2 设计方法的演变 ································· 15
1.3 面临的挑战 ····································· 20
1.4 四大基本原则 ··································· 22
1.4.1 硬件原则 ···································· 23
1.4.2 同步原则 ···································· 24
1.4.3 流水原则 ···································· 25
1.4.4 面积与速度的平衡与互换原则 ·················· 27
1.5 性能指标 ······································· 29
1.6 思考空间 ······································· 31

第2章 优化时钟网络 ································· 32
2.1 时钟资源 ······································· 32
2.1.1 7系列 FPGA 中的时钟资源 ····················· 32
2.1.2 UlatraScale/UltraScale+ FPGA 中的时钟资源 ··· 42
2.1.3 Versal ACAP 中的时钟资源 ···················· 47
2.2 时钟偏移 ······································· 52
2.3 时钟抖动 ······································· 64
2.4 安全的时钟启动方式 ····························· 71
2.5 时钟规划 ······································· 75
2.6 创建输出时钟 ··································· 79
2.7 思考空间 ······································· 80

第3章 优化组合逻辑 ································· 81
3.1 组合逻辑资源 ··································· 81
3.2 译码器与编码器 ································· 82
3.2.1 译码器代码风格 ······························ 82
3.2.2 编码器代码风格 ······························ 93
3.3 多路复用器与多路解复用器 ······················· 104
3.3.1 多路复用器代码风格 ·························· 104
3.3.2 多路解复用器代码风格 ························ 117
3.4 加法器与累加器 ································· 119

         3.4.1 加法器代码风格 119
         3.4.2 累加器代码风格 134
   3.5 其他组合逻辑电路 149
         3.5.1 移位器代码风格 149
         3.5.2 比较器代码风格 153
         3.5.3 奇偶校验电路代码风格 166
         3.5.4 二进制码与格雷码互转电路代码风格 167
   3.6 避免组合逻辑环路 170
   3.7 思考空间 171

第 4 章 优化触发器 172
   4.1 触发器资源 172
         4.1.1 7 系列 FPGA 中的触发器资源 172
         4.1.2 UltraScale/UltraScale+ FPGA 中的触发器资源 174
         4.1.3 Versal ACAP 中的触发器资源 175
   4.2 建立时间和保持时间 179
   4.3 亚稳态 181
   4.4 控制集 184
   4.5 复位信号的代码风格 189
         4.5.1 异步复位还是同步复位 189
         4.5.2 全局复位还是局部复位 192
         4.5.3 是否需要上电复位 195
   4.6 同步边沿检测电路代码风格 199
   4.7 串并互转电路代码风格 201
   4.8 避免意外生成的锁存器 206
   4.9 思考空间 209

第 5 章 优化移位寄存器 211
   5.1 移位寄存器资源 211
         5.1.1 7 系列 FPGA 中的移位寄存器资源 211
         5.1.2 UltraScale/UltraScale+ FPGA 中的移位寄存器资源 212
         5.1.3 Versal ACAP 中的移位寄存器资源 212
   5.2 移位寄存器的代码风格 216
   5.3 移位寄存器的应用场景 227
   5.4 管理时序路径上的移位寄存器 228
   5.5 思考空间 232

第 6 章 优化存储器 234
   6.1 存储器资源 234

		6.1.1	分布式 RAM	234
		6.1.2	BRAM	235
		6.1.3	UltraRAM	242
	6.2	单端口 RAM 代码风格		246
	6.3	简单双端口 RAM 代码风格		266
	6.4	真双端口 RAM 代码风格		276
	6.5	RAM 的初始化与 ROM 代码风格		284
	6.6	同步 FIFO 代码风格		287
	6.7	异步 FIFO 代码风格		301
	6.8	平衡 BlockRAM 的功耗与性能		310
	6.9	异构 RAM		312
	6.10	以 IP 方式使用 RAM 和 FIFO		312
	6.11	以 XPM 方式使用 RAM 或 FIFO		319
	6.12	管理时序路径上的 BRAM 和 UltraRAM		322
	6.13	思考空间		328

第 7 章 优化乘加运算单元 329

	7.1	乘加器资源		329
		7.1.1	7 系列 FPGA 中的乘加器资源	329
		7.1.2	UltraScale/UltraScale+ FPGA 中的乘加器资源	332
		7.1.3	Versal ACAP 中的乘加器资源	332
	7.2	以乘法为核心运算的代码风格		335
	7.3	复数乘法运算代码风格		363
	7.4	向量内积代码风格		378
	7.5	以加法为核心运算的电路结构		380
	7.6	管理时序路径上的乘加器		386
	7.7	思考空间		387

第 8 章 优化状态机 388

	8.1	基本概念	388
	8.2	状态机代码风格	390
	8.3	状态编码方式	410
	8.4	基于 ROM 的控制器	413
	8.5	思考空间	416

第 9 章 优化扇出 417

	9.1	生成扇出报告	417
	9.2	利用设计流程降低扇出	419
	9.3	利用约束降低扇出	421
	9.4	从代码层面降低扇出	424

  9.5 改善扇出的正确流程 ·············· 424
  9.6 思考空间 ·············· 425

第 10 章 优化布线拥塞 ·············· 426
  10.1 布线拥塞的三种类型 ·············· 426
  10.2 利用设计流程改善布线拥塞 ·············· 428
  10.3 利用约束缓解布线拥塞 ·············· 429
  10.4 从代码层面降低布线拥塞程度 ·············· 430
  10.5 缓解布线拥塞的正确流程 ·············· 430
  10.6 思考空间 ·············· 432

第 1 章

FPGA 技术分析

1.1 芯片架构的演变

FPGA，英文全称是 Field Programmable Gate Array，中文意思是现场可编程门阵列，是在 PAL（Programmable Array Logic，可编程阵列逻辑）、GAL（Generic Array Logic，通用阵列逻辑）、CPLD（Complex Programmable Logic Device，复杂可编程逻辑器件）等可编程器件的基础上进一步发展得来的。作为 ASIC（Application-Specific Integrated Circuit，专用集成电路）领域中的一种半定制电路，FPGA 既弥补了全定制电路的不足，又克服了原有可编程器件门电路数目有限的缺点。

FPGA 的两个主要特征就体现在其名称中，即"现场"和"可编程"。"现场"意味着工程师只需要一台电脑，安装了 FPGA 厂商提供的开发软件，焊接了 FPGA 芯片的板卡，即可完成从模型搭建到生成烧写文件的全过程。"可编程"意味着工程师可以根据设计需求，通过编程方式把 FPGA 内部逻辑块连接起来，这就好比把一个电路试验板放在一颗芯片里。一个出厂后的成品 FPGA 内部的逻辑块和连接方式可以按照工程师的需要而改变，所以 FPGA 可以完成特定的逻辑功能。更重要的是这种编程是可以反复操作的，即可以编程、除错、再编程。因此，FPGA 也称为"液体硬件"。

一般情况下，FPGA 比 ASIC 的速度要慢，若要实现同样的功能，FPGA 需要的面积更大，功耗也更高。但相比 ASIC，FPGA 的优势也很明显，具体如下。①快速成品：这有助于缩短产品上市时间；②反复编程：这使得工程师可以根据设计需求修改电路功能，同时有助于除错；③较低的调试成本：只要有厂商开发工具（如 Vivado）、下载器（如 Xilinx Platform Cable Usb）、FPGA 板卡，即可进行调试。

FPGA 芯片起源于 Xilinx（赛灵思）公司。1985 年，首款 FPGA（赛灵思 XC2064）采用双列直插式封装，也称为 DIP 封装（Dual In-line Package），如图 1-1 所示，只包含 64 个可配置逻辑模块，每个模块含有两个 3 输入查找表（Look-Up Table，LUT，后面介绍时，"查找表"和"LUT"会交替使用）和一个触发器（Flip-Flop，FF，后面介绍时，"触发器"和"FF"会交替使用）。按照现在的方式计算，该器件有 64 个逻辑单元，不足 1000 个逻辑门。这也构成了早期 FPGA 的基本架构：可配置逻辑模块（Configurable Logic Block，CLB）、开关矩阵（Switch Matrix，也称为 Switch Box）、互连线（Interconnect Wires）和输入/输出管脚，如图 1-2 所示。较为简单的结构决定了其主要扮演胶合逻辑的角色。

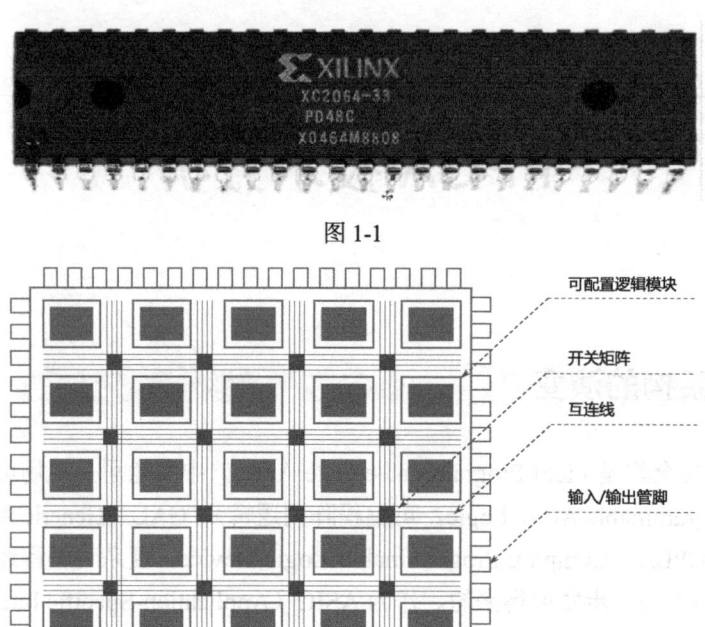

图 1-1

图 1-2

在摩尔定律的推动下，Xilinx 于 2001 年推出了基于 150nm 工艺制程的 Virtex-II FPGA，其最大亮点是嵌入了硬核的 18×18 的有符号数乘法器和 18Kb（1Kb=1024bit）的 Block RAM（简写为 BRAM），如图 1-3 所示（图中未包含布线资源和输入/输出管脚）。其中，XC2V8000 中包含 168 个乘法器和 168 个 BRAM，这大大提升了 FPGA 的计算能力。

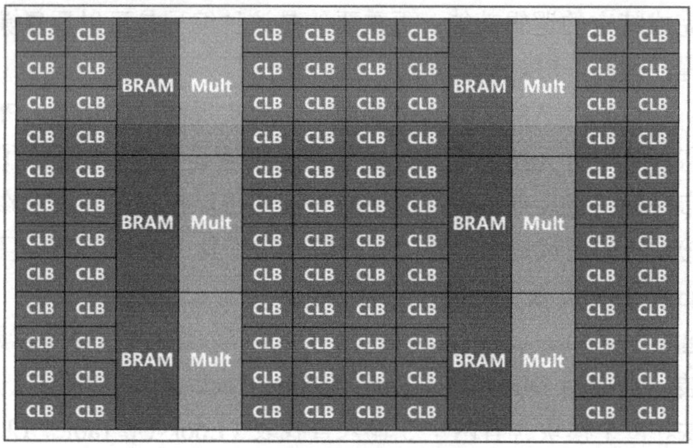

图 1-3

时隔一年，也就是 2002 年，Xilinx 推出了基于 130nm 工艺制程的 Virtex-II Pro FPGA，在 Virtex-II 的基础上嵌入了硬核 PowerPC 405 处理器（简写为 PPC405），这是当时 IBM 公司的 RISC CPU（Reduced Instruction Set Computer，精简指令集中央处理器），主频最高可达 400MHz。同时，在接口方面，Virtex-II Pro 集成了硬核高速收发器，传输速率最高可达 6.25Gbit/s，如图 1-4 所示。XC2VP70 所包含的高速收发器个数达到了 20 个。

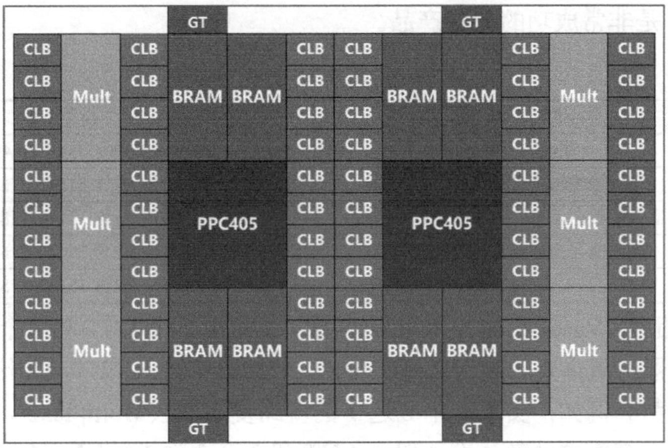

图 1-4

2004 年，Xilinx 推出了基于 90nm 工艺制程的 Virtex-4 FPGA。采用了基于列的 ASMBL（Advanced Silicon Modular Block）架构，如图 1-5 所示，大大提升了 FPGA 的可编程能力，使得其成为 ASIC 的强有力的替代者。FPGA 也由胶合逻辑的角色变为平台级产品。通过不同列（CLB 列、DSP 列、BRAM 列、GT 列等）的组合，Virtex-4 FPGA 形成了 3 个平台级家族：LX、FX 和 SX，给用户提供了多种选择，以应对不同的应用场景。其中，LX 面向高性能逻辑应用，CLB 列会更多一些，XC4VLX200 中的逻辑单元（Logic Cell，1 个 Logic Cell 等于 1 个 4 输入 LUT+1 个触发器+1 个进位链）达到了 200 448 个。SX 面向高性能数字信号处理，DSP 列会更多一些，XC4VSX55 中的乘加器多达 512 个。FX 面向嵌入式应用，增加了 GT 列，内部嵌入了硬核 PowerPC 处理器 PPC405，主频可达 450MHz。和 Virtex-II Pro 相比，Virtex-4 FPGA 扩展了乘法器的功能，将乘法器变为包含 48 位累加器的乘加器，可将其看作第一代 DSP48。DSP48 在数字信号处理应用中扮演着重要的角色。例如，在高性能 FIR（Finite Impulse Response）滤波器和 FFT（Fast Fourier Transform）中，DSP48 是重要的计算单元。此外，从功耗的角度看，Virtex-4 的核电压也由 Virtex-II 和 Virtex-II Pro 的 1.5V 降至 1.2V，这对于降低动态功耗是有益的，因为动态功耗与核电压的平方成正比。

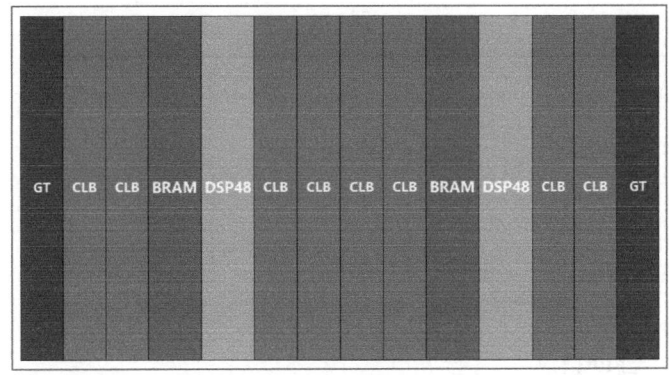

图 1-5

2006 年，Xilinx 推出了基于 65nm 工艺制程的 Virtex-5 FPGA。依托于第二代 ASMBL 架构，Virtex-5 形成了 5 个平台级家族系列，以满足不同领域的需求，如表 1-1 所示。实践证明，Virtex-5 是非常成功的一代产品。

表 1-1

Virtex-5 系列	应用场景	代表
LX	高性能逻辑应用	XC5VLX330：包含 51840 个 SLICE
LXT	高性能逻辑应用和高速数据传输	XC5VLX330T：包含 51840 个 SLICE 和 24 个 GT
SXT	高性能数字信号处理和高速数据传输	XC5VSX240T：包含 37440 个 SLICE 和 24 个 GT
TXT	高性能、高密度、高速数据传输	XC5VTX240T：包含 37440 个 SLICE 和 48 个 GT
FXT	高性能嵌入式系统和高速数据传输	XC5VFX200T：包含 30720 个 SLICE 和 24 个 GT

与 Virtex-4 相比，Virtex-5 的 SLICE 的容量扩大，由 2 个 LUT 和 2 个触发器变为 4 个 LUT 和 4 个触发器。更重要的是 Virtex-5 中的 LUT 是 6 输入 LUT，相比 Virtex-4 中的 4 输入 LUT 更有优势。例如，实现一个 8 选 1 的多路复用器（Multiplexer，MUX），如果使用 Virtex-4 FPGA，就需要消耗 4 个 4 输入 LUT 和 3 个 2 选 1 MUX（SLICE 里固有的）。如果使用 Virtex-5 FPGA，那么只需要消耗 2 个 6 输入 LUT 和 1 个 2 选 1 MUX（SLICE 里固有的），如图 1-6 所示，显然，后者的逻辑延迟会更小。分别选用 Virtex-4 和 Virtex-5 最快的 FPGA（对应速度等级分别为-12 和-3），在 ISE 8.1 版本下，实现一个 6 输入布尔表达式，Virtex-4 FPGA 的逻辑延迟为 1.1ns，而 Virtex-5 FPGA 的逻辑延迟为 0.9ns。从分布式 RAM 的角度看，实现一个 64×1（深度为 64，宽度为 1）的 RAM，Virtex-4 与 Virtex-5 的实现方式如图 1-7 所示，显然 Virtex-5 的实现方式具有更小的延迟。

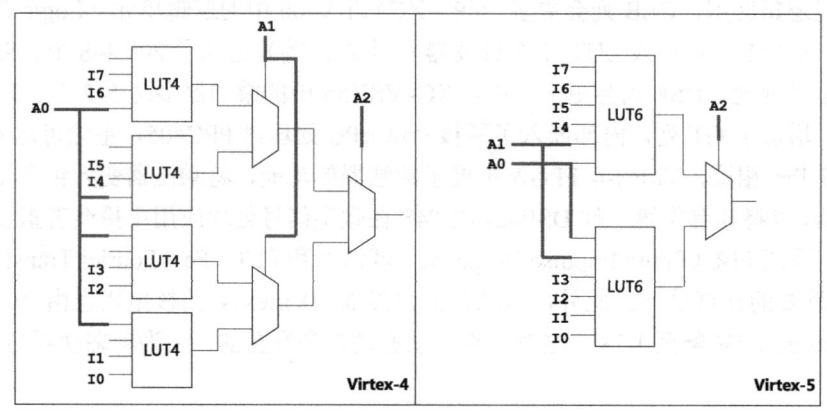

图 1-6

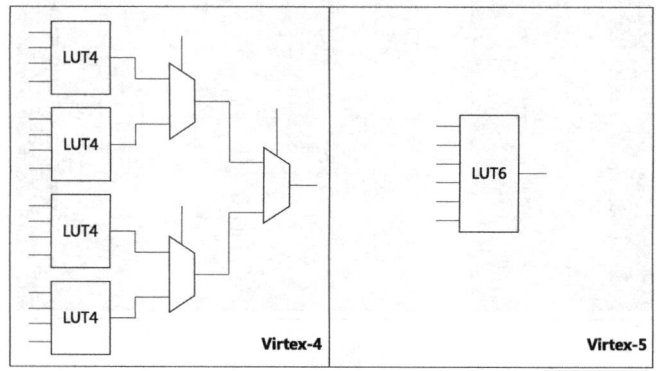

图 1-7

Virtex-5 中的 BRAM 也有所改进。Virtex-4 的每个 BRAM 大小为 18Kb，而 Virtex-5 的每个 BRAM 大小为 36Kb，由 2 个独立的 18Kb BRAM 构成。这对于搭建更大的存储空间非常有利。同时，对于每个独立的 18Kb BRAM，如果只使用了其中的 9Kb（或小于 9Kb），那么剩余的 9Kb（或大于 9Kb）会被自动关闭，从而可以有效降低功耗。

相比 Virtex-4，Virtex-5 中的乘加器 DSP48E 的功能更强大。不仅乘法器的位宽由 18×18 提升为 25×18，而且每个 DSP48E 的 C 端口是独立的（Virtex-4 中同列相邻的两个 DSP48 共享 C 端口），这增强了 DSP48E 的灵活性，也简化了数字信号处理算法的实现。此外，不得不提的是，DSP48E 中的累加器还可用来实现按位逻辑运算，如逻辑与、逻辑或、逻辑异或、逻辑非等。确切地说，DSP48E 中的乘法器后面跟随的是算术逻辑运算单元（Arithmetic Logic Unit，ALU），既可以实现加法运算，又可以实现逻辑运算。

在时钟方面，Virtex-5 在保留 Virtex-4 数字时钟管理单元（Digital Clock Manager，DCM）的基础上，引入了模拟锁相环（Phased Lock Loop，PLL），这对于改善时钟抖动非常有利。

此外，Virtex-5 FXT 系列嵌入了 PowerPC 440（简写为 PPC440）处理器，主频最高可达 550MHz，性能比 Virtex-4 中的 PPC405 有了进一步的提升。

2009 年，Xilinx 推出了基于 40nm 工艺制程的 Virtex-6 FPGA 和基于 45nm 工艺制程的 Spartan-6 FPGA。Virtex-6 采用了第三代 ASMBL 架构，通过不同资源列的配比，形成了三个系列：面向高性能逻辑应用和高速数据传输的 LXT 系列、面向高性能数字信号处理和高速数据传输的 SXT 系列、面向高带宽应用的 HXT 系列。与 Virtex-5 相比，所有的 Virtex-6 FPGA 都带有高速收发器，同时，Virtex-6 不再提供嵌入 PowerPC 处理器的 FPGA。Spartan-6 属于 Xilinx 的低端产品，与 Virtex-6 相比，性能略逊一筹。例如，Spartan-6 中的乘加器 DSP48A1 的乘法器位宽为 18×18，且其后的累加器不支持逻辑运算，而 Virtex-6 中的乘加器 DSP48E1 的乘法器位宽为 25×18，其后为 ALU。Spartan-6 中的 BRAM 大小为 18Kb，可当作两个独立的 9Kb BRAM 使用。Virtex-6 中的 BRAM 大小为 36Kb，由两个独立的 18Kb BRAM 构成。Spartan-6 中的时钟模块由 DCM 和 PLL 构成，而 Virtex-6 中的时钟模块为 MMCM（Mixed-Mode Clock Managers）。但是，Spartan-6 也有自己的特色，如封装尺寸小，CPG196 的封装尺寸只有 8mm×8mm，而 Virtex-6 的最小封装尺寸也要达到 23mm×23mm。此外，Spartan-6 还集成了硬核存储器控制器模块（Memory Controller Blocks，MCB）。从输入/输出管脚的角度看，Spartan-6 的管脚个数依据不同芯片和封装尺寸在 102～576 间变化，输出电平最大为 3.3V；Virtex-6 的管脚个数在 240～1200 间变化，输出电平最大为 2.5V。

2011 年，Xilinx 推出了基于 28nm 高性能低功耗（High Performance Low Power，HPL）工艺制程的 7 系列 FPGA，分为高端、中端和低端三类产品，对应 Virtex-7、Kintex-7 和 Artix-7。其中，Virtex-7 面向对系统性能要求比较苛刻的场合；Kintex-7 可提供业界最佳的价格/性能比；Artix-7 面向低功耗且需要使用高速收发器的场合。2017 年，Xilinx 又发布了 Spartan-7，具有封装尺寸小（CPGA196 的封装尺寸只有 8mm×8mm）、性能/功耗比高的特点。不同于 Virtex-6 和 Spartan-6，7 系列 FPGA 采用了统一的内部架构。图 1-8 对 7 系列 FPGA 从逻辑单元个数（K，1K=1024，下同）、BRAM 容量、DSP 个数、高速收发器个数、高速收发器速率和外部存储器速率几个方面进行了比较（这里显示的均为最大值），可进一步理解其中的差异及各类产品的应用场合。

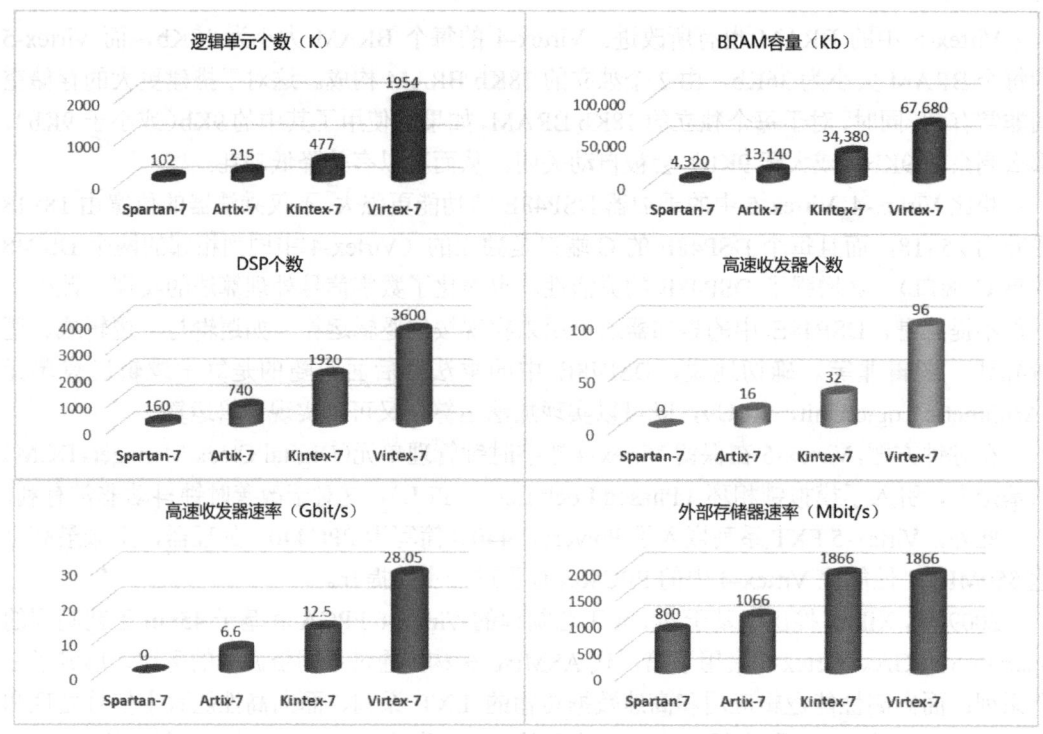

图 1-8

7 系列 FPGA 还首次使用了 SSI（Stacked Silicon Interconnect，堆叠硅片互联）技术，包括 XC7V2000T、XC7VX1140T、XC7VH580T 和 XC7VH870T 4 颗芯片。SSI 技术有效解决了采用传统方式互联多片 FPGA 时导致的互联资源有限（如输入/输出管脚个数不够）、互联延迟过大、高速串行互联的信号完整性等问题。以 XC7V2000T 为例，整个芯片包含 4 个 SLR（Super Logic Region，又称为 die），这 4 个 SLR 被并行放置在一个硅中介板上，如图 1-9 所示。再看 4 个 SLR 的内部布局，如图 1-10 所示。可以看到，4 个 SLR 的大小和内部结构是完全相同的。图 1-10 中的 GTX 代表的高速收发器的最大速率为 12.5Gbit/s。每个 SLR 由 3 个 CR（Clock Region，时钟区域）构成。每个 CR 的高度为 50 个 CLB，也可以表示为 20 个 DSP 或 10 个 36Kb BRAM，如图 1-11 所示。

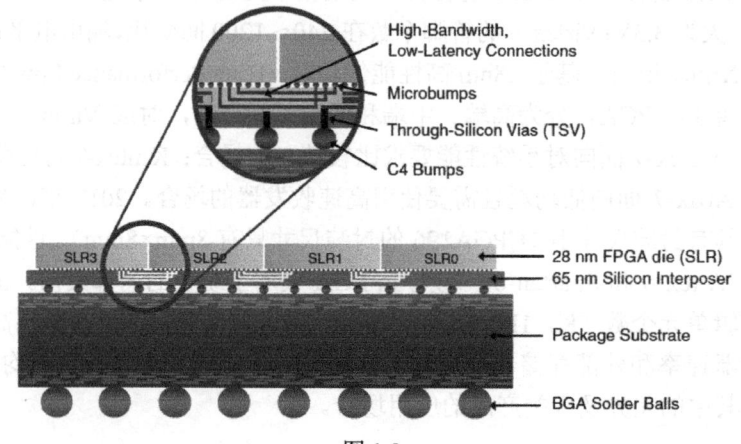

图 1-9

在图 1-9 中，英文对应释义如下。High-Bandwidth, Low-Latency Connections，高带宽低延迟互连线；Microbumps，微凸块；Through-Silicon Vias（TSV），硅穿孔；C4 Bumps，C4 凸块；28nm FPGA die，28nm FPGA 超级逻辑区域；65nm Silicon Interposer，65nm 硅中介板；Package Substrate，封装基板；BGA Solder Balls，BGA 锡球。

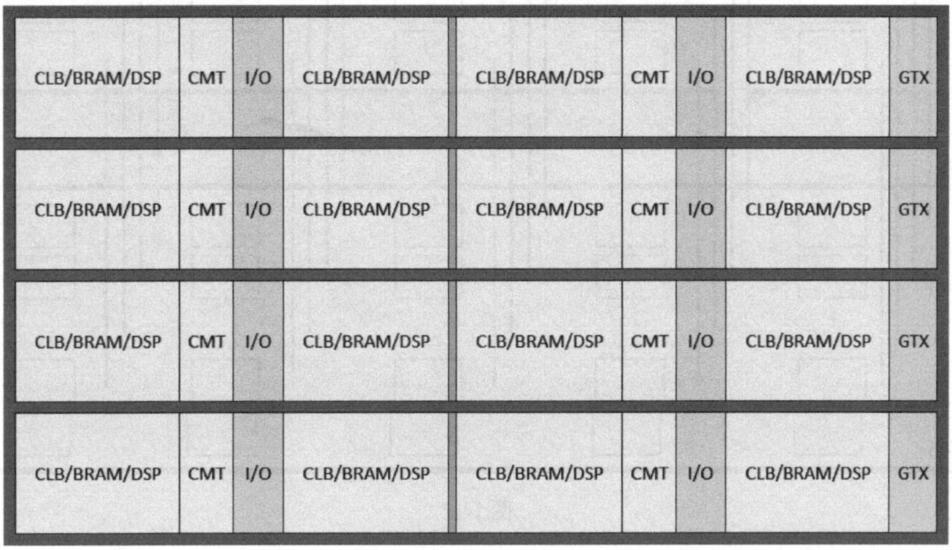

CLB/BRAM/DSP	CMT	I/O	CLB/BRAM/DSP	CLB/BRAM/DSP	CMT	I/O	CLB/BRAM/DSP	GTX
CLB/BRAM/DSP	CMT	I/O	CLB/BRAM/DSP	CLB/BRAM/DSP	CMT	I/O	CLB/BRAM/DSP	GTX
CLB/BRAM/DSP	CMT	I/O	CLB/BRAM/DSP	CLB/BRAM/DSP	CMT	I/O	CLB/BRAM/DSP	GTX
CLB/BRAM/DSP	CMT	I/O	CLB/BRAM/DSP	CLB/BRAM/DSP	CMT	I/O	CLB/BRAM/DSP	GTX

图 1-10

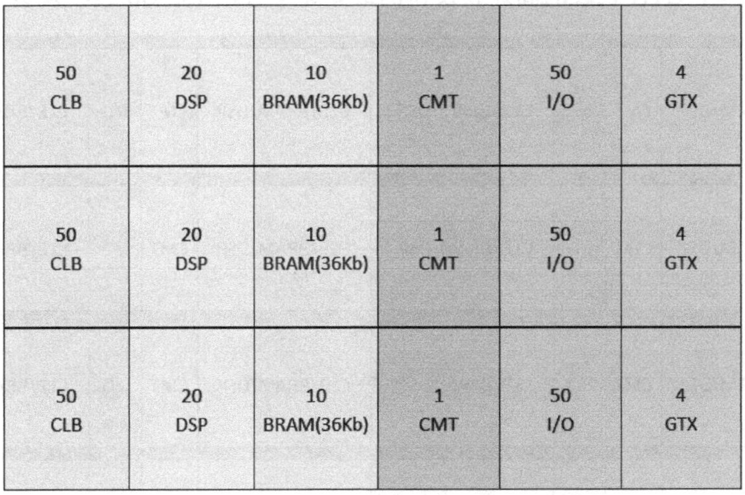

50 CLB	20 DSP	10 BRAM(36Kb)	1 CMT	50 I/O	4 GTX
50 CLB	20 DSP	10 BRAM(36Kb)	1 CMT	50 I/O	4 GTX
50 CLB	20 DSP	10 BRAM(36Kb)	1 CMT	50 I/O	4 GTX

图 1-11

SLR 之间通过（Super Long Line，SLL）互连，实现 die 间的数据传输。紧邻每个 CLB 会有一个互联单元（Interconnect Tile），SLL 的起点和终点正是这些位于相邻 SLR 的不同互联单元。每个互联单元列可提供 49 个 SLL，如图 1-12 所示，不同芯片互联单元列的个数有所不同，因此 7 系列 SSI 器件相邻 SLR 的 SLL 总个数为 10800 或 13700，从而保证了 die 间的高带宽数据传输。

图 1-12

XC7VX1140T 内部结构如图 1-13 所示，XC7VX1140T 由 4 个 SLR 构成，有更多的高速收发器（图中，GTH 代表的高速收发器的最高速率可达 13.1Gbit/s，共 96 个）。

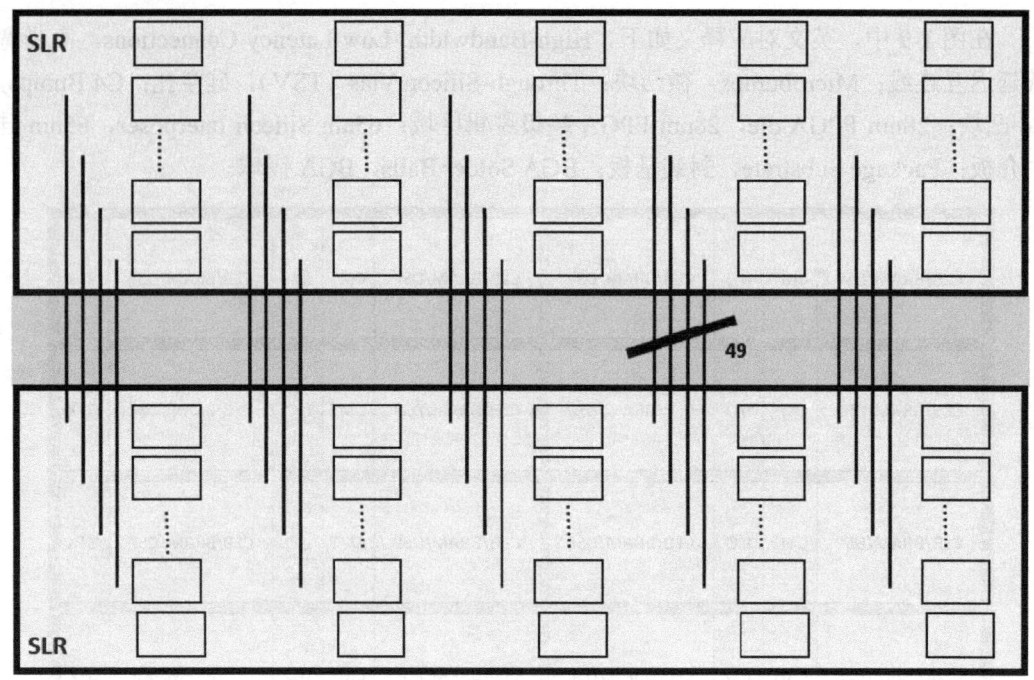

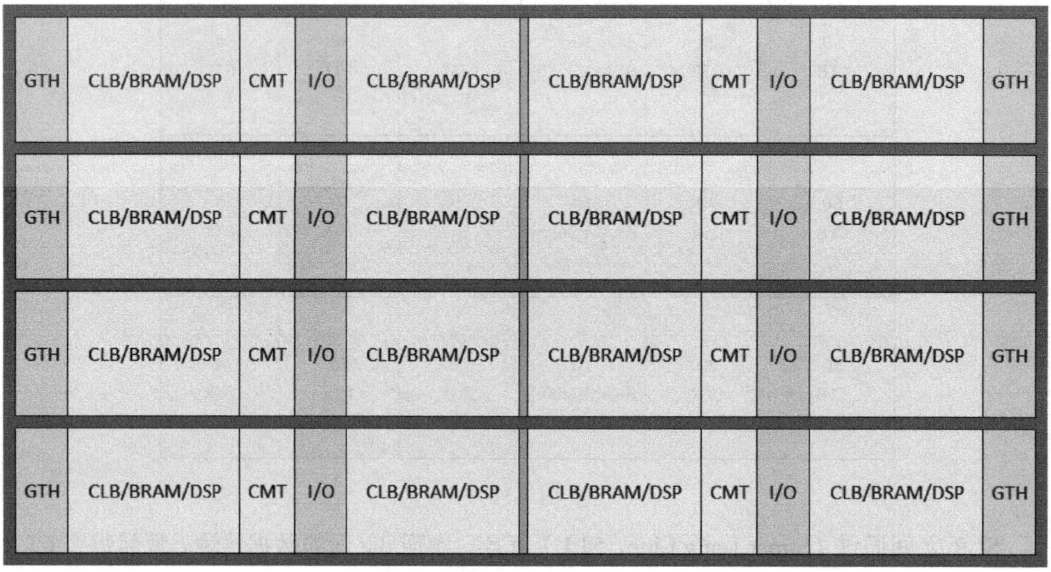

图 1-13

面向高带宽数据传输的 XC7VH870T 由 6 个 SLR 构成，其中，顶部和底部的 SLR 仅包含 GTZ（高速收发器，共 16 个，最高速率可达 28.05Gbit/s），如图 1-14 所示，因此，这是一颗由异构 SLR 构成的芯片。

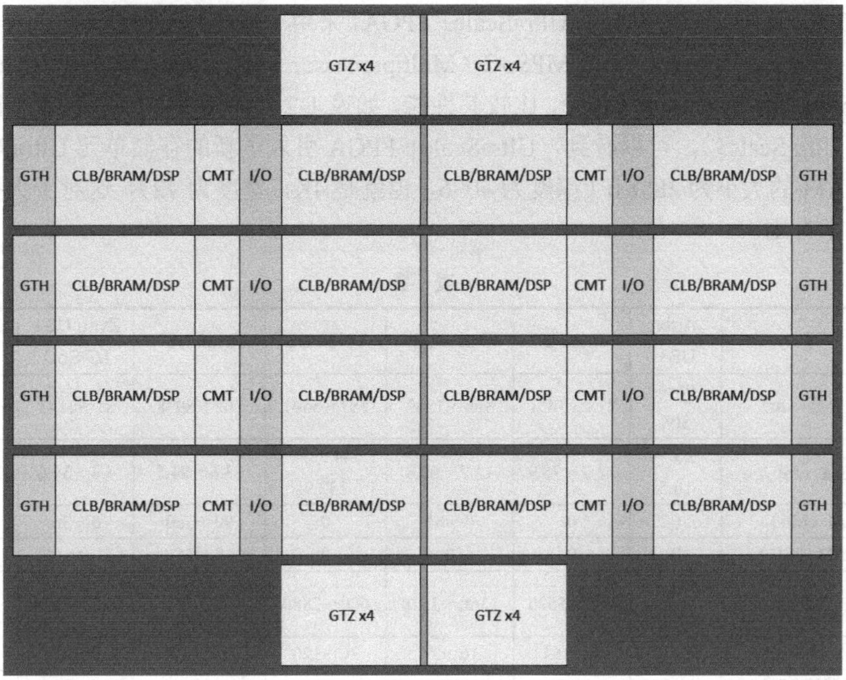

图 1-14

在推出传统 FPGA 的同时,2011 年,Xilinx 也发布了集成 ARM Cortex-A9 CPU 的 Zynq-7000 系列 SoC(System on Chip)。将 FPGA 分割为 PS(Processing System)和 PL(Programmable Logic)两部分,实现了全可编程:软件可编程、硬件可编程和 I/O 可编程。其中,PS 包含一个或两个 ARM Cortex-A9 核及丰富的硬核接口,CPU 主频最高可达 1GHz。PL 为传统的逻辑资源,如 CLB、BRAM、DSP 和 GT 等,如图 1-15 所示,图中,灰色部分表示 PL。Zynq-7000 因其异构特性在 Xilinx FPGA 发展史上留下了浓墨重彩的一笔。

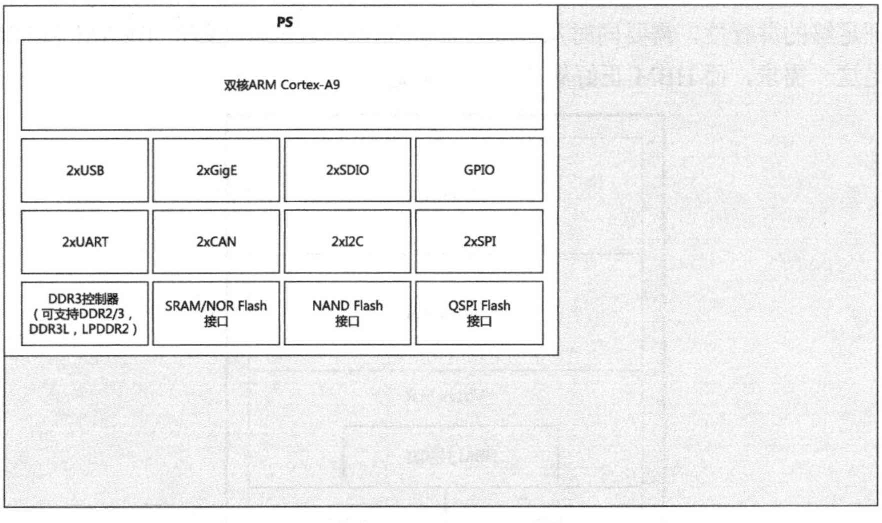

图 1-15

2014 年,Xilinx 推出了基于 20nm 工艺制程的 UltraScale FPGA,包括 Kintex 和 Virtex 两个系列。与 Kintex 相比,Virtex 无论是在规模上还是在性能上都更胜一筹。时隔两年,也就是2016年，Xilinx又推出了UltraScale+ FPGA，采用16nm工艺制程，包括Artix（2021年推出）、Kintex、Virtex、Zynq MPSoC（Multiprocessor System on Chip）和Zynq RFSoC（Radio Frequency System on Chip），比较其性能，如表1-2所示（表中，US代表UltraScale，US+代表UltraScale+）。可以看到，UltraScale+ FPGA引入了新的存储单元UltraRAM，每个UltraRAM的大小为288Kb（深度为4096，也就是4K，宽度为72），这对于大批量的数据存储非常有用。

表1-2

内 容	Artix US+	Kintex US	Kintex US+	Virtex US	Virtex US+	Zynq US+ MPSoC	Zynq US+ RFSoC
逻辑单元个数（K）	96～308	318～1451	356～1843	783～5541	862～8938	81～1143	489～930
BRAM容量（Mb）	3.5～10.5	12.7～75.9	12.7～60.8	44.3～132.9	23.6～94.5	3.8～34.6	22.8～38.0
UltraRAM（Mb）	0	0	0～81	0	90～360	0～36	13.5～45.0
HBM DRAM（GB）	0	0	0	0	0～16	0	0
DSP个数	400～1200	768～5520	1368～3528	600～2880	1320～12288	216～3528	1872～4272
高速收发器个数	8～12	12～64	16～76	36～120	32～128	0～72	8～16
高速收发器速率（Gbit/s）	16.30	16.30	32.75	30.50	58.00	32.75	32.75
外部存储器速率（Mbit/s）	2400	2400	2666	2400	2666	2666	2666
管脚个数	128～304	312～832	280～668	338～1456	208～2072	82～668	152～408

Vritex UltraScale+还引入了HBM（HighBandwidth Memory），有效地增大了数据存储空间及数据传输带宽，其内部布局如图1-16所示，仍然采用了SSI技术，HBM位于单独的一个SLR上。在与其紧邻的SLR内，集成了硬核的HBM控制器。Vritex UltraScale+ HBM FPGA内HBM的容量对比如图1-17所示，可以看到，HBM最小为4GB（也就是32Gb），最大为16GB（也就是128Gb）。在执行机器学习算法时，通常需要存储大量的权值，而且为了保证足够的并行性，需要同时从存储空间读出多个数据，传统的BRAM和UltraRAM很难满足这一需求，而HBM正好解决了这一问题。

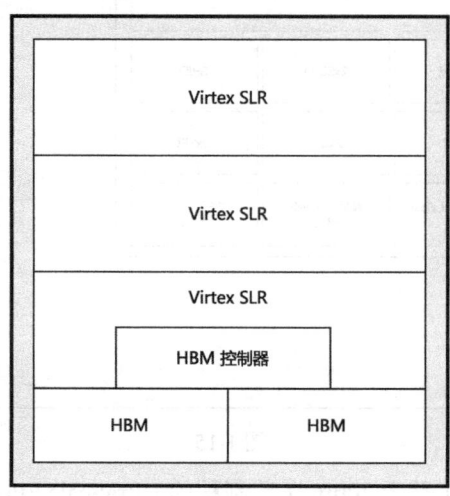

图1-16

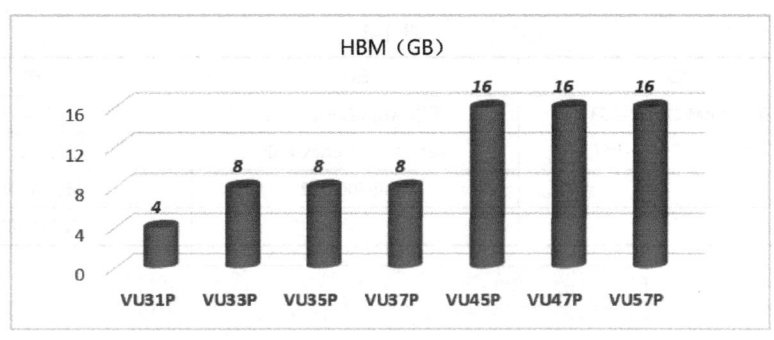

图 1-17

这里不得不提一下 Xilinx 第二代 SSI 芯片。Kintex UltraScale FPGA 只有 KU085 和 KU115 为 SSI 芯片，均包含 2 个 SLR。Virtex UltraScale FPGA 有 4 颗芯片为 SSI 芯片，分别为 VU125（2 个 SLR）、VU160（3 个 SLR）、VU190（3 个 SLR）和 VU440（3 个 SLR）。而 Virtex UltraScale+ FPGA 的芯片绝大多数为 SSI 芯片。与第一代 SSI 芯片相比，第二代 SSI 芯片不仅有专用跨 die 布线资源（SLL），还有专用跨 die 寄存器（LAGUNA 寄存器），如图 1-18 所示。在相邻两个 die 的边界，每个时钟区域内有 2 列 LAGUNA 寄存器，SLL 起点和 SLL 终点位于同列不同 die 的 LAGUNA 寄存器上。每 2 列 LAGUNA 寄存器，也就是 die 边界的每个时钟区域可提供 2880 个 SLL，在图 1-18 中，每个 SLR 位于边界的时钟区域共有 6 个，因此可提供 17280 个 SLL。

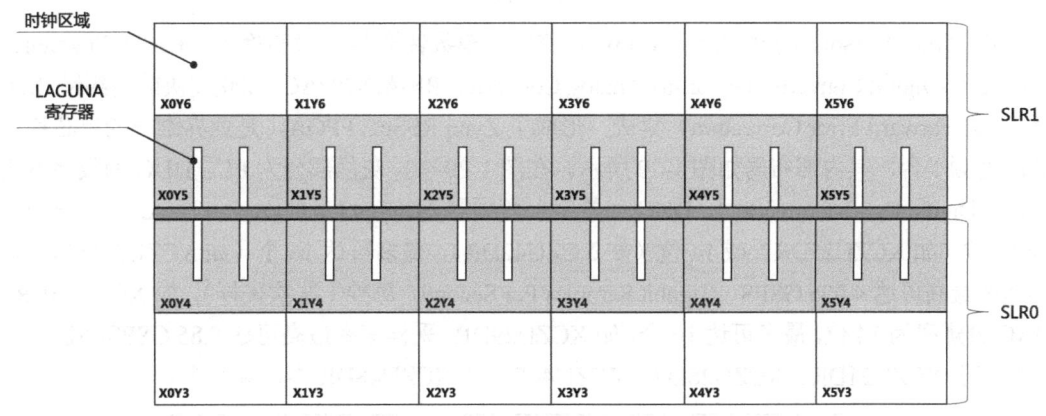

图 1-18

作为第二代嵌入了 ARM 核的 FPGA，Zynq UltraScale+MPSoC 的性能进一步提升，共有 3 个系列：CG、EG 和 EV。ARM 核由 Cortex-A9 升级为 Cortex-A53，采用双核（CG）或四核（EG 和 EV），扮演着 APU（Application Processing Unit，应用处理单元）的角色，主频可达 1.5GHz。同时，嵌入了 RPU（Real-time Processing Unit，实时处理单元），采用双核 ARM Cortex-R5F，主频最高可达 600MHz。EG 和 EV 系列还嵌入了 GPU（Graphics Processing Unit，图形处理单元）Mali-400MP2，主频最高可达 667MHz。在此基础上，EV 系列增强了视频处理功能，嵌入了硬核视频编解码处理单元 VCU（Video Encoder/Decoder），可支持 H.264/H.265，如表 1-3 所示。以 EV 系列为例，其内部布局如图 1-19 所示，图中，灰色部分为 PL。

表 1-3

	CG	EG	EV
APU	双核 ARM Cortex-A53	四核 ARM Cortex-A53	四核 ARM Cortex-A53
RPU	双核 ARM Cortex-R5F	双核 ARM Cortex-R5F	双核 ARM Cortex-R5F
GPU		Mali-400MP2	Mali-400MP2
VCU			H.264/H.265

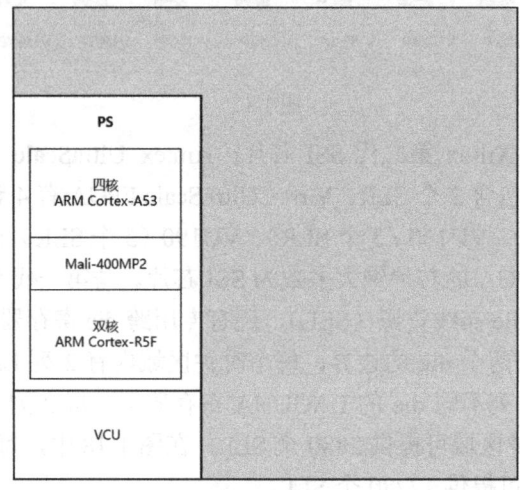

图 1-19

在 Zynq MPSoC 的基础上，Xilinx 还将射频模数转换器/数模转换器（Radio Frequency Analog-to-Digital Converter/Digital-to-Analog Converter，RF-ADC/DAC）和软判决前向纠错（Soft Decision Forward Error Correction）集成，形成了 Zynq RFSoC FPGA，是业界唯一的单芯片自适应射频平台，其内部布局如图 1-20 所示。在图 1-20 中，灰色部分为 PL。DDC 为数字下变频器（Digital Down Converter），DUC 为数字上变频器（Digital Up Converter）。ADC 的分辨率为 12 位（如 XCZU25DR）或 14 位（如 XCZU42DR），最多可达 16 个（如 XCZU29DR）。采样速率最高可达 4.096 GSPS（Gigabit Samples Per Second，每秒千兆次采样），如 XCZU25DR。DAC 分辨率为 14 位，最多可达 16 个，如 XCZU29DR，采样速率最高可达 9.85 GSPS。SD-FEC 仅存在于 XCZU21DR、XCZU28DR、ZCZU46DR 和 XCZU48DR 中，共 8 个。

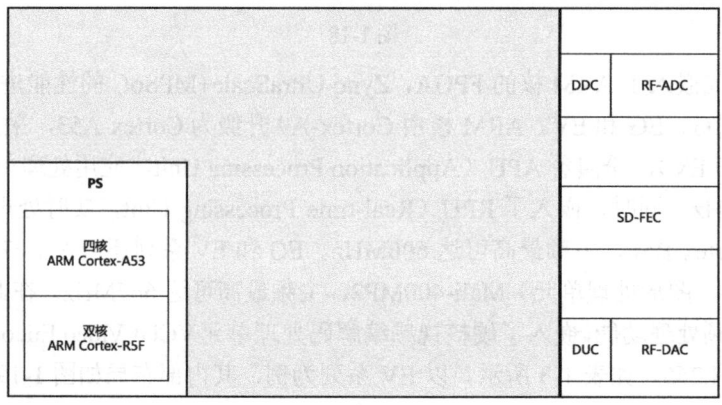

图 1-20

2019 年，Xilinx 推出了基于 7nm 工艺制程的 Versal 芯片，将其定义为 ACAP（Adaptive Compute Acceleration Platform，自适应计算加速平台），而不再是 FPGA。ACAP 包含三大引擎：标量引擎、自适应引擎和智能引擎，通过可编程片上网络（Programmable Network on Chip，可编程 NoC）将三者连接在一起，如图 1-21 所示。其中，标量引擎由双核 ARM Cortex-A72 和双核 ARM Cortex-R5F 构成，主频可分别达到 1.7GHz 和 800MHz；自适应引擎由传统的 FPGA 逻辑资源构成，包括 CLB、BRAM 和 UltraRAM；智能引擎主要提供计算功能，由 AIE（Artificial intelligence Engine）和 DSP58 构成。

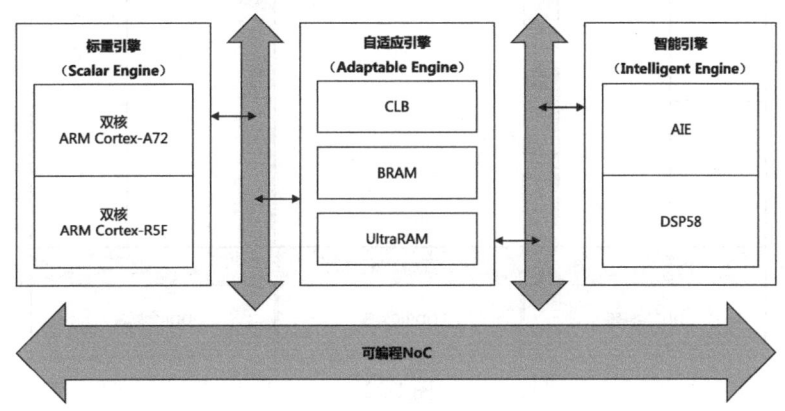

图 1-21

目前，Versal ACAP 已推出 5 个系列：AI Edge 系列、AI Core 系列、Prime 系列、Premium 系列和 HBM 系列，其性能对比如表 1-4 所示。表 1-4 中的存储空间是指分布式 RAM，即 LUTRAM、BRAM 和 UltraRAM 的总和。AI Edge 系列嵌入了针对机器学习推理应用而优化的 AIE-ML，AI Core 系列嵌入的则是第一代 AIE。其他系列没有嵌入 AIE。

表 1-4

内　　容	AI Edge 系列	AI Core 系列	Prime 系列	Premium 系列	HBM 系列
可编程 NoC	√	√	√	√	√
逻辑单元个数（K）	44～1139	540～1968	329～2233	1575～7352	3837～5631
存储空间（Mb）	40～177	90～191	54～282	198～994	509～752
DSP58 个数	90～1332	928～1968	464～3984	1904～14352	7392～10848
AIE/AIE-ML	8～304	128～400	0	0	0
高速收发器个数	0～44	8～44	8～48	48～168	88～128
管脚个数	114～530	478～770	316～770	586～780	780
硬核外部存储器控制器个数	1～3	2～4	1～4	3～4	4
HBM 容量（GB）	0	0	0	0	8～32

以 AI Core 系列为例，芯片内部布局如图 1-22 所示。其中，AIE 阵列位于芯片最上方，DDR 控制器位于最下方，两者通过可编程 NoC 实现数据传输。不难看出，与前一代 UltraScale+ FPGA 相比，Versal 中的硬核越来越多。为了发挥这些硬核的性能，需要在设计初期研究芯片架构，规划好整个系统的布局。

AIE 阵列中的每个 AIE 核的内部结构如图 1-23 所示，其核心部分是向量单元，可提供强大的计算能力。例如，其中的乘加器在一个时钟周期内可实现 128 个 8×8 有符号数乘法运算或 32 个 16×16 有符号数乘法运算。

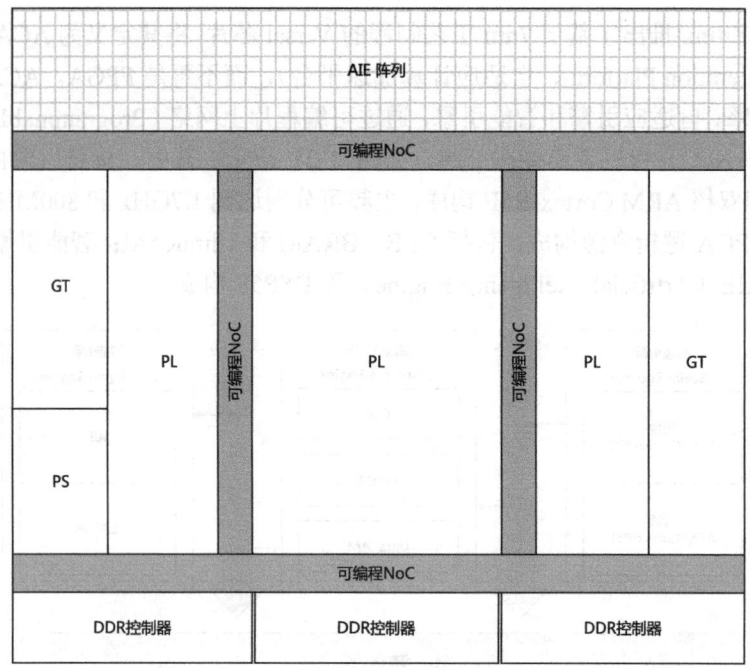

图 1-22

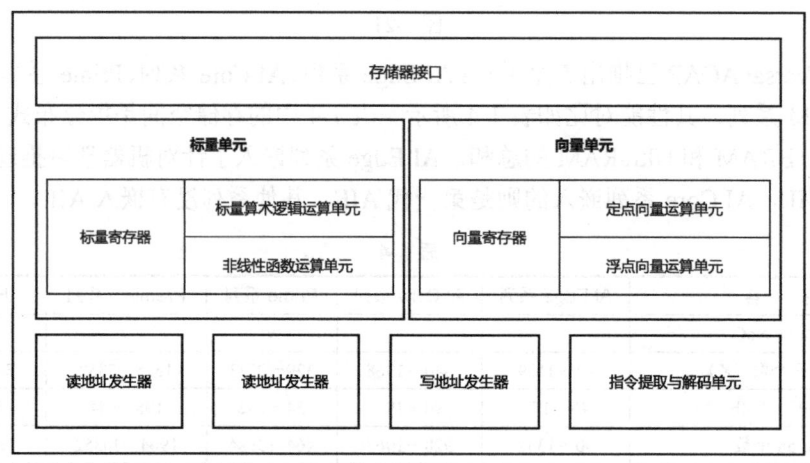

图 1-23

从 2001 年发布 Virtex-II FPGA 到 2019 年发布 Versal ACAP，在近 20 年的时间里，Xilinx 先后推出了 9 代产品，如表 1-5 所示，平均每两年就会有一代新产品诞生。产品性能不断突破：容量越来越大，速度越来越快。产品架构不断革新：从仅有 PL 到嵌入 CPU，再到嵌入 AIE。架构的不断演进使得 FPGA 除了在传统领域（如原型验证、通信、工业控制、航空航天等）继续保持一定优势，还使得 FPGA 进入新的领域，如数据中心，成为数据中心加速器、AI 加速器、SmartNIC（Smart Network Interface Card，智能网卡）及网络基础设施中的加速器。而 FPGA 厂商除了提供 FPGA 芯片，还生产制作了各种加速卡，如 Xilinx 的 Alveo 加速卡、Varium 加速卡，以期在最大限度上满足业务需求，帮助客户缩短研发周期。

表 1-5

内容	Virtex-II	Virtex-II Pro	Virtex-4	Virtex-5	Virtex-6	Virtex-7	UltraScale FPGA	UltraScale+ FPGA	Versal ACAP
发布时间（年）	2001	2002	2004	2006	2009	2011	2014	2016	2019
工艺制程（nm）	150	130	90	65	40	28	20	16	7
逻辑单元个数（K）	104	100	142	330	758	1955	5541	8938	7352
BRAM Fmax（MHz）	278	355	500	550	600	741	660	825	1,000
BRAM 容量（Mb）	2.95	7.80	9.70	18.14	37.41	66.09	132.89	94.50	174.00
DSP Fmax（MHz）	333	402	500	550	600	741	741	891	1,150
DSP 个数	168	444	512	1,056	2,016	3,600	5,520	12,288	14,352
GT（Gbit/s）	0	6.25	6.5	6.5	11	28.05	30.5	58	112
GT 个数	0	20	24	48	48	96	120	128	168
管脚个数	1108	1164	960	1200	1200	1200	1456	2072	780
CPU		PPC405	PPC405	PPC440		Cortex-A9 (Zynq-7000)		Cortex-A53	Cortex-A72
CPU 主频（MHz）		400	450	550		1000		1500	1700

1.2 设计方法的演变

FPGA 的设计方法是伴随着芯片架构的演变而演变的。在 FPGA 诞生初期，由于其内部资源较为单一，因此仅仅扮演着胶合逻辑的角色，在整个系统中只能起到协同作用。此时的设计方法也比较简单，如图 1-24 所示。设计输入使用 RTL（Register Transfer Level，寄存器传输级）描述方式，功能仿真也称为前仿或行为级仿真，以验证设计功能是否正确；综合后仿真（Post Synthesis Simulation）则用于验证综合后的电路功能是否正确；时序仿真又称为后仿，仿真时反标了门级延迟和布线延迟信息，用于验证布线后的电路功能是否正确。一般情况下，无论是综合后仿真还是时序仿真，相比于行为级仿真都更为耗时，尤其是当设计规模比较大时，因此，通常当设计出现问题时才会执行综合后仿真和时序仿真。

随着 FPGA 内部资源越来越丰富，嵌入了 BRAM，增大了存储空间，嵌入了乘加器，增强了计算能力，嵌入了高速收发器，提升了数据传输带宽，进行 FPGA 设计时要结合算法特征，分析哪些算法适合在 FPGA 上实现，以充分发挥 FPGA 的性能，从而形成如图 1-25 所示的开发流程。这里，软/硬件分割是设计的关键点。总体而言，数据流比较单一、运算密集但比较规整的算法（如 FIR 数字滤波器、FFT 等）就非常适合在 FPGA 上实现。进一步划分，则要采取"缓存-计算-缓存"的模式，以适配 FPGA 的架构。而分支较多、判断条件复杂、数据路径形成反馈回路的算法更适合在 CPU 上实现。同时，FPGA 更适合处理定点数据类型（尽管目前 FPGA 也支持浮点数据类型）。图 1-25 中 FPGA 部分的顶层功能验证过程与图 1-24 一致，可采用 Xilinx 开发工具 ISE（Integrated Software Environment）完成。

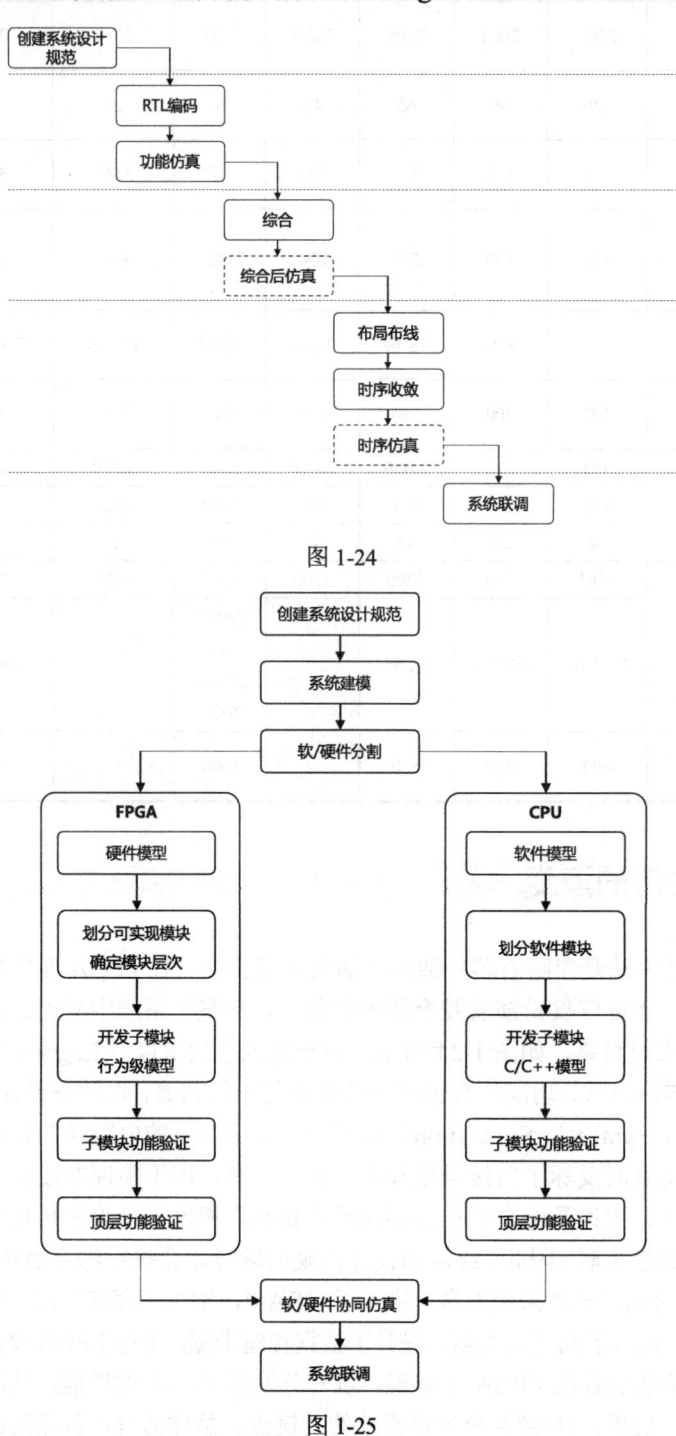

图 1-24

图 1-25

针对 FPGA 子系统，要从 3 个角度考虑，即物理级设计规范、时序设计规范和硬件设计规范，如图 1-26 所示。其中，物理级设计规范是为芯片选型服务的，根据资源评估、功耗预算和时钟频率确定芯片型号。时序设计规范是指在确定时钟网络拓扑结构（时钟管脚位置、输入时钟频率、输出时钟频率、全局时钟/区域时钟）的基础上，规划 I/O 时序（尤其是源同步设计和系统同步设计）和跨时钟域路径时序，基于此描述时序约束。硬件设计规范则是指根据数据流合理规划 RTL 代码层次结构，在此基础上进行各子模块的开发，最终为时序收敛服务。

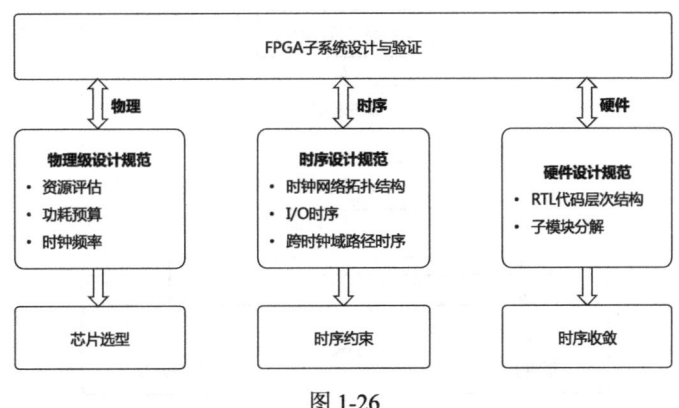

图 1-26

在硬件设计规范中，RTL 代码层次结构是重点，这对综合、布局布线和时序收敛都会产生直接影响，应遵循的原则如下。

- 需要实例化的输入/输出单元（如 IDDR、ODDR、ISERDES、OSERDES 等）应尽可能靠近设计顶层，尽管 IBUF、OBUF、IOBUF 和 OBUFT 可由综合工具自动推断出来，但要确保 IOBUF、OBUFT 的使能信号和输入/输出信号在同一层次，以确保工具正确推断。
- 将时钟生成模块（通常采用 IP Clocking Wizard 生成时钟，不建议使用 MMCM 或 PLL 原语）放在顶层，方便其他模块使用时钟。
- 在层次边界添加寄存器，确保关键模块是寄存器输出，这样可将关键路径隔离在单一层次或模块之内，对于修复时序违例及设计调试大有裨益。
- 确保需要手工布局的模块在同一层次之内。

图 1-27 给出了基于上述原则而形成的层次结构，图中，每个子模块的阴影部分表示输出寄存器。

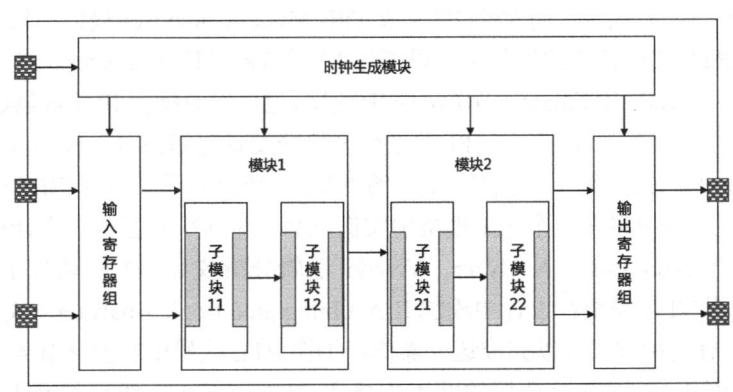

图 1-27

随着 FPGA 中嵌入 ARM 核构成 SoC 芯片，如 Zynq-7000，SoC 的设计方法应运而生，如图 1-28 所示。在软/硬件分割阶段，根据算法特征和系统需求（是否运行操作系统）将系统分为两大模块：硬件模块（在 PL 上实现）和软件模块（在 PS 上实现）。硬件开发依然采用传统的 FPGA 开发模式。软件开发则需要借助 Xilinx 开发工具 SDK（Software Development Kit）。

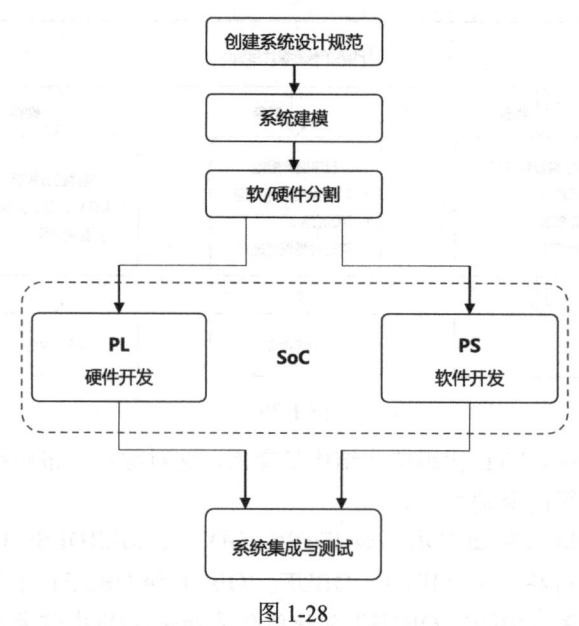

图 1-28

芯片架构和设计方法的演变也催生了 Xilinx 新一代开发工具 Vivado 的问世，随之问世的还有高层次综合工具 Vivado HLS（现更名为 Vitis HLS），Vivado 反过来又影响了设计方法。如图 1-29 所示，Vivado 提出了以 IP 为核心的设计理念，进一步强调了设计的可复用性。在设计输入阶段，设计源文件可以是传统的 RTL 代码，可以是 C/C++或 OpenCL 模型（采用 Vitis HLS 开发），也可以是 Simulink 下的 AIE 模型、HDL 模型（原 System Generator）或 HLS 模型（这 3 个模型隶属于同一个开发工具 Vitis Model Composer，该开发工具嵌入在 Simulink 中）。其中，以高级语言（C/C++或 OpenCL）描述的模型和以 Vitis Model Composer 搭建的模型均为高抽象度模型，需要借助相应的工具将其转化为 HDL（Hardware Description Language）代码。而 Vitis HLS 或 Vitis Model Composer 最终都将其封装为 IP，同时，传统的 RTL 代码描述的模型也可以通过 IP 封装器（IP Packager）封装为 IP，这些 IP 均可直接嵌入 Vivado IP Catalog 当作常规 IP 使用，也可以直接在 IP 集成器（IP Integrator，IPI）中以模块方式使用。这样，在 IPI 中进行设计开发就像搭积木一样。基于 IPI 开发设计形成的文件为.bd 文件，为了实现 IPI 模型的可复用，Vivado 又引入了 BDC（Block Design Container）功能，即可以在一个 IPI 模型中实例化另一个 IPI 模型。在设计调试方面，可以借助 VLA（Vivado Logic Analyzer，Vivado 逻辑分析仪）。VLA 取代了 ISE 时代的 ChipScope。使用 VLA 需要在设计中添加 ILA（Integrated Logic Analyzer）或 VIO（Virtual Input/Output）。有 3 种方式可以完成这一操作：①在 RTL 代码中实例化 ILA 和 VIO（VIO 仅支持代码实例化）；②在综合后的网表中插入 ILA；③在布线后的网表中采用 ECO（Engineering Change Order，工程变更命令）方式修改 ILA。

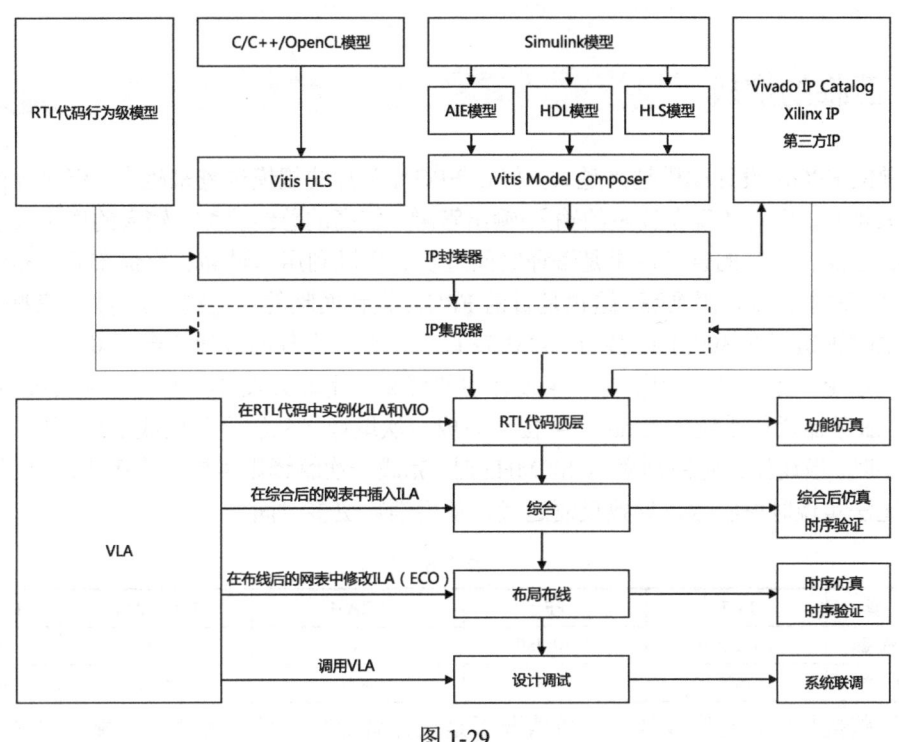

图 1-29

对比图 1-29 和图 1-24 不难发现，Vivado 要求在综合之后就要对设计进行时序分析，确保建立时间收敛或接近收敛，这实际上是一个重大变化。这意味着若综合后依然存在建立时间违例，那么布线后时序收敛的可能性也不会大。

在软件开发方面，Xilinx 于 2019 年推出了统一的软件开发平台 Vitis，取代了原有的 SDAccel，同时，其功能进一步增强。SoC、MPSoC、AIE 均可在 Vitis 下进行开发。

无论工具如何推陈出新，功能如何变化，到目前为止，FPGA 设计始终遵循的一个思路就是提高设计的可复用性。简而言之，对于一些常用模块，通过参数化处理，使其可以适配不同项目的需求，提高设计的可复用性。这样的好处是避免了重复开发，从而缩短开发周期，同时因为这些模块已经经过实际项目的验证，所以在功能和质量方面都有所保障。这其实就是用户自己开发的 IP。我们把"可复用性"的理念进一步扩大，如图 1-30 所示。设计输入阶段，从 RTL 代码到 C/C++代码，再到 Model Composer 模型，都可以做成参数化形式。相应的功能仿真阶段用到的 HDL 测试平台、C/C++测试平台和 Model Composer 测试平台也可以做成参数化形式。

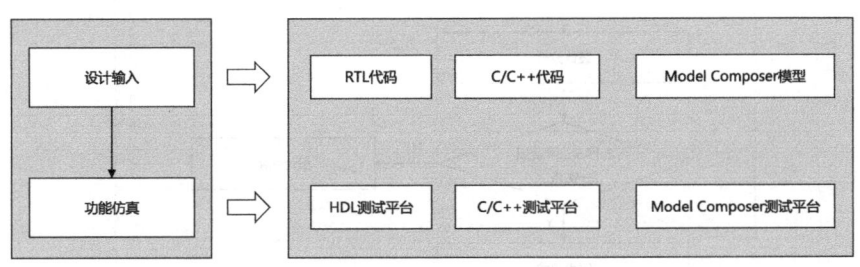

图 1-30

本书提供的很多 VHDL 代码均已做参数化处理，可以方便快捷地实现设计复用。

1.3 面临的挑战

当前的 FPGA 设计规模越来越大,好在 FPGA 芯片的规模也越来越大,但设计的复杂度也越来越高,主要体现在较多的输入/输出管脚、较高的总线位宽、较大的扇出及较高的逻辑级数方面。带来的直接后果是编译时间过长、资源利用率过高、次优布局、布线资源消耗太多、时序收敛较困难等,这也是目前 FPGA 设计面临的一些挑战。另外,有些 FPGA 设计要求时钟运行在 300MHz 甚至 400MHz 以上,这样面临的挑战将会更多。

以 VU9P 为例,其可用资源及某设计中的资源利用率如表 1-6 所示。Vivado 版本为 2018.3,服务器操作系统为 Linux 64 位,完成一次编译(从综合到生成.bit 文件)至少需要十几小时。当设计的时钟频率为 400MHz 时,完成一次编译则需要二十多小时甚至更长,有时还无法实现时序收敛,导致反复迭代,从而消耗更多时间。

表 1-6

类 型	LUT	FF	BRAM	UltraRAM	DSP48
可用资源	1 182 240	2 364 480	2160	960	6840
利 用 率	70%	50%	80%	80%	80%

在这些挑战中,时序收敛在多数情况下是设计者面临的最大挑战。传统的时序收敛流程如图 1-31 所示,分为两大阶段:综合阶段和布局布线阶段。综合阶段只要满足 WNS(Worst Negative Slack,最差建立时间裕量)等于 0 即可进行布局布线,这个目标通常比较容易实现。而布局布线阶段的时序收敛往往涉及多方因素,需要设计者对工具特性、设计本身深入理解,同时有一定的经验积累。因此,这个过程有可能出现反复迭代。

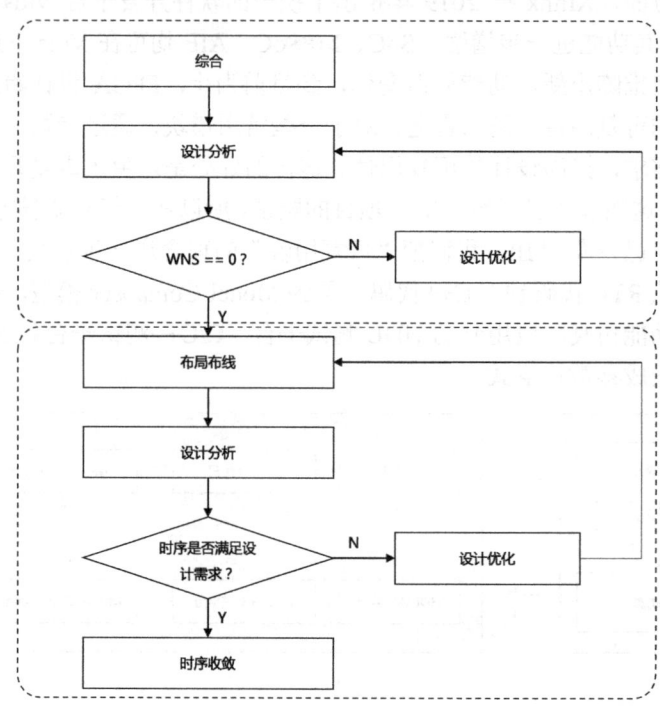

图 1-31

针对布局布线阶段的时序收敛,通常需要采用如图 1-32 所示的流程。这个流程的关键点是分析时序违例原因,分析的对象并不是所有的时序违例路径,而是最为关键的几条时序违例路径。实践表明,往往只用手工修复这些关键时序违例路径,工具会自动修复其他时序违例路径。找到原因之后才能有的放矢地找到解决办法。一般情况下,出现时序违例的原因包括:逻辑延迟太大、布线延迟太大、时钟偏移或时钟抖动太大。若确定原因是逻辑延迟太大,那么可能的解决办法有两种:插入流水寄存器和采用重定时。前者需要修改代码,由于增加流水寄存器会导致从输入到输出所需要的时钟周期个数发生改变,多条路径的数据需要重新对齐,改动量可能会比较大;后者无须改动代码,但需要确定是在综合阶段设置重定时还是在布局布线阶段设置重定时,是全局设置(使用全局设置选项)还是局部设置(使用模块化综合技术)。这两种解决办法到底哪个是最优解?这需要尝试,也需要设计者积累足够的经验,同时比较耗时。尤其是在多个解决办法无法并行执行时,耗时更多。

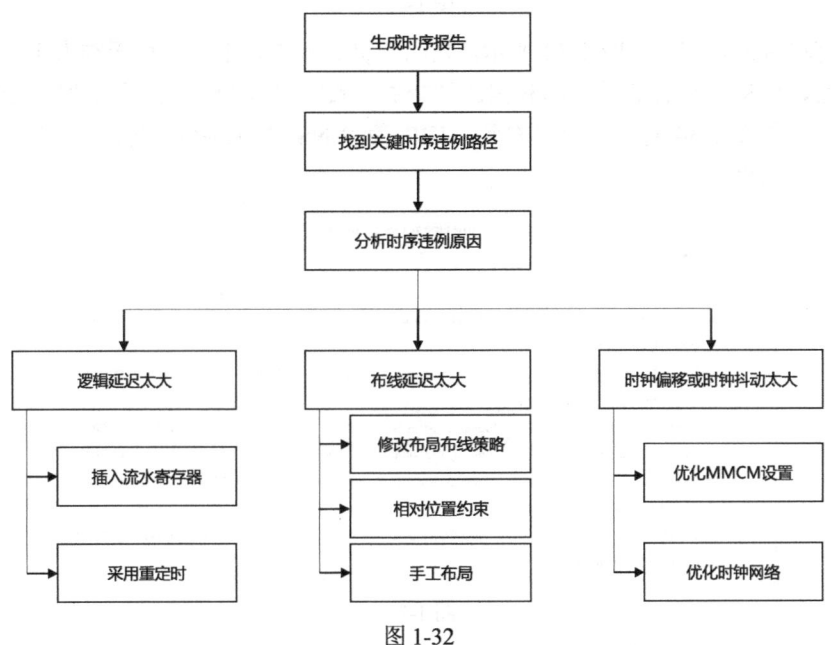

图 1-32

为了应对这些挑战,我们不仅要从设计本身入手,还要从 EDA 工具的角度入手,以期工具能够帮助我们缩短编译时间、提高编译质量。这就要提到 Vivado ML(Machine Learning)版本。

Vivado ML 的首个版本是 2021.1。其在设计的多个阶段引入了机器学习算法。例如,在综合阶段,引入了基于机器学习算法的门级优化,从而大大提高了综合质量并且将综合时间平均缩短了 30%;在布局布线阶段,引入了基于机器学习算法的拥塞评估和延迟评估,从而有效减少了迭代次数并将布局布线时间平均缩短了 20%;其最大亮点在于引入了智能设计流程(Intelligent Design Run,IDR)。该设计流程专门为时序收敛服务,共包含 3 个阶段,如图 1-33 所示。阶段 1,工具会从 5 个方面,即时序约束、资源利用率、时钟网络、布线拥塞和时序违例路径进行优化,生成相应的优化建议(通常以 Tcl 脚本的方式呈现),并自动将这些建议施加到对应的子步骤(实现阶段 1 的 4 个子步骤,对应的 Tcl 命令分别为 opt_design、place_design、phys_opt_design 和 route_design）中。该阶段结束后，工具会对所有的布线后的网表进行比较，选取 WNS 最小者，获取其中有效的优化建议供阶段 2 或阶段 3 使用。阶段 2 在阶段 1 的基础上（继承了阶段 1 的结果，即继承了有效的优化建议），工具会采用机器学习算法生成 3 个布局布线策略。阶段 3 为增量编译阶段。实验数据表明，IDR 可以给设计带来平均 10%的收益，即 Fmax（设计可运行的最高时钟频率）可提升 10%。

```
┌─────────────┐    ┌─────────────┐    ┌─────────────┐
│    阶段1    │    │    阶段2    │    │    阶段3    │
│  时序约束   │    │             │    │             │
│ 资源利用率  │    │   生成3个   │    │   增量编译  │
│  时钟网络   │    │ 布局布线策略│    │             │
│  布线拥塞   │    │             │    │             │
│时序违例路径 │    │             │    │             │
└─────────────┘    └─────────────┘    └─────────────┘
```

图 1-33

IDR 有单独的窗口，如图 1-34 所示。采用一键式操作，中间过程无须人工干预。从这个角度而言，IDR 对设计者的经验积累的要求并不高。图 1-35 显示了未使用 IDR 时最终的时序性能。与图 1-34 对比，可以看到，IDR 使 WNS 从-1.040ns 变为-0.158ns，同时修复了保持时间违例。

图 1-34

图 1-35

1.4 四大基本原则

尽管 FPGA 设计越来越复杂，但只要遵循一些基本的原则，就可以提高综合质量、减轻布局布线压力、减少设计迭代次数、加速时序收敛进程。

1.4.1 硬件原则

FPGA 设计采用的主要编程语言包括 VHDL、Verilog 和 SystemVerilog，三者均属于硬件描述语言（Hardware Description Language，HDL），其中，SystemVerilog 因其在验证方面显示的强大功能又被称为硬件描述与验证语言（Hardware Description and Verification Language，HDVL）。硬件描述语言，顾名思义，其描述的对象是硬件电路，是实实在在存在的硬件电路，这要求我们必须采用 HDL 可综合的部分进行电路设计，遵循可综合的代码规范。设计者需要了解所用的 HDL 哪些是可综合的，哪些是不可综合的。例如，延迟语句，无论是 VHDL 中的 wait for 语句，还是 Verilog 或 SystemVerilog 中的"#延迟时间"（延迟时间为具体数字），均是不可综合的。同样，除法运算"/"也是不可综合的（只有当除数为 2 的整数次幂时才可综合，此时等效于右移操作）。从数据类型的角度而言，实数（Real）是不可综合的。

相比软件编程语言（如 C/C++），HDL 具有两大特征：并发性和时序性。并发性体现在同一模块中的不同进程（VHDL 中的 process，Verilog 中的 always，SystemVerilog 中的 always_ff、always_comb 和 always_latch）是同时执行的，这反映了硬件电路的特征：一旦上电，同时工作。时序性则体现了不同数据路径在时钟作用下的相互关系及时序路径与控制路径的关系。时序是设计出来的，不是凑出来的，更不是测出来的。就这点而言，在开始编程之前，要有待设计电路的基本雏形。例如，要设计一个复数乘法器，就要先画出如图 1-36 所示的硬件电路，图中，D 表示 D 触发器，第一个复数的实部和虚部分别为 ar 和 ai，第二个复数的实部和虚部分别为 br 和 bi。从图 1-36 中也可以看出各数据路径之间的时序关系。"先有电路，再写代码"，RTL 代码的每一条语句都有与之对应的电路单元。

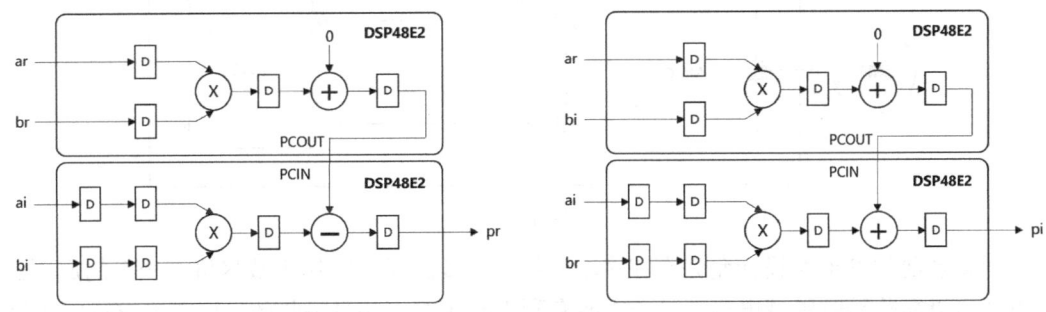

图 1-36

硬件原则还要求我们对所选用的 FPGA 的内部结构有所认识和了解，以确保 RTL 代码风格与 FPGA 内部结构相匹配，这样才能保证综合电路的质量。例如，7 系列 FPGA 中的触发器复位信号只支持高电平有效，如果代码中的复位是低电平有效，就会消耗额外的查找表（用于逻辑取反），而 UltraScale/UltraScale+ FPGA 中的触发器既支持高电平有效，又支持低电平有效。7 系列 FPGA 中的 DSP48 不支持异步复位，因此，如果代码中描述的乘法器使用了异步复位，那么相应的触发器是无法被吸收到 DSP48 内部的。

综上所述，硬件原则如下。

（1）使用可综合的 HDL 代码描述电路。
（2）先有电路，再写代码。
（3）RTL 代码风格与 FPGA 内部结构相匹配。

1.4.2 同步原则

同步原则指的是整个设计采用同步时序电路,"同步"意味着设计中的所有逻辑单元"步调保持一致",即所有电路在同一时钟沿的触发下同步处理数据。这样看来,设计中只允许存在一个时钟,我们把这种"同步"称为狭义的同步。毕竟,随着芯片规模增大和设计复杂度增加,设计中的时钟个数也越来越多。例如,外部存储器接口时钟、以太网接口时钟、PCIE 接口时钟、PS 侧时钟、PL 主时钟等。因此,我们又引入了广义的同步。这里我们先给同步时钟下定义,所谓同步时钟,是指时钟组(一个时钟组至少有两个时钟)内的时钟之间有明确的相位关系。反之,若时钟组内的时钟之间没有明确的相位关系,则认为它们是异步时钟。最典型的同步时钟场景是时钟组内的时钟是由同一个 MMCM/PLL 生成的,而异步时钟场景是时钟组内的时钟由不同 MMCM/PLL 生成,如图 1-37 所示。在图 1-37 中,clk0 和 clk1 是同步时钟,clka 和 clkb 是同步时钟,但{clk0, clk1}和{clka, clkb}是异步时钟。从而,模块 1 内存在同步跨时钟域电路,而模块 2 内存在异步跨时钟域电路。显然,后者无论是电路设计还是时序约束都更为复杂。因此,广义的同步是指电路在同步时钟的作用下处理数据。

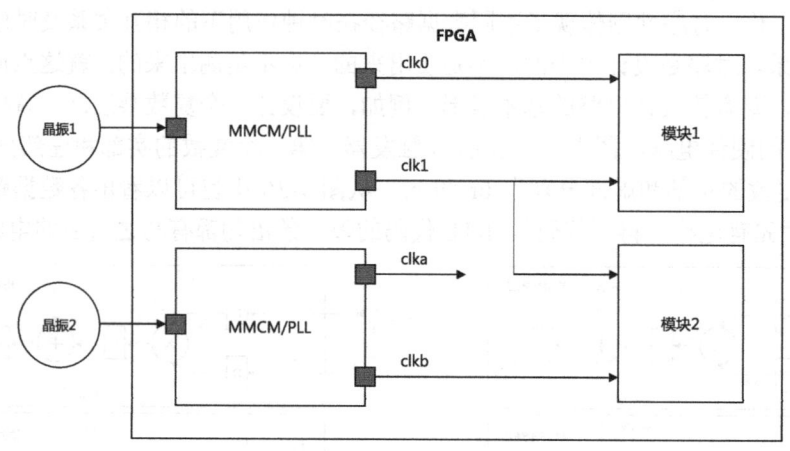

图 1-37

尽管当前的 FPGA 设计允许存在多个时钟,仍要遵循"时钟个数尽可能少"的原则。一方面可以简化跨时钟域电路,另一方面有利于时序约束和时序收敛。这就要求我们在设计初期制定时钟方案时就要明确设计中有哪些时钟,它们之间的关系如何,是否可以由同一个 MMCM/PLL 生成。

之所以遵循同步原则,还因为相比异步设计,同步设计有着明显的优点。首先,同步设计可以有效地避免毛刺的影响,增强设计的稳定性。组合逻辑可能会产生毛刺,如果毛刺仅存在于同步时序的数据路径中,那么受时钟驱动的触发器就可以过滤掉毛刺,因为触发器仅在时钟有效沿才会有动作,从而消除其对电路的影响。如果将该组合逻辑的输出连接到触发器的控制端,如作为触发器的异步复位/置位信号,那么当毛刺足够宽时,就可能导致触发器误动作。即使毛刺的宽度不足以驱动异步复位/置位端,也会造成触发器不稳定,甚至激发其产生亚稳态。其次,同步设计可以减少外部环境对芯片的影响。芯片的实际工作环境可能要比实验环境更为恶劣,这也是我们做高低温实验的一个主要原因,即检测当前设计是否能在不同环境下正常工作。此外,长时间运行也会使芯片自身温度升高,使电压变得不稳定，芯片内部延时可能会发生微小变化。如果采用异步设计，对时序要求比较严苛的电路将无法正常工作，这是因为异步设计的时序正确与否完全依赖于每个逻辑元件的逻辑延时和布线延时。最后，同步设计更有利于静态时序分析（Static Timing Analysis，STA）和验证设计的时序性能。这得益于同步设计的时序约束更为简单。即使对于同步跨时钟域路径，工具也可以自动对其进行约束，当然有时会出现约束不合理的情形。

1.4.3 流水原则

流水原则是指要求我们在设计中使用流水线设计方法将数据处理流程分割为若干个子步骤，使数据在这些子步骤中流动起来。

如图 1-38 所示，假定某处理流程可分解为读操作、计算操作和写操作 3 个步骤，采用流水线方式就是在这 3 个步骤之间插入流水寄存器。这样处理流程就由顺序方式变为流水线方式，顺序方式和流水线方式如图 1-39 所示，这样带来的好处也是显而易见的，不仅提高了处理速度，而且降低了从输入到输出的总时钟周期个数（Latency）。使用流水线技术是有要求的：数据流是单向流动，不存在反馈支路。流水线技术也体现了 FPGA 处理数据的特征：动态处理。

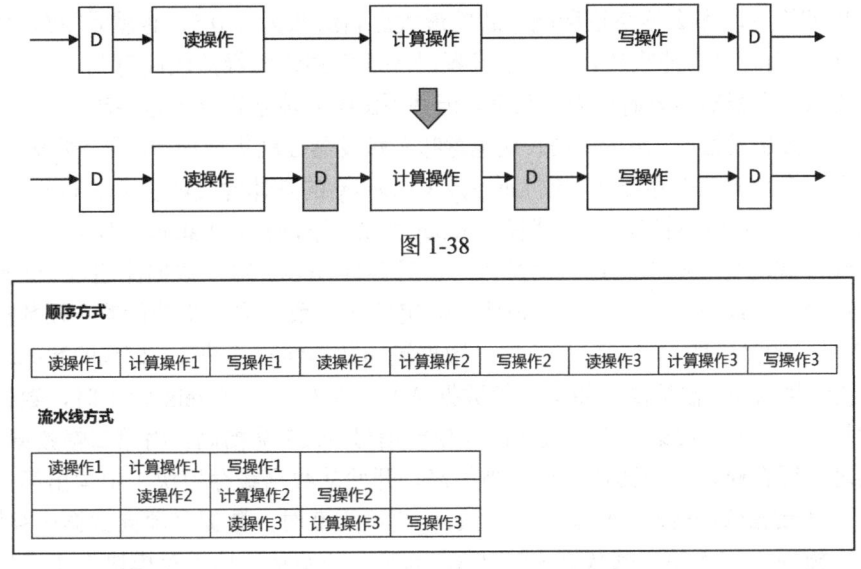

图 1-38

图 1-39

利用这个思想，从微观角度而言，我们可以把一段长路径切割为多段短路径，在每段短路径之间插入流水寄存器以暂存中间数据，目的仍是将一个大操作分解为若干个小操作，而每个小操作比大操作的延时要小，因此可以提高时钟频率，同时，各个小操作可以并行执行，因此又能提高数据吞吐率。如图 1-40 所示，图中，上部路径有 6 个 LUT，仅从逻辑延时的角度来看，总逻辑延时为 $6T_{ilo}$（T_{ilo} 为 LUT 从输入到输出的延时）。现将这 6 个 LUT 分为两组，在两组 LUT 之间插入寄存器，从而形成两条时序路径，每条时序路径的逻辑延时降低至 $3T_{ilo}$。这实际上是将路径的逻辑级数由 6 降至 3。从这个角度而言，流水线技术也是修复时序违例的一种方法。

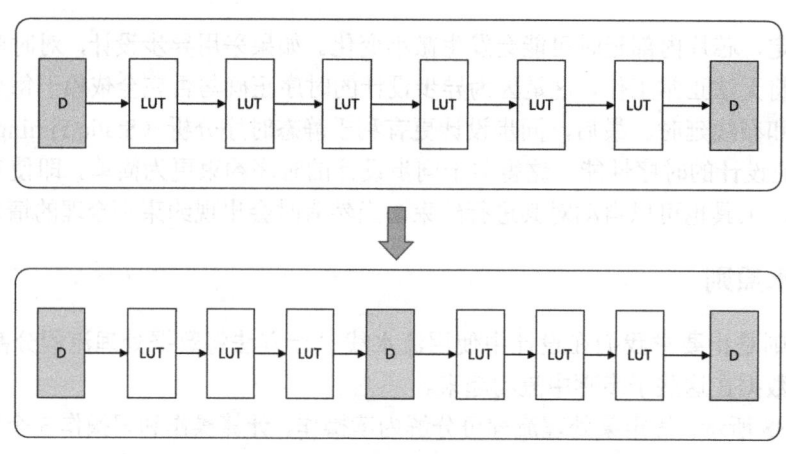

图 1-40

FPGA 芯片内触发器的个数远远多于 LUT 的个数，如 UltraScale SLICE 内有 8 个 LUT，但有 16 个触发器。因此，对于逻辑级数较高的路径，可通过插入流水寄存器的方法改善其时序性能。那么，如何判断逻辑级数是否过高呢？通常，一个"LUT+一根走线"的延迟为 0.5ns（经验值，适用于 Xilinx 7 系列 FPGA 和 UltraScale/UltraScale+ FPGA），假定寄存器时钟周期为 T，那么该路径所能承载的最大逻辑级数为 $T/0.5$，也就是 $2T$。如果逻辑级数大于 $2T$，而时序未能收敛，就可以怀疑时序违例跟逻辑级数较高有关。

尽管流水寄存器对改善时序有所帮助，但并不意味着流水寄存器越多越好。事实上，过重的流水会导致触发器的利用率增加，这也意味着触发器控制集增加，从而会引发布线拥塞。同时应注意，增加流水寄存器会导致 Latency（Latency 的具体含义请参考本书 1.5 节）发生变化。从图 1-40 中不难看出，上部路径 Latency 为 2，下部路径 Latency 为 3。

流水线方式的另一种形式是乒乓操作，如图 1-41 所示。输入数据流通过 1:2 解复用器分时流向两个数据缓存模块。为便于说明，假定每 4 个数据构成待处理的一帧数据，形成如图 1-42 所示的时序图。当 selx 为 0 时，数据流向数据缓存模块 1，相应的数据缓存模块 1 的写使能信号 wen1 被抬高，将第一帧数据 A0~A3 写入；当 selx 为 1 时，数据流向数据缓存模块 2，相应的数据缓存模块 2 的写使能信号 wen2 被抬高，将第二帧数据 B0~B3 写入，同时，缓存模块 1 的读使能信号被抬高，开始从数据缓存模块 1 中读出第一帧数据 A0~A3，并传送给数据预处理模块 1。当 selx 再次为 0 时，数据又流向数据缓存模块 1，向数据缓存模块 1 写入第三帧数据 C0~C3，同时，开始从数据缓存模块 2 中读出第二帧数据 B0~B3 并传送给数据预处理模块 2，如此循环。从而，数据缓存模块 1 中存储的永远是奇数帧数据，数据缓存模块 2 中存储的是偶数帧数据。数据缓存模块 1 在执行写操作时，数据缓存模块 2 在执行读操作，而数据缓存模块 1 在执行读操作时，数据缓存模块 2 在执行写操作。这样，对两个数据预处理模块而言，只要在 8 个时钟周期内处理 4 个数据即可。这实际上减轻了数据预处理模块的时序压力。两个数据预处理模块的输出端连接 2:1 复用器，在 sely 的控制下即可将预处理结果不间断地传送给下游处理单元（在图 1-42 中，数据预处理模块的 Latency 为 6）。

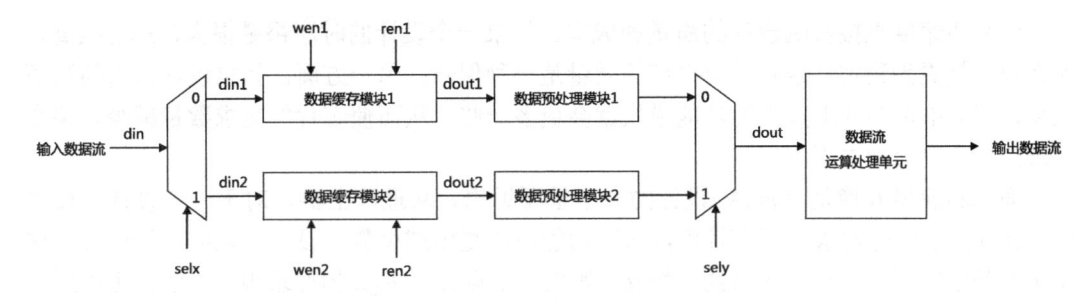

图 1-41

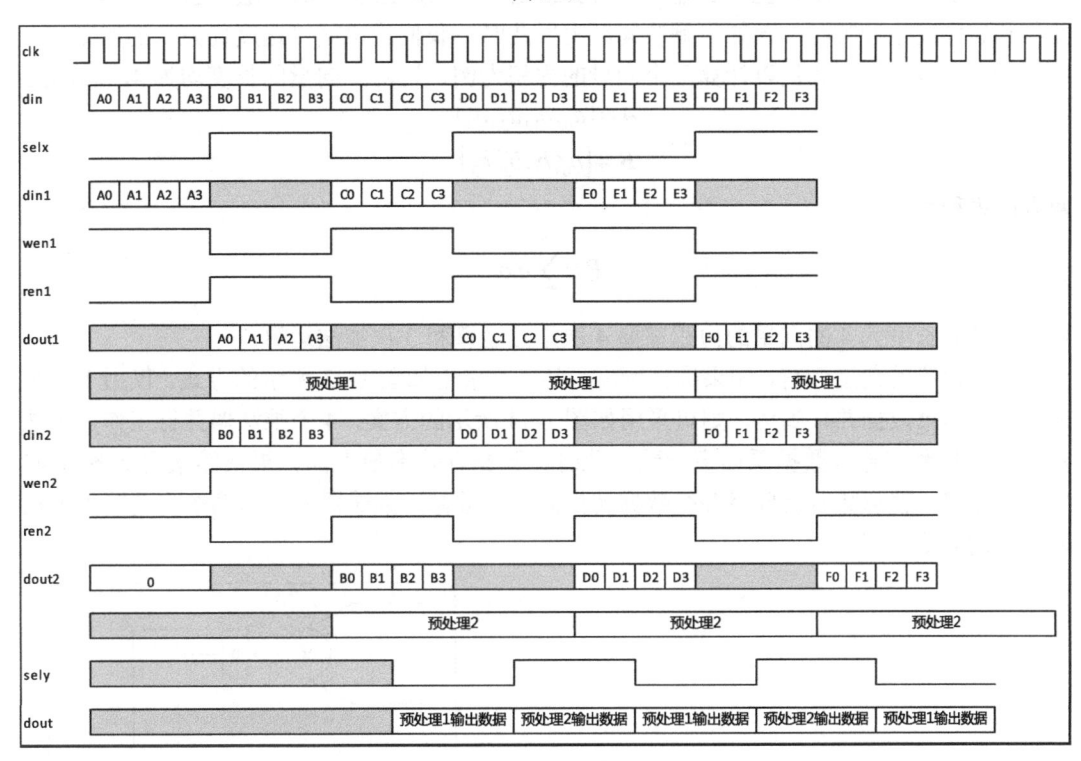

图 1-42

1.4.4 面积与速度的平衡与互换原则

这里的面积指的是一个设计所消耗的 FPGA 的逻辑资源数量，其直观的体现就是资源利用率报告。速度指的是设计在 FPGA 上稳定运行时所能达到的最高频率，也就是我们常说的 F_{max}，可间接地通过时序报告中的 WNS（Worst Negative Slack，建立时间最小裕量，若此值大于等于 0，说明建立时间收敛；若此值小于 0，说明建立时间存在时序违例）换算获得。面积和速度这两个指标贯穿 FPGA 设计的始终，是设计质量评价的终极标准。

面积和速度是对立统一的。要求一个设计以最少的资源为代价运行在最高的时钟频率下是不可行的。科学的设计目标应该是在保证设计满足时序性能（能够达到预期的 F_{max}）的前提下，尽可能地减少设计所消耗的资源；或者在限定的资源用量下，使设计的时序裕量更大，即使 F_{max} 更高。这两种目标充分体现了面积和速度的平衡思想。

面积和速度直接影响设计的质量和成本。如果一个设计的时序裕量很大，F_{max} 很高，那么设计就更为稳定，这对整个系统的质量是一种保障；另一方面，如果设计占用的资源很少，那么单位芯片上实现的功能模块就会更多一些，从而使芯片的需求数量减少，整个系统的成本也会随之下降。

面积和速度互换是 FPGA 设计的一个重要思想。从理论上讲，对于一个设计，如果 F_{max} 远远高于实际需求，那么就可以通过功能模块复用减少整个设计消耗的逻辑资源，这其实就是用速度优势换面积节约。相反，如果一个设计对 F_{max} 的要求很高，那么可以通过并行复制多个操作模块来提高 F_{max}，这其实就是通过增加面积换取 F_{max} 的提升。那么，当面积和速度发生冲突时，该如何解决问题呢？此时，我们应遵循"速度优先"的原则。

为进一步说明，我们以计算两个向量的内积为例，这两个向量的长度均为 4，表示为

$$A = [a_0, a_1, a_2, a_3]$$
$$B = [b_0, b_1, b_2, b_3]$$

两者的内积表示为

$$P = \sum_{i=0}^{3} a_i b_i$$

可见，计算长度为 4 的向量的内积需要 4 次乘法运算和 3 次加法运算。

从数据流的角度而言，如果数据顺序到达，可采用如图 1-43 所示的方案，仅用 1 个乘加单元。如果数据并行到达，则可采用如图 1-44 所示的方案，4 个乘法器并行工作，乘法器的工作频率与输入数据采样率一致，此时，数据吞吐率最大。如果依然采用如图 1-43 所示的方案，就意味着需要将并行数据转换为串行数据，乘法器的工作频率将是输入数据采样率的 4 倍。

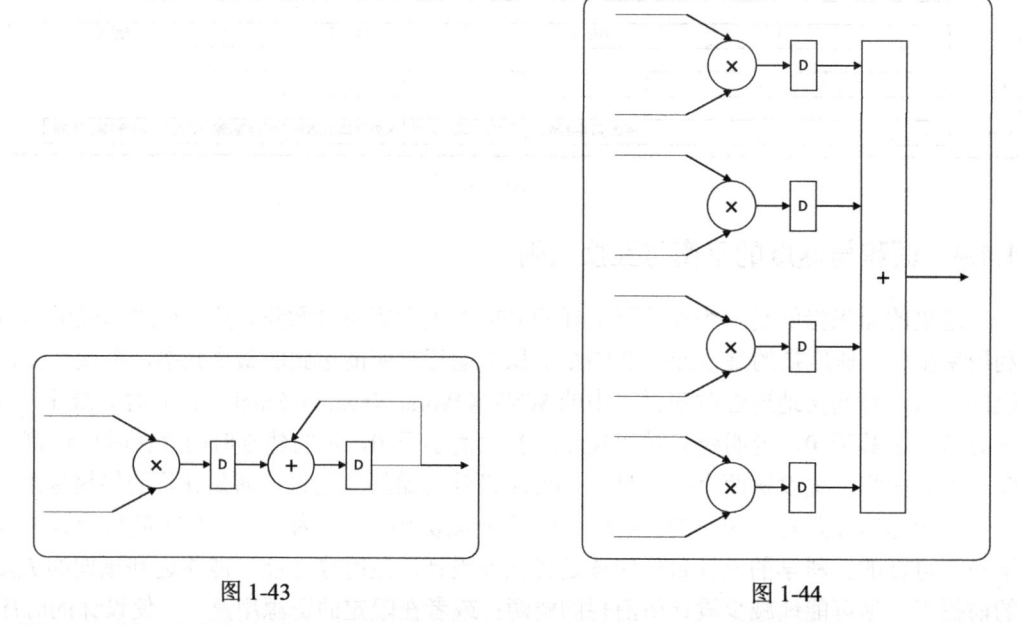

图 1-43

图 1-44

图 1-43 所示为串行方案，图 1-44 所示为全并行方案，前者消耗的资源最少，但并行数据流 F_{max} 将受限，后者 F_{max} 最高，但消耗的资源也最多。能否在面积与速度之间取得折中呢？答案是肯定的，这就是半并行方案，如图 1-45 所示。此时，对于 4 路并行数据，将

其转为 2 路并行数据，即 a_0、a_2 和 b_0、b_2 使用上部乘法器，a_1、a_3 和 b_1、b_3 使用下部乘法器。乘法器的工作频率将是输入数据采样率的 2 倍。

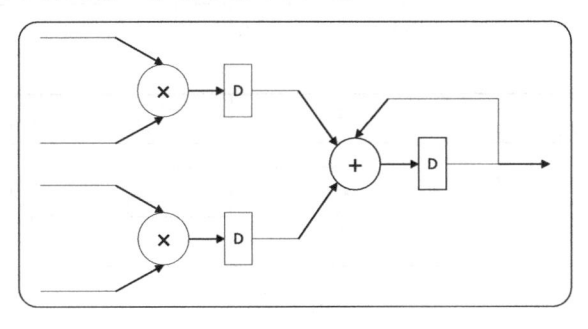

图 1-45

1.5 性能指标

对于一个 FPGA 设计，我们该如何评估其性能呢？通常会用到以下几个指标：设计可运行的最高频率（F_{max}）、输入到输出的时钟周期数（Latency）、吞吐率（Throughput）、资源利用率和功耗（Power）。

F_{max} 可通过时序报告计算得到。在 Vivado 中，可通过命令 report_timing_summary 生成时序报告，如图 1-46 所示。当 WNS、WHS 和 WPWS 均大于或等于 0 时，表明时序已收敛。

图 1-46

在图 1-46 中，WNS 为 0.171ns，若时钟周期为 10ns，那么 F_{max} 为

$$F_{max} = \frac{1}{10 - 0.171} \times 1000 = 101.74 \text{MHz}$$

显然，WNS 越大越好。

输入到输出的延迟通常用时钟周期个数来表示，称为 Latency，该指标也反映了设计的流水级数。如图 1-47 所示，输入 x0 对应的输出为 y0，输入 x1 对应的输出为 y1，从输入到输出需要 3 个时钟周期，因此 Latency 为 3。相邻两个输入之间间隔的时钟周期个数反映了该设计的吞吐率。显然，Latency 越小越好。但 Latency 小意味着流水级数低，这可能会导致 F_{max} 降低。

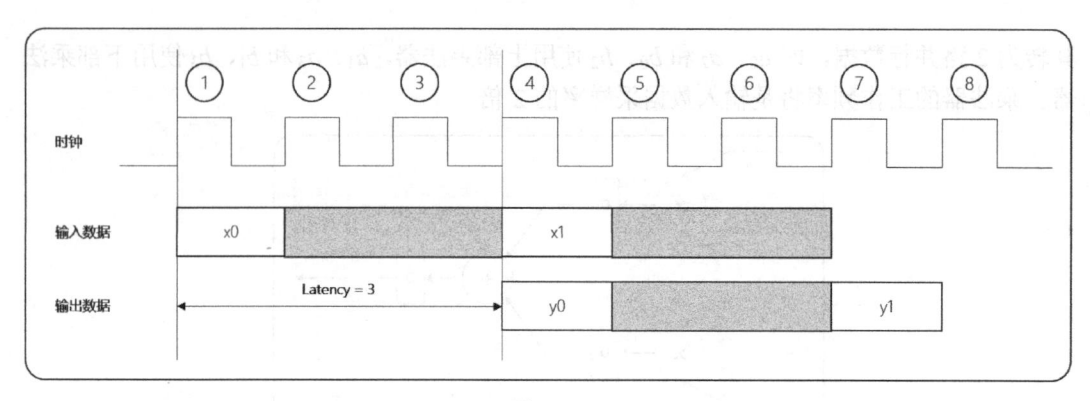

图 1-47

在 Vivado 下,可通过命令 xilinx::designutils::report_failfast(从 Vivado 2022.1 开始,可直接使用命令 report_qor_assessment)查看资源利用率的指导值和实际值,如图 1-48 所示,图中,Guideline 列对应指导值,Actual 列对应实际值,Status 列若为 REVIEW,则表明其所在行对应的资源利用率超过指导值。

```
+-----------------------------------------------------------+
| Design Summary                                            |
| checkpoint_top_routed                                     |
| xc7k70tfbg676-2                                           |
+-----------------------------------------------------------+
| Criteria                              | Guideline | Actual | Status |
+-----------------------------------------------------------+
| LUT                                   | 70%       | 48.08% | OK     |
| FD                                    | 50%       | 19.13% | OK     |
| LUTRAM+SRL                            | 25%       | 0.07%  | OK     |
| MUXF7                                 | 15%       | 3.38%  | OK     |
| DSP                                   | 80%       | 28.33% | OK     |
| RAMB/FIFO                             | 80%       | 80.74% | REVIEW |
| DSP+RAMB+URAM (Avg)                   | 70%       | 54.53% | OK     |
| BUFGCE* + BUFGCTRL                    | 24        | 12     | OK     |
| DONT_TOUCH (cells/nets)               | 0         | 1      | REVIEW |
| MARK_DEBUG (nets)                     | 0         | 0      | OK     |
| Control Sets                          | 769       | 453    | OK     |
| Average Fanout for modules > 100k cells | 4       | 0      | OK     |
| Non-FD high fanout nets > 10k loads   | 0         | 0      | OK     |
```

图 1-48

将图 1-48 中的信息提取出来,形成表 1-7,表中,LUTRAM 表示分布式 RAM,SRL 表示用 LUT 实现的移位寄存器。根据此表,我们可以在设计初期进行芯片选型。需要注意的是 BRAM、UltraRAM 和 DSP48 三者的资源利用率都不能超过 80%,若超过了 80%,则要保证三者的平均利用率低于 80%。

表 1-7

查找表(LUT)	LUTRAM + SRL	MUXF7	触发器(FF)	BRAM	UltraRAM	DSP48
70%	25%	15%	50%	80%	80%	80%

此外,我们还要看设计功耗是否达到预期目标。在 Vivado 下,可通过命令 report_power 生成当前设计的功耗报告,如图 1-49 所示。报告提供的信息越多(如通过仿真提供.saif 文件等),可信度越高。

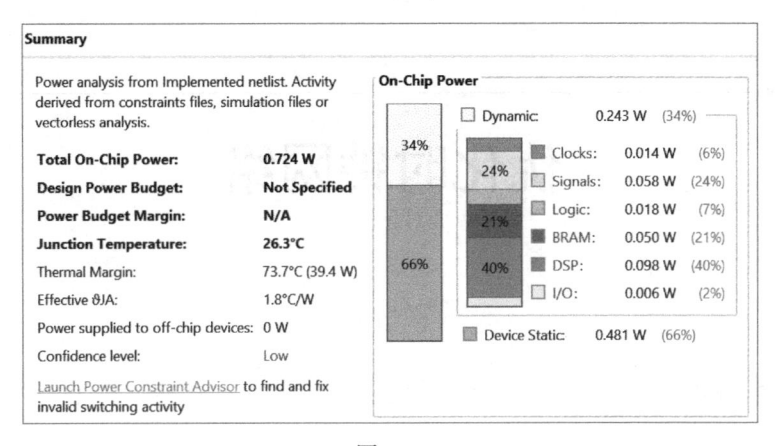

图 1-49

这些性能指标不是独立的，而是相互影响的。有时为了提升 F_{max} 而增加 Latency 或资源利用率；有时为了降低资源利用率而牺牲 F_{max}。无论如何，最终的目标是实现时序和功耗均收敛。时序收敛意味着设计达到了预期的 F_{max}，功耗收敛意味着设计的功耗在目标范围内。

1.6 思考空间

1．FPGA 内部的主要资源有哪些？以 Xilinx UltraScale FPGA 为例，试在 Vivado 下观察这些逻辑资源的分布状况。

第一步，打开 Vivado。

第二步，在 Vivado Tcl Console 中输入如下两条 Tcl 命令：

```
set part [lindex [get_parts xcku040*] 0]
link_design -part $part
```

2．创建一个 Vivado 工程，试给出 Vivado 的设计流程（设计中仅包含 RTL 代码）。

3．创建一个 Vitis HLS 工程，试给出 Vitis HLS 的设计流程（设计中仅包含 C++ 代码）。

4．创建一个 Model Composer 工程，试给出 Model Composer 的设计流程（设计中仅包含 HDL 模型）。

5．试解释为什么流水线技术可以提高系统处理速度。

6．某设计中用 HDL 代码描述了一个可支持异步复位的移位寄存器，目标芯片为 UltraScal 系列 FPGA，试判断该移位寄存器会映射为哪类逻辑资源。

7．试给出同步设计相比异步设计的优势。

8．试解释乒乓操作为什么会降低数据预处理模块的时序压力。

9．什么是逻辑级数？如何判断逻辑级数过高是造成时序违例的主要原因？

10．如何根据时序报告判断当前设计的 F_{max}？

11．如何计算设计中指定模块的 Latency？

12．试描述 Latency 和 F_{max} 之间是如何相互影响的。

第 2 章

优化时钟网络

2.1 时钟资源

时钟资源主要包括全局时钟管脚、时钟缓冲器和时钟管理单元。本节重点介绍 7 系列 FPGA、UltraScale/UltraScale+ FPGA 和 Versal ACAP 中的时钟资源。

2.1.1 7 系列 FPGA 中的时钟资源

1. 全局时钟管脚

7 系列 FPGA 是基于 28nm 工艺制程的芯片。在 7 系列 FPGA 中，每个输入/输出区域（I/O Bank）包含 50 个输入/输出管脚，其中有 4 对（8 个）全局时钟管脚，称为 CCIO（Clock-Capable IO）。每对 CCIO 包含一个 P 端和一个 N 端，用于构成差分 IO，从而使得输入时钟支持差分时钟。若外部时钟是单端时钟，则需要连接到 CCIO 的 P 端。此时，相应的 N 端只能用作通用 IO，而不能再当作另一个全局单端时钟管脚。在这 4 对 CCIO 中，有 2 对 MRCC（Multi-Region Clock-Capable）和 2 对 SRCC（Single-Region Clock-Capable）。当它们未被当作时钟管脚使用时，可用作通用 IO。

> 设计规则 1：确保外部时钟由全局时钟管脚进入 FPGA 内部。

2. 时钟缓冲器

7 系列 FPGA 中的时钟缓冲器分为全局时钟缓冲器（Global Clock Buffer）和区域时钟缓冲器（Regional Clock Buffer）。全局时钟缓冲器是指由 BUFGCTRL 配置成的 BUFG、BUFGCE、BUFGMUX 和 BUFGMUX_CTRL。7 系列 FPGA 有 32 个全局时钟缓冲器。这 32 个全局时钟缓冲器位于全局时钟列，不属于任何时钟区域（Clock Region）。其中，有 16 个全局时钟缓冲器位于全局时钟列的上侧，另外 16 个全局时钟缓冲器位于全局时钟列的下侧，如图 2-1 所示。在图 2-1 中，XmYn 是时钟区域坐标，每个时钟区域的高度为 50 个 CLB（Configurable Logic Block），宽度为整个芯片宽度的一半。同时，可以看到全局时钟列位于芯片的中心位置，垂直时钟布线资源也位于此。水平时钟布线资源则位于每个时钟区域水平方向的中心位置。

之所以称为全局时钟缓冲器，是因为它们输出的时钟可经全局时钟树（Global Clock Tree）到达 FPGA 内部逻辑以及输入/输出逻辑（指 ILOGIC 和 OLOGIC，位于输入/输出区域）的时钟端口，而其输入时钟可来自与之同侧的 MRCC、SRCC、CMT（时钟管理带，Clock Management Tile，一个 CMT 包含一个 MMCM 和一个 PLL）、BUFG 或高速收发器的输出时钟 RXOUTCLK/TXOUTCLK，如图 2-2 所示。

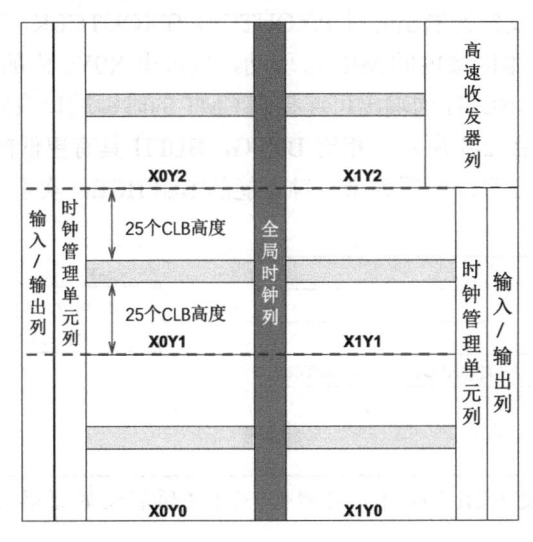

图 2-1

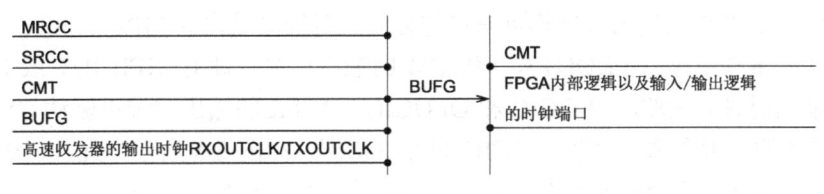

图 2-2

区域缓冲器也称为局部缓冲器，包括 BUFH、BUFIO、BUFR 和 BUFMR。之所以称之为区域缓冲器，是因为它们输出的时钟经区域时钟树（Regional Clock Tree）之后可到达特定区域内的逻辑时钟端口。区域缓冲器隶属于时钟区域。每个时钟区域包括 12 个 BUFH、4 个 BUFIO、4 个 BUFR 和 2 个 BUFMR，如图 2-3 所示。

4个BUFIO，4个BUFR，2个BUFMR，12个BUFH

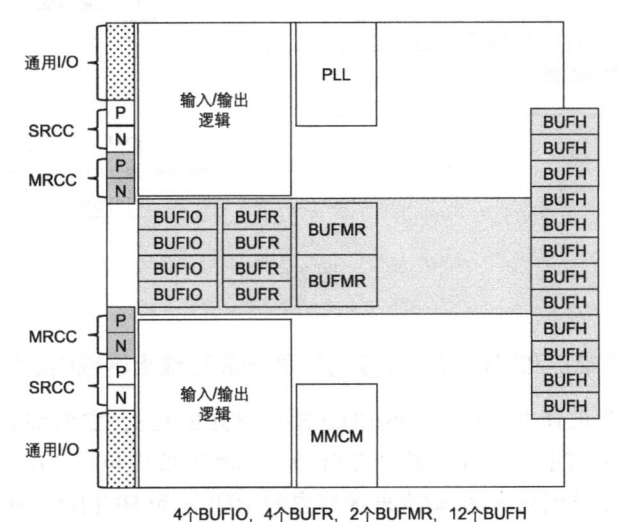

图 2-3

BUFH 的输入时钟可来自与之同一时钟区域或相邻水平时钟区域内的 MRCC、SRCC、CMT、BUFG 或高速收发器的输出时钟 RXOUTCLK/TXOUTCLK。以图 2-1 为例,时钟区域 X1Y0 的 BUFH 可由本区域内的 MRCC 驱动,也可由 X0Y0 内的 MRCC 驱动,这就是相邻水平时钟域的概念。BUFH 的输出可连接到 CMT 的时钟端口或同一时钟区域内 BUFG 可到达的时钟端口,如图 2-4 所示。相比 BUFG,BUFH 具有更低的功耗,其输出时钟抖动更小。此外,BUFH 还可以配置为带时钟使能的 BUFHCE,实现门控时钟功能。

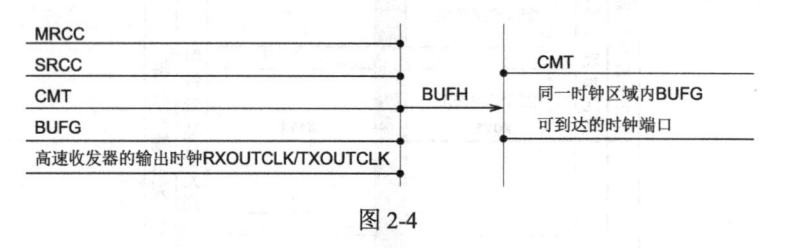

图 2-4

> 设计规则 2:使用 BUFH 时,要确保 BUFH 所驱动的逻辑资源可以放置在一个时钟区域内。

BUFIO 和 BUFR 的输入时钟均可来自与之同一时钟区域内的 MRCC、SRCC、MMCM 的输出时钟 CLKOUT0~CLKOUT3、MMCM 的输出反馈时钟 CLKFBOUT 或 BUFMR。BUFIO 的输出时钟只能驱动 ILOGIC 和 OLOGIC。BUFR 的输出时钟可驱动 CMT 和与之同一时钟区域的逻辑资源,包括 ILOGIC 和 OLOGIC,如图 2-5 所示。此外,BUFR 还具备分频功能,可支持的分频因子为 1~8(包含 1 和 8)的整数。

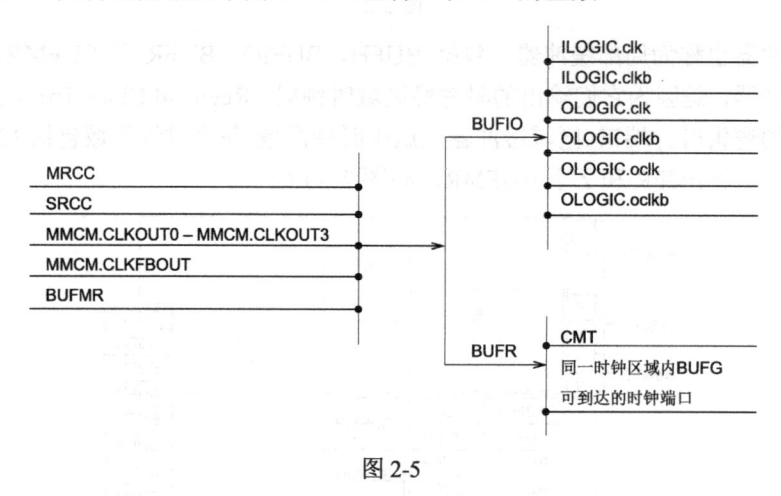

图 2-5

应用案例 1:数据端口和时钟端口可容纳于同一输入/输出区域内的源同步设计

通常,在源同步设计中会用到 OSERDES 或 ISERDES,它们都有时钟端口 CLK 和 CLKDIV。其中,CLKDIV 是 CLK 的分频时钟。此时可采用如图 2-6 所示的时钟方案。时钟由 CCIO 进入,连接到同一输入/输出区域内的 BUFR 和 BUFIO,借助 BUFR 的分频功能产生 CLKDIV,而 BUFIO 的输出直接连接到 CLK 端口。

图 2-6

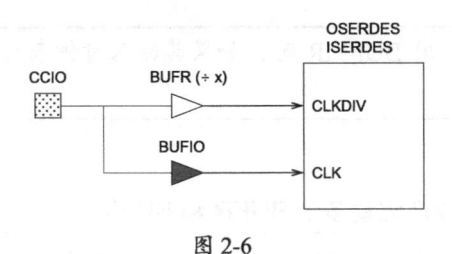

BUFMR 的输入时钟可以是 MRCC，也可以是高速收发器的输出时钟 RXOUTCLK 或 TXOUTCLK，而其输出时钟可以是与之同一时钟区域内的 BUFIO/BUFR，或与之相邻同列上方/下方的时钟区域内的 BUFIO/BUFR，如图 2-7 所示。

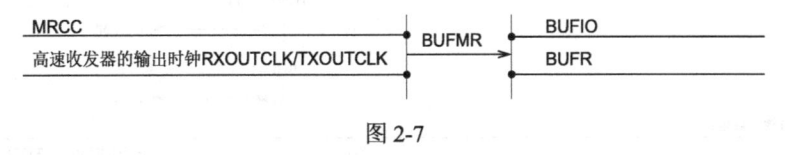

图 2-7

所谓"相邻同列上方/下方的时钟区域"，可结合图 2-8 进行理解。图中，BUFMR 位于时钟区域 X0Y1，与之相邻同列上方的时钟区域是 X0Y2，与之相邻同列下方的时钟区域是 X0Y0，因此，BUFMR 可驱动这三个时钟区域内的 BUFIO/BUFR。此外，BUFMR 还可配置为 BUFMRCE，以保证提供给 BUFR 和 BUFIO 的时钟是同步且同相位的。

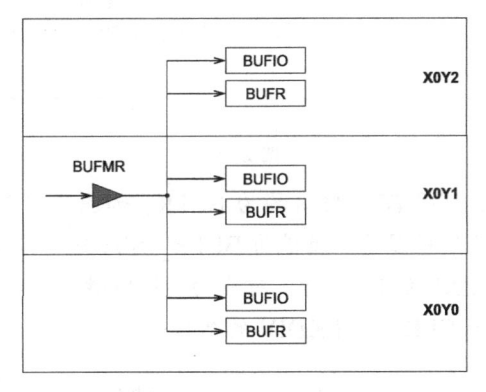

图 2-8

BUFMR 的输入时钟可以来自 MRCC，但不能来自 SRCC，这正是 MRCC 和 SRCC 的区别，如图 2-9 所示。两者的输入时钟均来自 FPGA 外部，输出时钟的差异仅在于 MRCC 可驱动的对象多了与之同一时钟区域内的两个 BUFMR。

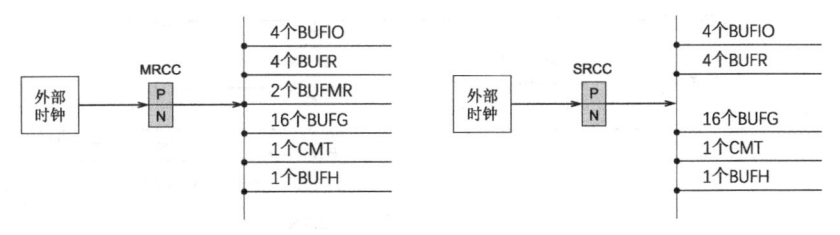

图 2-9

💡 **设计规则 3**：在使用 BUFMR 时，如果其输入时钟来自 FPGA 外部，那么要确保该时钟由 MRCC 进入。

应用案例 2：使用 BUFMR 驱动多个 BUFR 和 BUFIO

在使用多个 ISERDES/OSERDES，如进行源同步设计时，如果它们不在同一个输入/输出区域，但在相邻同列的输入/输出区域，则可行的一种时钟方案就是使用 BUFMR 驱动多个 BUFR 和 BUFIO，借助 BUFR 的分频功能产生 CLKDIV，如图 2-10 所示。

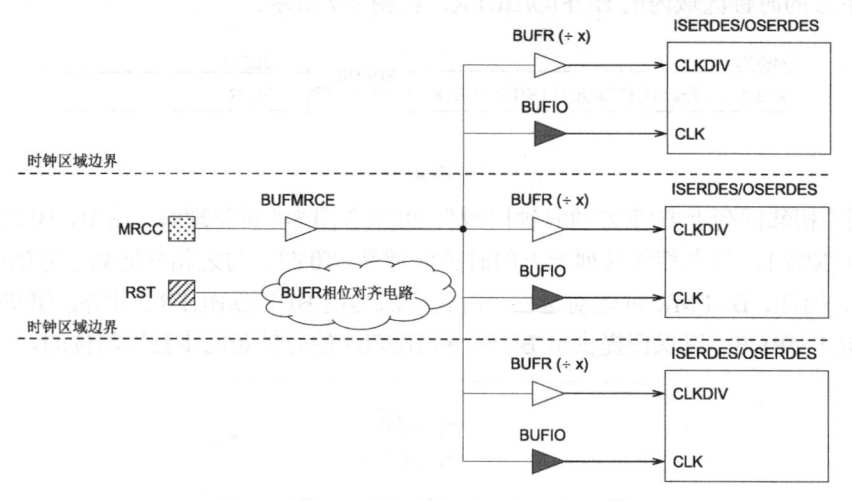

图 2-10

如果 BUFIO 不够用，可将 BUFIO 替换为 BUFR，如图 2-11 所示。需要注意的是连接到 CLK 端口的 BUFR，其分频因子（属性为 BUFR_DIVIDE）应设置为 1，而不能设置为 BYPASS（当 BUFR_DIVIDE 为 1～8 时，对应的延迟特性是一致的，但与 BYPASS 对应的延迟特性不同），以保证 CLK 和 CLKDIV 的相位一致。

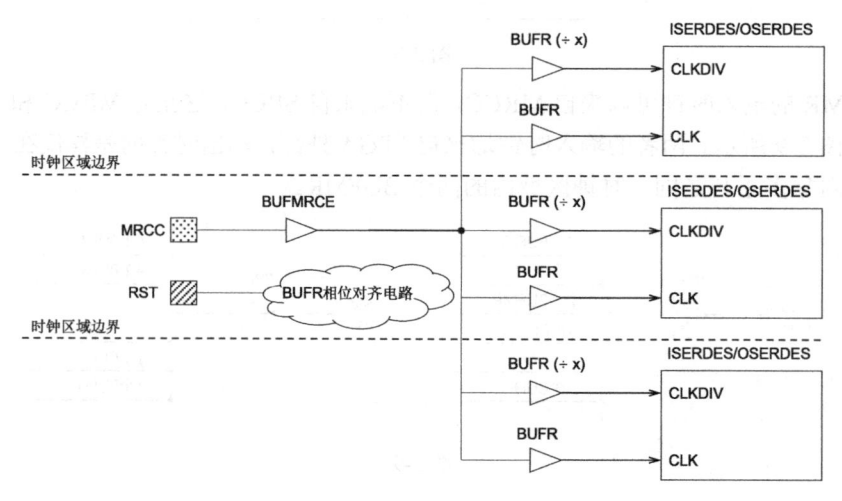

图 2-11

7 系列 FPGA 的每个输入/输出区域都包含 4 个高性能时钟 HPC（High Performance Clocks），它们具有最低的抖动和最小的占空比失真。这 4 个时钟只能由同一时钟区域内 MMCM 的 CLKOUT0～CLKOUT3 驱动，且它们与同一时钟区域内的 BUFIO 和 BUFR 之间有专用布线资源，因此，在使用 MMCM 生成时钟驱动 BUFIO/BUFR 时，要确保使用的是由 CLKOUT0～CLKOUT3 端口输出的时钟。

3. 时钟管理单元

7 系列 FPGA 最多可包含 24 个时钟管理带（CMT），位于时钟管理单元列，紧邻输入/输出列。每个时钟管理带由一个 MMCM（Mixed-Mode Clock Manager）和一个 PLL（Phase-Locked Loop）构成，可实现大范围的频率合成、抖动过滤和改善时钟偏移的功能。两者的输入/输出端口如图 2-12 所示。可以看到，相比 PLL，MMCM 多了 PSCLK、PSEN、PSINCDEC 和 PSDONE 4 个和动态相位调整相关的端口，表明当使用动态相位调整功能时，只能用 MMCM，而不能用 PLL。同时，MMCM 的输出时钟端口有 CLKOUT0～CLKOUT6，且 CLKOUT0～CLKOUT3 有对应的反相时钟 CLKOUT0B～CLKOUT3B。PLL 只有 CLKOUT0～CLKOUT5 这 6 个输出时钟端口。

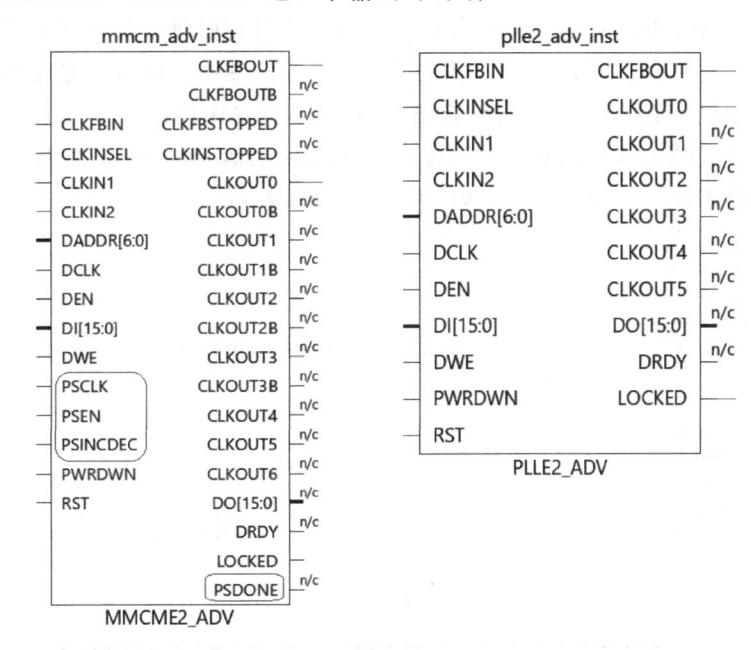

图 2-12

就频率合成而言，MMCM 和 PLL 的工作原理是一样的。其内部均包含一个压控振荡器（Voltage Controlled Oscillator，VCO），用于产生高频时钟。每个输出时钟端口都有分频器，VCO 的输出经分频器产生各端口需要的时钟。例如，输入时钟为 100MHz，输出时钟为 400MHz，采用 MMCM 时，配置参数如图 2-13 所示。图中，CLKIN1_PERIOD 等于 10.000，意味着输入时钟周期为 10ns，对应频率为 100MHz。CLKFBOUT_MULT_F 等于 10.000，DIVCLK_DIVIDE 等于 1，表明 VCO 的频率为 100×10/1MHz，也就是 1000MHz。端口 CLKOUT0 分频器的分频因子为 2.500，因此 CLKOUT0 的输出频率为 1000/2.5MHz，即 400MHz。

如果使用 PLL，配置如图 2-14 所示。图 2-13 与图 2-14 中参数的含义是一致的。因此，不难得出，PLL 中的 VCO 频率为 100×12/1MHz，即 1200MHz，CLKOU0 的输出频率为 1200/3MHz，即 400MHz。

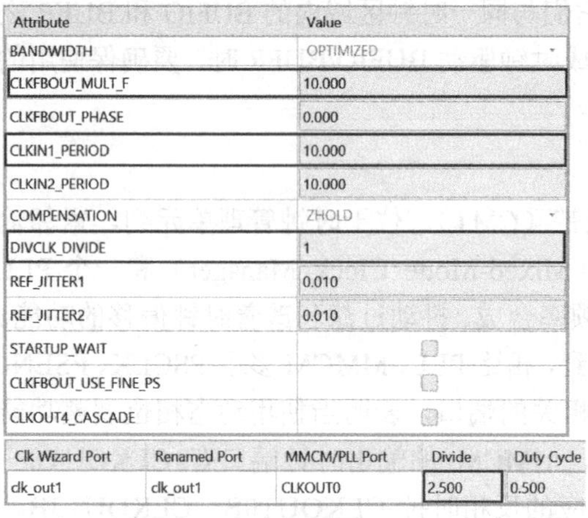

图 2-13

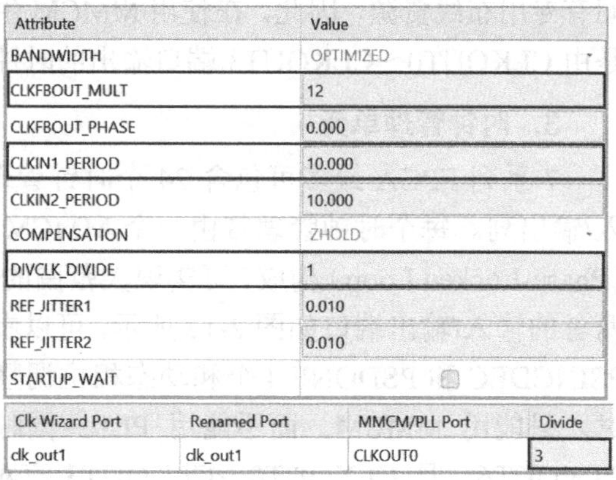

图 2-14

既然在这种情况下，MMCM 和 PLL 都可以生成目标时钟，那么两者是否完全等效呢？这里从生成时钟的性能角度看，如图 2-15 所示。图中，Divide Counter 对应参数 DIVCLK_DIVIDE，Mult Counter 对应参数 CLKFBOUT_MULT_F。不难看出，PLL 输出时钟的峰峰抖动（Pk-to-Pk Jitter）和相位误差（Phase Error）更小。这是因为此时 PLL 的 VCO 为 1200MHz，而 MMCM 的 VCO 为 1000MHz。

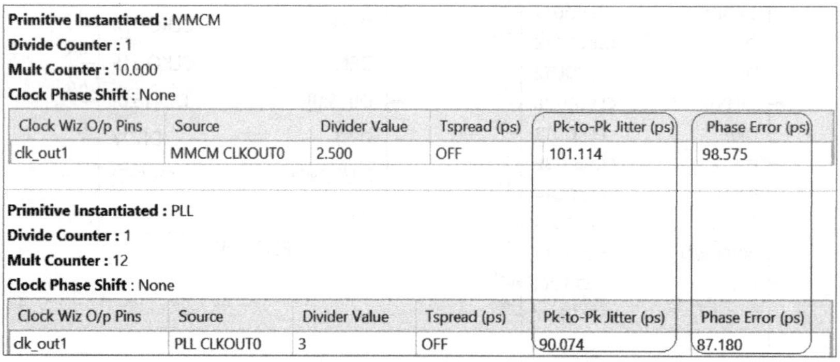

图 2-15

💡 **设计规则 4**：在使用 MMCM 或 PLL 时，应尽可能使 VCO 频率更高一些，这样输出时钟的抖动会更低一些，相位误差会更小一些。

Clocking Wizard IP 会自动计算得出相应数据，但这些参数也是可以手工修改的，如图 2-16 所示，只需要勾选图中方框圈起来的选项即可。但需要注意，VCO 的频率是有限制的。例如，对于 Virtex-7 FPGA，速度等级为-1 的芯片，VCO 频率介于 600～1200MHz，速度等级为-2 的芯片，VCO 频率介于 600～1440MHz，速度等级为-3 的芯片，VCO 频率介于 600～1600MHz。

Attribute	Value
BANDWIDTH	OPTIMIZED
CLKFBOUT_MULT_F	10.000
CLKFBOUT_PHASE	0.000
CLKIN1_PERIOD	10.000
CLKIN2_PERIOD	10.000
COMPENSATION	ZHOLD

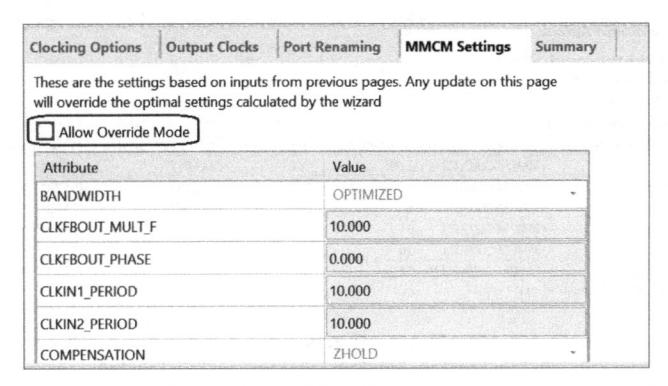

图 2-16

> 💡 **设计规则 5**：在使用 MMCM 或 PLL 时，应尽可能通过 Clocking Wizard IP 的方式实例化，避免使用原语（Primitive）。因为原语中有很多参数需要手工填写，极易导致错误。

应用案例 3：使用 MMCM 生成多个时钟

外部时钟为 33MHz，需要通过此时钟生成 6 个时钟，其频率分别为 528MHz、264MHz、176MHz、132MHz、66MHz 和 33MHz。MMCM 的参数配置分别如图 2-17 和图 2-18 所示。从图 2-17 中可以看到，CLKFBOUT_MULT_F 为 32.000，DIVCLK_DIVIDE 为 1，因此 VCO 频率为 33×32/1MHz，即 1056MHz。图 2-19 显示了 6 个输出时钟的基本性能指标。

Attribute	Value
BANDWIDTH	OPTIMIZED
CLKFBOUT_MULT_F	32.000
CLKFBOUT_PHASE	0.000
CLKIN1_PERIOD	30.303
CLKIN2_PERIOD	10.0
COMPENSATION	ZHOLD
DIVCLK_DIVIDE	1
REF_JITTER1	0.010
REF_JITTER2	0.010
STARTUP_WAIT	☐
CLKFBOUT_USE_FINE_PS	☐
CLKOUT4_CASCADE	☐

图 2-17

Clk Wizard Port	Renamed Port	MMCM/PLL Port	Divide	Duty Cycle	Phase	Use Fine Ps
clk_out1	clk_out1	CLKOUT0	2.000	0.500	0.000	☐
clk_out2	clk_out2	CLKOUT1	4	0.500	0.000	☐
clk_out3	clk_out3	CLKOUT2	6	0.500	0.000	☐
clk_out4	clk_out4	CLKOUT3	8	0.500	0.000	☐
clk_out5	clk_out5	CLKOUT4	16	0.500	0.000	☐
clk_out6	clk_out6	CLKOUT5	32	0.500	0.000	☐

图 2-18

Clocking Primitive Attributes					
Primitive Instantiated : MMCM					
Divide Counter : 1					
Mult Counter : 32.000					
Clock Phase Shift : None					
Clock Wiz O/p Pins	Source	Divider Value	Tspread (ps)	Pk-to-Pk Jitter (ps)	Phase Error (ps)
clk_out1	MMCM CLKOUT0	2.000	OFF	147.017	191.781
clk_out2	MMCM CLKOUT1	4	OFF	162.151	191.781
clk_out3	MMCM CLKOUT2	6	OFF	171.967	191.781
clk_out4	MMCM CLKOUT3	8	OFF	179.394	191.781
clk_out5	MMCM CLKOUT4	16	OFF	211.956	191.781
clk_out6	MMCM CLKOUT5	32	OFF	262.264	191.781
clk_out7	OFF	OFF	OFF	OFF	OFF

图 2-19

MMCM 的 CLKOUT0 和 CLKFBOUT 端口对应的分频器可实现小数分频，分频因子的分辨率为 0.125。而相应的 PLL 是不具备此功能的。

应用案例 4：借助 MMCM 小数分频功能输出目标时钟

已知输入时钟为 100MHz，现在需要生成频率为 320MHz 和 400MHz 的两个时钟。使用 Clocking Wizard IP 调用 MMCM。在 Output Clocks 页面下，clk_out1（对应 MMCM 的 CLKOUT0 端口）的输出频率填写 320，clk_out2（对应 MMCM 的 CLKOUT1 端口）的输出频率填写 400，最终在 Actual 列显示的实际值分别为 320.00000 和 400.00000，达到了预期效果，如图 2-20 所示。此时，MMCM 的参数配置情况如图 2-21 所示，由此可推断出 VCO 的频率为 800MHz。同时可以看到 CLKOUT0 端口的分频器的分频因子为 2.500，CLKOUT1 端口的分频器的分频因子为 2。

Clocking Options	Output Clocks	Port Renaming	MMCM Settings	Summary			
The phase is calculated relative to the active input clock.							
Output Clock	Port Name	Output Freq (MHz)		Phase (degrees)			
		Requested	Actual	Requested		Actual	
☑ clk_out1	clk_out1	320	320.00000	0.000		0.000	
☑ clk_out2	clk_out2	400	400.00000	0.000		0.000	

图 2-20

BANDWIDTH	OPTIMIZED
CLKFBOUT_MULT_F	8.000
CLKFBOUT_PHASE	0.000
CLKIN1_PERIOD	10.000
CLKIN2_PERIOD	10.000
COMPENSATION	ZHOLD
DIVCLK_DIVIDE	1
REF_JITTER1	0.010
REF_JITTER2	0.010
STARTUP_WAIT	☐
CLKFBOUT_USE_FINE_PS	☐
CLKOUT4_CASCADE	☐

Clk Wizard Port	Renamed Port	MMCM/PLL Port	Divide	Duty Cycle
clk_out1	clk_out1	CLKOUT0	2.500	0.500
clk_out2	clk_out2	CLKOUT1	2	0.500

图 2-21

如果将 320 和 400 的位置互换一下，如图 2-22 所示，可以看到，在 Actual 列显示的实际输出值将分别变成 400.00000 和 316.66667，与实际值不符。

图 2-22

如果使用 PLL，如图 2-23 所示，也无法达到预期效果。

图 2-23

使用 Clocking Wizard IP 时，若勾选图 2-24 中的 Phase Alignment 复选框，则 CLKFBOUT 端口会连接 BUFG，BUFG 的输出会连接到 CLKFBIN 端口，如图 2-25 所示。从而使得标记①、④、⑥处的时钟相位一致，标记②、③、⑤处的时钟相位一致，但两组之间的相位是不同的。如果没有勾选 Phase Alignment 复选框，那么 CLKFBOUT 会直接连接到 CLKFBIN，所有输出端口（CLKOUT0～CLKOUT6）若使用了同类型的时钟缓冲器，则输出时钟的相位一致，但与标记①、②处的相位不再一致。由此可见，选项 Phase Alignment 的目的是保证 MMCM 输出时钟与输入时钟的相位一致。如果不需要此功能，MMCM 反馈支路就不会插入 BUFG，这样也可以节省一个 BUFG。

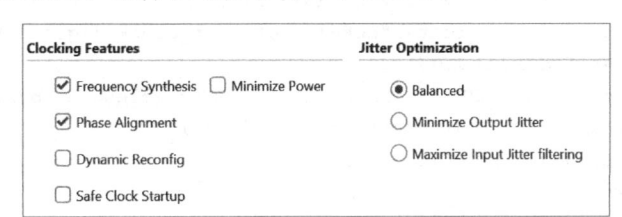

图 2-24

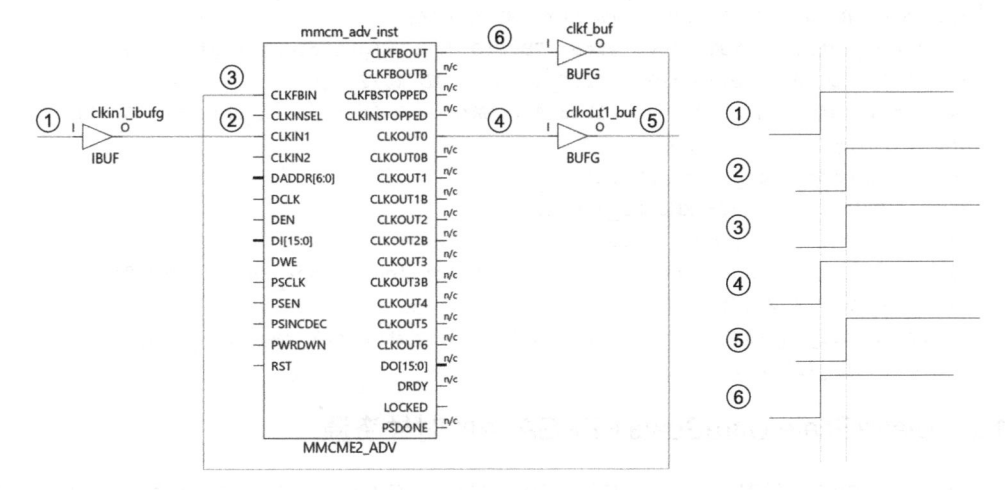

图 2-25

在芯片选型阶段，可在 Vivado 下借助 Tcl 脚本查看指定芯片中的时钟资源分布情况，如 Tcl 代码 2-1 所示。只需要打开 Vivado，在 Tcl Console 中执行此脚本即可。

Tcl 代码 2-1

```
1.  #File: check_clk_resource_7fpga.tcl
2.  set mypart [get_parts xc7k70tfbg484-3]
3.  => xc7k70tfbg484-3
4.  link_design -part $mypart
5.  => design_1
6.  set mrcc_in_bank13 [get_package_pins \
7.  -filter "IS_CLK_CAPABLE==1 && PIN_FUNC =~ *MRCC*" -of [get_iobanks 13]]
8.  => V19 W19 Y18 Y19
9.  set srcc_in_bank13 [get_package_pins \
10. -filter "IS_CLK_CAPABLE==1 && PIN_FUNC =~ *SRCC*" -of [get_iobanks 13]]
11. => V20 W20 W17 Y17
12. set mybufg [get_sites *BUFG*]
13. => BUFGCTRL_X0Y16 BUFGCTRL_X0Y17 ...
14. puts "The number of BUFG in $mypart: [llength $mybufg]"
15. => The number of BUFG in xc7k70tfbg484-3: 32
16. highlight_objects $mybufg -color red
17. set mybufh [get_sites "BUFH*"]
18. => BUFHCE_X0Y36 BUFHCE_X0Y37 ...
19. puts "The number of BUFH in $mypart: [llength $mybufh]"
20. => The number of BUFH in xc7k70tfbg484-3: 96
21. set myclock_region [get_clock_regions]
22. => X0Y0 X1Y0 X0Y1 X1Y1 X0Y2 X1Y2 X0Y3 X1Y3
23. set bufh_in_X0Y0 [get_sites "BUFH*" -of [get_clock_regions X0Y0]]
24. => BUFHCE_X0Y0 BUFHCE_X0Y1 ...
25. puts "The number of BUFH in clock region X0Y0: [llength $bufh_in_X0Y0]"
26. => The number of BUFH in clock region X0Y0: 12
27. mark_objects $bufh_in_X0Y0 -color red
28. get_iobanks
29. => 0 13 14 15 16 33 34 115
30. set mybufr_in_X0Y0 [get_sites "BUFR*" -of [get_clock_regions X0Y0]]
31. => BUFR_X0Y1 BUFR_X0Y0 BUFR_X0Y3 BUFR_X0Y2
32. set mybufio_in_X0Y0 [get_sites "BUFIO*" -of [get_clock_regions X0Y0]]
33. => BUFIO_X0Y1 BUFIO_X0Y0 BUFIO_X0Y3 BUFIO_X0Y2
34. set mybufmr_in_X0Y0 [get_sites "BUFMR*" -of [get_clock_regions X0Y0]]
35. => BUFMRCE_X0Y1 BUFMRCE_X0Y0
36. set mymmcm [get_sites *MMCM*]
37. => MMCME2_ADV_X0Y3 MMCME2_ADV_X0Y2 ...
38. mark_objects $mymmcm -color blue
39. set mymmcm_in_X0Y0 [get_sites "MMCM*" -of [get_clock_regions X0Y0]]
40. => MMCME2_ADV_X0Y0
41. set mypll_in_X0Y0 [get_sites "PLL*" -of [get_clock_regions X0Y0]]
42. => PLLE2_ADV_X0Y0
```

2.1.2　UlatraScale/UltraScale+ FPGA 中的时钟资源

UltraScale FPGA 是基于 20nm 工艺制程的芯片，而 UltraScale+ FPGA 是基于 16nm 工

艺制程的芯片。尽管两者的工艺制程不同，但其内部结构是一致的。若无特殊声明，后面阐述的 UltraScale FPGA 架构也适用于 UltraScale+ FPGA。

1. 全局时钟管脚

在 UltraScale FPGA 中，每个输入/输出区域（I/O Bank）都位于单一的时钟区域 CR（Clock Region）内，且都包含 52 个输入/输出管脚。在这 52 个管脚中，有 4 对（8 个）全局时钟管脚 GC（Global Clock）I/O。其使用方法与 7 系列 FPGA 是一致的。不同之处在于，这 4 对 GC 的地位是相等的，不再有 MRCC 和 SRCC 之分。UltraScale+ FPGA 新增了高密度输入/输出区域（High Density I/O Bank，HD I/O Bank），位于此区域内的全局时钟管脚 HDGC 只能通过 BUFGCE 连接到 MMCM 或 PLL。

> 💡 **设计规则 6**：使用 UltraScale+ FPGA 时，如果外部时钟需要借助 MMCM 生成新的时钟，就要避免外部时钟由 HDGC 进入。如果外部时钟只能由 HDGC 进入，那么需要将属性 CLOCK_DEDICATED_ROUTE 的值设置为 FALSE。

2. 时钟缓冲器

7 系列 FPGA 既包含全局时钟缓冲器，又包含区域时钟缓冲器。UltraScale FPGA 简化了时钟缓冲器，即只有全局时钟缓冲器。包含输入/输出列的时钟区域内有 24 个 BUFGCE、4 个 BUFGCE_DIV 和 8 个 BUFGCTRL，同时只能使用其中的 24 个，如图 2-26 所示。

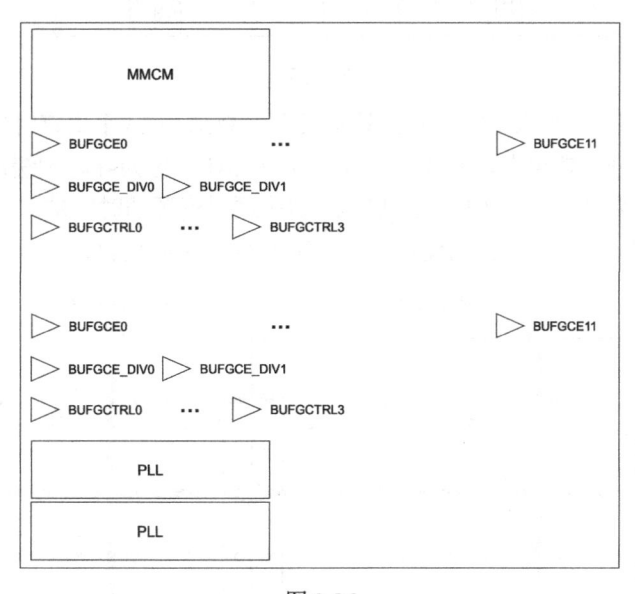

图 2-26

这些全局时钟缓冲器位于时钟列，可驱动水平时钟布线/分发轨道和垂直时钟布线/分发轨道。其中，分发轨道是 7 系列 FPGA 所没有的。这些轨道均位于时钟区域的中心位置，如图 2-27 所示（有的芯片只有一侧有高速收发器）。不难看出，每个时钟区域的宽度相比 7 系列 FPGA 有所缩减，不再是半个芯片的宽度，高度由 7 系列 FPGA 中的 50 个 CLB 变为 60 个 CLB。时钟区域的粒度更加细化。无论是水平时钟布线轨道和分发轨道，还是垂直时钟布线轨道和分发轨道，都以时钟区域作为边界。这意味着如果某个时钟区域内的资源未使用时钟，工具就会关闭相应的轨道，节省功耗。布线轨道可驱动相邻时钟区域内的布线轨道和分发轨道。分发轨道只能驱动相邻时钟区域内的水平分发轨道。布线轨道的目的是将时钟从全局时钟缓冲器布线到某个中心点。在这个中心点，时钟经分发轨道连接到其负载的时钟端口。分发轨道可进一步移动这个点，以改善时钟的局部偏移。这个点被称为时钟根节点（Clock Root）。

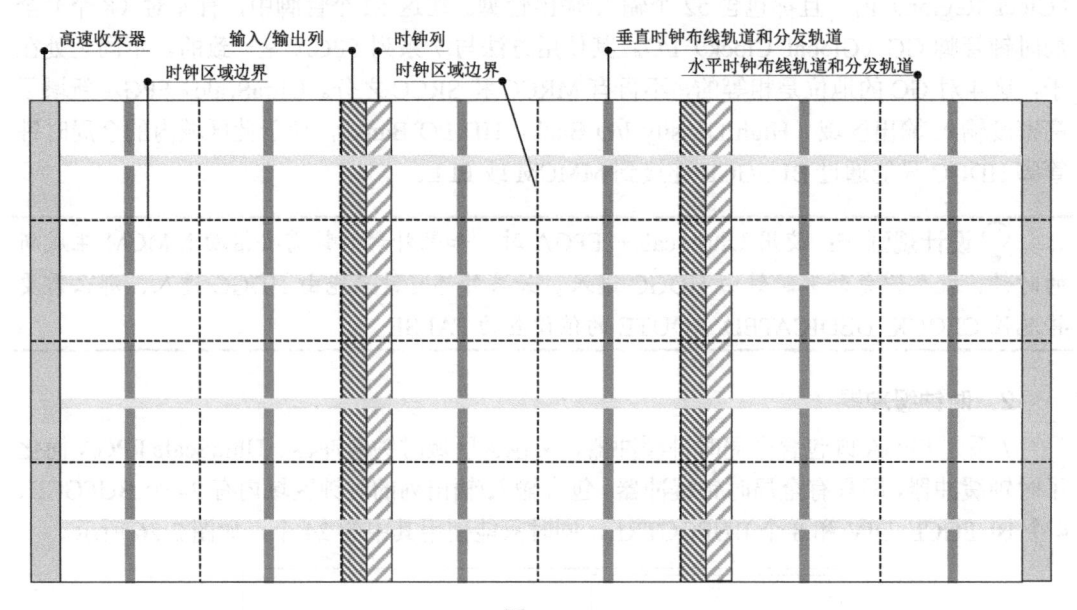

图 2-27

每个时钟区域有 24 个水平时钟布线轨道及分发轨道和 24 个垂直时钟布线轨道和分发轨道。在水平时钟分发轨道上，有 32 个 BUFCE_LEAF，称为叶时钟缓冲器。时钟从水平时钟分发轨道上下来之后，经 BUFCE_LEAF 到达逻辑资源的时钟端口，如图 2-28 所示。BUFCE_LEAF 只能由 Vivado 自动使用，而不能在代码中实例化。

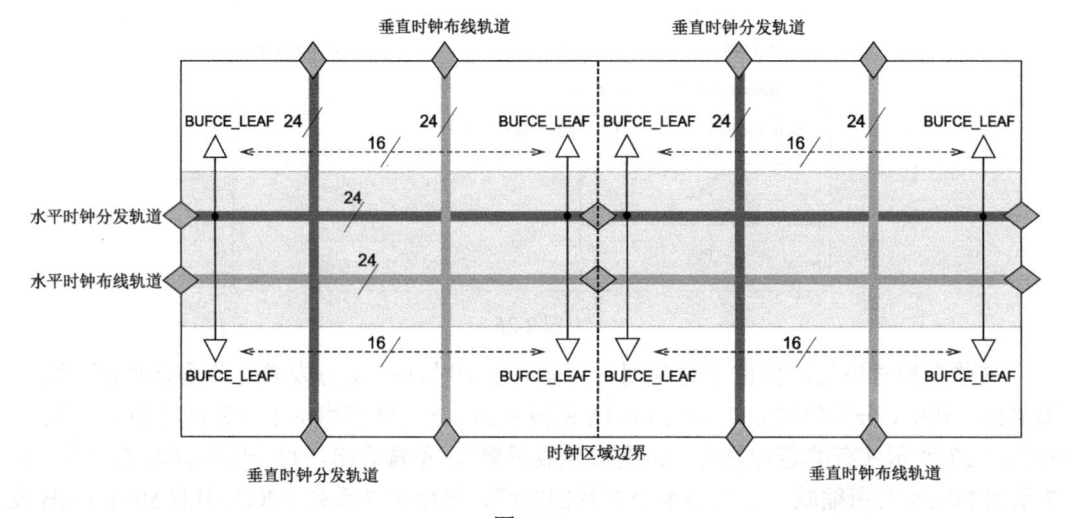

图 2-28

进一步理解水平时钟布线轨道和分发轨道、垂直时钟布线轨道和分发轨道，如图 2-29

所示。图中有两个时钟网络，左侧的时钟根节点在时钟区域 X0Y2，右侧的时钟根节点在时钟区域 X4Y1。换言之，时钟根节点其实就是时钟负载在芯片中布局的中心位置，其值为时钟区域。

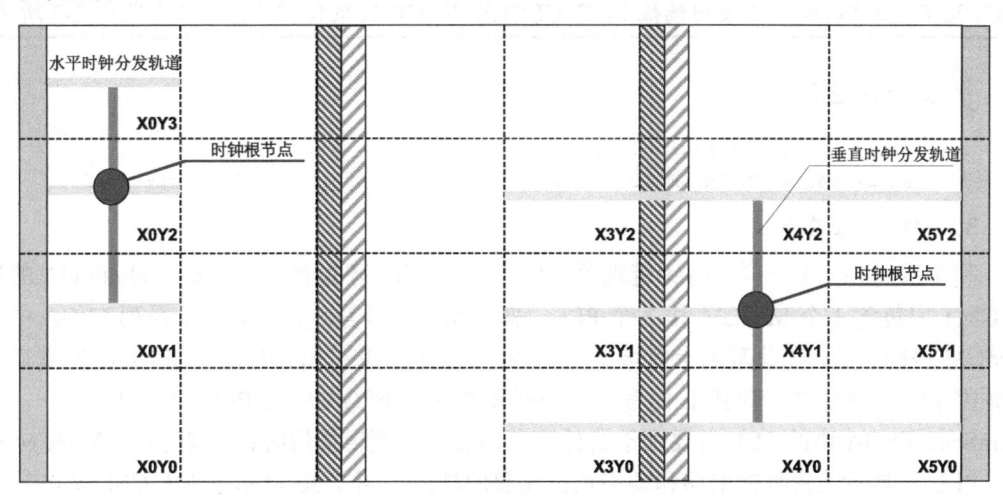

图 2-29

UltraScale FPGA 有独立的 BUFGCE，无须通过 BUFGCTRL 配置而成。但 BUFGCTRL 仍是可配置的，BUFGCE_1、BUFGMUX 和 BUFGMUX_1 都是通过 BUFGCTRL 配置生成的。BUGCE_DIV 取代了 BUFR，但比 BUFR 具有更强大的驱动能力，因为它已成为全局时钟缓冲器。同时，BUFGCE_DIV 具有分频功能，分频因子可以是 1~8（包含 1 和 8）的整数。只是当分频因子为奇数时，输出时钟的占空比将不再是 50%。UltraScale FPGA 新增了 BUFG_GT。BUFG_GT 只可以由高速收发器或 RFSoC 中的 ADC/DAC 模块驱动。BUFG_GT_SYNC 是 BUFG_GT 的同步器，当 Vivado 推断出 BUFG_GT 时会自动在设计中插入 BUFG_GT_SYNC。和 BUFGCE_DIV 类似，BUFG_GT 也具有分频功能，可用的分频因子为 1~8（包含 1 和 8）的整数。分频因子由 DIV 端口输入。DIV 位宽为 3，当其为 3'b000 时，对应的分频因子为 1。在包含高速收发器的时钟区域内有 24 个 BUFG_GT。

在 Zynq UltraScale+ MPSoC 中新增了一种全局时钟缓冲器 BUFG_PS（在 Zynq 7000 系列 FPGA 中是没有的）。该缓冲器位于内部 ARM 处理器的旁边。PS 侧的输出时钟需要经此缓冲器访问 PL（Programmatic Logic）侧的时钟布线资源，从而驱动 PL 侧的逻辑资源。BUFG_PS 的个数因不同的芯片而异。例如，ZU4EG 有 96 个 BUFG_PS，而 ZU2CG 有 72 个 BUFG_PS。可借助 Tcl 代码 2-2 查看芯片中 BUFG_PS 的个数和位置。

Tcl 代码 2-2

```
1.  #File: check_bufg_ps.tcl
2.  link_design -part [get_parts xczu4eg-fbvb900-1-e]
3.  set bufg_ps [get_sites "BUFG_PS*"]
4.  => BUFG_PS_X0Y72 BUFG_PS_X0Y73 BUFG_PS_X0Y74 ...
5.  llength $bufg_ps
6.  => 96
7.  mark_objects $bufg_ps -color red
```

> 💡 **设计规则 7**：应由 Vivado 自动决定设计中全局时钟缓冲器的位置，尽量避免通过约束的方式进行干预。如果需要指定设计中某些全局时钟缓冲器的位置，应使用属性 CLOCK_REGION，而不要使用属性 LOC。CLOCK_REGION 的使用方法如 Tcl 代码 2-3 所示。

📝 Tcl 代码 2-3

```
1.  #File: set_clock_region.tcl
2.  set_property CLOCK_REGION X4Y6 [get_cells {sys_clk_pll/inst/clkf_buf}]
```

3. 时钟管理单元

和 7 系列 FPGA 一样，时钟管理单元位于 CMT 内。不同的是，UltraScale FPGA 的每个 CMT 内包含 1 个 MMCM 和 2 个 PLL。就 MMCM 而言，其功能和 7 系列 FPGA 中的 MMCM 保持一致，在此基础上新增了一个特性：动态调整 CLKOUT0～CLKOUT6 端口的分频因子时无须复位。就 PLL 而言，其功能和 7 系列 FPGA 中的 PLL 相比更加专一，即 UltraScale FPGA 中的 PLL 主要用于给外部存储器接口逻辑提供时钟。同时，与 MMCM 相比，PLL 不具备改善时钟偏移和动态相位调整的功能。图 2-30 显示了 MMCM 和 PLL 的输入/输出端口。不难看出，PLL 的输出时钟个数与 7 系列 FPGA 相比明显减少。

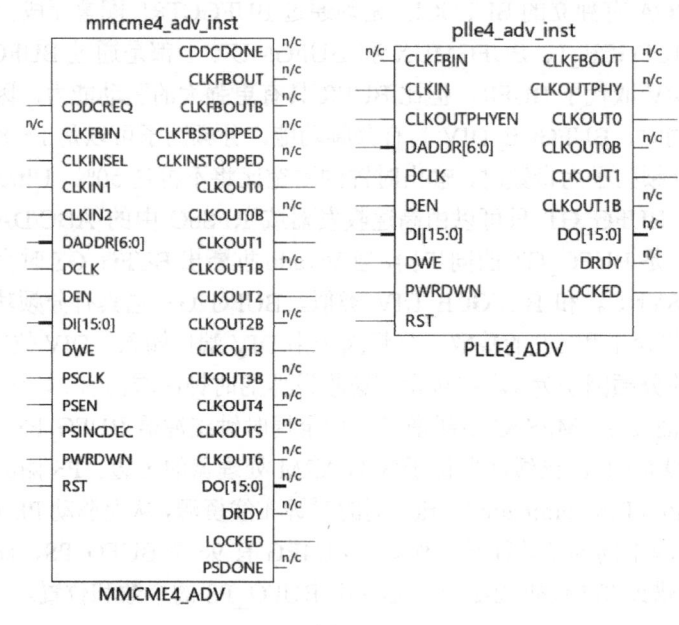

图 2-30

可借助 Tcl 代码 2-4 分析 UltraScale/UltraScale+ FPGA 芯片中的时钟资源，只需要打开 Vivado，在 Tcl Console 中执行此脚本即可。

📝 Tcl 代码 2-4

```
1.  #File: check_clk_resource_us.tcl
2.  set mypart [get_parts xcku040-fbva676-1-c]
3.  link_design -part $mypart
4.  set gc_in_bank44 [get_package_pins -of [get_iobanks 44]\
```

```
5.      -filter "IS_GLOBAL_CLK"]
6.  puts "The number of global clock pins in each bank: [llength $gc_in_bank44]"
7.  => The number of global clock pins in each bank: 8
8.  mark_objects $gc_in_bank44 -color red
9.  set cr [get_clock_regions -of [get_iobanks 44]]
10. set bufgce_in_x0y0 [get_sites -filter "SITE_TYPE==BUFGCE" \
11.     -of [get_clock_regions $cr]]
12. set bufgce_div_in_x0y0 [get_sites -filter "SITE_TYPE==BUFGCE_DIV" \
13.     -of [get_clock_regions $cr]]
14. set bufgctrl_in_x0y0 [get_sites -filter "SITE_TYPE==BUFGCTRL" \
15.     -of [get_clock_regions $cr]]
16. set bufce_leaf_in_x0y0 [get_sites "BUFCE_LEAF*" \
17.     -of [get_clock_regions $cr]]
18. set bufg_gt_in_x3y0 [get_sites -filter "SITE_TYPE==BUFG_GT" \
19.     -of [get_clock_regions X3Y0]]
20. set mmcm_in_x0y0 [get_sites "MMCM*" -of [get_clock_regions $cr]]
21. set pll_in_x0y0 [get_sites "PLL*" -of [get_clock_regions $cr]]
22. puts "#BUFGCE in each clock region: [llength $bufgce_in_x0y0]"
23. => #BUFGCE in each clock region: 24
24. puts "#BUFGCE_DIV in each clock region: [llength $bufgce_div_in_x0y0]"
25. => #BUFGCE_DIV in each clock region: 4
26. puts "#BUFGCTRL in each clock region: [llength $bufgctrl_in_x0y0]"
27. => #BUFGCTRL in each clock region: 8
28. puts "#BUFCE_LEAF in each clock region: [llength $bufce_leaf_in_x0y0]"
29. => #BUFCE_LEAF in each clock region: 32
30. puts "#BUFG_GT in clock region containing GT: [llength $bufg_gt_in_x3y0]"
31. => #BUFG_GT in clock region containing GT: 24
32. puts "#MMCM in each clock region: [llength $mmcm_in_x0y0]"
33. => #MMCM in each clock region: 1
34. puts "#PLL in each clock region: [llength $pll_in_x0y0]"
35. => #PLL in each clock region: 10
36. mark_object $bufgce_in_x0y0 -color green
37. mark_object $bufgce_div_in_x0y0 -color blue
38. mark_object $bufgctrl_in_x0y0 -color yellow
39. mark_object $bufce_leaf_in_x0y0 -color magenta
40. mark_object $bufg_gt_in_x3y0 -color red
41. mark_object $mmcm_in_x0y0 -color cyan
42. mark_object $pll_in_x0y0 -color orange
```

2.1.3 Versal ACAP 中的时钟资源

1. 全局时钟管脚

Versal 是基于 7nm 工艺制程的芯片。和 UltraScale FPGA 一样，其内部被分割为矩阵形式的时钟区域（Clock Region，CR）。不同的是，每个 CR 的高度由 60 个 CLB 变为 96 个 CLB，同时，Versal 中的 CR 还分为满 CR（高度为 96 个 CLB）和半 CR（高度为 48 个 CLB）。

Versal 中的全局时钟布线轨道和分发轨道既有水平方向的，也有垂直方向的，且都以时钟区域为边界。对于满 CR，水平时钟布线轨道和分发轨道位于 CR 水平方向的中心位置；对于半 CR，水平时钟布线轨道和分发轨道位于 CR 的底部。垂直时钟布线轨道和分发轨道位于两个背靠背 CR 的临界处，如图 2-31 所示。

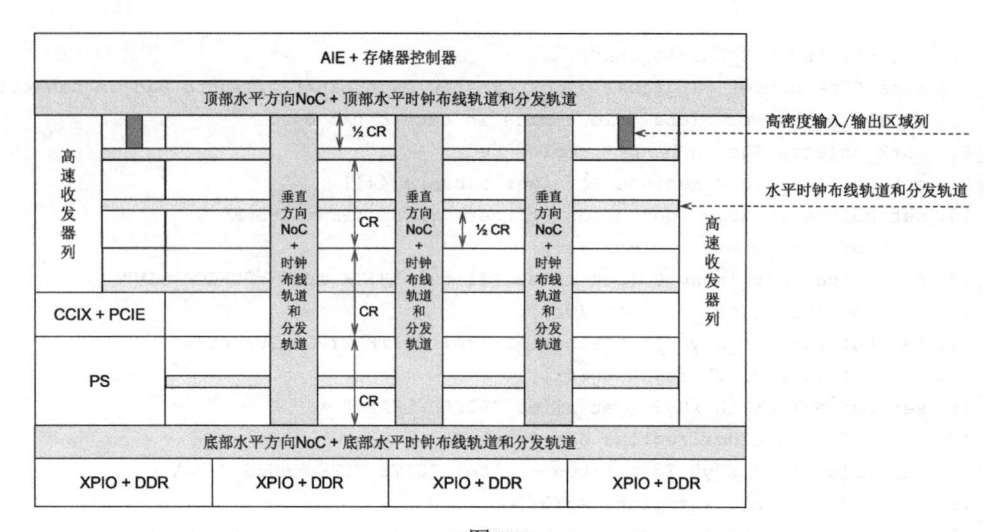

图 2-31

Versal 有两种输入/输出区域，分别是高性能输入/输出区域（XPIO Bank）和高密度输入/输出区域（HDIO Bank）。XPIO 主要用于实现高性能接口，支持的电平范围为 1.0～1.5V，同时，硬核存储单元控制器也在其中。HDIO 可实现的接口速率比 XPIO 低一些，可支持的电平范围为 1.8～3.3V。XPIO 和 HDIO 支持的电平标准是没有重叠的。每个 XPIO 有 54 个管脚，其中有 4 对（8 个）全局时钟管脚 GCIO；每个 HDIO 有 24 个管脚，其中有 2 对（4 个）全局时钟管脚 HDGC。这些全局时钟管脚的用法与 UltraScale FPGA 中的保持一致。

2. 时钟缓冲器

与 UltraScale FPGA 一样，Versal 中的时钟缓冲器均为全局时钟缓冲器。在 XPIO 对应的 CR 中，每个 CR 有 24 个 BUFGCE、8 个 BUFGCTRL 和 4 个 BUFGCE_DIV，但同时只能使用其中的 24 个。在 HDIO 对应的 CR 中，每个 CR 只有 4 个 BUFGCE。紧邻高速收发器的 CR 会分布一些 BUFG_GT，紧邻 PS 的 CR 会分布一些 BUFG_PS。这些时钟缓冲器的功能和 UltraScale FPGA 中的保持一致。Versal 中新增了一种全局时钟缓冲器 BUFG_FABRIC，紧邻垂直方向的 NoC（Network on Chip）。时钟缓冲器在芯片中的分布情况如图 2-32 所示，同时，可以借助 Tcl 代码 2-5 进行分析。

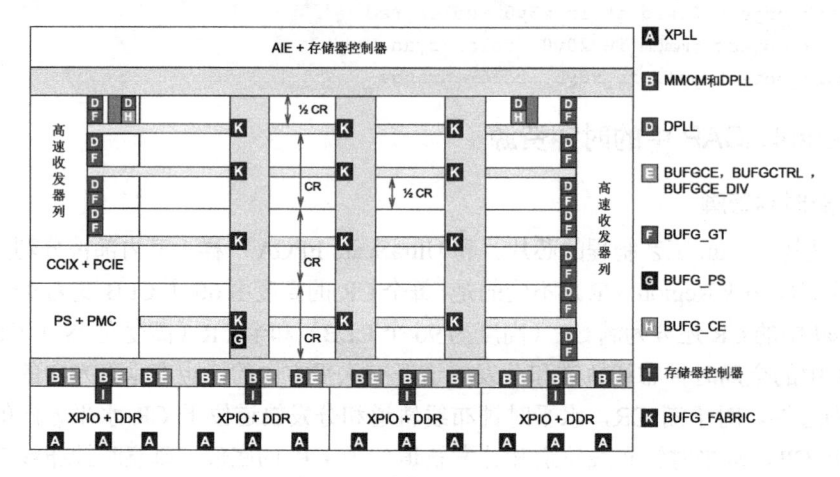

图 2-32

Tcl 代码 2-5

```
1.  #File: check_clock_buffer_versal.tcl
2.  link_design -part [get_parts xcvc1902-viva1596-1LHP-i-L]
3.  set hdio_banks [get_iobanks -filter "BANK_TYPE==BT_HIGH_DENSITY"]
4.  set xpio_banks [get_iobanks -filter "BANK_TYPE==BT_XP"]
5.  set gc_in_bank700 [get_package_pins -filter "PIN_FUNC=~*GC_XCC*" \
6.  -of [get_iobanks 700]]
7.  set gc_in_bank306 [get_package_pins -filter "PIN_FUNC=~*HDGC*" \
8.  -of [get_iobanks 306]]
9.  puts "#GC in each XPIO: [llength $gc_in_bank700]"
10. => #GC in each XPIO: 8
11. puts "#GC in each HDIO: [llength $gc_in_bank306]"
12. => #GC in each HDIO: 4
13. set xpcr [get_clock_regions X4Y0]
14. set bufgce_xpio  [get_sites -filter "SITE_TYPE==BUFGCE" -of $xpcr]
15. set bufgce_div   [get_sites -filter "SITE_TYPE==BUFGCE_DIV" -of $xpcr]
16. set bufgctrl    [get_sites -filter "SITE_TYPE==BUFGCTRL" -of $xpcr]
17. set hdcr [get_clock_regions X0Y4]
18. set bufgce_hdio  [get_sites *BUFGCE_HDIO* -of $hdcr]
19. set bufg_gt [get_sites -filter "SITE_TYPE==BUFG_GT" \
20. -of [get_clock_regions X0Y4]]
21. set bufg_ps [get_sites "BUFG_PS*" -of [get_clock_regions X1Y1]]
22. set bufg_fabric  [get_sites *BUFG_FABRIC* -of [get_clock_regions X1Y4]]
23. puts "#BUFGCE in each CR corresponding to XPIO: [llength $bufgce_xpio]"
24. => #BUFGCE in each CR corresponding to XPIO: 24
25. puts "#BUFGCE_DIV in each CR corresponding to XPIO: [llength $bufgce_div]"
26. => #BUFGCE_DIV in each CR corresponding to XPIO: 4
27. puts "#BUFGCTRL in each CR corresponding to XPIO: [llength $bufgctrl]"
28. => #BUFGCTRL in each CR corresponding to XPIO: 8
29. puts "#BUFGCE in each CR corresponding to HDIO: [llength $bufgce_hdio]"
30. => #BUFGCE in each CR corresponding to HDIO: 4
31. puts "#BUFG_GT in each CR adjacent to GT: [llength $bufg_gt]"
32. => #BUFG_GT in each CR adjacent to GT: 24
33. puts "#BUFG_PS in each CR adjacent to PS: [llength $bufg_ps]"
34. => #BUFG_PS in each CR adjacent to PS: 12
35. puts "#BUFG_FABRIC available in some CRs: [llength $bufg_fabric]"
36. => #BUFG_FABRIC available in some CRs: 24
37. mark_objects $bufgce_xpio -color red
38. mark_objects $bufgce_div -color yellow
39. mark_objects $bufgctrl -color green
40. mark_objects $bufgce_hdio -color magenta
41. mark_objects $bufg_gt -color orange
42. mark_objects $bufg_ps -color blue
43. mark_objects $bufg_fabric -color cyan
```

设计规则 8：虽然 BUFG_FABRIC 是全局时钟缓冲器，但是只能用于驱动设计中的高扇出网线，不能用于驱动时钟信号。

在 Versal 中也有叶时钟缓冲器 BUFDIV_LEAF，取代了 UltraScale FPGA 中的

BUFCE_LEAF，功能更加丰富，支持静态分频（分频因子由属性 BUFLEAF_DIVIDE 确定，一旦确定，不可动态更改），可支持的分频因子为 1、2、4 和 8。借助 BUFDIV_LEAF，可形成以下 5 种 MBUFG（Multi-clock Buffer）。

- MBUFGCTRL（BUFGCTRL+BUFDIV_LEAF）。
- MBUFGCE（BUFGCE+BUFDIV_LEAF）。
- MBUFGCE_DIV（BUFGCE_DIV+BUFDIV_LEAF）。
- MBUFG_PS（BUFG_PS+BUFDIV_LEAF）。
- MBUFG_GT（BUFG_GT+BUFDIV_LEAF）。

由此可见，MBUFG 并非物理存在，而是逻辑存在。MBUFG 有 4 个输出端口 O1~O4，两种工作模式，由模式控制字 MODE 设定。MODE 与输入时钟频率的关系如表 2-1 所示。其中，FIN 为 MBUFG 的输入时钟频率。

表 2-1

MODE	O1	O2	O3	O4
PERFORMANCE	FIN	FIN/2	FIN/4	FIN/8
POWER	FIN×2	FIN	FIN/2	FIN/4

3．时钟管理单元

Versal 中的时钟管理单元带分为 3 种：1 个 XPLL；1 个 MMCM（原语为 MMCME5）和 1 个 DPLL；1 个 DPLL。MMCM 的功能和 UltraScale FPGA 中的保持一致，即在模拟锁相环的基础上增加了通用时钟功能。XPLL 是 Versal 中新增的时钟管理单元，主要功能是给 XPHY 提供时钟，时钟输出个数会比 MMCM 少一些。DPLL 是数字锁相环，可将其看作轻量版的 MMCM。三者的端口如图 2-33 所示。

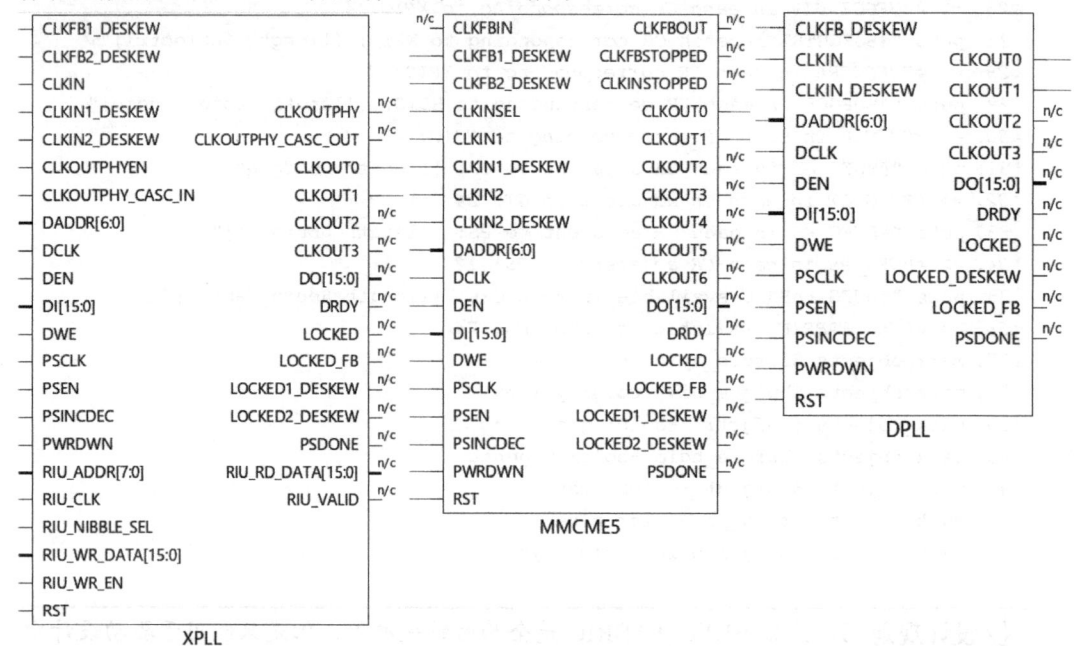

图 2-33

XPLL 位于 XPIO 内。每个 XPIO 内有两个 XPLL。MMCM 和 DPLL 分布在紧邻 XPIO

的水平时钟布线轨道和分发轨道上，与其在同一时钟区域内的全局时钟管脚可直接连接这些位置上的 MMCM 和 DPLL。同时，在紧邻高速收发器的时钟区域内还分布着一些 DPLL，位于 HDIO 的全局时钟管脚可与之直接连接，但如果需要连接 XPIO 的 MMCM，则需要先将其连接到 BUFGCE 上，再由 BUFGCE 的输出连接到 MMCM 上。XPLL、MMCM 和 DPLL 的分布状况也可查看图 2-32。借助 Tcl 代码 2-6 可查看指定芯片中的时钟管理单元。

Tcl 代码 2-6

```
1.  #File: check_pll_mmem_versal.tcl
2.  link_design -part [get_parts xcvc1902-viva1596-1LHP-i-L]
3.  set mmcm [get_sites "MMCM*"]
4.  => MMCM_X0Y0 MMCM_X1Y0 ...
5.  set dpll [get_sites "DPLL*"]
6.  => DPLL_X1Y8 DPLL_X14Y8 ...
7.  set xpll [get_sites "XPLL*"]
8.  => XPLL_X0Y0 XPLL_X1Y0 ...
9.  mark_objects $dpll -color green
10. mark_objects $mmcm -color yellow
11. mark_objects $xpll -color cyan
12. set mmcm_in_x0y0 [get_sites "MMCM*" -of [get_clock_regions X0Y0]]
13. => MMCM_X0Y0
14. set dpll_in_x0y0 [get_sites "DPLL*" -of [get_clock_regions X0Y0]]
15. => DPLL_X0Y0
16. set xpll_in_x0y0 [get_sites "XPLL*" -of [get_clock_regions X0Y0]]
17. => XPLL_X0Y0 XPLL_X1Y0
18. set dpll_in_x0y4 [get_sites "DPLL*" -of [get_clock_regions X0Y4]]
19. => DPLL_X1Y8
20. puts "#MMCM in each XPIO bank: [llength $mmcm_in_x0y0]"
21. => #MMCM in each XPIO bank: 1
22. puts "#DPLL in each XPIO bank: [llength $dpll_in_x0y0]"
23. => #DPLL in each XPIO bank: 1
24. puts "#XPLL in each XPIO bank: [llength $xpll_in_x0y0]"
25. => #XPLL in each XPIO bank: 2
26. puts "#DPLL in each CR adjacent to GT: [llength $dpll_in_x0y4]"
27. => #DPLL in each CR adjacent to GT: 1
```

从 28nm 工艺制程的 7 系列 FPGA 到 20nm 工艺制程的 UltraScale FPGA 和 16nm 工艺制程的 UltraScale+ FPGA，再到 7nm 工艺制程的 Versal ACAP，芯片架构发生了很大的变化。表 2-2 从时钟资源的角度对其进行了对比。可以看到，时钟区域的规模越来越大，时钟缓冲器逐渐趋向单一。

表 2-2

比较条目	7 系列 FPGA	UltraScale/UltraScale+ FPGA	Versal ACAP	备注
CR 大小	50 个 CLB 高度	60 个 CLB 高度	96 个 CLB 高度	
	BRAM 列高度为 10	BRAM 列高度为 12	BRAM 列高度为 24	
	DSP 列高度为 20	DSP 列高度为 24	DSP 列高度为 48	
全局时钟管脚	MRCC 和 SRCC	GCIO	GCIO 和 HDGC	
BUFGCTRL	√	√	√	全局时钟缓冲器
BUFGCE		√	√	全局时钟缓冲器

续表

比较条目	7系列FPGA	UltraScaleI/UltraScale+ FPGA	Versal ACAP	备注
BUFGCE_DIV		√	√	全局时钟缓冲器
BUFG_GT		√	√	全局时钟缓冲器，高速收发器输出时钟使用
BUFG_PS		√	√	全局时钟缓冲器，PS输出时钟使用
BUFG_FABRIC			√	全局时钟缓冲器，非时钟高扇出网线使用
BUFIO	√			区域时钟缓冲器
BUFR	√			区域时钟缓冲器
BUFMR	√			区域时钟缓冲器，仅驱动BUFR和BUFIO
BUFCE_LEAF		√		不可实例化，由工具自动使用
BUFDIV_LEAF			√	不可实例化，由工具自动使用

2.2 时钟偏移

FPGA 设计中的绝大部分电路为同步时序电路，基本模型为"寄存器+组合逻辑+寄存器"。同步意味着时序路径上的所有寄存器在时钟信号的驱动下步调一致地运作。这就要求时钟信号在同一时间点到达所有寄存器的时钟端口。为此，FPGA 内部提供了专用的时钟布线资源。然而，即便如此，实际情形也是时钟信号往往在不同时间点到达寄存器的时钟端口，这种现象就是时钟偏移。

时钟偏移反映了时钟信号到达同一时序路径上不同寄存器时钟端口之间的时间差异，如图 2-34 所示。时钟从源端到达寄存器 FF1 的时间点为 T_{clk1}，到达寄存器 FF2 的时间点为 T_{clk2}，因此时钟偏移为 T_{clk2} 与 T_{clk1} 的差。若 clk 源端记为零时刻点，那么 T_{clk1} 和 T_{clk2} 分别对应发送时钟路径延迟和捕获时钟路径延迟。

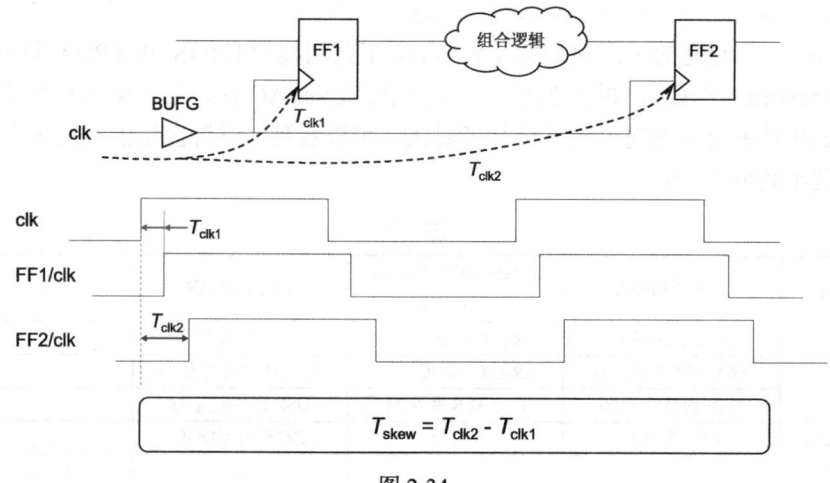

图 2-34

时钟偏移可正可负。通常,若数据流向与时钟前进方向相同,则时钟偏移为正,否则,时钟偏移为负,如图 2-35 所示。

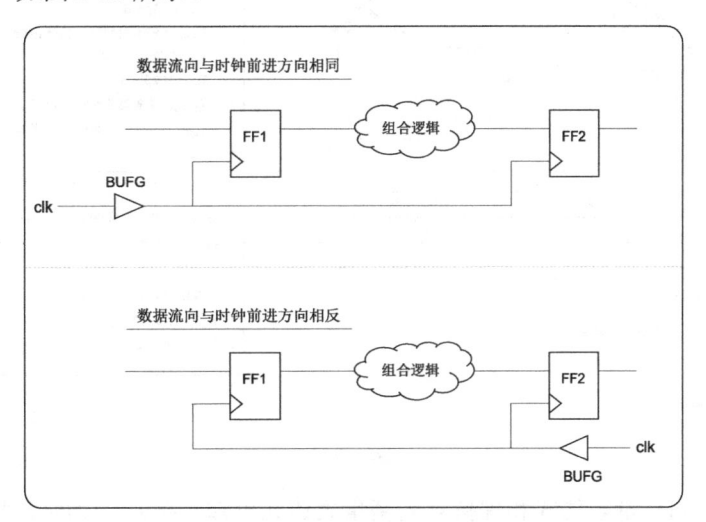

图 2-35

时钟偏移对时序收敛有什么影响呢?从建立时间裕量和保持时间裕量两个角度分析。先以正向的时钟偏移为例。建立时间裕量分析,如图 2-36 所示,发起沿和捕获沿相差一个时钟周期。由图 2-36 中的建立时间裕量表达式可以得出结论:正向的时钟偏移对建立时间收敛是有利的,相当于捕获寄存器的建立时间由 T_{su} 减小至 $T_{su}-T_{skew}$。保持时间裕量分析如图 2-37 所示,保持时间检查的发起沿和捕获沿为同一时钟沿(保持时间检查是基于建立时间检查的,要求当前发送沿发送的数据不能被前一个捕获沿捕获;下一个发送沿发送的数据不能被当前捕获沿捕获)。由图 2-37 中的保持时间裕量表达式可以得出结论:正向的时钟偏移不利于保持时间收敛,相当于数据在有效沿到达之后还要稳定保持的时间变长了,由原来的 T_h 变为 T_h+T_{skew}。这显然阻碍了保持时间收敛。

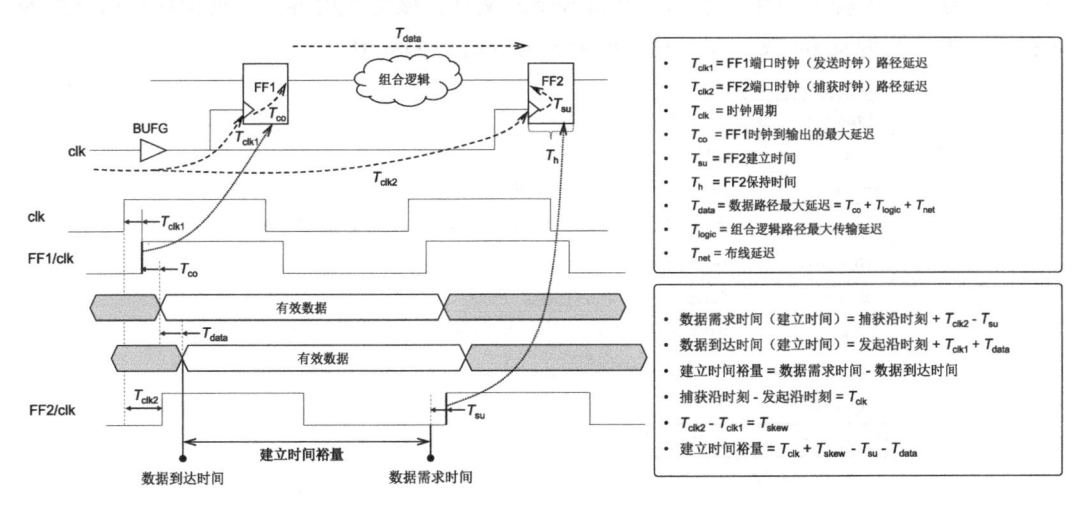

- T_{clk1} = FF1端口时钟(发送时钟)路径延迟
- T_{clk2} = FF2端口时钟(捕获时钟)路径延迟
- T_{clk} = 时钟周期
- T_{co} = FF1时钟到输出的最大延迟
- T_{su} = FF2建立时间
- T_h = FF2保持时间
- T_{data} = 数据路径最大延迟 = T_{co} + T_{logic} + T_{net}
- T_{logic} = 组合逻辑路径最大传输延迟
- T_{net} = 布线延迟

- 数据需求时间(建立时间)= 捕获沿时刻 + T_{clk2} - T_{su}
- 数据到达时间(建立时间)= 发起沿时刻 + T_{clk1} + T_{data}
- 建立时间裕量 = 数据需求时间 - 数据到达时间
- 捕获沿时刻 - 发起沿时刻 = T_{clk}
- T_{clk2} - T_{clk1} = T_{skew}
- 建立时间裕量 = T_{clk} + T_{skew} - T_{su} - T_{data}

图 2-36

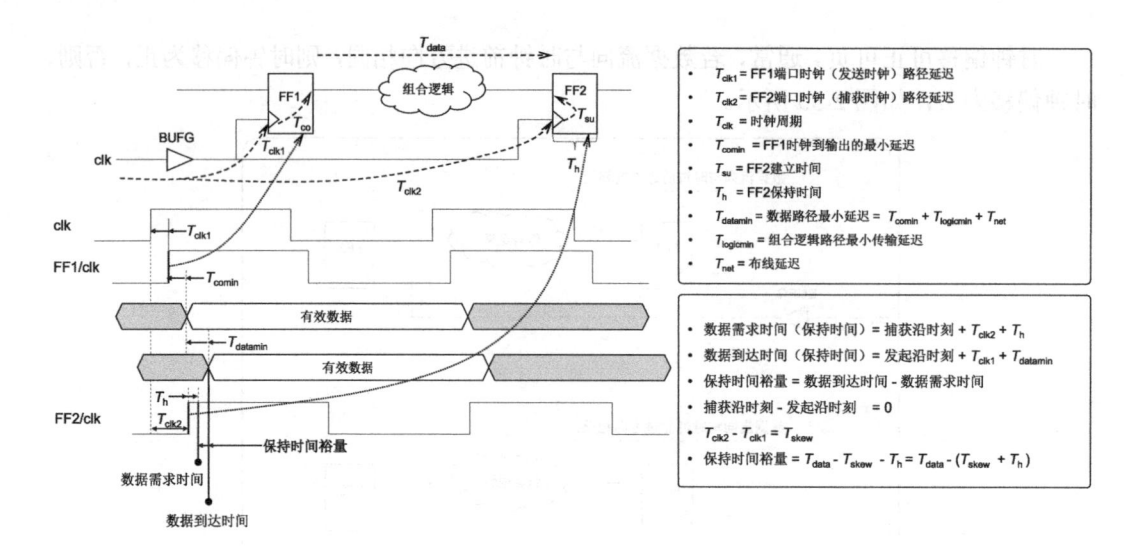

图 2-37

结合建立时间裕量表达式和保持时间裕量表达式可知，若 T_{skew} 为负，则建立时间收敛更加困难，保持时间收敛更加容易。

> 💡 **设计规则 9**：在设计早期规划阶段，就要做到系统时钟输入管脚和系统数据输入管脚在 FPGA 芯片同侧，这样能最大限度地保证时钟前进方向与数据流向是一致的。

哪些因素会导致时钟偏移过大呢？诸如不合理的时钟结构（如级联的 BUFG 或时钟路径上出现组合逻辑，可通过 report_methodology 命令生成的报告进一步分析）、时钟同时驱动 I/O 资源和 SLICE 中的资源、时钟跨 die 等都会增大时钟偏移。通常，如果时钟偏移超过 0.5ns，就要格外关注。时序报告的总结部分会显示时钟偏移的具体数值，如图 2-38 所示，但该数值未必是对应时钟最糟糕的时钟偏移情形。此时，可通过 Tcl 代码 2-7 的方式获得这一数值。该脚本最终会生成一个 skew.csv 文件。该文件内容由三列构成：第一列为时钟名；第二列为 setup 或 hold；第三列为相应的时钟偏移值。

Name	Path 41
Slack	0.135ns
Source	rst_gen_i0/reset_bridge_clk_tx_i0/rst_dst_reg/C (rising edge-triggere
Destination	lb_ctl_i0/debouncer_i0/meta_harden_signal_in_i0/signal_meta_reg/R
Path Group	clk_tx_clk_core
Path Type	Setup (Max at Slow Process Corner)
Requirement	6.000ns (clk_tx_clk_core rise@6.000ns - clk_tx_clk_core rise@0.000ns)
Data Path Delay	5.118ns (logic 0.308ns (6.018%) route 4.810ns (93.982%))
Logic Levels	0
Clock Path Skew	-0.318ns
Clock Uncertainty	0.062ns

图 2-38

Tcl 代码 2-7

```
1.  #File: report_max_clock_skew.tcl
2.  proc report_max_clock_skew {max_paths} {
3.      set clks [get_clocks]
4.      set f [open skew.csv w]
5.
6.      foreach ana_ele [list setup hold] {
7.          foreach i_clks $clks {
8.              if { [llength [get_timing_paths \
9.                  -from $i_clks -to  $i_clks]] == 0} {
10.                 puts "No timing paths found for clock $i_clks"
11.                 continue
12.             }
13.             set skew_val [get_property SKEW \
14.                 [get_timing_paths -from $i_clks \
15.                 -to $i_clks -max $max_paths -$ana_ele]]
16.             set skew_val_abs [list]
17.             foreach val $skew_val {
18.                 lappend skew_val_abs [expr abs($val)]
19.             }
20.             set max_skew_abs [lindex \
21.                 [lsort -decreasing $skew_val_abs] 0]
22.             if {[lsearch $skew_val $max_skew_abs]== -1} {
23.                 set max_skew_val [expr $max_skew_abs * (-1)]
24.             } else {
25.                 set max_skew_val $max_skew_abs
26.             }
27.             puts "The max clock skew of Clock \
28.                 $i_clks ($ana_ele) = $max_skew_val"
29.             puts $f [join "$i_clks $ana_ele $max_skew_val" ,]
30.         }
31.     }
32.     close $f
33. }
```

那么有哪些方法可以降低时钟偏移呢？这里介绍一些常见方法（注意：以下方法若无特殊说明，则为通用方法，适用于 Xilinx 7 系列 FPGA、UltraScale/UltraScale+ FPGA 和 Versal ACAP）。

方法 1：移除时钟路径上多余的时钟缓冲器

之所以会出现多余的时钟缓冲器，是因为设计中出现了级联时钟缓冲器。如图 2-39 所示，对于 BUFGCE 级联的情形可优化为 BUFGCE 并联，但更好的方法是移除不必要的 BUFGCE。目前，Vivado 已相当智能，在 opt_design 阶段可自动移除时钟路径上多余的缓冲器。

方法 2：合并并联的时钟缓冲器为单一的时钟缓冲器

并联时钟缓冲器常见的情形是两个并联的 BUFGCE，一个使能端恒接高电平，另一个使能端受其他信号控制，如图 2-40 所示。此时，可将这两个 BUFGCE 合并，将原本的使能信号连接到寄存器的使能端。

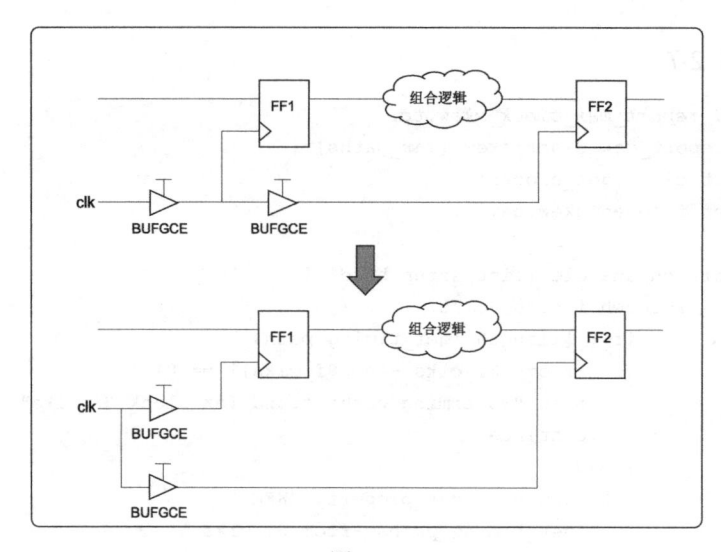

图 2-39

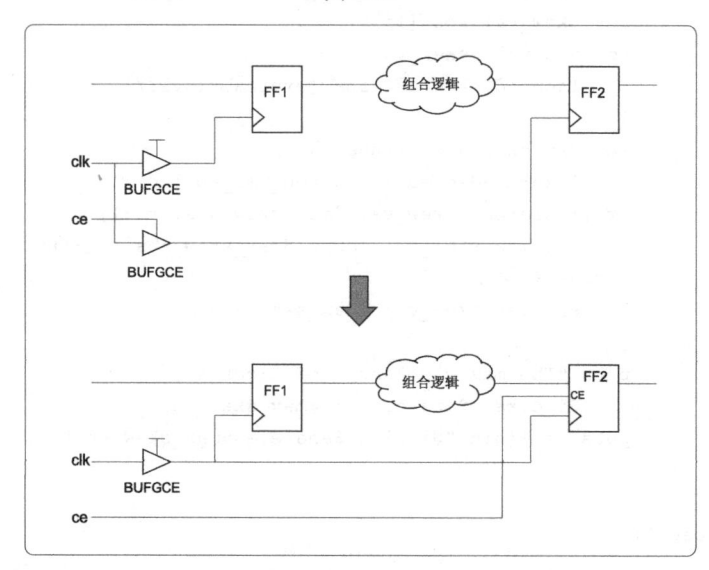

图 2-40

方法 3：移除时钟路径上的组合逻辑

一旦时钟路径上出现了组合逻辑，就意味着时钟布线采用了"常规布线资源+专用时钟布线资源"的组合形式，从而显著增大时钟延迟且使得时钟偏移无法被预测。同时，相比专用时钟布线资源，常规布线资源对噪声更加敏感，这使得时钟质量显著下降。时钟路径上出现组合逻辑的常见场景是工程师试图通过组合逻辑的形式实现门控时钟的功能，以节省功耗，如图 2-41 所示。此时，可将组合逻辑从时钟路径上移除，将其输出连接到寄存器的时钟使能端口。

方法 4：避免使用约束 CLOCK_DEDICATED_ROUTE=FALSE

CLOCK_DEDICATE_ROUTE=FALSE 意味着时钟布线采用了"常规布线资源+专用时钟布线资源"的组合形式，这在很大程度上会增大时钟偏移，如图 2-42 所示。图中，右侧使用了专用时钟布线资源，而左侧没有使用。之所以 Vivado 会要求设置此约束，是因为设计中的某个时钟网线不是由全局时钟缓冲器驱动的,而是直接由时钟网络的某部分驱动的,如 MMCM/PLL 的时钟输出端口或输入缓冲器 IBUF 的输出端口,如 Tcl 代码 2-8 所示。一个典型的设计失误是时钟信号未从全局时钟管脚进入 FPGA,而从通用输入端口进入 FPGA。此时,只能添加此约束,否则在布局布线阶段会报错。

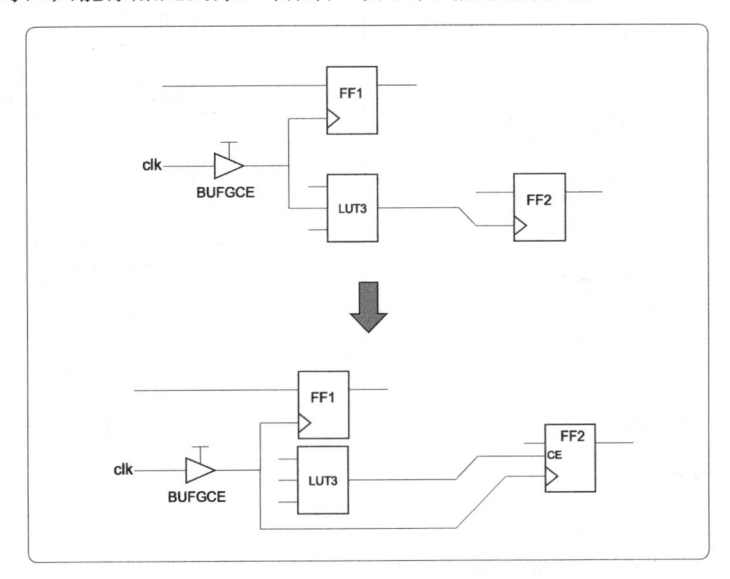

图 2-41

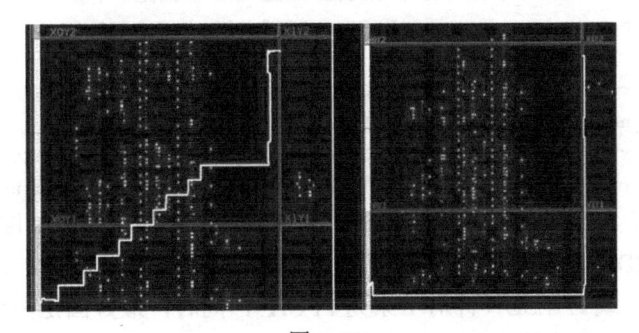

图 2-42

Tcl 代码 2-8

```
1.  #File: clk_dedicated_route_7fpga.tcl
2.  set_property CLOCK_DEDICATED_ROUTE FALSE \
3.  [get_nets -of [get_pins MMCME4_ADV_inst/CLKOUT0]]
4.  set_property CLOCK_DEDICATED_ROUTE FALSE \
5.  [get_nets -of [get_pins IBUF_inst/O]]
```

方法 5:当设计中出现并联的 MMCM/PLL 时,应合理设置 CLOCK_DEDICATED_ROUTE 的值

当设计中出现了一个 BUFGCE 输出驱动两个 MMCM 或 PLL 时,如图 2-43 所示,应根据三者的位置合理设置 CLOCK_DEDICATED_ROUTE 约束,以保证时钟布线仅使用专用时钟布线资源。

如果 BUFGCE 和两个 MMCM 或 PLL 位于同列相邻的时钟区域,如图 2-44 所示,此时应将 CLOCK_DEDICATED_ROUTE 设置为 SAM_CMT_COLUMN（针对 UltraScale/UltraScale+ FPGA）或 BACKBONE（针对 7 系列 FPGA），同时，最好固定两个 MMCM 或 PLL 的位置（通过属性 LOC 实现），如 Tcl 代码 2-9 所示。

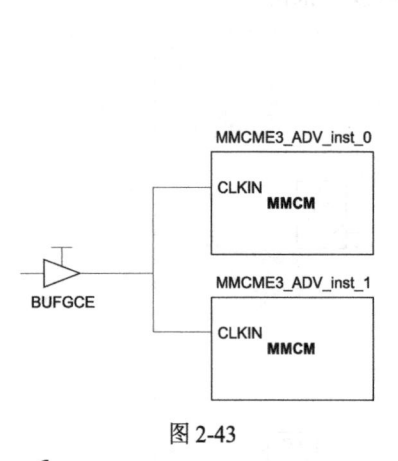

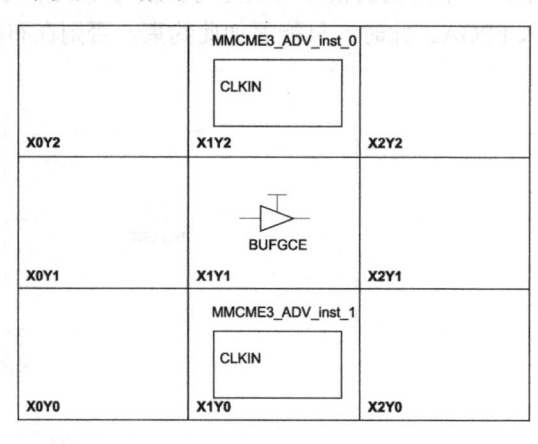

图 2-43

图 2-44

Tcl 代码 2-9

```
1. #File: same_cmt_col.tcl
2. #For UltraScale/UltraScale+
3. set_property CLOCK_DEDICATED_ROUTE SAME_CMT_COLUMN \
4.     [get_nets -of [get_pins BUFG_inst_0/O]]
5. #For 7 Series FPGA
6. set_property CLOCK_DEDICATED_ROUTE BACKBONE\
7.     [get_nets -of [get_pins BUFG_inst_0/O]]
8. set_property LOC MMCME3_ADV_X1Y2 [get_cells MMCME3_ADV_inst_0]
9. set_property LOC MMCME3_ADV_X1Y0 [get_cells MMCME3_ADV_inst_1]
```

如果 BUFGCE 和两个 MMCM 位于不同列但相邻的时钟区域，如图 2-45 所示，此时应将 CLOCK_DEDICATED_ROUTE 设置为 ANY_CMT_COLUMN（针对 UltraScale/UltraScale+ FPGA）或 FALSE（针对 7 系列 FPGA），同时，最好固定两个 MMCM 或 PLL 的位置（通过属性 LOC 实现），如 Tcl 代码 2-10 所示。

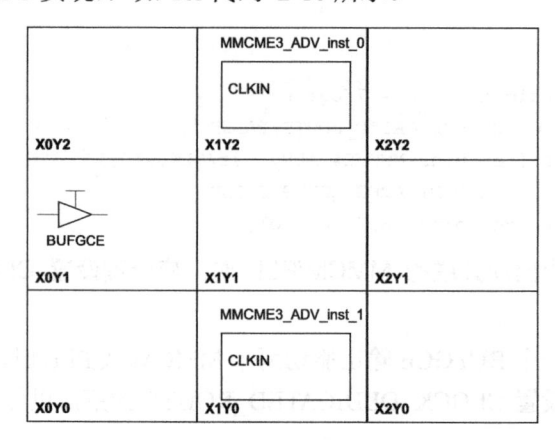

图 2-45

Tcl 代码 2-10

```
1.  #File: any_cmt_col.tcl
2.  #For UltraScale/UltraScale+
3.  set_property CLOCK_DEDICATED_ROUTE ANY_CMT_COLUMN \
4.      [get_nets -of [get_pins BUFG_inst_0/O]]
5.  #For 7 Series FPGA
6.  set_property CLOCK_DEDICATED_ROUTE FALSE\
7.      [get_nets -of [get_pins BUFG_inst_0/O]]
8.  set_property LOC MMCME3_ADV_X1Y2 [get_cells MMCME3_ADV_inst_0]
9.  set_property LOC MMCME3_ADV_X1Y0 [get_cells MMCME3_ADV_inst_1]
```

方法 6（针对 7 系列 FPGA）：避免使用区域时钟缓冲器（如 BUFIO、BUFR 或 BUFH）驱动分散在不同时钟区域内的逻辑

使用 7 系列 FPGA 中的区域时钟缓冲器时要格外小心。首先，要明确区域时钟缓冲器的作用域。例如，BUFR 只能驱动其所在时钟区域内的逻辑资源。其次，要评估设计中区域时钟缓冲器负载的个数，以保证这些负载可以被放置在一个时钟区域内。最后，添加合理的位置约束，保证区域时钟缓冲器与其负载在同一个时钟区域内。这可通过手工布局（画PBLOCK）的方式实现。

方法 7（针对 UltraScale/UltraScale+ FPGA 和 Versal ACAP）：避免使用 MMCM/PLL 对来自 BUFG_GT 的输出时钟执行简单的分频

BUFG_GT 具有分频功能，可支持的分频因子为 1～8（包含 1 和 8）的整数，分频因子控制字由位宽为 3 的输入端口 DIV 提供。当 DIV 为 3'b000 时，对应的分频因子为 1。借助 BUFG_GT 的分频功能可生成分频时钟，节省了 MMCM，如图 2-46 所示。

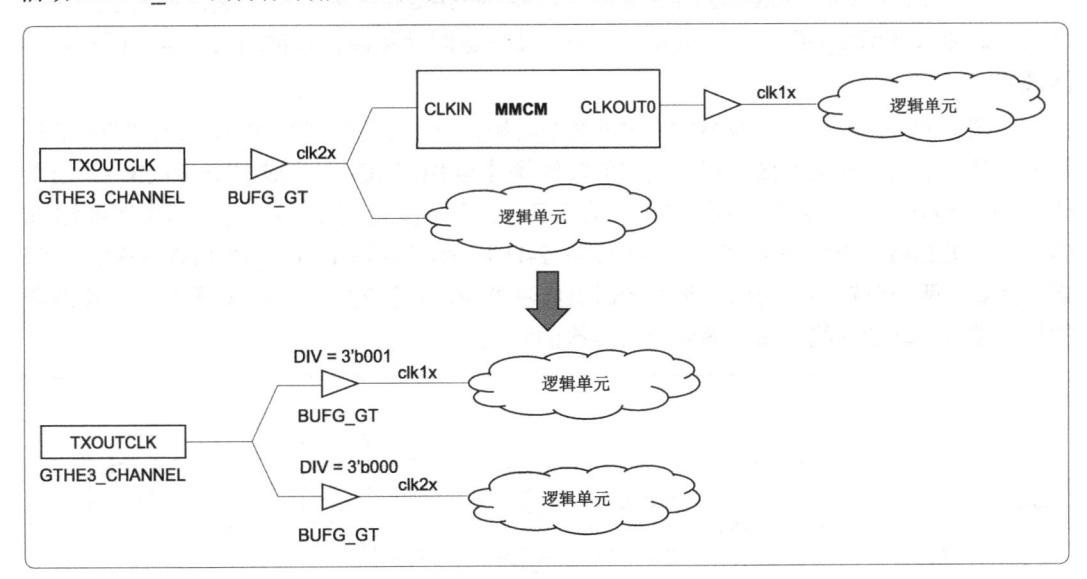

图 2-46

此外，具有分频功能的时钟缓冲器还包括 UltraScale/UltraScale+ FPGA 系列中的 BUFGCE_DIV，以及 Versal ACAP 中的 BUFGCE_DIV 和 MBUFG，如图 2-47 所示。图中，clk2x 的频率为 clk1x 的 2 倍。

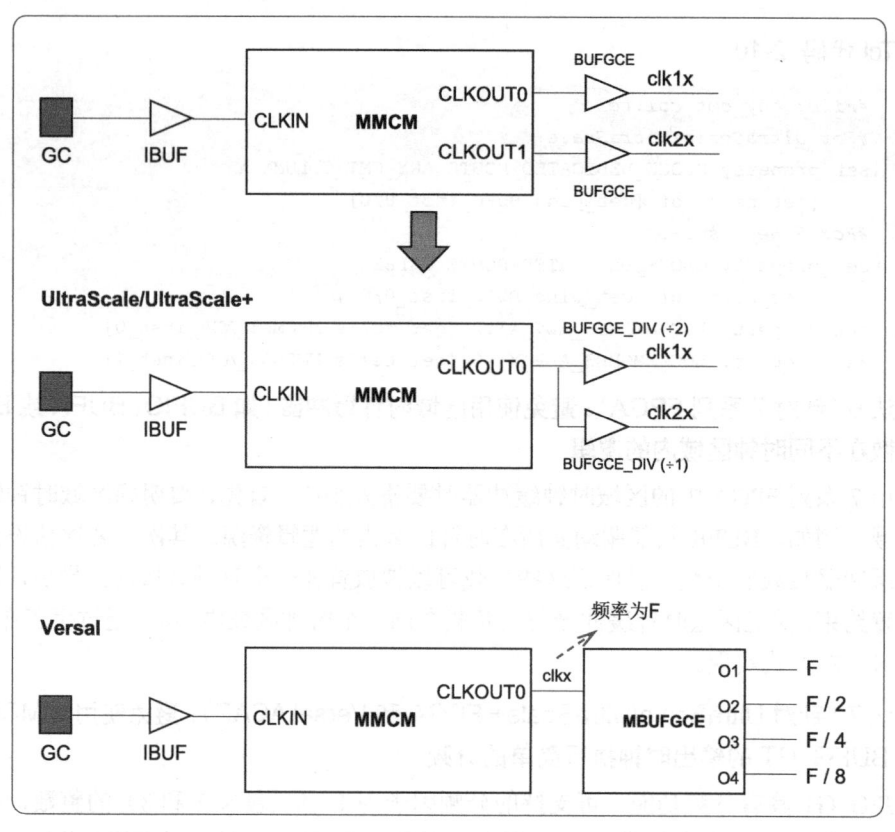

图 2-47

方法 8（针对 UltraScale/UltraScale+ FPGA 和 Versal ACAP）：对关键的同步跨时钟域路径添加 CLOCK_DELAY_GROUP 约束，使发送时钟和接收时钟的根节点在同一时钟区域内

以图 2-48 为例，图中，MMCM 输出两个时钟，分别由 CLKOUT0 和 CLKOUT1 端口输出，且两个端口输出连接了相同类型的时钟缓冲器 BUFGCE，形成 clktx 和 clkrx。由于 clktx 和 clkrx 之间存在数据传输，从而形成了同步跨时钟域路径，因此可添加 CLOCK_DELAY_GROUP 约束，如 Tcl 代码 2-11 所示。使用 CLOCK_DELAY_GROUP 时需要注意：两个时钟必须由同一类型的时钟缓冲器驱动；同时，其约束对象是时钟缓冲器的输出端口驱动的网线，而非输入端口连接的网线。

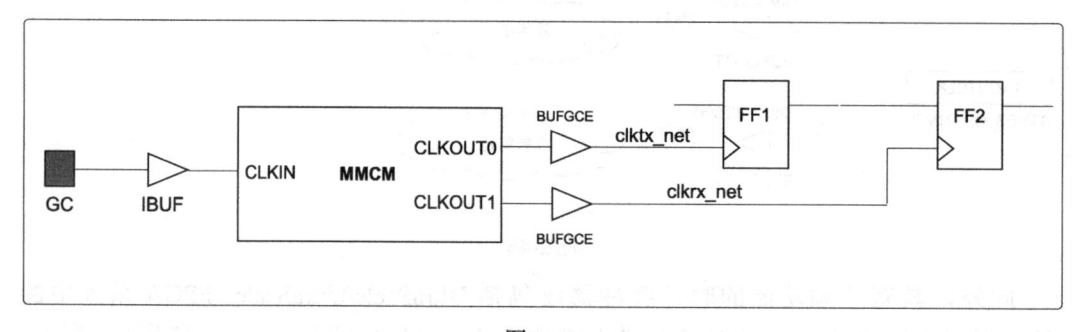

图 2-48

Tcl 代码 2-11

```
1. #File: set_clock_delay_group.tcl
2. set_property CLOCK_DELAY_GROUP group0 [get_nets {clktx_net clkrx_net}]
```

方法 9（针对 UltraScale/UltraScale+ FPGA）：修改时钟根节点的位置

在默认情况下，Vivado 在布局阶段会自动给每个时钟分配根节点的位置，以获取最佳的时序性能。打开 place_design 或 route_design 生成的 .dcp 文件，找到目标时钟对应的网线。该网线一定是全局时钟缓冲器驱动的网线，如图 2-49 所示。借助属性 CLOCK_ROOT 可获得指定时钟的根节点位置，如 Tcl 代码 2-12 所示。代码第 2 行表示获取时钟网线，代码第 3 行表示获取 CLOCK_ROOT 值，通常为一时钟区域坐标。也可以借助命令 report_clock_utilization 添加 -clock_roots_only 生成时钟根节点报告，如代码第 5 行所示，报告样例如图 2-50 所示。此外，还可以找到时钟网线驱动的所有负载，使其以高亮方式显示，如代码第 6 行~第 8 行所示。

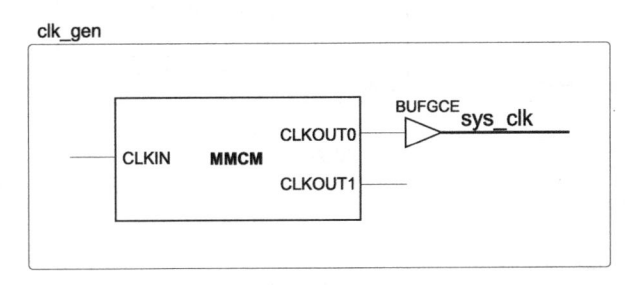

图 2-49

Tcl 代码 2-12

```
1. #File: get_clk_root.tcl
2. set clk_net [get_nets clk_gen/sys_clk]
3. set clk_root [get_property CLOCK_ROOT $clk_net]
4. puts "Clock root: $clk_root"
5. report_clock_utilization -clock_roots_only -name clock_root
6. set loads [get_cells -of [get_pins -of $clk_net -leaf]]
7. puts "#Loads of $clk_net: [llength $loads]"
8. highlight_objects $loads -color red
```

Global Id	Root	Net
g0	X3Y7	u_f3_shell_top/u_static_slr1/u_base_clock/base_clking/clkwiz_kernel/inst/CLK_CORE_DRP_I/clk_inst/clk_out1
g9	X3Y7	u_f3_shell_top/u_static_slr1/u_base_clock/base_clking/clkwiz_sysclks/inst/clk_out2
g13	X3Y7	dbg_hub/inst/BSCANID.u_xsdbm_id/itck_i
g14	X3Y7	u_f3_shell_top/u_static_slr1/u_base_clock/base_clking/clkwiz_kernel2/inst/CLK_CORE_DRP_I/clk_inst/clk_out1
g15	X3Y7	u_f3_shell_top/u_static_slr1/u_base_clock/base_clking/clkwiz_sysclks/inst/clk_out1

图 2-50

通常情况下，时钟根节点位于其所驱动负载的中心位置，以图 2-51 为例。图中，时钟网线 sys_clk 所驱动的负载以小方格的形式表示。如果 sys_clk 的根节点位于时钟区域 X2Y1，那么应该是合理的；若不是，则可通过属性 USER_CLOCK_ROOT 进行调整，如

Tcl 代码 2-13 所示。该属性需要在 place_design 之前部署。若是直接打开 route_design 生成的.dcp，则设置该属性之后，需要先撤销原时钟网线的布线（对应代码第 4 行），再重新布线（对应代码第 5 行），这样 USER_CLOCK_ROOT 才能生效。对于基于 SSI（Stacked Silicon Interconnect）芯片的设计，如果时钟穿越多个 SLR（Super Logic Region），有时需要借助此方法调整时钟根节点的位置。

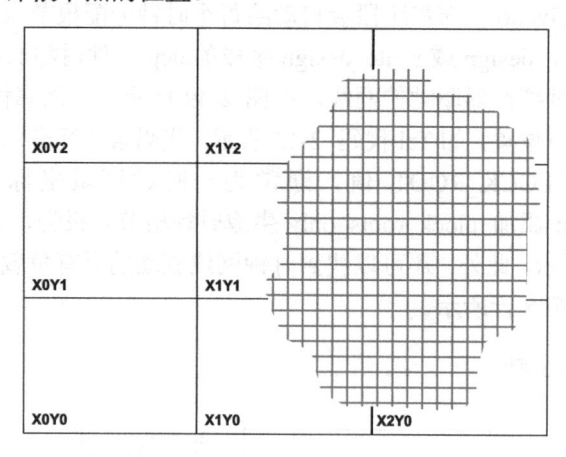

图 2-51

Tcl 代码 2-13

```
1.  #File: change_clk_root.tcl
2.  set clk_net [get_nets clk_gen/sys_clk]
3.  set_property USER_CLOCK_ROOT X2Y1 [get_nets $clk_net]
4.  route_design -unroute -nets [get_nets $clk_net]
5.  update_clock_routing
```

方法 10：避免关键路径穿越 SLR 或输入/输出列

如果时序违例的根本原因在于时钟偏移过大，那么首先检查该路径是否跨 die（穿越多个 SLR）或是否穿过输入/输出列。若是，则可采用手工布局的方式将关键路径封闭在同一个 SLR 内或避免其穿过输入/输出列。

方法 11（针对 UltraScale/UltraScale+ FPGA）：使用 CLOCK_LOW_FANOUT 属性使时钟网线驱动的负载位于同一个时钟区域内

对于低扇出的时钟网线（负载个数小于 2000），可通过属性 CLOCK_LOW_FANOUT 使这些负载最终位于同一个时钟区域内，如 Tcl 代码 2-14 所示。这里需要特别注意的是命令 get_nets 的对象必须是全局时钟缓冲器输出端连接的网线。最终结果如图 2-52 左侧所示。此外，CLOCK_LOW_FANOUT 的优先级低于 CLOCK_DEDICATED_ROUTE、CLOCK_DELAY_GROUP、USER_CLOCK_ROOT、LOC 和 PBLOCK，因此，如果发生约束冲突，CLOCK_LOW_FANOUT 可能不会生效。

Tcl 代码 2-14

```
1.  #File: set_clk_low_fanout.tcl
2.  set_property CLOCK_LOW_FANOUT TRUE [get_nets -of \
3.     [get_pins clkOut0_bufg_inst/O]]
```

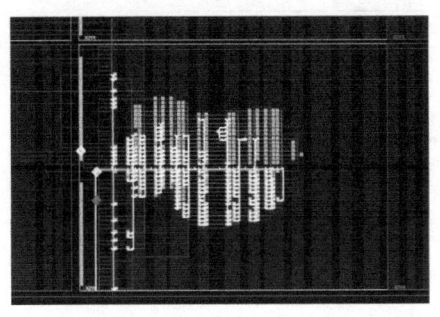

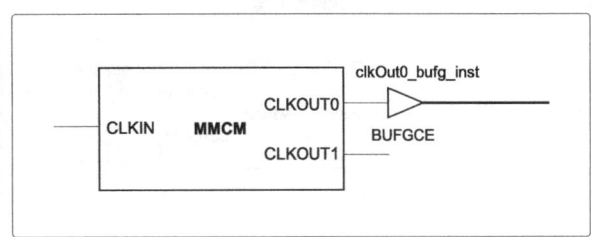

图 2-52

应用案例 5：UltraScale/UltraScale+ FPGA 芯片中 ISERDESE3 和 IDDRE1 的时钟方案

ISERDESE3 有时钟端口 CLK 和 CLK_B。其中，CLK_B 是 CLK 的反相时钟。IDDRE1 有时钟端口 C 和 CB。其中，CB 是 C 的反相时钟。ISERDESE3 和 IDDRE1 对这两个时钟之间的时钟偏移有严格的要求。尽管 MMCM 的 CLKOUT0～CLKOUT3 有专门的反相时钟输出端口，但仍建议直接使用 ISERDESE3 和 IDDRE1 自带的时钟反相器，如图 2-53 所示。

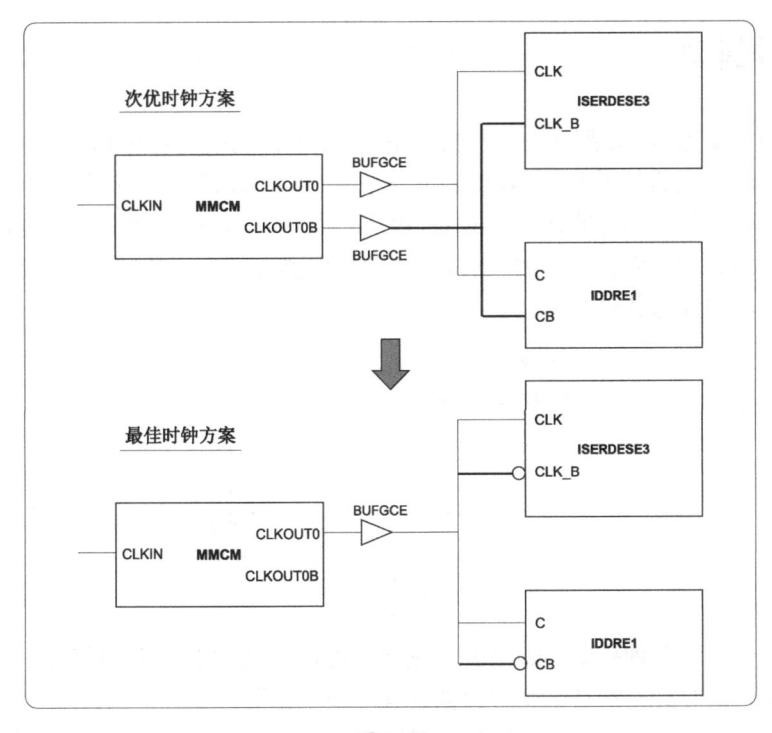

图 2-53

应用案例 6：UltraScale/UltraScale+ FPGA 芯片中 OSERDESE3 的时钟方案

OSERDESE3 有时钟端口 CLK 和 CLKDIV。其中，CLK 是高速时钟，CLKDIV 是其分频时钟。OSERDESE3 对这两个时钟的时钟偏移有严格的要求。尽管可以利用 MMCM 生成 CLK 及其分频时钟，但更为有效的方式是采用 BUFGCE_DIV 实现分频，如图 2-54 所示。

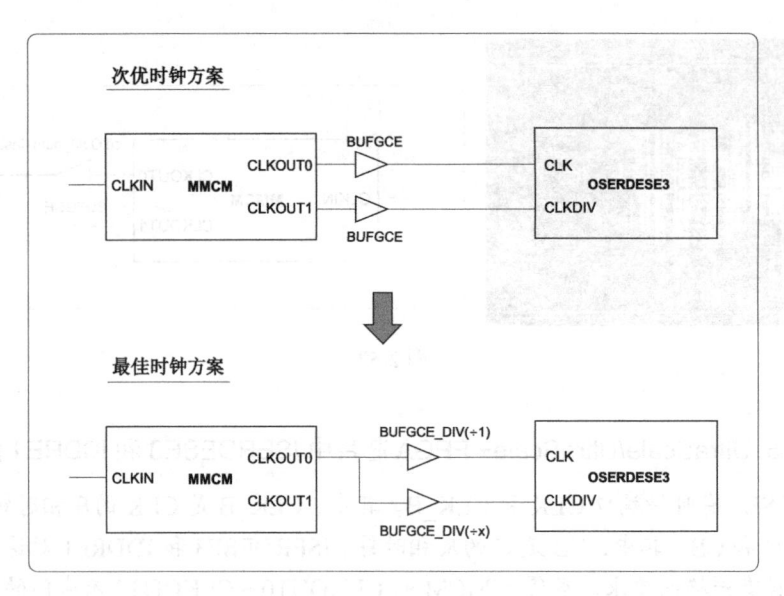

图 2-54

2.3 时钟抖动

假定时钟周期为 10ns，那么在理想情况下，其上升沿将会出现在 0ns,10ns,20ns…时刻。已知某个上升沿出现的时刻为 30ns，那么下一个上升沿出现的时刻应该为 40ns。但实际上，下一个上升沿出现的时刻可能是 40.1ns 或 39.9ns。这个在时钟周期中出现的 0.1ns 的偏差为时钟抖动，如图 2-55 所示。时钟抖动是在时钟发生器内部产生的，和晶振或 MMCM/PLL 内部电路有关，布线对其没有影响。

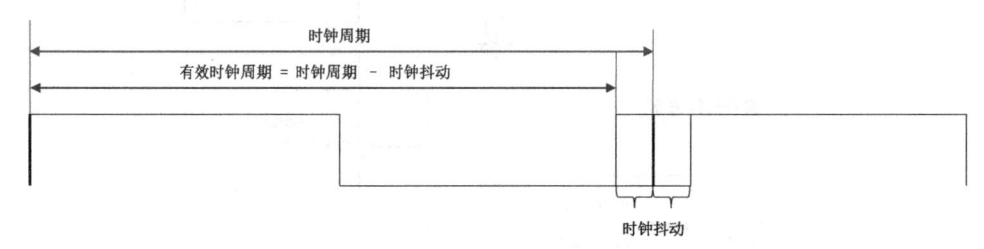

图 2-55

对比时钟偏移和时钟抖动，如图 2-56 所示。可以看出，时钟偏移反映的是两个时钟之间的相位关系，而时钟抖动是每个时钟都具有的基本特征。

时钟抖动对建立时间裕量和保持时间裕量有什么影响呢？根据静态时序分析理论，以图 2-57 为例，对于单周期时序路径，发起沿为图中标记①对应的上升沿，捕获沿为图中标记③对应的上升沿。若标记①对应 0ns，则标记③可能对应 T_{clk}+Jitter 或 T_{clk}-Jitter（Jitter 表示时钟抖动）。考虑到最坏情况，应按 T_{clk}-Jitter 进行分析。这意味着有效时钟周期为 T_{clk}-Jitter。由此可见，时钟抖动进一步减小了建立时间裕量，增加了建立时间收敛的难度。对于保持时间裕量，由于发送沿与捕获沿均为图中标记①对应的上升沿，因此时钟抖动对保持时间裕量没有影响。

图 2-56

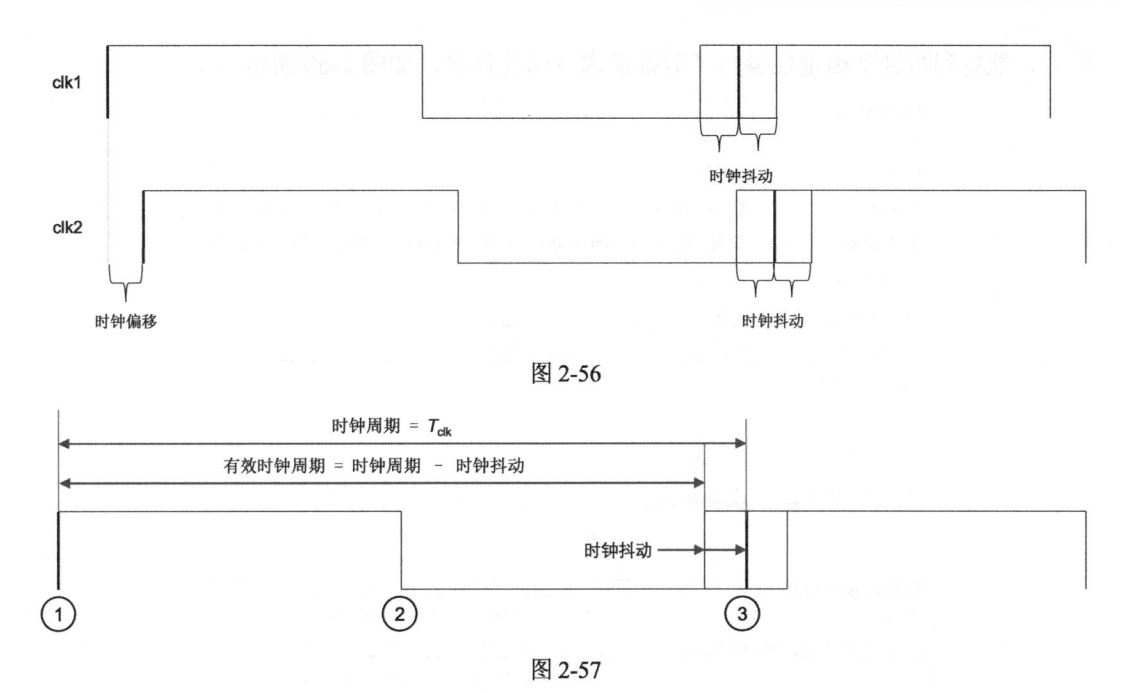

图 2-57

在建立时间时序报告总结部分，有一项 Clock Uncertainty（时钟不确定性），如图 2-58 所示。单击其后的数字 0.062ns，会弹出一个对话框，显示为 Clock Uncertainty Equation，即 Clock Uncertainty 的计算公式。从这个计算公式中可以看出，时钟抖动是时钟不确定性的重要组成部分。

$$((TSJ^2 + DJ^2)^{1/2}) / 2 + PE$$

图 2-58

在建立时间时序报告的总结部分，可以看到路径类型（Path Type）为建立时间（Setup），时钟周期为 6ns，时钟不确定性为 0.062ns。在源时钟路径和目的时钟路径部分，可以看到发起沿出现在 0ns 时刻，捕获沿出现在 6ns 时刻，两者相差一个时钟周期，如图 2-59 所示。

同时，能看到时钟不确定性参与了数据需求时间的计算，如图 2-60 所示。

Summary	
Name	Path 41
Slack	0.135ns
Source	rst_gen_i0/reset_bridge_clk_tx_i0/rst_dst_reg/C (rising edge-triggere
Destination	lb_ctl_i0/debouncer_i0/meta_harden_signal_in_i0/signal_meta_reg/R
Path Group	clk_tx_clk_core
Path Type	Setup (Max at Slow Process Corner)
Requirement	6.000ns (clk_tx_clk_core rise@6.000ns - clk_tx_clk_core rise@0.000ns)
Data Path Delay	5.118ns (logic 0.308ns (6.018%) route 4.810ns (93.982%))
Logic Levels	0
Clock Path Skew	-0.318ns
Clock Uncertainty	0.062ns

图 2-59

Source Clock Path			
Delay Type	Incr (ns)	Path (ns)	Location
(clock clk_tx_clk_core rise edge)	(r) 0.000	0.000	
	(r) 0.000	0.000	Site: AA3
net (fo=0)	0.000	0.000	
IBUFDS (Prop_ibufds_I_O)	(r) 1.033	1.033	Site: AA3
net (fo=1, routed)	1.230	2.263	
MMCME2_ADV (Prop_mmc...adv_CLKIN1_CLKOUT1)	(r) 382	-5.119	Site: MMC...ADV_X1Y1
net (fo=1, routed)	1.556	-3.563	
BUFG (Prop_bufg_I_O)	(r) 0.120	-3.443	Site: BUFGCTRL_X0Y0
net (fo=248, routed)	1.656	-1.787	
FDPE			Site: SLICE_X6Y47
Data Path			
Destination Clock Path			
Delay Type	Incr (ns)	Path (ns)	Location
(clock clk_tx_clk_core rise edge)	(r) 6.000	6.000	
	(r) 0.000	6.000	Site: AA3
net (fo=0)	0.000	6.000	
IBUFDS (Prop_ibufds_I_O)	(r) 0.915	6.915	Site: AA3
net (fo=1, routed)	1.149	8.064	
MMCME2_ADV (Prop_mmc...adv_CLKIN1_CLKOUT1)	(r) 458	1.606	Site: MMC...ADV_X1Y1
net (fo=1, routed)	1.476	3.082	
BUFG (Prop_bufg_I_O)	(r) 0.113	3.195	Site: BUFGCTRL_X0Y0
net (fo=248, routed)	1.338	4.533	
FDRE			Site: SLICE_X0Y108
clock pessimism		-0.638	3.895
clock uncertainty		-0.062	3.833

图 2-60

在保持时间时序报告总结部分，会显示时钟偏移，但没有时钟不确定性，如图 2-61 所示。这进一步证明了时钟不确定性对保持时间裕量没有影响。在源时钟路径部分和目的时钟路径部分，可以看到发起沿的出现时刻和捕获沿的出现时刻均为 0ns，如图 2-62 所示，这再次证明了保持时间检查的发起沿和捕获沿为同一上升沿。

Summary	
Name	Path 51
Slack (Hold)	0.100ns
Source	lb_ctl_i0/debouncer_i0/meta_harden_signal_in_i0/signal_meta_reg/C
Destination	lb_ctl_i0/debouncer_i0/meta_harden_signal_in_i0/signal_dst_reg/D
Path Group	clk_tx_clk_core
Path Type	Hold (Min at Fast Process Corner)
Requirement	0.000ns (clk_tx_clk_core rise@0.000ns - clk_tx_clk_core rise@0.000ns)
Data Path Delay	0.358ns (logic 0.100ns (27.920%) route 0.258ns (72.080%))
Logic Levels	0
Clock Path Skew	0.218ns

图 2-61

Source Clock Path

Delay Type	Incr (ns)	Path (...)	Location
(clock clk_tx_clk_core rise edge)	(r) 0.000	0.000	
	(r) 0.000	0.000	Site: AA3
net (fo=0)	0.000	0.000	
IBUFDS (Prop_ibufds_I_O)	(r) 0.390	0.390	Site: AA3
net (fo=1, routed)	0.503	0.893	
MMCME2_ADV (Prop_mmc...adv_CLKIN1_CLKOUT1)	(r) 468	-1.575	Site: MMC...ADV_X1Y1
net (fo=1, routed)	0.523	-1.052	
BUFG (Prop_bufg_I_O)	(r) 0.026	-1.026	Site: BUFGCTRL_X0Y0
net (fo=248, routed)	0.566	-0.460	
FDRE			Site: SLICE_X0Y108

Data Path
Destination Clock Path

Delay Type	Incr (ns)	Path (...)	Location
(clock clk_tx_clk_core rise edge)	(r) 0.000	0.000	
	(r) 0.000	0.000	Site: AA3
net (fo=0)	0.000	0.000	
IBUFDS (Prop_ibufds_I_O)	(r) 0.471	0.471	Site: AA3
net (fo=1, routed)	0.553	1.024	
MMCME2_ADV (Prop_mmc...adv_CLKIN1_CLKOUT1)	(r) 872	-1.848	Site: MMC...ADV_X1Y1
net (fo=1, routed)	0.581	-1.267	
BUFG (Prop_bufg_I_O)	(r) 0.030	-1.237	Site: BUFGCTRL_X0Y0
net (fo=248, routed)	0.784	-0.453	
FDRE			Site: SLICE_X1Y97

图 2-62

结合图 2-36，考虑到时钟抖动，可以进一步得到建立时间裕量更为精确的表达式

$$建立时间裕量 = T_{clk} - T_x + T_{skew} - T_{su} - T_{data}$$

其中，T_{clk} 为时钟周期，T_x 为时钟不确定性，T_{skew} 为时钟偏移，T_{su} 为捕获寄存器的建立时间，T_{data} 为数据路径延迟。可见，时钟不确定性会减小建立时间裕量。

为了降低时钟不确定性，一个重要手段就是降低时钟抖动。在 Clocking Wizard IP 中有针对抖动优化的选项，如图 2-63 所示。其中，Balanced 用于在功耗和抖动之间获得折中。Minimize Output Jitter 则是以牺牲功耗（有时也包括输出时钟的相位误差）为代价换取最小的输出时钟抖动。Maximize Input Jitter filtering 则允许输入时钟有较大的抖动，但对输出时钟抖动有负面影响。

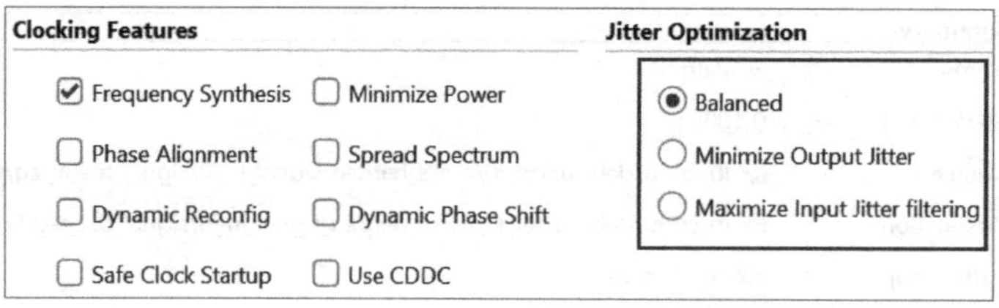

图 2-63

应用案例 7：利用 Minimize Output Jitter 优化输出时钟抖动

输入时钟为 62.5MHz，输出时钟为 300MHz，若 Jitter Optimization 选择 Balanced（见图 2-63），则 MMCM 配置参数如图 2-64 所示。据此可推断出 VCO 频率为 62.5×96/5MHz，即 1200MHz，输出时钟频率为 1200/4MHz，即 300MHz，峰峰抖动为 188.156ps。若 Jitter Optimization 选择 Minimize Output Jitter，则 MMCM 配置参数如图 2-65 所示。据此可推断出 VCO 频率为 62.5×24/1MHz，即 1500MHz，输出时钟频率为 1500/5MHz，即 300MHz，峰峰抖动为 88.936ps，减小了 99.22ps，几乎是原来的一半。由此可见，VCO 频率越高，输出时钟抖动越小。

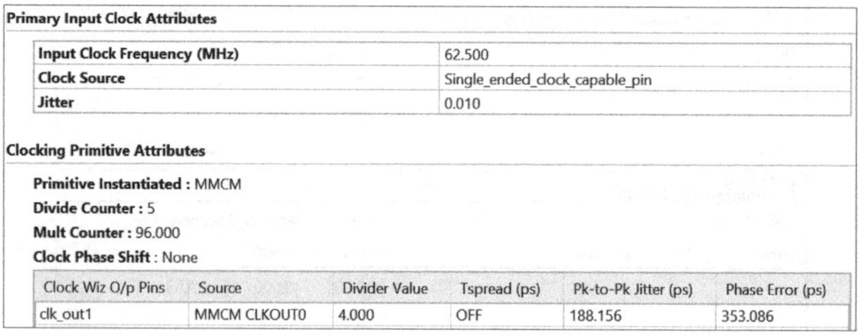

Primary Input Clock Attributes

Input Clock Frequency (MHz)	62.500
Clock Source	Single_ended_clock_capable_pin
Jitter	0.010

Clocking Primitive Attributes

Primitive Instantiated : MMCM
Divide Counter : 5
Mult Counter : 96.000
Clock Phase Shift : None

Clock Wiz O/p Pins	Source	Divider Value	Tspread (ps)	Pk-to-Pk Jitter (ps)	Phase Error (ps)
clk_out1	MMCM CLKOUT0	4.000	OFF	188.156	353.086

图 2-64

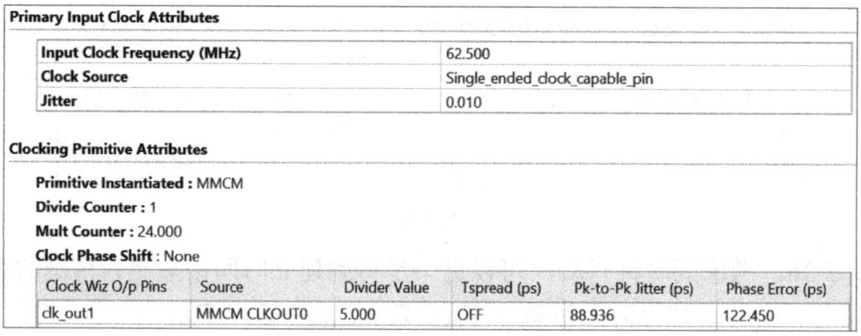

Primary Input Clock Attributes

Input Clock Frequency (MHz)	62.500
Clock Source	Single_ended_clock_capable_pin
Jitter	0.010

Clocking Primitive Attributes

Primitive Instantiated : MMCM
Divide Counter : 1
Mult Counter : 24.000
Clock Phase Shift : None

Clock Wiz O/p Pins	Source	Divider Value	Tspread (ps)	Pk-to-Pk Jitter (ps)	Phase Error (ps)
clk_out1	MMCM CLKOUT0	5.000	OFF	88.936	122.450

图 2-65

实际上，Jitter Optimization 与 MMCM 的属性 BANDWIDTH 是对应的。换言之，当 Jitter Optimization 选择 Balanced 时，对应的 BANDWIDTH 是 OPTIMIZED；选择 Minimize Output Jitter 时对应 HIGH；选择 Maximize Input Jitter filtering 时对应 LOW，如图 2-66 所示。这就不难理解 Minimize Output Jitter 对输出时钟抖动的优化力度最大。

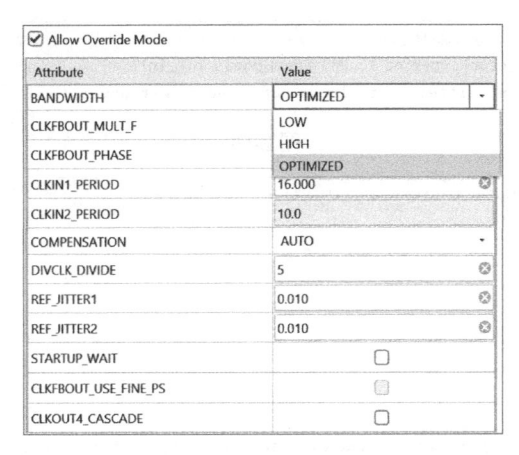

图 2-66

表 2-3 给出了 UltraScale/UltraScale+ FPGA 中 MMCM/PLL VCO 所允许的最小频率和最大频率。不难看出，MMCM VCO 的频率范围更大一些，可承受的最大频率更高一些。

表 2-3

类 别	指 标	UltraScale FPGA（速度等级-3）		UltraScale+ FPGA（速度等级-1/-2/-3）	
		Virtex	Kintex	Virtex	Kintex
MMCM	VCO 最小频率（MHz）	600	600	800	800
	VCO 最大频率（MHz）	1600	1600	1600	1600
PLL	VCO 最小频率（MHz）	600	600	750	750
	VCO 最大频率（MHz）	1335	1335	1500	1500

💡 **设计规则 10（针对 UltraScale/UltraScale+ FPGA）**：使用 Clocking Wizard 生成时钟时，优先选择 MMCM：一方面，可以利用 MMCM VCO 能承受更高的频率来改善输出时钟抖动；另一方面，可以利用 MMCM CLKOUT0 端口的分频器支持小数分频来获得更精确的时钟频率。若生成时钟频率较高（大于或等于 300MHz），则要勾选 Minimize Output Jitter 选项，以获得较低的时钟抖动。

在 Clocking Wizard IP 中还可以指定输入时钟峰峰抖动，如图 2-67 所示。对于应用案例 7，如果输入时钟峰峰抖动为 160ps（0.16ns），那么在最终生成的 .xdc 文件中会体现出时钟周期与输入时钟峰峰抖动，如图 2-68 所示。

图 2-67

UltraScale/UltraScale+ FPGA 中的时钟缓冲器 BUFGCE_DIV 具备分频功能，对一些简单分频可以采用 BUFGCE_DIV 实现，具体电路可参考图 2-47。同时，Vivado 2021.1 版本增强了 Clocking Wizard IP 的功能，提供了选项 Optimize Clocking Structure，如图 2-69 所示。一旦勾选该选项，IP 就会根据输入/输出时钟之间的关系优化时钟网络架构。

```
55  # Connect to input port when clock capable pin is selected for input
56  create_clock -period 16.000 [get_ports clk_in1]
57  set_input_jitter [get_clocks -of_objects [get_ports clk_in1]] 0.160
```

图 2-68

图 2-69

例如，输入时钟的频率为 61.44MHz，4 个输出时钟的频率分别为 491.52MHz、245.76MHz、122.88MHz 和 61.44MHz（相邻两个频率为 2 分频关系），时钟缓冲器选择 Buffer，如图 2-70 所示。最终的时钟网络架构如图 2-71 所示。可以看到，该架构利用了 BUFGCE_DIV 的分频功能。

图 2-70

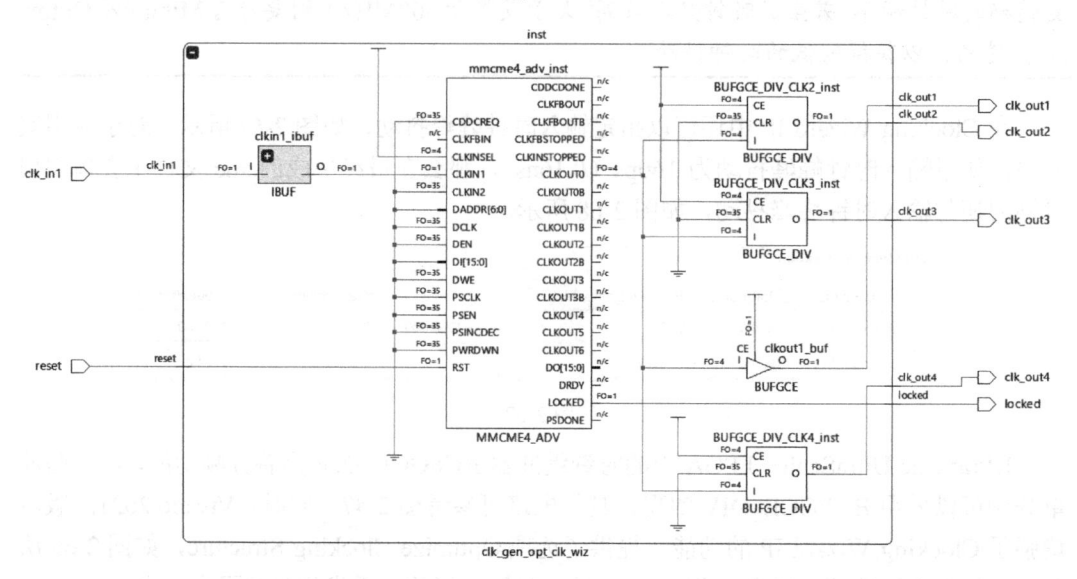

图 2-71

2.4 安全的时钟启动方式

通常情况下,在 MMCM/PLL 的 LOCKED 信号抬高之后(由 0 变为 1),MMCM/PLL 就处于锁定状态,输出时钟已保持稳定。在此之前,输出时钟会发生持续翻转且不稳定,如图 2-72 所示。一旦时钟发生翻转,时序逻辑就可能将其判定为一个有效沿,进而执行相应的操作。在某些场合下,需要 LOCKED 信号为 1 之后才发生时钟翻转,这就是所谓的安全时钟启动。Clocking Wizard IP 提供了 Safe Clock Startup 这个选项,用于实现此功能,如图 2-73 所示。

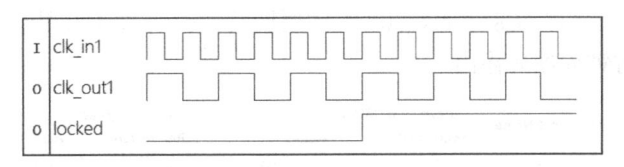

图 2-72

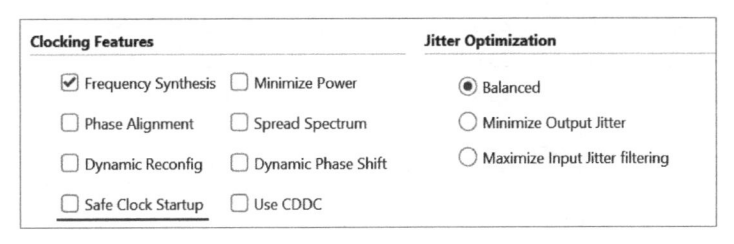

图 2-73

以 UltraScale/UltraScale+ FPGA 为例,一旦勾选 Safe Clock Startup 选项,Clocking Wizard IP 会生成如图 2-74 所示的电路。可以看到每个时钟输出端口会出现并联的时钟缓冲器 BUFGCE(此处只能是 BUFGCE)和一个深度为 8 的移位寄存器。移位寄存器的数据输入端口由 MMCM 的 LOCKED 信号提供。最终输出时钟 clka 和 clkb 对应的全局时钟缓冲器 BUFGCE 的时钟使能端口由移位寄存器的数据输出端口提供。因此,对于 clka,只有当 LOCKED 信号为高电平,且经过 8 个 clka 周期后,clka 才稳定输出,clkb 亦是如此。这可由如图 2-75 所示的仿真波形进一步验证。图中,"I"代表输入;"O"代表输出。

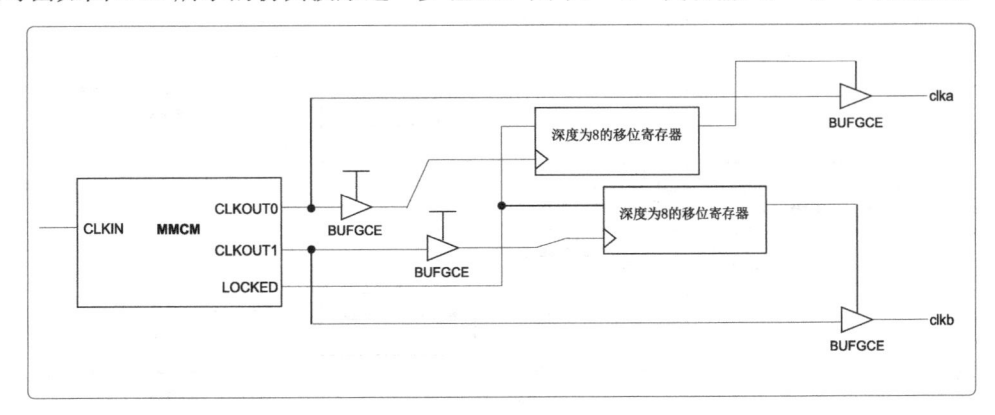

图 2-74

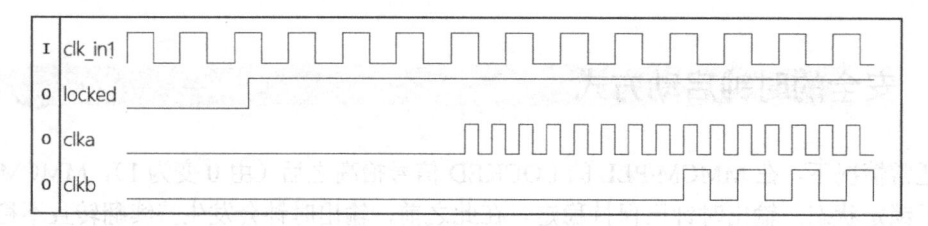

图 2-75

勾选 Safe Clock Startup 之后,在 Output Clocks 页面可勾选 USE CLOCK SEQUENCING,如图 2-76 所示。该选项的功能是使输出时钟按指定顺序依次输出,标记为 1 的先输出,类似图 2-75。clk_300 是在 LOCKED 为高电平之后,经过 8 个 clk_300 时钟周期后输出;clk_200 是在 clk_300 有效之后,经过 8 个 clk_200 时钟周期后输出;clk_100 是在 clk_200 有效之后,经过 8 个 clk_100 时钟周期后输出。

图 2-76

之所以会形成顺序时钟,是因为勾选 USE CLOCK SEQUENCING 选项之后会形成如图 2-77 所示的电路,对比图 2-74 可以看出,其最大的变化在于移位寄存器由并联关系变为级联关系。

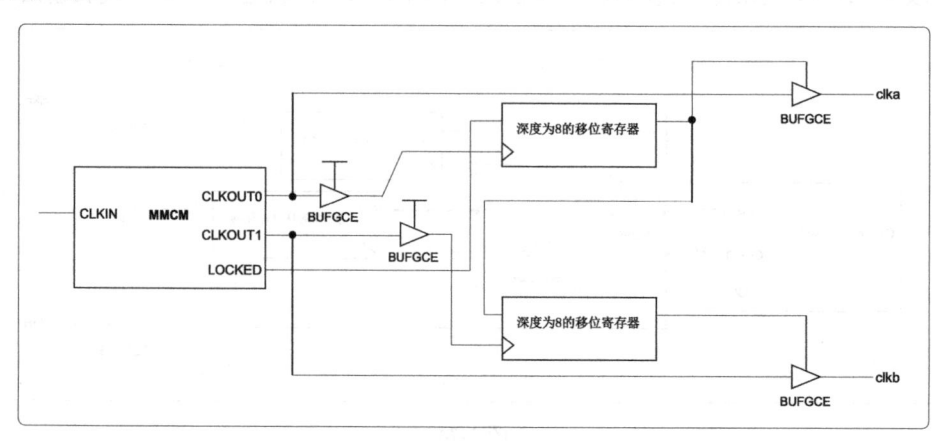

图 2-77

应用案例 8（针对 UltraScale/UltraScale+ FPGA）：当输出时钟频率大于 300MHz 时，勾选 Safe Clock Startup 选项，如何保证时序收敛

在使用 UltraScale/UltraScale+ FPGA 芯片时，若勾选 Safe Clock Startup 选项，则当输出时钟频率大于 300MHz 时，可能会出现时序违例。此时，需要人工搭建安全时钟启动电路，如图 2-78 所示。在原有电路的基础上插入一个 BUFGCE，并对此 BUFGCE 设置 DONT_TOUCH 约束，以保证 Vivado 在 opt_design 阶段不会将其移除。同时，还要添加 CLOCK_DELAY_GROUP 和 CLOCK_REGION 约束。在图 2-78 中，CLOCK_REGION 的值 XmYn 意在表明这三个 BUFGCE 需要处于同一时钟区域内，该时钟区域也是 MMCM 所在的时钟区域。

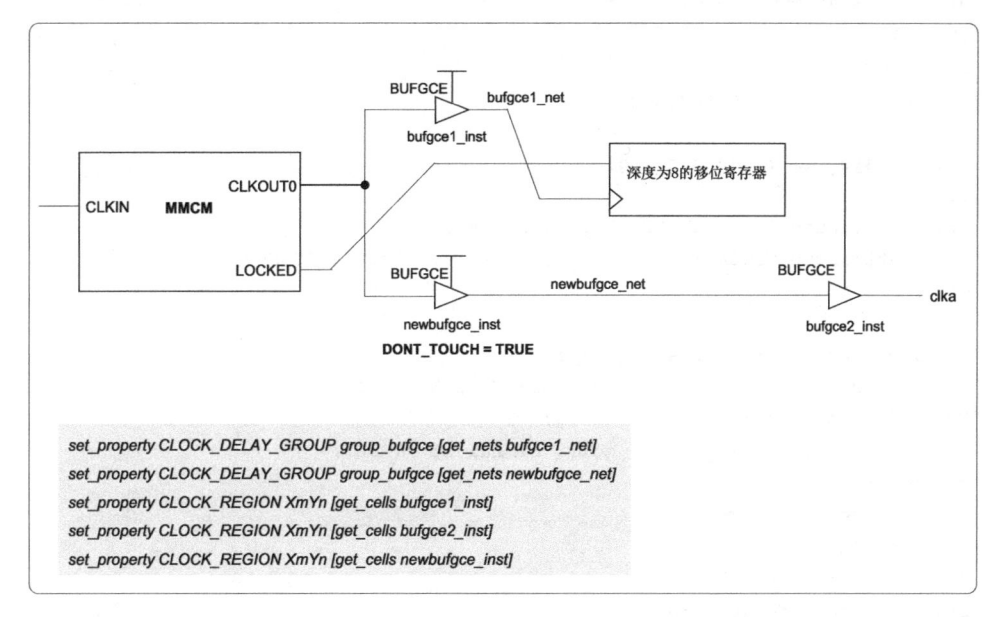

图 2-78

安全启动时钟的另外一种方式是通过外部复位信号控制，具体电路如图 2-79 所示。这里，输入时钟 clk 因扇出很小而不用插入全局时钟缓冲器，这可以通过在代码或约束文件中添加属性 CLOCK_BUFFER_TYPE 实现。同时，对这 4 个异步复位寄存器添加属性 ASYNC_REG，以保证它们最终在一个 SLICE 内，如 VHDL 代码 2-1 所示，代码第 9 行中的参数 DELAY_NUM 应小于或等于 8（7 系列 FPGA 的一个 SLICE 内有 8 个触发器）。

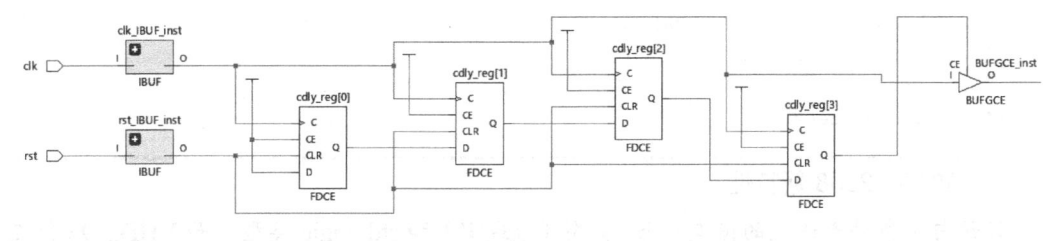

图 2-79

对于 VHDL 代码，端口的属性应写在 entity 域内，如 VHDL 代码 2-1 第 16 行和第 17

行所示，信号的属性应写在 architecture 域内，如第 22 行和第 23 行所示。描述异步复位触发器时，时钟和复位信号都应出现在 process 的敏感变量列表里，如代码第 25 行所示。

> **VHDL 代码 2-1**

```vhdl
1.  --File: safe_clk.vhd
2.  library ieee;
3.  use ieee.std_logic_1164.all;
4.  library UNISIM;
5.  use UNISIM.vcomponents.all;
6.  
7.  entity safe_clk is
8.    generic (
9.      DELAY_NUM : positive := 4
10.   );
11.   port (
12.     clk     : in std_logic;
13.     rst     : in std_logic;
14.     sys_clk : out std_logic
15.   );
16.   attribute CLOCK_BUFFER_TYPE : string;
17.   attribute CLOCK_BUFFER_TYPE of clk : signal is "NONE";
18. end entity;
19. 
20. architecture rtl of safe_clk is
21.   signal cdly : std_logic_vector(DELAY_NUM - 1 downto 0) := (others => '0');
22.   attribute ASYNC_REG : string;
23.   attribute ASYNC_REG of cdly: signal is "TRUE";
24. begin
25.   process(clk, rst)
26.   begin
27.     if rst then
28.       cdly <= (others => '0');
29.     elsif rising_edge(clk) then
30.       cdly <= cdly(DELAY_NUM - 2 downto 0) & '1';
31.     end if;
32.   end process;
33. 
34.   i_bufgce: BUFGCE
35.   port map (
36.     O  => sys_clk,
37.     CE => cdly(DELAY_NUM-1),
38.     I  => clk
39.   );
40. end architecture;
```

> **VHDL-2008 新特性**
>
> if 语句中条件表达式的值支持 bit 类型（'0'或'1'）和 std_logic 类型，而 VHDL-93 仅支持 boolean 类型（false, true）。因此 VHDL 代码 2-1 第 27 行如果采用 VHDL-93 描述，应写为 if rst = '1' then；如果采用 VHDL-2008 描述，也可以写为 if (rst) then。

2.5 时钟规划

时钟规划在设计初期就要完成，与芯片选型紧密相关。时钟规划要解决以下几个问题。
- 设计中需要的时钟个数。
- 每个时钟的频率。
- 时钟之间的相位关系。
- 外部时钟的管脚位置与电平标准。

尽管目前 Xilinx 主流的 FPGA 芯片中都有多个 MMCM/PLL，每个 MMCM/PLL 都可以生成多个时钟，同时有很多全局时钟缓冲器可供使用，但是时钟个数越少越好仍然是一个黄金准则。这对于降低设计复杂度和减少跨时钟域路径是很有帮助的。

有些时钟伴随着相应的硬核，一旦硬核指标和位置确定，相应的时钟频率和位置也就确定了，如高速收发器的参考时钟。因此，确定硬核的位置就显得尤为重要。通常硬核的位置与系统数据流是息息相关的。例如，某设计中要用到三类硬核 CMAC、PCIe 和 Interlaken。其中，CMAC 和 PCIe 之间有数据交互，Interlaken 和 CMAC 之间有数据交互。根据硬核在芯片中的位置可形成三种方案，如图 2-80 所示：方案 1 走线很长，同时存在数据流冲突；方案 3 走线最短且不存在数据流冲突；方案 2 没有数据流冲突，走线长度介于方案 1 和方案 3 之间。

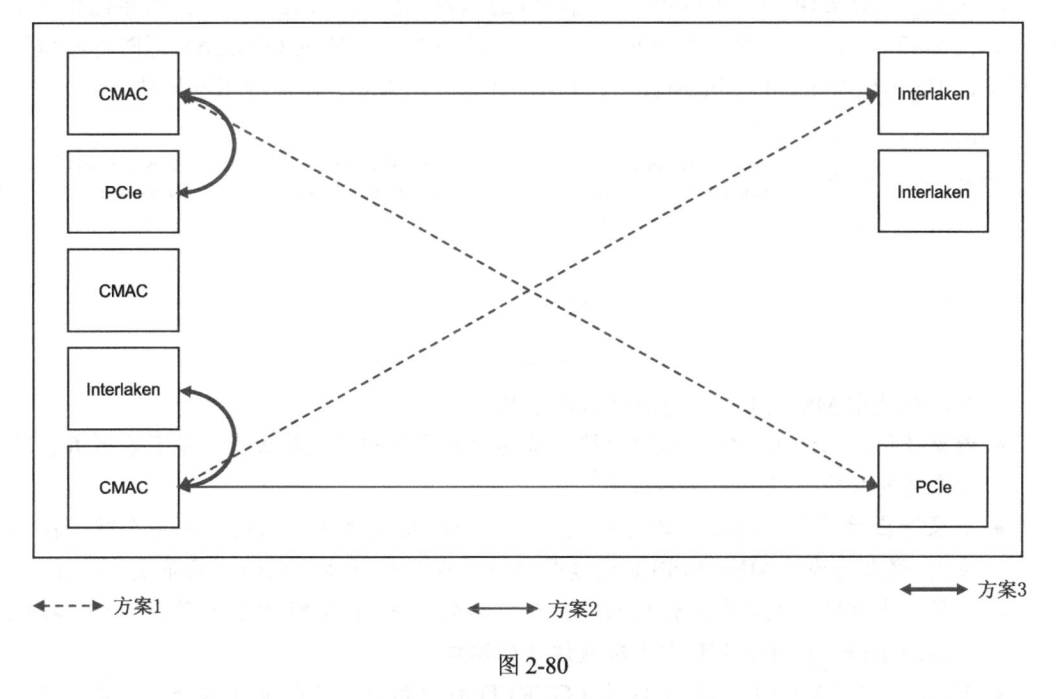

图 2-80

FPGA 内部逻辑时钟若需要由外部时钟经 MMCM/PLL 生成，则要确保外部时钟从全局时钟管脚进入。同时，一些复杂的 IP 本身就包含特定的时钟架构，如 MIG IP。工程师最好能够充分理解这些 IP 内部的时钟架构，判断其是否可以给设计的其他部分使用。

通常，对于时钟网线，Vivado 会自动插入全局时钟缓冲器 BUFG。但是，如果设计中需要额外的时钟缓冲器，那么实例化几乎是唯一的方式。借助语言模板（Tools→Language

Templates），搜索 BUFG，可快速找到各类时钟缓冲器的实例化模板，如图 2-81 所示，这样可以有效加速设计进程。

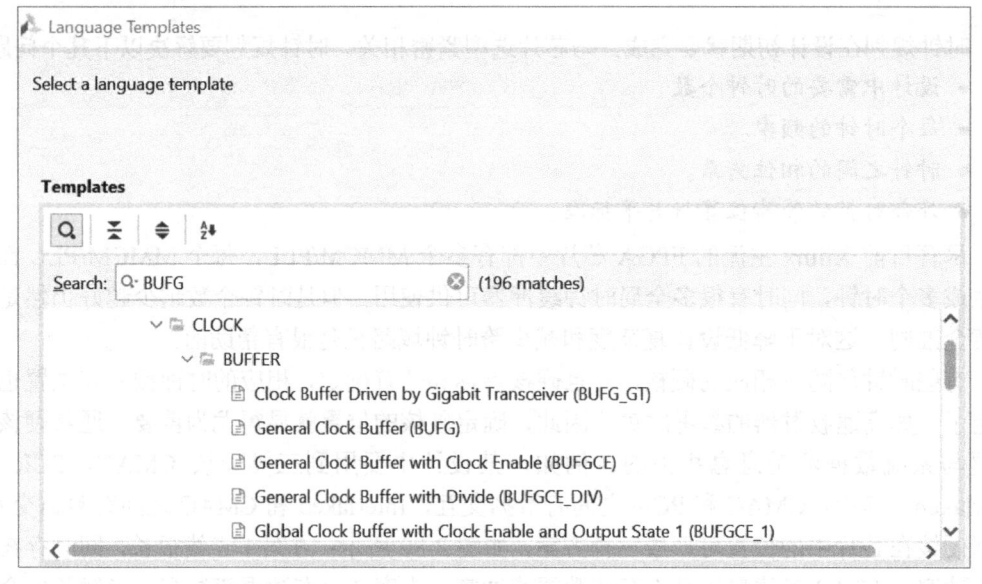

图 2-81

Clocking Wizard IP 不仅可以控制输出时钟的频率，还可以管理输出时钟的相位和占空比，如图 2-82 所示。对于 7 系列 FPGA，还可以使用 IDELAY 或 ODELAY 对时钟相位进行移动。但对于 UltraScale/UltraScale+ FPGA，建议使用 MMCM 实现相位移动。

图 2-82

同时，在使用 MMCM/PLL 时应注意以下几点。
- 避免任何输入管脚悬空。不同的综合工具对悬空管脚的处理方式可能存在差异，有的会将其接地，有的会将其拉高。
- 如果输出时钟是断断续续的（在设计的某个时间段存在，在设计的某个时间段消失），那么应确保 MMCM/PLL 的复位端口是由用户逻辑控制的，而不是直接接地。
- 如果输出时钟一直存在且保持频率不变，那么应将 MMCM/PLL 的复位端口恒接地（或在 Clocking Wizard IP 中不勾选该复位端口）。
- 可以将 MMCM/PLL 的输出信号 LOCKED 作为用户逻辑的复位信号，但要先将其同步化到相应的时钟域。
- MMCM/PLL 应放置在设计的顶层，以保证生成时钟可以通畅地传送给各个模块，而不用在层次之间穿梭，如图 2-83 所示。

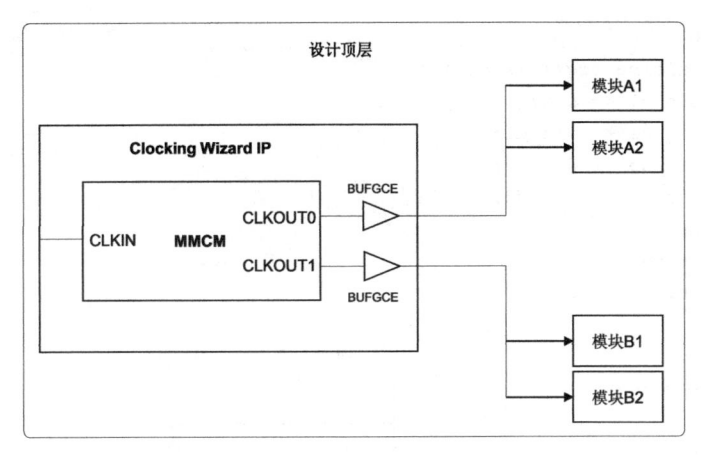

图 2-83

对于由寄存器或查找表生成的门控时钟，尽管其可以用来降低功耗，但在设计中应避免使用门控时钟：一是因为 Vivado 不会对门控时钟自动插入 BUFG；二是因为时钟周期约束无法覆盖到相应的路径。以 VHDL 代码 2-2 为例，其对应的电路如图 2-84 所示。寄存器 clk_div_reg 经反相器（LUT1）生成二分频时钟 clk_div。该设计通过 create_clock 设置源时钟的频率为 200MHz。在时钟网络报告（由命令 report_clock_networks 生成）中，可以看到 clk_div 所驱动的路径没有被约束，如图 2-85 所示。

📄 **VHDL 代码 2-2**

```
1.  --File: gated_clk.vhd
2.  library ieee;
3.  use ieee.std_logic_1164.all;
4.
5.  entity gated_clk is
6.    port (
7.      clk : in std_logic;
8.      din : in std_logic;
9.      dout: out std_logic
10.   );
11. end entity;
12.
13. architecture rtl of gated_clk is
14.   signal clk_div : std_logic := '0';
15.   signal din_d1  : std_logic := '0';
16.   signal din_d2  : std_logic := '0';
17.   signal din_d3  : std_logic := '0';
18. begin
19.   process(clk)
20.   begin
21.     if rising_edge(clk) then
22.       clk_div <= not clk_div;
23.     end if;
24.   end process;
25.
26.   process(clk_div)
```

```
27.   begin
28.     if rising_edge(clk_div) then
29.       din_d1 <= din;
30.       din_d2 <= din_d1;
31.       din_d3 <= din_d2;
32.       dout   <= din_d3;
33.     end if;
34.   end process;
35. end architecture;
```

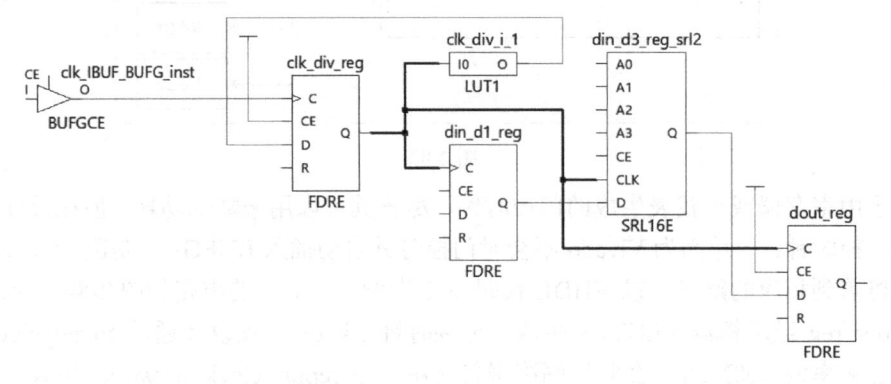

图 2-84

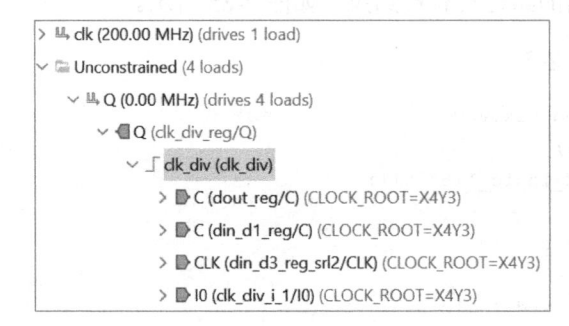

图 2-85

在 Vivado 综合选项设置中，有一个选项-gated_clock_conversion，将其值设置为 on，如图 2-86 所示，可将图 2-84 中的门控时钟信号转换为时钟使能信号，如图 2-87 所示。但该选项并不一定能百分之百地生效，同时，对于一些综合属性，如 KEEP_HIERARCHY、DONT_TOUCH 和 MARK_DEBUG，若其对象与门控时钟相关或就是门控时钟信号本身，那么也会对门控时钟转换造成干扰。因此，对于一些复杂的门控时钟，有时需要手工将其变为时钟使能信号。

图 2-86

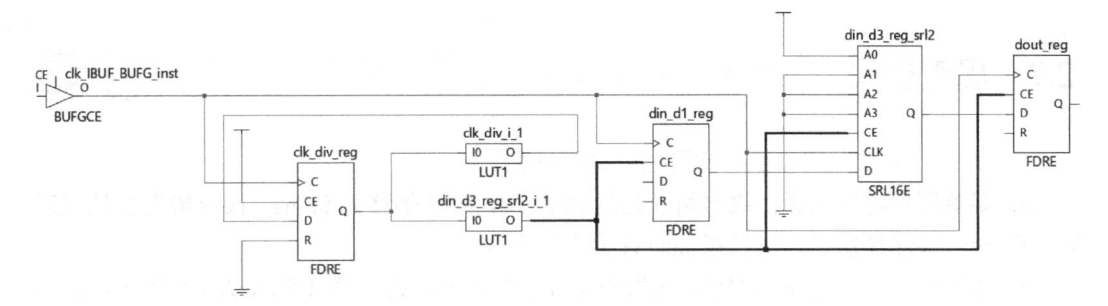

图 2-87

应避免使用局部时钟。局部时钟是指使用非专用全局时钟资源生成的时钟，如通过计数器分频生成的时钟。这类时钟通常时钟偏移较大，给时序收敛带来了压力。进行时钟规划时，需要了解对应芯片的 MMCM 所能输出的最小频率，如表 2-4 所示，以确保设计所需要的时钟频率在 MMCM 可生成的范围之内。并且，若设计中的确需要计数器通过分频生成频率很低的时钟，则要确保该时钟的负载较低（不超过 30 个），同时，要通过手工布局的方式使这些负载在较小的区域内。

表 2-4

指标	7 系列 FPGA	UltraScale FPGA	UltraScale+ FPGA
MMCM 输出最小频率（MHz）	4.69	4.69	6.25

2.6 创建输出时钟

如果设计中需要将 FPGA 内部时钟传递到芯片外部，最有效的方式是使用 ODDR 或 OSERDES，而不是将时钟直接连接到管脚上。以 UltraScale FPGA 为例，若将 ODDRE1 的 D1 端口恒接地，将 D2 端口恒接高电平，则输出时钟与输入时钟同频反相；若将 D1 端口恒接高电平，将 D2 端口恒接地，则输出时钟与输入时钟同频同相，相应的电路如图 2-88 所示（ODDRE1 映射为 OSERDESE3）。

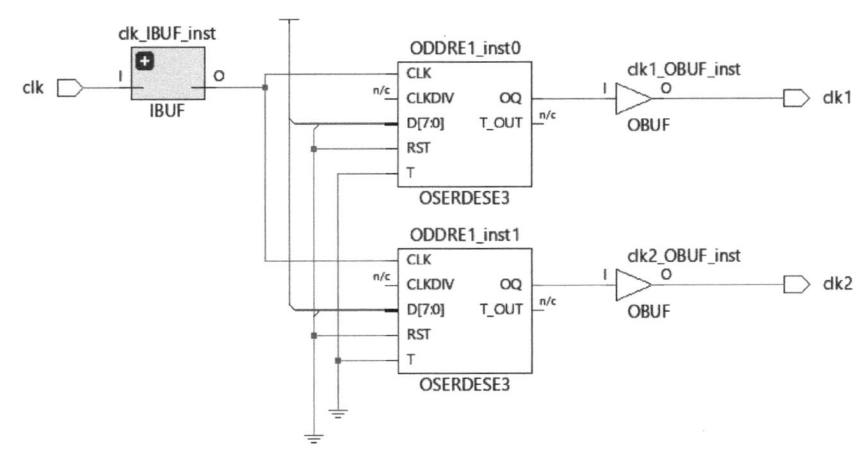

图 2-88

2.7 思考空间

1. 请列出 7 系列 FPGA 中的区域时钟缓冲器。
2. 某单周期时序路径，数据路径延迟为 2.6ns，时钟偏移为 0.2ns，接收触发器建立时间为 0.1ns，该路径所能达到的最高时钟频率是多少？
3. 某单周期时序路径，时钟周期为 4ns，T_{co} 为 0.1ns，T_{logic} 为 1.8ns，T_{net} 为 2.0ns，T_{skew} 为 -0.2ns，T_{su} 为 0.1ns，该路径是否发生建立时间违例？
4. 某设计采用 7 系列 FPGA，输入时钟频率为 100MHz，由外部晶振提供，内部需要频率为 100MHz 和 300MHz 的两个时钟，请给出可行的时钟方案。
5. 某源同步设计采用 7 系列 FPGA，随路时钟为 clk，还需要生成 clk 的 4 分频时钟，并行输入数据可放置在同一个 I/O Bank 内，请给出可行的时钟方案。如果采用 UltraScale FPGA，又该如何设计时钟方案？
6. 某设计采用 UltraScale FPGA，输入时钟频率为 150MHz，内部需要一个频率为 300MHz 的时钟和一个频率为 600MHz 的时钟，请给出可行的时钟方案。
7. 某设计采用 Versal ACAP，输入时钟频率为 200MHz，内部需要 4 个时钟，频率分别为 400MHz、200MHz、100MHz 和 50MHz，请给出可行的时钟方案。
8. 某设计中存在一个高扇出的网线，可采用哪些方法插入 BUFG 并将其引入全局时钟网络？
9. 就时钟缓冲器而言，Versal ACAP 和 UltraScale+ FPGA 有哪些差异？
10. 如何发现设计中的门控时钟？

第 3 章

优化组合逻辑

3.1 组合逻辑资源

组合逻辑主要由查找表（Look-Up Table，LUT）实现，因此，LUT 也被称为逻辑函数发生器。无论是 SLICEL 中的 LUT 还是 SLICEM 中的 LUT，均可用作逻辑函数发生器。此外，SLICE 中的数据选择器（Multiplex，MUX）和进位链（Carry Chain）也可用于构成组合逻辑电路。表 3-1 显示了不同芯片的 SLICE 内部逻辑单元的分布情况。就 LUT 个数而言，UltraScale/UltraScale+ FPGA 和 Versal ACAP 一致，均为 8 个，且是 7 系列 FPGA 的 2 倍。就 MUX 而言，UltraScale/UltraScale+ FPGA 中的 SLICE 不仅有 F7MUX 和 F8MUX，还增加了 F9MUX，当然，三者均是 2 选 1 的数据选择器。7 系列 FPGA 则只有 F7MUX 和 F8MUX。而 Veral ACAP 中的 SLICE 不再包含任意一种 MUX。就进位链而言，7 系列 FPGA 和 UltraScale/UltraScale+ FPGA 中的进位链为 CARRY8，而 Versal ACAP 中为 LOOKAHEAD8。

表 3-1

SLICE	LUT	FF	F7MUX	F8MUX	F9MUX	进位链
7 系列 FPGA SLICE	4	8	2	1	0	1
UltraScale/UltraScale+ SLICE	8	16	4	2	1	1
Versal ACAP SLICE	8	16	0	0	0	1

借助 Tcl 代码 3-1 可以验证如表 3-1 所示的结论。代码第 2 行表明器件为 7 系列 FPGA，但可以将其改为 UltraScale/UltraScale+ FPGA 或 Versal ACAP。

Tcl 代码 3-1

```
1.  #File: check_bel_in_slice.tcl
2.  set mypart [get_parts xc7k70tfbg484-3]
3.  link_design -part $mypart
4.  set slice [get_sites SLICE_X0Y0]
5.  set lut6 [get_bels -of $slice -filter "NAME =~ *6LUT"]
6.  => SLICE_X0Y0/A6LUT SLICE_X0Y0/B6LUT SLICE_X0Y0/C6LUT SLICE_X0Y0/D6LUT
7.  set f7mux [get_bels -of $slice -filter "NAME =~ *F7*"]
8.  =>
SLICE_X0Y0/F7MUX_AB SLICE_X0Y0/F7MUX_CD SLICE_X0Y0/F7MUX_EF SLICE_X0Y0/F7MUX_GH
9.  set f8mux [get_bels -of $slice -filter "NAME =~ *F8*"]
10. => SLICE_X0Y0/F8MUX_BOT SLICE_X0Y0/F8MUX_TOP
11. set f9mux [get_bels -of $slice -filter "NAME =~ *F9*"]
12. => SLICE_X0Y0/F9MUX
13. puts "#N LUT6 in one SLICE: [llength $lut6]"
14. => #N LUT6 in one SLICE: 8
```

```
15.  puts "#N F7MUX in one SLICE: [llength $f7mux]"
16.  => #N F7MUX in one SLICE: 4
17.  puts "#N F8MUX in one SLICE: [llength $f8mux]"
18.  => #N F8MUX in one SLICE: 2
19.  puts "#N F9MUX in one SLICE: [llength $f9mux]"
20.  => #N F9MUX in one SLICE: 1
```

3.2 译码器与编码器

3.2.1 译码器代码风格

先看一个 2 输入 4 输出的译码器，其输出与输入的关系如表 3-2 所示。本质上，这就是一个真值表，而组合逻辑总是可以用真值表的形式表示。

表 3-2

输	入	输			出
0	0	0	0	0	1
0	1	0	0	1	0
1	0	0	1	0	0
1	1	1	0	0	0

对于如表 3-2 所示的真值表，用 HDL 描述，如 VHDL 代码 3-1 所示。这里增加了一个输入端口 en，当其值为 1 时，译码器才可正常工作。同时，这里使用了 case 语句，如代码第 20~26 行所示。不难看出，2-4 译码器其实完成了从二进制码到独热码（所有位中只有某一位为 1）的转换。

VHDL 代码 3-1

```
1.  --File: decoder_2to4_high_v1a.vhd
2.  library ieee;
3.  use ieee.std_logic_1164.all;
4.
5.  entity decoder_2to4_high_v1a is
6.    port (
7.      a  : in std_logic_vector(1 downto 0);
8.      en : in std_logic;
9.      y  : out std_logic_vector(3 downto 0)
10.   );
11. end entity;
12.
13. architecture rtl of decoder_2to4_high_v1a is
14. begin
15.   process(all)
16.   begin
17.     if (not en) then
18.       y <= (others => '0');
19.     else
20.       case a is
21.         when "00" => y <= "0001";
```

```
22.            when "01" => y <= "0010";
23.            when "10" => y <= "0100";
24.            when "11" => y <= "1000";
25.            when others => y <= (others => '0');
26.        end case;
27.    end if;
28. end process;
29. end architecture;
```

> 🔍 **VHDL-2008 新特性**
>
> 若 process 中描述的是纯组合逻辑，则可以用关键字 all 表征涉及的所有敏感变量，这使得敏感变量列表更为简洁，同时避免了 VHDL-93 版本因遗漏敏感变量而导致的仿真错误。

在 Vivado 下，打开 Elaborated Design，可查看其 Schematic 视图，如图 3-1 所示，两个 MUX 级联。最左侧的 MUX 对应 VHDL 代码 3-1 中的 case 语句，右侧的 MUX 对应 VHDL 代码 3-1 第 17 行～第 19 行的 if-else 语句。由此可以进一步理解 case 和 MUX 的关系：VHDL 中的 case 语句要求选择分支选项全面（when others 选项不可或缺）且彼此互斥。综合工具认为它们具有相等的优先级，并综合为一个 MUX 而不是优先级结构。与 case 语句不同，if 语句隐式地指定第一个分支具有更高的优先级，但当 if 语句的条件分支互斥时，如当 en 为 0 时，执行 if 分支，否则执行 else 分支，综合工具会将其综合为 MUX。

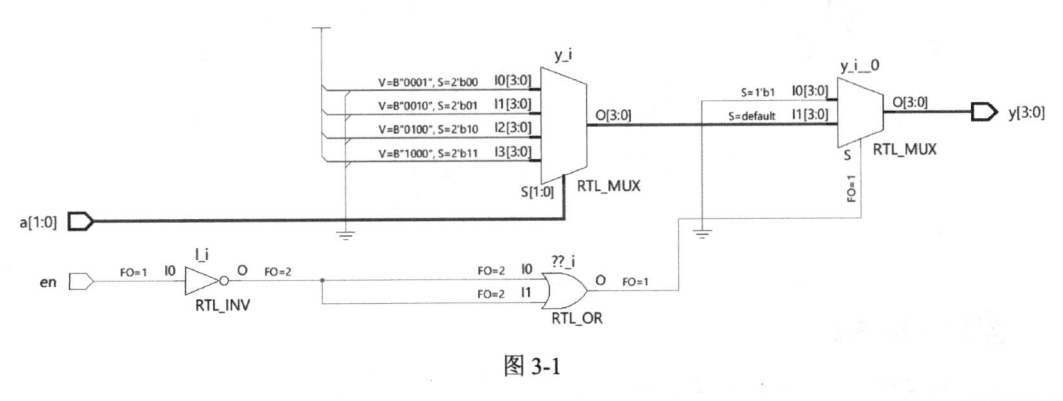

图 3-1

> 💡 **设计规则 1**：VHDL 中的 case 语句选择分支选项彼此互斥，因此综合工具会将其综合为一个 MUX，而不是优先级结构。同时，when others 选项不可或缺，以保证 case 语句可以覆盖所有可能的情况。

进一步分析 VHDL 代码 3-1，输入共 3 位（其中，a 为 2 位，en 为 1 位），输出共 4 位，且输出的每一位均与 3 位输入有关，因此可推断出 2-4 译码器需要消耗 4 个 LUT。而实际综合后的结果显示只消耗了 2 个 LUT。这是为什么呢？这是因为一个 6 输入 LUT（LUT6）是由 2 个 5 输入 LUT（LUT5）加上 1 个 2 选 1 的 MUX 构成的。这个 LUT6 有 2 个输出，分别为 O5 和 O6。其中，O5 来自其中一个 LUT5，O6 来自 MUX 的输出。如果两个布尔表达式有共同因子，是有可能将它们放置在一个 LUT6 中实现的。这就是所谓的 LUT 整合（LUT Combining）。对于 2-4 译码器，每个输出对应一个布尔表达式。不难发现，这 4 个布尔表达式的 3 个变量是完全相同的，因此只用 2 个 LUT6 就可以实现 4 个布尔表达式。

打开综合后的网表文件，生成资源利用率报告，可以看到，实际 LUT 使用个数为 2，如图 3-2 所示。

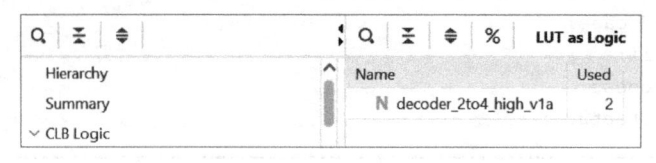

图 3-2

图 3-3 为 VHDL 代码 3-1 所描述的 2-4 译码器综合后的结果。右侧部分为顶部 LUT3 的真值表。可以看到，综合后的 Schematic 视图不会显示整合 LUT。对于被整合的 LUT，其属性 SOFT_HLUTNM 是非空字符串，据此，可以很容易地找到此类 LUT，如 Tcl 代码 3-2 所示。代码第 2 行用于获取图 3-3 中所有被整合的 LUT，代码第 6 行用于获取这些 LUT 的 SOFT_HLUTNM 值，可以看到，其值为 soft_lutpair0 或 soft_lutpair1。

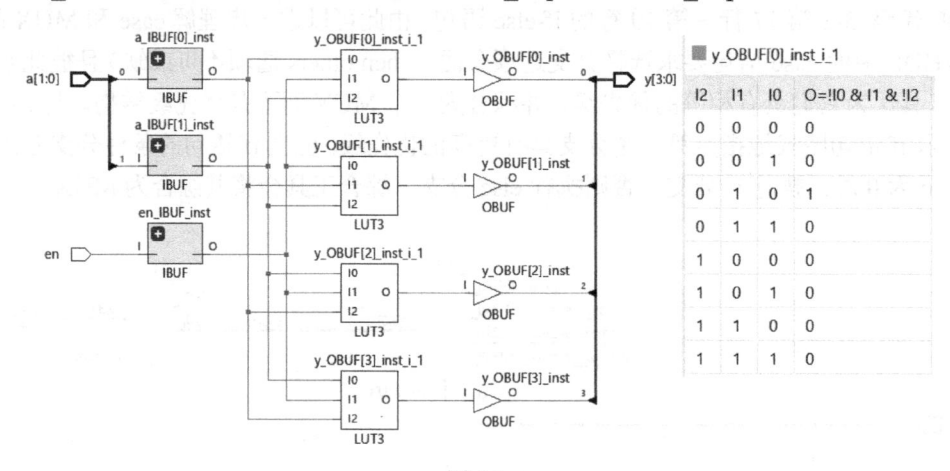

图 3-3

Tcl 代码 3-2

```
1. #File: get_combined_lut.tcl
2. set combined_lut [get_cells -hier -filter {SOFT_HLUTNM != "" || HLUTNM != ""}]
3. => y_OBUF[0]_inst_i_1 y_OBUF[1]_inst_i_1 y_OBUF[2]_inst_i_1 y_OBUF[3]_inst_i_1
4. puts "Number of Combined LUT: [llength $combined_lut]"
5. => Number of Combined LUT: 4
6. get_property SOFT_HLUTNM $combined_lut
7. => soft_lutpair0 soft_lutpair0 soft_lutpair1 soft_lutpair1
8. get_property HLUTNM $combined_lut
9. => {} {} {} {}
```

在 VHDL 中，with-select 语句和 case 语句具有相同的功能，因此，也可以用 with-select 语句描述 2-4 译码器，如 VHDL 代码 3-2 所示。这里只给出结构体部分的代码，其与 VHDL 代码 3-1 具有完全相同的综合结果。

VHDL 代码 3-2

```
1.  --File: decoder_2to4_high_v1b.vhd
2.  architecture rtl of decoder_2to4_high_v1b is
3.  begin
4.    process(all)
5.    begin
6.      if (not en) then
7.        y <= (others => '0');
8.      else
9.        with a select y <=
10.          "0001" when "00",
11.          "0010" when "01",
12.          "0100" when "10",
13.          "1000" when "11",
14.          "0000" when others;
15.      end if;
16.    end process;
17. end architecture;
```

🔍 VHDL-2008 新特性

选择信号赋值语句 with-select 既可以在进程 process 内部使用，也可以在其外部使用。

2-4 译码器的另一种描述方式是采用 if-elsif-else 语句（注意是 elsif，不是 else if），如 VHDL 代码 3-3 所示。代码第 17 行和第 19 行构成一个 if-else 语句。第 19 行的 else 分支内又嵌套了 if-elsif 语句。其综合后的结果与 VHDL 代码 3-1 所示的 case 语句的描述方式完全相同。通常，我们认为 if-elsif-else 语句是有优先级的，但在这里并没有体现出来，这是因为每个 if-elsif-else 分支的条件变量完全相同，在这里均为输入信号 a，同时，条件之间是互斥的。

VHDL 代码 3-3

```
1.  --File: decoder_2to4_high_v2a.vhd
2.  library ieee;
3.  use ieee.std_logic_1164.all;
4.
5.  entity decoder_2to4_high_v2a is
6.    port (
7.      a  : in std_logic_vector(1 downto 0);
8.      en : in std_logic;
9.      y  : out std_logic_vector(3 downto 0)
10.   );
11. end entity;
12.
13. architecture rtl of decoder_2to4_high_v2a is
14. begin
15.   process(all)
16.   begin
```

```
17.     if (not en) then
18.       y <= (others => '0');
19.     else
20.       if a = 2b"00" then
21.         y <= 4b"0001";
22.       elsif a = 2b"01" then
23.         y <= 4b"0010";
24.       elsif a = 2b"10" then
25.         y <= 4b"0100";
26.       elsif a = 2b"11" then
27.         y <= 4b"1000";
28.       else
29.         y <= 4X"0";
30.       end if;
31.     end if;
32.   end process;
33. end architecture;
```

🔍 VHDL-2008 新特性

在 VHDL-2008 中，常数可以采用<位宽><进制><值>的形式表示。进制支持二进制（用 b 或 B 表示）、八进制（用 o 或 O 表示）、十进制（用 d 或 D 表示）和十六进制（用 x 或 X 表示均可）。<值>应与进制匹配，如 2b"20" 是非法的（二进制可使用的数字符号只有 0 和 1），2d"2" 是合法的。当<位宽>被省略时，可以通过<值>中数字的位数来推断其数据宽度。例如 X"0" 是 4 位（VHDL 代码 3-3 第 29 行可以将 4X"0" 写为 X"0"），而 X"00" 是 8 位。d"0" 是 1 位（VHDL 代码 3-3 第 29 行若将 4X"0" 写为 d"0"，则会报错，因为赋值操作符<=左右两边的操作数位宽不同）。若<位宽>和<进制>均被省略，则默认是二进制数，如 VHDL 代码 3-2 中的 "0001" 表示一个 4 位二进制常数，其值为 1。此外，为了增强代码的可读性，对于<值>表征的字符串，可以插入下划线，如 "00_11" 与 "0011" 均表示十进制数 3。对于可综合的 VHDL 代码，建议使用二进制形式，并给出明确的位宽。

2-4 译码器的 if-elsif-else 语句也可以采用 when-else 语句替代，如 VHDL 代码 3-4 所示。when-else 语句与 if-elsif-else 语句一样，都具有优先级，但当分支的条件变量完全相同且彼此互斥时，这种优先级就不存在了，转变为全并行结构。

📄 VHDL 代码 3-4

```
1.  --File: decoder_2to4_high_v2b.vhd
2.  architecture rtl of decoder_2to4_high_v2b is
3.  begin
4.    process(all)
5.    begin
6.      if (not en) then
7.        y <= (others => '0');
8.      else
9.        y <= "0001" when a = "00" else
10.            "0010" when a = "01" else
```

```
11.             "0100" when a = "10" else
12.             "1000" when a = "11" else
13.             "0000";
14.     end if;
15.   end process;
16. end architecture;
```

> 🔍 **VHDL-2008 新特性**
> 条件信号赋值语句 when-else 既可以在进程 process 内部使用，也可以在其外部使用。

> 💡 **设计规则 2**：尽管当用互斥条件指定 if-elsif-else 语句的每个分支时，其功能和 case 语句等效，将被综合为 MUX，不再具有优先级，但仍然建议此时采用 case 语句描述，这样更为合理，不仅可以增强代码的可读性，也便于代码的管理和维护。

VHDL 代码 3-1 至 VHDL 代码 3-4 采用的都是行为级描述方式，而实际上，VHDL 还支持另外两种描述方式：结构化描述方式和数据流描述方式。采用结构化描述方式时需要先构建非门和与门，分别如 VHDL 代码 3-5 和 VHDL 代码 3-6 所示，在此基础上构建 2-4 译码器，如 VHDL 代码 3-7 所示。VHDL 代码 3-7 第 16~21 行采用了 entity 实例化方法。这种方法是 VHDL-93 引入的，相比 component 实例化方法更为简洁。采用 entity 实例化方法时，若一个 entity 内只有一个 architecture，则可以不指定 architecture 名，如 VHDL 代码 3-7 第 18~21 行所示。结构化描述方式就像在面包板上搭建数字电路一样，如图 3-4 所示，唯一的不同点在于我们是通过 HDL 的形式来描述数字电路所需要的元器件及它们之间的连接关系的。显然，随着 FPGA 芯片集成度的提高和项目复杂性的增大，纯粹使用这种描述方式完成 FPGA 设计是完全行不通的。

📄 **VHDL 代码 3-5**

```
1.  --File: not_gate.vhd
2.  library ieee;
3.  use ieee.std_logic_1164.all;
4.
5.  entity not_gate is
6.    port (
7.      a     : in std_logic;
8.      a_inv : out std_logic
9.    );
10. end entity;
11.
12. architecture gate of not_gate is
13. begin
14.   a_inv <= not a;
15. end architecture;
```

VHDL 代码 3-6

```vhdl
1.  --File: and_gate.vhd
2.  library ieee;
3.  use ieee.std_logic_1164.all;
4.
5.  entity and_gate is
6.    port (
7.      a0 : in std_logic;
8.      a1 : in std_logic;
9.      a2 : in std_logic;
10.     y  : out std_logic
11.   );
12. end entity;
13.
14. architecture gate of and_gate is
15. begin
16.   y <= a0 and a1 and a2;
17. end architecture;
```

VHDL 代码 3-7

```vhdl
1.  --File: decoder_2to4_high_structure.vhd
2.  library ieee;
3.  use ieee.std_logic_1164.all;
4.
5.  entity decoder_2to4_high_structure is
6.    port (
7.      a  : in std_logic_vector(1 downto 0);
8.      en : in std_logic;
9.      y  : out std_logic_vector(3 downto 0)
10.   );
11. end entity;
12.
13. architecture rtl of decoder_2to4_high_structure is
14.   signal a0_inv, a1_inv : std_logic;
15. begin
16.   g0 : entity work.not_gate(gate) port map (a => a(0), a_inv => a0_inv);
17.   g1 : entity work.not_gate(gate) port map (a => a(1), a_inv => a1_inv);
18.   g2 : entity work.and_gate port map
(a0 => en, a1 => a0_inv, a2 => a1_inv, y => y(0));
19.   g3 : entity work.and_gate port map
(a0 => en, a1 => a1_inv, a2 => a(0),   y => y(1));
20.   g4 : entity work.and_gate port map
(a0 => en, a1 => a0_inv, a2 => a(1),   y => y(2));
21.   g5 : entity work.and_gate port map
(a0 => en, a1 => a(0),   a2 => a(1),   y => y(3));
22. end architecture;
```

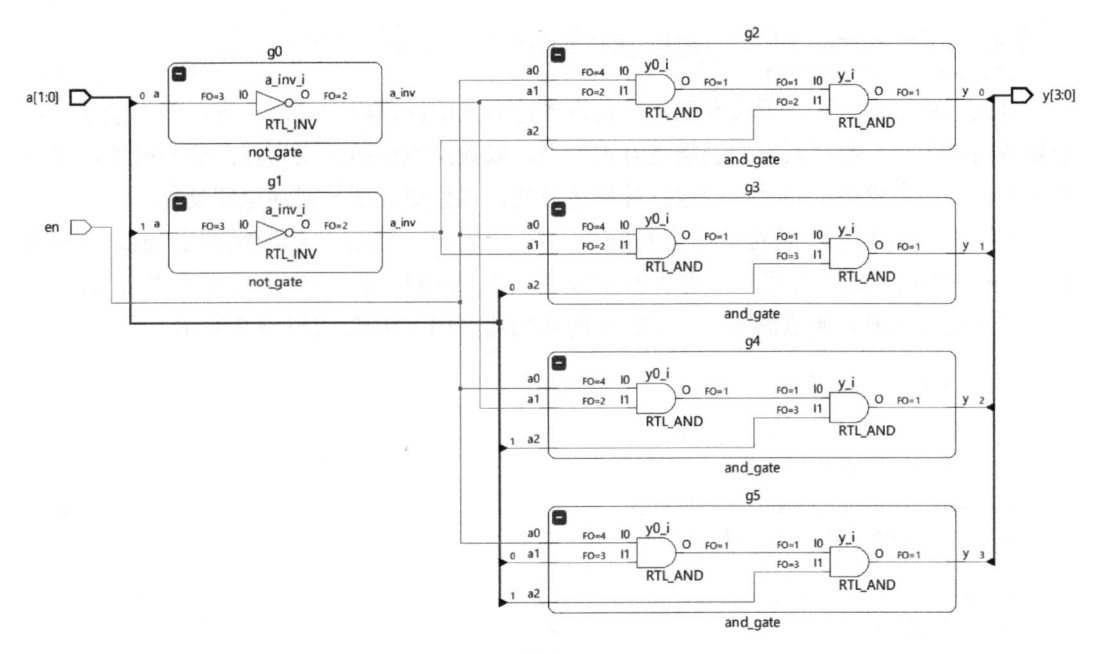

图 3-4

数据流描述方式如 VHDL 代码 3-8 所示。数据流描述方式又可称为 RTL（Register Transfer Level）描述方式，即寄存器传输级描述，因为它主要从数据的变换和传送的角度来描述设计模块，并且使用的语句多为和硬件行为一致的并行语句。例如，在这里，就把真值表转换为布尔表达式，直接用布尔表达式描述电路功能，这样既显式地表达了模块的行为，又隐式地刻画了模块的电路结构。与结构化描述方式相比，数据流描述方式的抽象级别更高一些，因为它不再需要清晰地刻画出具体的数字电路结构，而是以比较直观的方式表达底层的逻辑行为。显然，对于大规模电路设计，采用纯数据流描述方式也是行不通的。

📄 VHDL 代码 3-8

```
1.  --File: decoder_2to4_high_dataflow.vhd
2.  library ieee;
3.  use ieee.std_logic_1164.all;
4.
5.  entity decoder_2to4_high_dataflow is
6.    port (
7.      a  : in std_logic_vector(1 downto 0);
8.      en : in std_logic;
9.      y  : out std_logic_vector(3 downto 0)
10.   );
11. end entity;
12.
13. architecture rtl of decoder_2to4_high_dataflow is
14. begin
15.   y(0) <= en and (not a(1)) and (not a(0));
16.   y(1) <= en and (not a(1)) and a(0);
17.   y(2) <= en and      a(1)  and (not a(0));
```

```
18.     y(3) <= en and          a(1) and a(0);
19. end architecture;
```

对比三种描述方式，不难发现：行为级描述方式的抽象级别最高，概括能力最强。对于复杂的模块，一般都以行为级描述方式为主，以数据流描述方式为辅；更一般地，我们可能会在同一个模块中同时使用这三种描述方式，以便更灵活地描述电路功能。

根据 2-4 译码器的特征，很容易描述 3-8 译码器。但是，若仍然采用 case 语句或 if-elsif-else 语句，则显得冗长。观察表 3-2，可以发现相邻的两个输出是相互移位的关系，结合此特征可以很方便地描述一个参数化的译码器，如 VHDL 代码 3-9 所示。

VHDL 代码 3-9

```
1.  --File: decoder_m2n_high_v1.vhd
2.  library ieee;
3.  use ieee.std_logic_1164.all;
4.  use ieee.numeric_std.all;
5.
6.  entity decoder_m2n_high_v1 is
7.    generic (
8.      M : positive := 2;
9.      N : positive := 2 ** M
10.   );
11.   port (
12.     a  : in std_logic_vector(M - 1 downto 0);
13.     en : in std_logic;
14.     y  : out std_logic_vector(N - 1 downto 0)
15.   );
16. end entity;
17.
18. architecture rtl of decoder_m2n_high_v1 is
19.   constant ONE : std_logic_vector(N - 1 downto 0) := (0 => '1', others => '0');
20.   signal shift : integer range 0 to 2 ** M - 1;
21. begin
22.   shift <= to_integer(unsigned(a));
23.   y <= (ONE sll shift) when en else
24.        (others => '0');
25. end architecture;
```

VHDL 代码 3-9 第 8 行将参数 M 的数据类型定义为 positive。实际上，VHDL 代码中的整数数据类型为 integer，它有两个预定义的子类型，分别为 positive 和 natural。其中，positive（正整数）的最小值为 1，natural（自然数）的最小值为 0。代码第 9 行使用了指数运算操作符 "**"。代码第 22 行使用了数据类型转换函数 to_integer，使用此函数时需要声明 VHDL 的算术运算包（Package）numeric_std，如代码第 4 行所示。使用聚合符号 "=>" 进行切片式赋值时，如果用到关键字 others，要将其放置在最后，如代码第 19 行所示，否则 Vivado 在编译时会报错。代码第 23 行使用了逻辑左移操作符 sll（shift left logic），其他移位操作符将在 3.5.1 节介绍。

图 3-5 中使用了 VHDL 预定义属性（Attribute）length（也可以写为 LENGTH）。表 3-3 总结了常用的可综合的 5 个属性。需要注意的是，当作用于多维数组时，若未指定数组维度，则默认为第一维度（行）。可用数字"1"表示第一维度行，用数字"2"表示第二维度列。

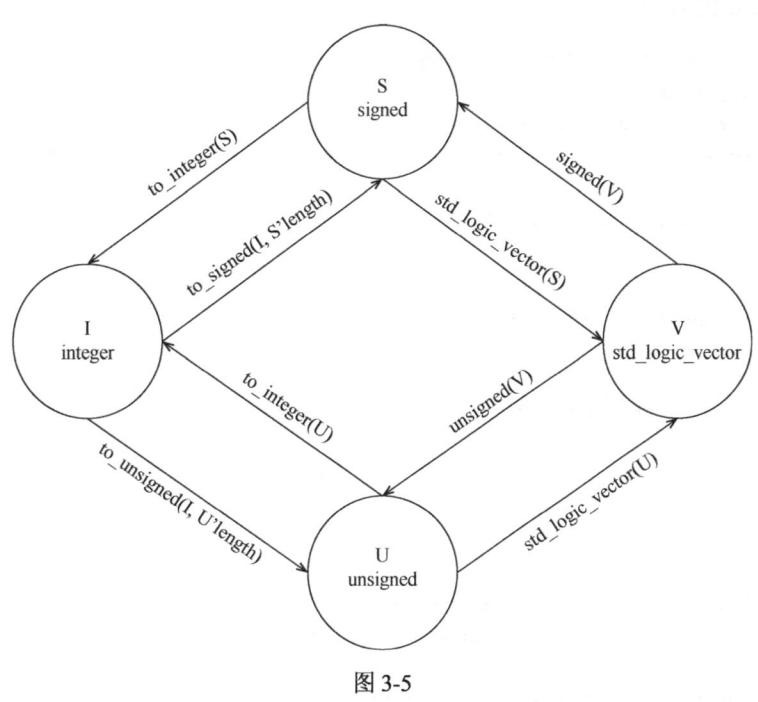

图 3-5

表 3-3

signal data : std_logic_vector(3 downto 0);				
data'LEFT	data'RIGHT	data'HIGH	data'LOW	data'LENGTH
3	0	3	0	4
type mem is array (0 to 15) of std_logic_vector(7 downto 0); signal mem16x8 : mem;				
mem16x8'LEFT	mem16x8'RIGHT	mem16x8'HIGH	mem16x8'LOW	mem16x8'LENGTH
0	15	15	0	16
type x2d_mem is array(0 to 7,0 to 15) of integer; signalx_2d : x2d_mem;				
x_2d'LEFT	x_2d'RIGHT	x_2d'HIGH	x_2d'LOW	x_2d'LENGTH
0	7	7	0	8
x_2d'LEFT(1)	x_2d'RIGHT(1)	x_2d'HIGH(1)	x_2d'LOW(1)	x_2d'LENGTH(1)
0	7	7	0	8
x_2d'LEFT(2)	x_2d'RIGHT(2)	x_2d'HIGH(2)	x_2d'LOW(2)	x_2d'LENGTH(2)
0	15	15	0	16

🔍 VHDL-2008 新特性

在 generic 中定义的参数是可以被引用的，如 VHDL 代码 3-9 第 8 行定义了参数 M，第 9 行定义的参数 N 引用了 M，从而表征了两者之间的关系。

另一方面，从表 3-2 中也能看出输出"1"的位置取决于输入，当输入为 0（2'b00）时，"1"在输出的 0 号位置；当输入为 1（2'b01）时，"1"在输出的 1 号位置，以此类推，输入就是输出"1"的地址。据此，可形成如 VHDL 代码 3-10 所示的描述方式。

📄 **VHDL 代码 3-10**

```vhdl
1.  --File: decoder_m2n_high_v2.vhd
2.  library ieee;
3.  use ieee.std_logic_1164.all;
4.  use ieee.numeric_std.all;
5.
6.  entity decoder_m2n_high_v2 is
7.    generic (
8.      M : positive := 2;
9.      N : positive := 2 ** M
10.   );
11.   port (
12.     a  : in std_logic_vector(M - 1 downto 0);
13.     en : in std_logic;
14.     y  : out std_logic_vector(N - 1 downto 0)
15.   );
16. end entity;
17.
18. architecture rtl of decoder_m2n_high_v2 is
19.   signal index : integer range 0 to 2 ** M - 1;
20. begin
21.   index <= to_integer(unsigned(a));
22.   process(all)
23.   begin
24.     if not en then
25.       y <= (others => '0');
26.     else
27.       y <= (others => '0');
28.       y(index) <= '1';
29.     end if;
30.   end process;
31. end architecture;
```

也可以使用 for 循环描述译码器，如 VHDL 代码 3-11 所示。再看资源利用率，M-N（其中，$N=2^M$）译码器会消耗 $N/2$ 个 LUT。

📄 **VHDL 代码 3-11**

```vhdl
1.  --File: decoder_m2n_high_v3.vhd
2.  library ieee;
3.  use ieee.std_logic_1164.all;
4.  use ieee.numeric_std.all;
5.
6.  entity decoder_m2n_high_v3 is
7.    generic (
```

```
8.       M : positive := 3;
9.       N : positive := 2 ** M
10.    );
11.    port (
12.      a  : in std_logic_vector(M - 1 downto 0);
13.      en : in std_logic;
14.      y  : out std_logic_vector(N - 1 downto 0)
15.    );
16. end entity;
17.
18. architecture rtl of decoder_m2n_high_v3 is
19. begin
20.   process(all)
21.     variable index : std_logic_vector(M - 1 downto 0) := (others => '0');
22.   begin
23.     for i in 0 to N - 1 loop
24.       index := std_logic_vector(to_unsigned(i, M));
25.       y(i) <= (a ?= index) and en;
26.     end loop;
27.   end process;
28. end architecture;
```

🔍 **VHDL-2008 新特性**

VHDL-2008 引入了匹配关系运算符，如 VHDL 代码 3-11 第 25 行的操作符 "?="，a ?= index 等效于：

if a = index then
    k <= '1';
else
    k <= '0';
end if;

其返回值可以是 bit，也可以是 std_logic。其他匹配关系运算符将在第 3.5 节介绍。

3.2.2 编码器代码风格

编码器所执行的功能可看作译码器的逆操作。以 4-2 编码器为例，其输入与输出的关系如表 3-4 所示。不难看出，该编码器还有一些未定义的情形，如当输入为 4'b1100 时，输出是未知的。

表 3-4

输			入	输	出
0	0	0	1	0	0
0	0	1	0	0	1
0	1	0	0	1	0
1	0	0	0	1	1

采用数据流描述方式，如 VHDL 代码 3-12 所示，其原理是根据表 3-4 推断出输出与输入的布尔表达式。不难发现，输出的最低位没有被用到。反过来，仅看代码第 14 行和第 15 行的布尔表达式，其对应的真值表与表 3-4 并不完全等效，如当输入为"1001"时，根据布尔表达式所得，输出为"11"，而表 3-4 中并未对此输入进行定义。

📄 VHDL 代码 3-12

```vhdl
1.  --File: encoder_4to2_dataflow.vhd
2.  library ieee;
3.  use ieee.std_logic_1164.all;
4.
5.  entity encoder_4to2_dataflow is
6.    port (
7.      a : in std_logic_vector(3 downto 0);
8.      y : out std_logic_vector(1 downto 0)
9.    );
10. end entity;
11.
12. architecture dataflow of encoder_4to2_dataflow is
13. begin
14.   y(0) <= a(1) or a(3);
15.   y(1) <= a(2) or a(3);
16. end architecture;
```

VHDL 代码 3-12 综合后的电路如图 3-6 所示，图中，每个 LUT2 都是一个逻辑函数发生器，均实现二输入逻辑或运算。表面上看，该电路会消耗 2 个 LUT，但由于 LUT 整合，最终只会消耗 1 个 LUT。同时，该代码的行为级仿真（Behavioral Simulation）结果和综合后的功能仿真结果是一致的，如图 3-7 所示，图中，"I"代表输入，"O"代表输出。

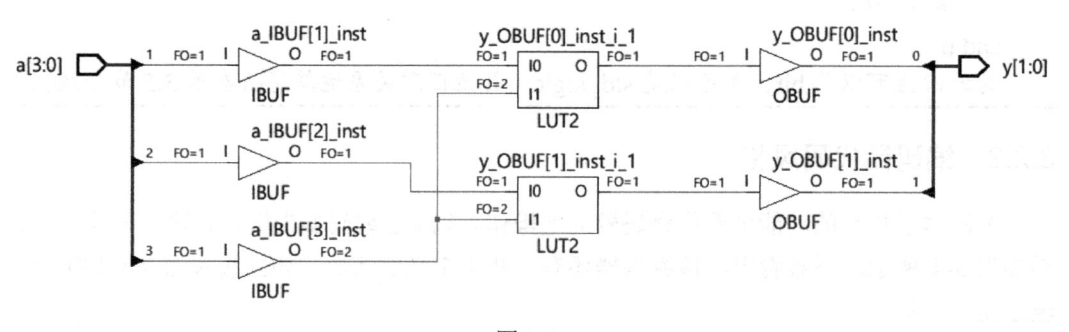

图 3-6

图 3-7

同样，我们可以采用行为级描述方式，可以用 if-elsif-else 语句，如 VHDL 代码 3-13 所示，也可以用 when-else 语句，如 VHDL 代码 3-14 所示（这里仅给出结构体）。代码所描述的功能与表 3-4 是完全一致的，对于表 3-4 中未定义的输入，在代码中将其输出设置为"-"（VHDL 中表示无关态，Don't care），如 VHDL 代码 3-13 第 25 行或 VHDL 代码 3-14 第 8 行所示。两者综合后的结果完全相同，如图 3-8 所示，可以看出，该电路结构与图 3-6 是不同的。

VHDL 代码 3-13

```vhdl
1.  --File: encoder_4to2_v1a.vhd
2.  library ieee;
3.  use ieee.std_logic_1164.all;
4.  
5.  entity encoder_4to2_v1a is
6.    port (
7.      a : in std_logic_vector(3 downto 0);
8.      y : out std_logic_vector(1 downto 0)
9.    );
10. end entity;
11. 
12. architecture rtl of encoder_4to2_v1a is
13. begin
14.   process(all)
15.   begin
16.     if a = "0001" then
17.       y <= "00";
18.     elsif a = "0010" then
19.       y <= "01";
20.     elsif a = "0100" then
21.       y <= "10";
22.     elsif a = "1000" then
23.       y <= "11";
24.     else
25.       y <= (others => '-');
26.     end if;
27.   end process;
28. end architecture;
```

VHDL 代码 3-14

```vhdl
1.  --File: encoder_4to2_v1b.vhd
2.  architecture rtl of encoder_4to2_v1b is
3.  begin
4.    y <= "00" when a = "0001" else
5.         "01" when a = "0010" else
6.         "10" when a = "0100" else
7.         "11" when a = "1000" else
8.         "--";
9.  end architecture;
```

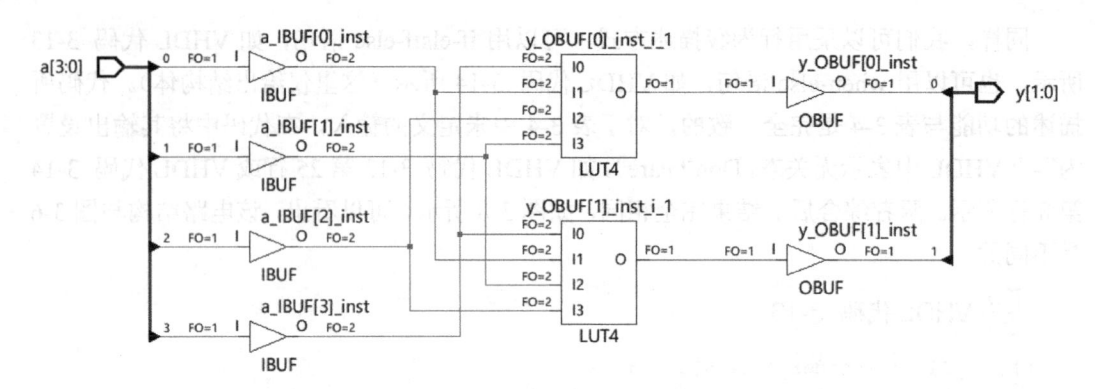

图 3-8

从仿真结果来看，在图 3-9 中，上部为行为级仿真结果，下部为综合后的功能仿真结果，两者是不同的。前者将未定义输入对应的输出设置为 x，后者则将其设置为 3，即"11"。使用 if 语句描述组合逻辑时，如果条件分支或条件赋值语句中出现无关态 "-"，那么行为级模型和综合后的网表可能会得到不同的仿真结果。

I	a[3:0]	0	1	2	3	4	5	6	7	8	9	A	B	C	D	E	F
O	y[1:0]	X	0	1	X	2	X	3	X

I	a[3:0]	0	1	2	3	4	5	6	7	8	9	A	B	C	D	E	F
O	y[1:0]	3	0	1	3	2	3

图 3-9

对于 VHDL 代码 3-13，如果删除第 24 行和第 25 行，即 if-elsif 语句没有匹配的 else 分支，那么会形成一个不完备的 if 语句，从而导致工具推断出锁存器，如图 3-10 中的 LDCE（锁存器原语）所示。

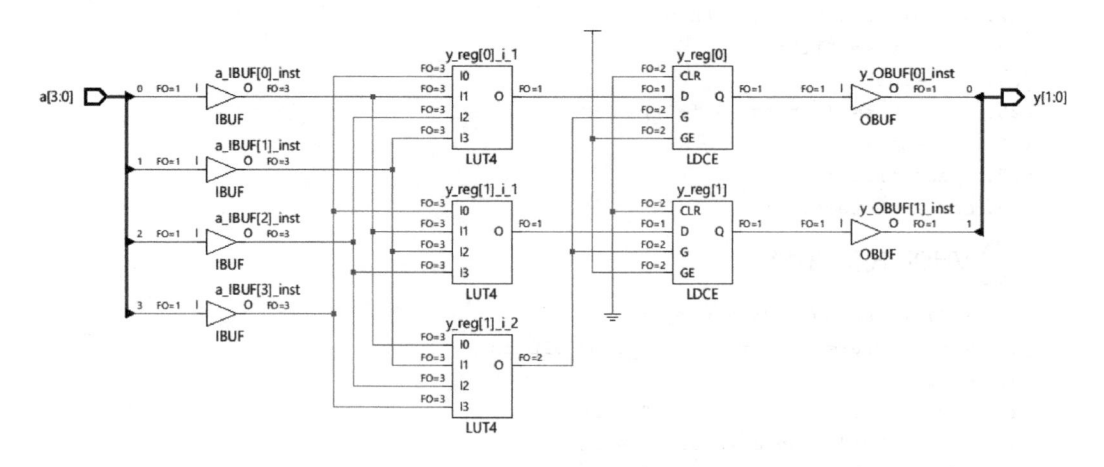

图 3-10

💡 **设计规则 3**：采用 if 语句描述组合逻辑时，对于不完备的 if 语句，即缺少 else 分支的 if 语句要格外关注，因为工具会将其视为锁存器。

对于 when-else 语句，如果也想得到如图 3-10 所示的电路，需要采用如 VHDL 代码 3-15 所示的描述方式。

📄 **VHDL 代码 3-15**

```
1.  --File: encoder_4to2_v1b_latch.vhd
2.  architecture rtl of encoder_4to2_v1b_latch is
3.  begin
4.    y <= "00" when a = "0001" else
5.         "01" when a = "0010" else
6.         "10" when a = "0100" else
7.         "11" when a = "1000" else
8.         y;
9.  end architecture;
```

🔍 **VHDL-2008 新特性**

在之前的 VHDL 版本中，若需要读取输出端口（端口声明为 out 类型）给内部信号使用，则要将其端口类型由 out 更改为 buffer。VHDL-2008 版本则删除了这一限制，可直接读取输出端口，如 VHDL 代码 3-15 第 8 行所示。

行为级描述方式也可采用 case 语句，如 VHDL 代码 3-16 所示。同样，对于未定义的输入，代码第 21 行将其输出设置为 "--"。也可以采用 with-select 语句，如 VHDL 代码 3-17 所示，对于未定义的输入，代码第 9 行将其输出设置为 "--"。这两段代码的 Elaborated Design 阶段对应的电路如图 3-11 所示，综合后的电路如图 3-12 所示，这个结果与图 3-6 是一致的。这也表明 case 语句描述的功能与数据流描述方式描述的功能是一致的，而与 if-elsif-else 是不一致的。图 3-11 与图 3-12 进一步佐证了 case 语句各条件分支选项是并行的、没有优先级的。

📄 **VHDL 代码 3-16**

```
1.  --File: encoder_4to2_v2a.vhd
2.  library ieee;
3.  use ieee.std_logic_1164.all;
4.
5.  entity encoder_4to2_v2a is
6.    port (
7.      a : in std_logic_vector(3 downto 0);
8.      y : out std_logic_vector(1 downto 0)
9.    );
10. end entity;
11.
12. architecture rtl of encoder_4to2_v2a is
13. begin
14.   process(all)
15.   begin
16.     case a is
17.       when "0001" => y <= "00";
18.       when "0010" => y <= "01";
19.       when "0100" => y <= "10";
```

```
20.        when "1000" => y <= "11";
21.        when others => y <= "--";
22.     end case;
23.   end process;
24. end architecture;
```

📄 VHDL 代码 3-17

```
1.  --File: encoder_4to2_v2b.vhd
2.  architecture rtl of encoder_4to2_v2b is
3.  begin
4.    with a select
5.      y <= "00" when "0001",
6.           "01" when "0010",
7.           "10" when "0100",
8.           "11" when "1000",
9.           "--" when others;
10. end rtl;
```

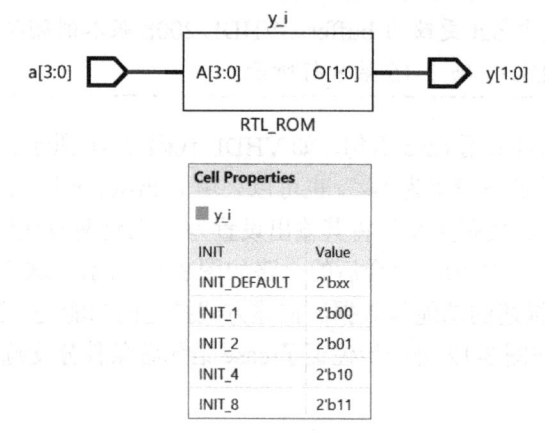

图 3-11

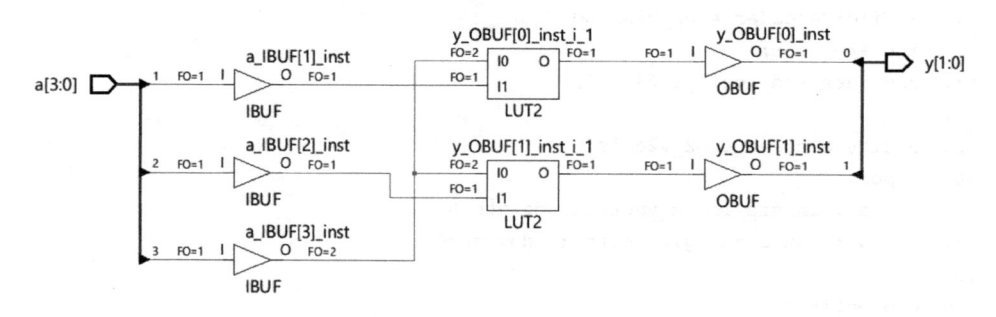

图 3-12

VHDL 代码 3-16 的行为级仿真结果如图 3-13 上部所示，综合后的功能仿真结果如图 3-13 下部所示。显然，两者是不同的。从综合后的仿真结果可以看出，对于未定义的输入，综合后的电路将其设置为 0，这与图 3-7 中的数据流描述方式的仿真结果是相同的。

I	a[3:0]	0	1	2	3	4	5	6	7	8	9	A	B	C	D	E	F
O	y[1:0]	X	0	1	X	2	X			3				X			

I	a[3:0]	0	1	2	3	4	5	6	7	8	9	A	B	C	D	E	F
O	y[1:0]	0		1		2				3							

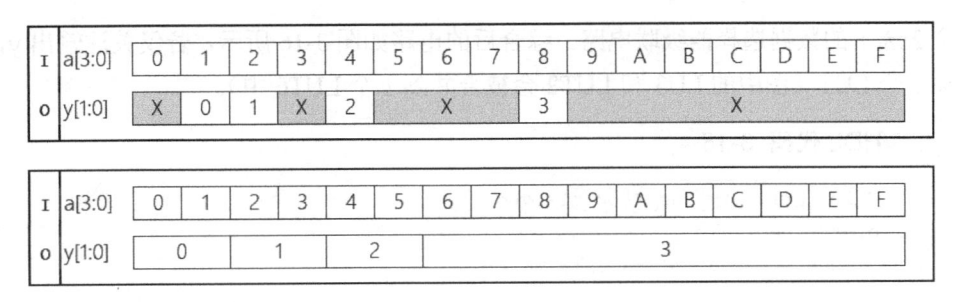

图 3-13

不同于 SystemVerilog，VHDL 要求 case 语句必须包含 when others 分支，否则会报错。例如，将 VHDL 代码 3-16 第 21 行删除，Vivado 在综合时会显示如图 3-14 所示的错误信息。

```
General Messages (5 errors)
  [Synth 8-426] missing choice(s) 32'b00000000000000000000000000000011 in case statement [encoder_4to2_v2a.vhd:16] (2 more like this)
  [Synth 8-285] failed synthesizing module 'encoder_4to2_v2a' [encoder_4to2_v2a.vhd:12]
  [Vivado_Tcl 4-5] Elaboration failed - please see the console for details
```

图 3-14

对比图 3-8 和图 3-12 可以看出，尽管 if-elsif-else 的条件分支变量与 case 的分支变量相同，均为 a，但两者综合后的电路是不同的，这一点在设计中要特别注意。

💡 **设计规则 4**：对于 RTL 代码中有无关态的情形，要分别进行行为级仿真（也称为功能仿真）和综合后的功能仿真，对比两者的差异，并观察综合后的仿真结果中未知状态的变化值（0 或 1）是否符合实际需求，如果符合，才能将其正视为无关态。此外，如果确定要在 RTL 代码中对输出信号使用"-"，那么要确保此时的输入对应的输出确实为"-"，这可以通过真值表来验证。

对于 4-2 优先编码器，其输出与输入之间的关系如表 3-5 所示，表中，"-"表示无关态。这里，"-"可代表 0、1、z（高阻）和 x，换言之，四值逻辑均可表示。

表 3-5

输		入		输	出
0	0	0	1	0	0
0	0	1	-	0	1
0	1	-	-	1	0
1	-	-	-	1	1

采用 if-elsif-else 语句描述，如 VHDL 代码 3-18 所示。此时，条件分支里的每个变量是不同的，意味着条件分支并非互斥的。例如，代码第 18 行的变量为 a[3]，而第 20 行的变量为 a[2]，从而形成了优先级。采用 when-else 语句描述，如 VHDL 代码 3-19 所示，同样构成了优先级。两者对应的 Elaborated Design Schematic 视图一致，如图 3-15 所示，这

是一个 2 选 1 的数据选择器级联电路。综合后的电路如图 3-16 所示，若仅关注输出 y，则只消耗 1 个 LUT（图中的 LU3 和 LUT2 会被合并到 1 个 LUT6 中）。

📄 **VHDL 代码 3-18**

```vhdl
1.  --File: priority_enc_4to2_v1a.vhd
2.  library ieee;
3.  use ieee.std_logic_1164.all;
4.
5.  entity priority_enc_4to2_v1a is
6.    port (
7.      a        : in std_logic_vector(3 downto 0);
8.      y        : out std_logic_vector(1 downto 0);
9.      valid_in : out std_logic
10.    );
11. end entity;
12.
13. architecture rtl of priority_enc_4to2_v1a is
14. begin
15.   valid_in <= or(a);
16.   process(all)
17.   begin
18.     if a(3) then
19.       y <= "11";
20.     elsif a(2) then
21.       y <= "10";
22.     elsif a(1) then
23.       y <= "01";
24.     elsif a(0) then
25.       y <= "00";
26.     else
27.       y <= "--";
28.     end if;
29.   end process;
30. end architecture;
```

📄 **VHDL 代码 3-19**

```vhdl
1.  --File: priority_enc_4to2_v1b.vhd
2.  architecture rtl of priority_enc_4to2_v1b is
3.  begin
4.    valid_in <= or(a);
5.    y <= "11" when a(3) else
6.         "10" when a(2) else
7.         "01" when a(1) else
8.         "00" when a(0) else
9.         "--";
10. end architecture;
```

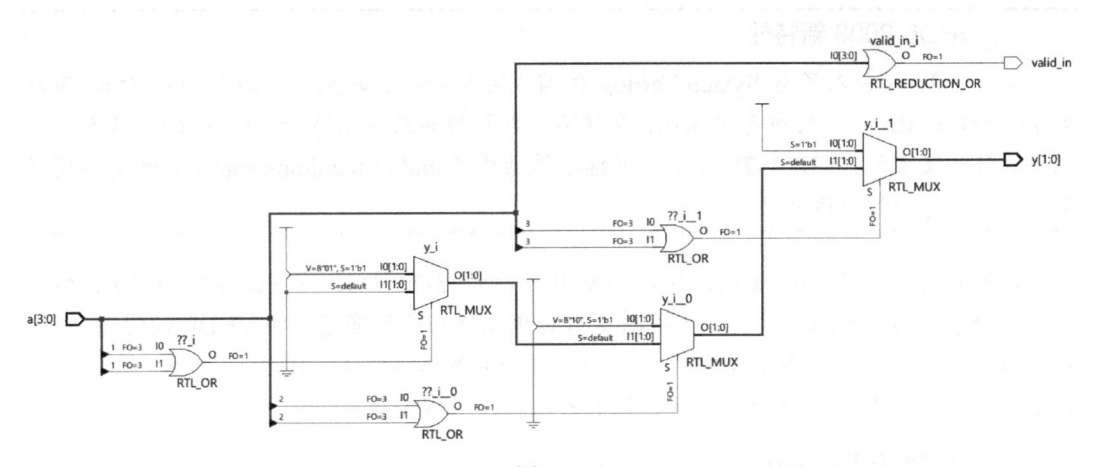

图 3-15

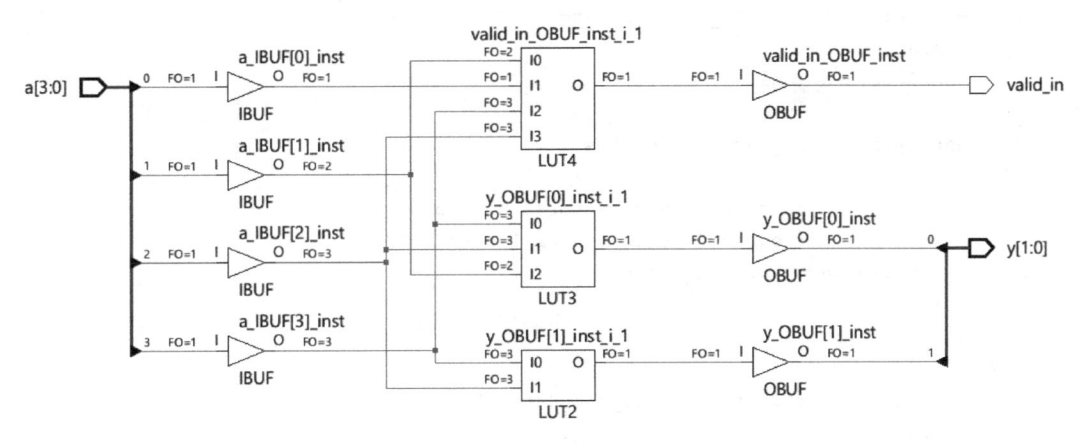

图 3-16

从仿真的角度来看，VHDL 代码 3-18 和 VHDL 代码 3-19 综合前和综合后的功能仿真如图 3-17 所示（上半部分为综合前的仿真结果，下半部分为综合后的仿真结果）。两者的差异在于当输入为 0 时，前者的输出为 x，而后者的输出为 0。

I	a[3:0]	0	1	2	3	4	5	6	7	8	9	A	B	C	D	E	F	0
o	y[1:0]	X	0	1			2					3						0
o	valid_in																	

I	a[3:0]	0	1	2	3	4	5	6	7	8	9	A	B	C	D	E	F	0
o	y[1:0]	0		1			2					3						0
o	valid_in																	

图 3-17

🔍 VHDL-2008 新特性

VHDL-2008 引入了与 SystemVerilog 位缩减运算相对应的操作，如 VHDL 代码 3-18 第 15 行的 or(a)表示 a 的所有位相或，这样不用再写为 a(3) or a(2) or a(1) or a(0)，从而使代码更简洁。换言之，VHDL-2008 中的逻辑运算操作符 and/or/nand/nor/xor/xnor 既支持逻辑运算，也支持位缩减运算。

优先编码器也可使用 case?语句，如 VHDL 代码 3-20 所示。注意，代码第 19 行至第 21 行的条件分支中包含了"-"。也可以使用 with-select?语句描述，如 VHDL 代码 3-21 所示。综合后的电路与图 3-16 一致，综合前和综合后的仿真结果与图 3-17 一致。可见，此时 case?、with-select?和 if-elsif-else 语句等效，隐含了优先级。

📄 **VHDL 代码 3-20**

```vhdl
1.  --File: priority_enc_4to2_v2a.vhd
2.  library ieee;
3.  use ieee.std_logic_1164.all;
4.
5.  entity priority_enc_4to2_v2a is
6.    port (
7.      a        : in std_logic_vector(3 downto 0);
8.      y        : out std_logic_vector(1 downto 0);
9.      valid_in : out std_logic
10.   );
11. end entity;
12.
13. architecture rtl of priority_enc_4to2_v2a is
14. begin
15.   valid_in <= or(a);
16.   process(all)
17.   begin
18.     case? a is
19.       when "1---" => y <= "11";
20.       when "01--" => y <= "10";
21.       when "001-" => y <= "01";
22.       when "0001" => y <= "00";
23.       when others => y <= "--";
24.     end case?;
25.   end process;
26. end architecture;
```

📄 **VHDL 代码 3-21**

```vhdl
1.  --File: priority_enc_4to2_v2b.vhd
2.  architecture rtl of priority_enc_4to2_v2b is
3.  begin
4.    valid_in <= or(a);
5.    with a select?
6.    y <= "11" when "1---",
7.         "10" when "01--",
```

```
8.          "01" when "001-",
9.          "00" when "0001",
10.         "--" when others;
11. end architecture;
```

观察 VHDL 代码 3-20，不难看出，y 的输出结果反映的是 a 中由高位至低位第一个"1"所出现的位置，如当 a 为"1000"～"1111"时，第一个 1 均出现在 a(3)，因此 y 输出"11"。这样，该电路可用于检测 4 位二进制数由高位至低位第一个"1"所出现的位置。

🔍 **VHDL-2008 新特性**

VHDL-2008 引入了匹配 case 语句 case?和匹配 select 语句 select?，这大大简化了描述条件分支选项中包含无关态"-"的场景。

从代码可复用性的角度而言，我们可以描述一个参数化优先编码器，如 VHDL 代码 3-22 所示。参数 M 定义了输入数据位宽，N 定义了输出数据位宽，M 和 N 的关系可表示为

$$N = \log_2^M$$

这在代码第 10 行有所体现，因此，需要添加 math_real 包，如代码第 5 行所示。注意 ceil 返回值仍然为实数。代码第 25 行～第 29 行通过 for 循环结合 if 语句给输出赋值，其中，第 25 行使用了属性 REVERSE_RANGE，其返回值为 0～M-1。这里要特别注意，代码第 24 行是必不可少的，否则就会生成锁存器，同时该行内容应位于 for 循环之前。与前面描述的优先编码器不同的是，这里将未定义输入所对应的输出设置为 0 而不是"-"，从而避免综合前和综合后的仿真结果不匹配。

📄 **VHDL 代码 3-22**

```
1.  --File: priority_enc_m2n.vhd
2.  library ieee;
3.  use ieee.std_logic_1164.all;
4.  use ieee.numeric_std.all;
5.  use ieee.math_real.all;
6.
7.  entity priority_enc_m2n is
8.    generic (
9.      M : positive := 4;
10.     N : positive := integer(ceil(log2(real(M))))
11.   );
12.   port (
13.     a        : in std_logic_vector(M - 1 downto 0);
14.     y        : out std_logic_vector(N - 1 downto 0);
15.     valid_in : out std_logic
16.   );
17. end entity;
18.
19. architecture rtl of priority_enc_m2n is
20. begin
21.   valid_in <= or(a);
22.   process(all)
```

```
23.    begin
24.      y <= (others => '0');
25.      for i in a'REVERSE_RANGE loop
26.        if a(i) then
27.          y <= std_logic_vector(to_unsigned(i, N));
28.        end if;
29.      end loop;
30.    end process;
31. end architecture;
```

> 设计规则 5：对于有优先级的编码器，可以使用 if-elsif-else 语句描述，也可以使用 case?或 with-select?语句描述，采用后两者会使代码更为简洁。

3.3 多路复用器与多路解复用器

3.3.1 多路复用器代码风格

多路复用器（Multiplexer，后面统一用 MUX 表示）又称为数据选择器，分为独热码 MUX 和二进制码 MUX。两者的区别在于控制端（也称为选择端）的编码类型不同。以 2 选 1 独热码 MUX 为例，其基本电路结构如图 3-18 所示。如果 a0 和 a1 的位宽均为 1，那么该电路就需要 2 个与门和 1 个或门。如果 a0 和 a1 的位宽均为 W，那么就需要把 s(0)和 s(1) 复制 W 份，每一份都对应 2 个与门和 1 个或门，因此最终需要2W 个与门和 W 个或门。这个电路也给我们一个启示：与门可用于实现数据选通。同时，若选择信号位宽为 N，则意味着 MUX 有 N 个输入数据源。反过来，输入数据源的个数决定了选择信号的位宽。

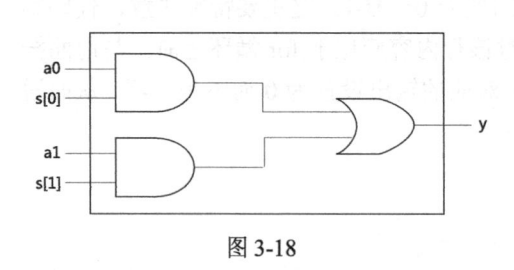

图 3-18

基于此电路可形成 VHDL 代码 3-23，相应的仿真结果如图 3-19 所示，图中 s 以二进制形式显示。可以看到，当输入不是独热码时，输出是不正确的。

VHDL 代码 3-23

```
1.  --File: one_hot_mux2.vhd
2.  library ieee;
3.  use ieee.std_logic_1164.all;
4.
5.  entity one_hot_mux2 is
6.    generic ( W : positive := 4 );
7.    port (
8.      s  : in std_logic_vector(1 downto 0);
9.      a0 : in std_logic_vector(W - 1 downto 0);
10.     a1 : in std_logic_vector(W - 1 downto 0);
11.     y  : out std_logic_vector(W - 1 downto 0)
12.   );
13. end entity;
```

```
14.
15. architecture rtl of one_hot_mux2 is
16. begin
17.   y <= (s(1) and a1) or (s(0) and a0);
18. end architecture;
```

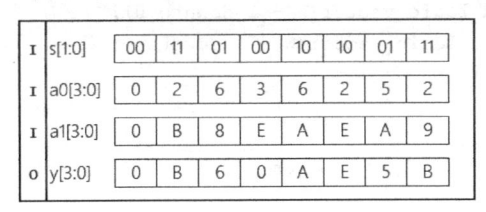

I	s[1:0]	00	11	01	00	10	10	01	11
I	a0[3:0]	0	2	6	3	6	2	5	2
I	a1[3:0]	0	B	8	E	A	E	A	9
O	y[3:0]	0	B	6	0	A	E	5	B

图 3-19

🔍 VHDL-2008 新特性

对于逻辑运算符 and/or/nand/nor/xor/xnor，VHDL-2008 对其进行了重载，可支持向量与标量逻辑运算，如 VHDL 代码 3-23 第 17 行所示，s(1) and a1，假定 a1 的位宽为 4。那么 s(1) and a1 的结果也是 4 位，这 4 位由低位至高位分别为

s(1) and a1(0);
s(1) and a1(1);
s(1) and a1(2);
s(1) and a1(3)。

2 个 2 选 1 独热码 MUX 可构成 1 个 4 选 1 独热码 MUX，如图 3-20 所示。采用行为级描述方式，如 VHDL 代码 3-24 所示，这里采用了 with-select 语句，当然，也可以用 case 语句。

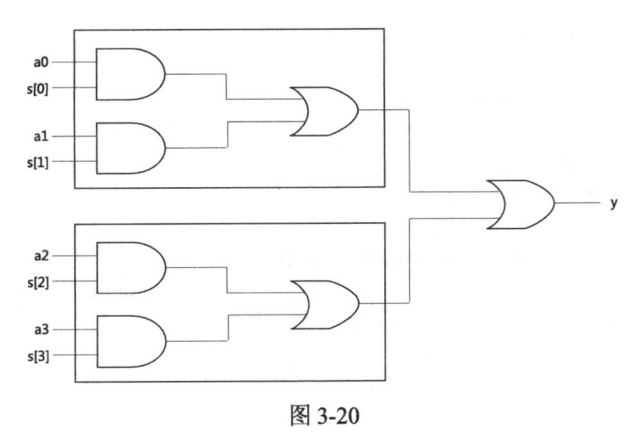

图 3-20

📄 VHDL 代码 3-24

```
1. --File: one_hot_mux4.vhd
2. library ieee;
3. use ieee.std_logic_1164.all;
4.
5. entity one_hot_mux4 is
6.   generic ( W : positive := 2 );
```

```
7.    port (
8.      s  : in std_logic_vector(3 downto 0);
9.      a0 : in std_logic_vector(W - 1 downto 0);
10.     a1 : in std_logic_vector(W - 1 downto 0);
11.     a2 : in std_logic_vector(W - 1 downto 0);
12.     a3 : in std_logic_vector(W - 1 downto 0);
13.     y  : out std_logic_vector(W - 1 downto 0)
14.   );
15. end entity;
16.
17. architecture rtl of one_hot_mux4 is
18. begin
19.   with s select
20.     y <= a0 when "0001",
21.          a1 when "0010",
22.          a2 when "0100",
23.          a3 when "1000",
24.          (others => '0') when others;
25. end architecture;
```

对于一个 N 选 1 的二进制码 MUX，其选择信号的位宽为

$$M = \lceil \log_2^N \rceil$$

这里，$\lceil \ \rceil$ 表示向上取整。我们可以使用已有电路搭建二进制码 MUX。例如，可以用一个 2-4 译码器和一个 4 选 1 的独热码 MUX 构造一个 4 选 1 的二进制码 MUX，如 VHDL 代码 3-25 所示，其综合后的电路如图 3-21 所示（数据位宽为 1），可以看到，只消耗了一个 LUT6。不难推测，如果数据位宽为 W，将消耗 W 个 LUT6。

📄 **VHDL 代码 3-25**

```
1.  --File: bin_mux4_v1.vhd
2.  library ieee;
3.  use ieee.std_logic_1164.all;
4.
5.  entity bin_mux4_v1 is
6.    generic ( W : positive := 4 );
7.    port (
8.      sel : in std_logic_vector(1 downto 0);
9.      a0, a1, a2, a3 : in std_logic_vector(W - 1 downto 0);
10.     y   : out std_logic_vector(W - 1 downto 0)
11.   );
12. end entity;
13.
14. architecture rtl of bin_mux4_v1 is
15.   signal one_hot_sel : std_logic_vector(3 downto 0);
16. begin
17.   i_decoder_m2n_high_v1 : entity work.decoder_m2n_high_v1
18.     generic map ( M => 2 )
19.     port map (a => sel, en => '1', y => one_hot_sel);
20.
21.   i_one_hot_mux4 : entity work.one_hot_mux4
22.     generic map (W => W)
```

```
23.     port map (s => one_hot_sel, a0 => a0, a1 => a1, a2 => a2, a3 => a3, y => y);
24. end architecture;
```

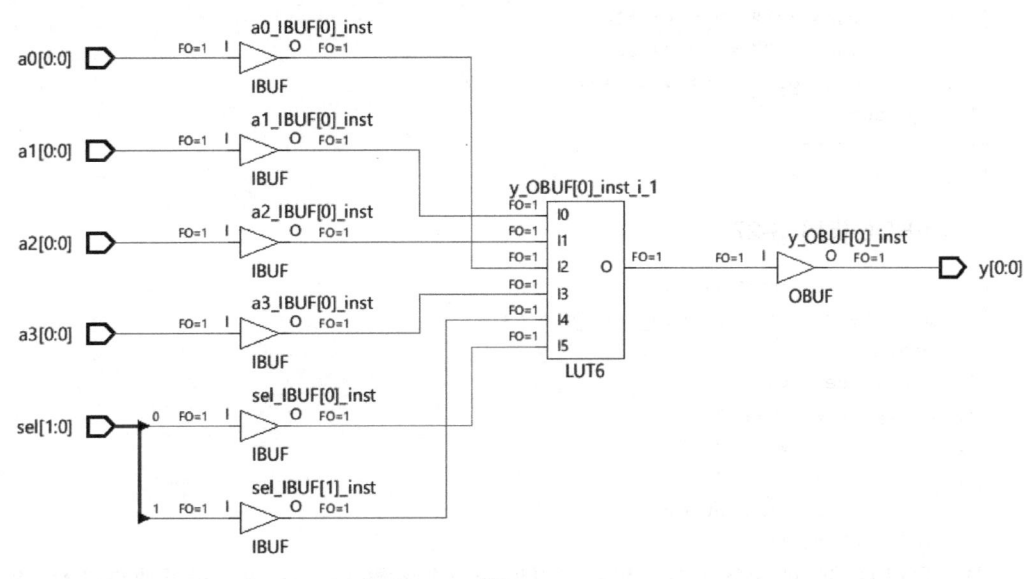

图 3-21

当然，我们也可以用 3 个 2 选 1 二进制码 MUX 搭建 1 个 4 选 1 二进制码 MUX，其电路结构如图 3-22 所示。类似地，也可以用 2 个 4 选 1 二进制码 MUX 加 1 个 2 选 1 二进制码 MUX 搭建 1 个 8 选 1 二进制码 MUX。

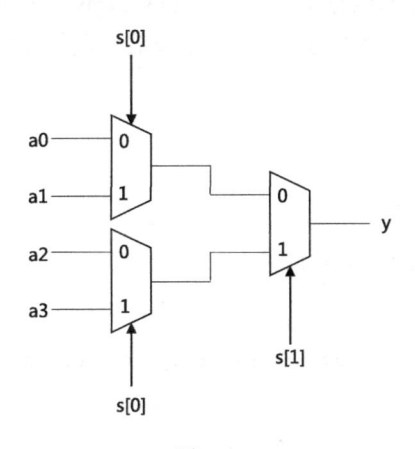

图 3-22

我们也可以直接采用 case 语句（见 VHDL 代码 3-26）描述一个 4 选 1 二进制码 MUX，也可以采用 with-select 语句（见 VHDL 代码 3-27）描述一个 4 选 1 二进制码 MUX，两者综合后的电路与图 3-21 一致。

📄 VHDL 代码 3-26

```
1.  --File: bin_mux4_v2a.vhd
2.  architecture rtl of bin_mux4_v2a is
3.  begin
4.    process(all)
```

```
5.    begin
6.      case sel is
7.        when "00" => y <= a0;
8.        when "01" => y <= a1;
9.        when "10" => y <= a2;
10.       when others => y <= a3;
11.     end case;
12.   end process;
13. end architecture;
```

VHDL 代码 3-27

```
1. --File: bin_mux4_v2b.vhd
2. architecture rtl of bin_mux4_v2b is
3. begin
4.   with sel select
5.     y <= a0 when "00",
6.          a1 when "01",
7.          a2 when "11",
8.          a3 when others;
9. end architecture;
```

从代码可复用性的角度而言，我们可以设计一个参数化的 MUX。对于独热码 MUX，其参数化形式如 VHDL 代码 3-28 所示。代码第 5 行至第 7 行声明 package（这里将 package 和 entity 放在一个文件里，尽管如此，代码第 2 行和第 3 行的内容仍然需要在 entity 前声明），其中定义了数据类型 bus_array 为一维数组。代码第 12 行表明需要使用 package 中的内容。代码第 21 行将端口 a 的数据类型声明为 bus_array。参数 W 为数据位宽，N 为选择信号位宽。对于参数化二进制码 MUX，只需要释放代码第 22 行并关闭代码第 21 行（在行首添加注释符 "--"）即可。

VHDL 代码 3-28

```
1.  --File: one_hot_mux.vhd
2.  library ieee;
3.  use ieee.std_logic_1164.all;
4.
5.  package bus_mux_pkg is
6.    type bus_array is array(natural range <>) of std_logic_vector;
7.  end package;
8.
9.  library ieee;
10. use ieee.std_logic_1164.all;
11. use ieee.numeric_std.all;
12. use work.bus_mux_pkg.all;
13.
14. entity one_hot_mux is
15.   generic (
16.     W : positive := 1; -- data width
17.     N : positive := 4  -- sel width
18.   );
19.   port (
```

```
20.      s : in std_logic_vector(N - 1 downto 0);
21.      a : in bus_array(0 to N - 1)(W - 1 downto 0);
22. -- a : in bus_array(0 to 2 ** N - 1)(W - 1 downto 0);
23.      y : out std_logic_vector(W - 1 downto 0)
24.    );
25. end entity;
26.
27. architecture rtl of one_hot_mux is
28. begin
29.    y <= a(to_integer(unsigned(s)));
30. end architecture;
```

> **VHDL-2008 新特性**
>
> VHDL-2008 支持非约束数组（Unconstrained Array），即数组的规模是未知的，如 VHDL 代码 3-28 第 6 行所示，这极大地增强了定义数据类型的灵活性。例如，在仅声明数组 bus_array 的情况下，可以定义：
>
> signal a1 : bus_array(0 to 3)(1 downto 0);
>
> signal a2 : bus_array(0 to 7)(3 downto 0)。

对于这两种 MUX，我们从资源利用率的角度对其进行比较，如表 3-6 所示。这里，输入数据位宽均为 1，目标芯片为 UltraScale+ FPGA。可以看到，二进制码 MUX 可能会消耗 MUFX7 或 MUXF8，如 32 选 1 二进制码 MUX 或 64 选 1 二进制码 MUX。独热码 MUX 则只消耗 LUT。

表 3-6

二进制码 MUX	LUT6	MUXF7	MUXF8	独热码 MUX	LUT6	LUT5	LUT4	LUT3	LUT 总消耗量
4 选 1	1	0	0	4 选 1	1	0	0	1	2
8 选 1	2	1	0	8 选 1	3	0	1	1	5
16 选 1	4	2	1	16 选 1	8	1	0	2	11
32 选 1	9	4	0	32 选 1	15	4	0	5	24
64 选 1	17	8	4	64 选 1	28	13	0	14	55

应用案例 1：优化代码中的 MUX

在实际应用中，独热码 MUX 和二进制码 MUX 两者哪个更适合应用场景呢？这里我们先看一个包含二进制码 MUX 的电路，如图 3-23 所示，相应的 RTL 代码如 VHDL 代码 3-29 所示。其电路功能很简单：计数器在时钟使能信号 en 的作用下计数，当计数到 144 时，执行加法操作，否则执行减法操作。MUX 的选择控制端由计数器译码获取。显然，这是一个二进制码 MUX。由代码第 19 行可知，该 MUX 的选择信号位宽为 8。目标芯片为 xcvc1902-vsva2197-1LP-i-S，时钟频率为 500MHz。在这种情况下，由 cnt_reg 出发经 RTL_MUX 到 dout_reg 这段路径成为关键路径，这是由 8 位选择信号对应的译码电路造成的。这在综合后的电路中也可看出，如图 3-24 所示。

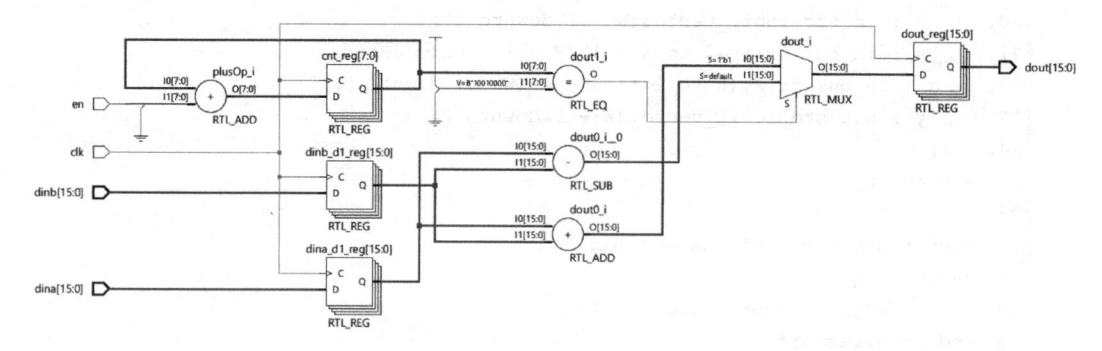

图 3-23

📄 **VHDL 代码 3-29**

```vhdl
1.  --File: mux_opt_before.vhd
2.  library ieee;
3.  use ieee.std_logic_1164.all;
4.  use ieee.numeric_std.all;
5.
6.  entity mux_opt_before is
7.    port (
8.      clk  : in std_logic;
9.      en   : in std_logic;
10.     dina : in unsigned(15 downto 0);
11.     dinb : in unsigned(15 downto 0);
12.     dout : out unsigned(15 downto 0)
13.   );
14. end entity;
15.
16. architecture rtl of mux_opt_before is
17. signal dina_d1 : unsigned(15 downto 0);
18. signal dinb_d1 : unsigned(15 downto 0);
19. signal cnt     : unsigned(7 downto 0) := (others => '0');
20. begin
21.   process(clk)
22.   begin
23.     if rising_edge(clk) then
24.       cnt <= cnt + en;
25.     end if;
26.   end process;
27.
28.   process(clk)
29.   begin
30.     if rising_edge(clk) then
31.       dina_d1 <= dina;
32.       dinb_d1 <= dinb;
33.       dout    <= dina_d1 + dinb_d1 when cnt = 8D"144" else
34.                  dina_d1 - dinb_d1;
35.     end if;
36.   end process;
37. end architecture;
```

图 3-24

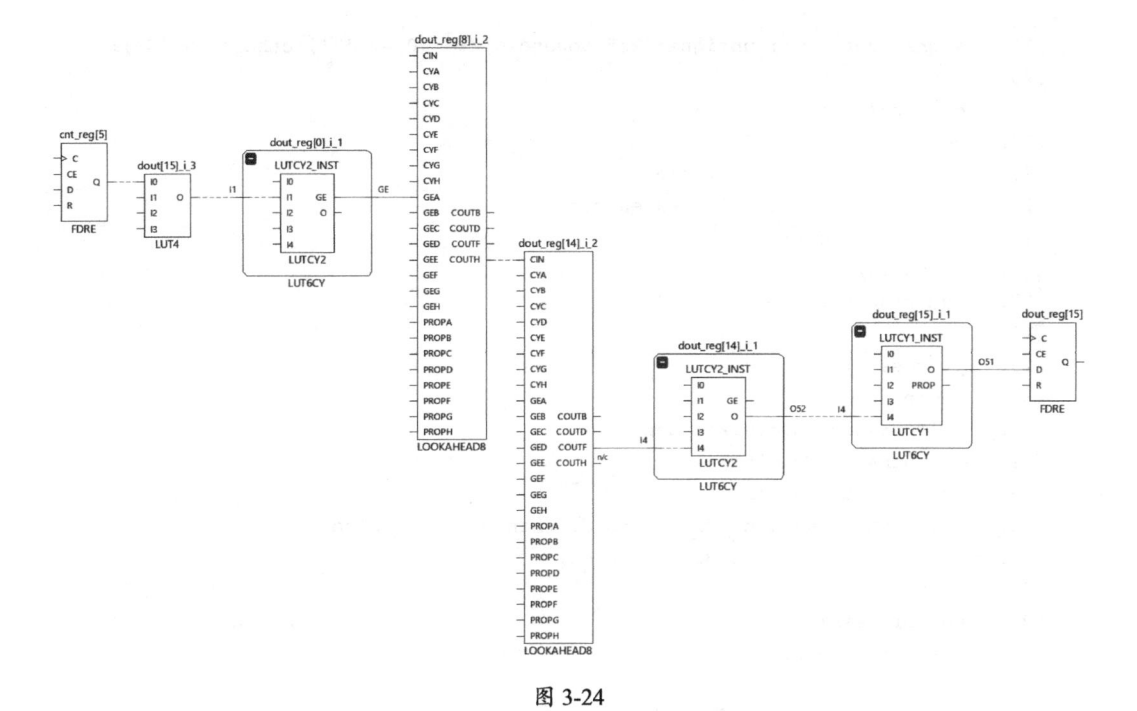

现在，我们对其中的二进制码 MUX 进行优化，如 VHDL 代码 3-30 所示。为保证电路功能不变，将 cnt 位宽调整为 256（原始 8 位计数器模值为 256），并将其初始值设置为 1，如代码第 19 行所示，同时将计数功能调整为循环移位功能，如代码第 24 行所示（操作符 rol 表示循环左移）。这样代码第 34 行和第 35 行就是一个简化的独热码 MUX，也可将其看作二进制码 MUX，因为选择信号位宽为 1。看似 cnt 位宽增大，但不会消耗太多资源，因为其移位功能可通过 SLICEM 中的 LUT 实现。综合后可看到关键路径如图 3-25 所示，逻辑级数由 6 降至 4。

📄 VHDL 代码 3-30

```vhdl
1.  --File: mux_opt_after.vhd
2.  library ieee;
3.  use ieee.std_logic_1164.all;
4.  use ieee.numeric_std.all;
5.
6.  entity mux_opt_after is
7.    port (
8.      clk  : in std_logic;
9.      en   : in std_logic;
10.     dina : in unsigned(15 downto 0);
11.     dinb : in unsigned(15 downto 0);
12.     dout : out unsigned(15 downto 0)
13.   );
14. end entity;
15.
16. architecture rtl of mux_opt_after is
17.   signal dina_d1 : unsigned(15 downto 0);
18.   signal dinb_d1 : unsigned(15 downto 0);
```

```vhdl
19.   signal cnt          : unsigned(255 downto 0) := (0 => '1', others => '0');
20. begin
21.   process(clk)
22.   begin
23.     if rising_edge(clk) then
24.       cnt <= cnt rol 1 when en else
25.              cnt;
26.     end if;
27.   end process;
28.
29.   process(clk)
30.   begin
31.     if rising_edge(clk) then
32.       dina_d1 <= dina;
33.       dinb_d1 <= dinb;
34.       dout    <= dina_d1 + dinb_d1 when cnt(144) else
35.                  dina_d1 - dinb_d1;
36.     end if;
37.   end process;
38. end architecture;
```

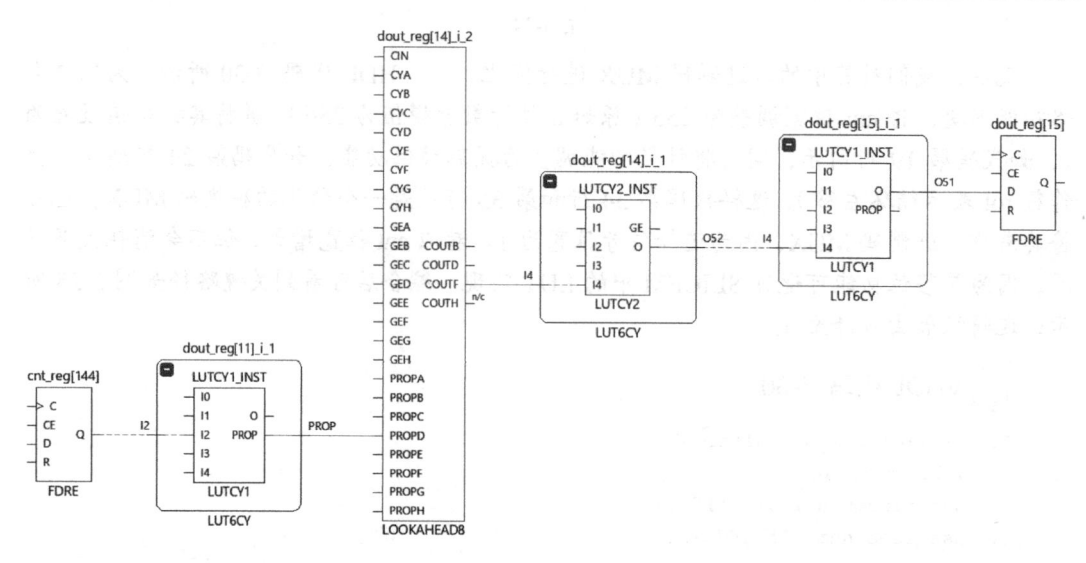

图 3-25

进一步对优化前后的性能进行比较（使用 Vivado 2021.1），从时序的角度看，如图 3-26 所示，可以看到优化后（impl_2）的 WNS 可以达到 0.539，意味着 Fmax 还可以提升到 684MHz。这得益于逻辑级数的改善，如图 3-27 所示，逻辑级数最大值由 6 降到 5。从资源利用率的角度看，优化前后，LUT 个数保持不变，而优化后寄存器个数减少，如图 3-28 所示。

Report	WNS	TNS	WHS	THS	TPWS
impl_1, Timing Summary - Route Design	0.090	0	0.086	0	0
impl_2, Timing Summary - Route Design	0.539	0	0.148	0	0

图 3-26

General Information	End Point Clock	Requirement	1	2	3	4	5	6
Logic Level Distribution	clk	2.000ns	7	2	1	1	7	6

General Information	End Point Clock	Requirement	0	1	2	3	4	5
Logic Level Distribution	clk	2.000ns	2	1	3	7	4

图 3-27

Name	Registers (1799680)	CLB LUTs (899840)	LUT as Logic (899840)	LOOKAHEAD8 (112480)	SLICE (112480)	CLB Registers (1799680)
mux_opt_before	56	24	24	2	12	56

Name	Registers (1799680)	CLB LUTs (899840)	LUT as Logic (899840)	LUT as Memory (449920)	LOOKAHEAD8 (112480)	SLICE (112480)	CLB Registers (1799680)
mux_opt_after	49	24	16	8	2	9	49

图 3-28

还有一种优先 MUX，也称为优先多路选择器，以 4 选 1 优先 MUX 为例，如 VHDL 代码 3-31 所示。从代码第 7 行和第 8 行可以看到，输入数据源的个数与选择信号宽度是一致的。代码第 18 行至第 21 行为 if 语句，但每条 if 语句的执行条件是不同的。表面上看，每条 if 语句的地位是相同的，但实际上，第一条 if 语句的优先级最低，最后一条 if 语句的优先级最高，这可以通过仿真验证，如图 3-29 所示。当 sel 为"1111"时，q 等于 1，为输入 d 的值。进一步查看其对应的电路，如图 3-30 所示。这是由 4 个 2 选 1 数据选择器级联构成的电路。从输入 a 到输出 q 经过的逻辑门最多，因此延迟最大，从输入 d 到输出 q 经过的逻辑门最少，因此延迟也最小。此外，这里虽然 if 语句没有对应的 else 分支，但是在代码第 17 行可以看到对 q 进行了赋值，因此工具不会推断出锁存器。

VHDL 代码 3-31

```
1.  --File: priority_mux_v1a.vhd
2.  library ieee;
3.  use ieee.std_logic_1164.all;
4.
5.  entity priority_mux_v1a is
6.    port (
7.      a, b, c, d : in std_logic;
8.      sel        : in std_logic_vector(3 downto 0);
9.      q          : out std_logic
10.   );
11. end entity;
12.
13. architecture rtl of priority_mux_v1a is
14. begin
15.   process(all)
16.   begin
17.     q <= '0';
18.     if sel(0) then q <= a; end if;
```

```
19.       if sel(1) then q <= b; end if;
20.       if sel(2) then q <= c; end if;
21.       if sel(3) then q <= d; end if;
22.    end process;
23. end architecture;
```

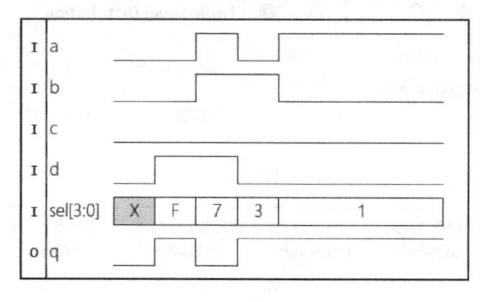

图 3-29

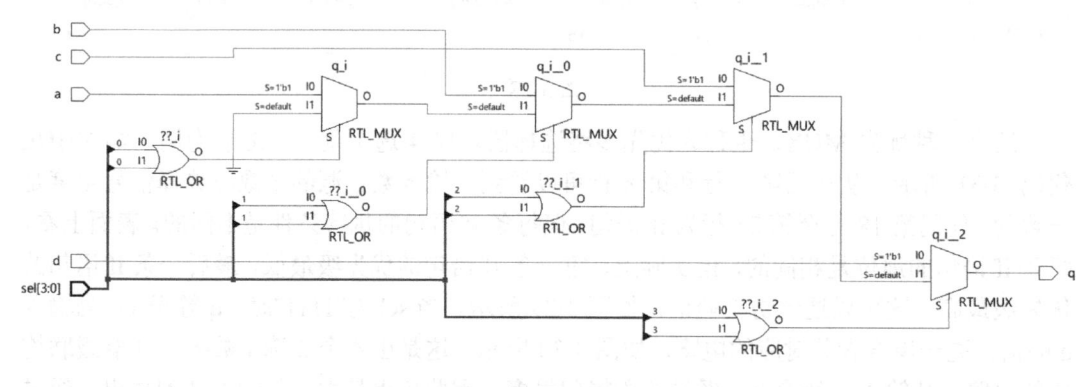

图 3-30

也可以采用 when-else 语句描述如图 3-30 所示的电路，如 VHDL 代码 3-32 所示。这里只给出结构体部分。在 when-else 语句中，第一个条件分支（也就是代码第 4 行）的优先级最高，最后一个条件分支（也就是代码第 7 行）的优先级最低。

VHDL 代码 3-32

```
1.  --File: priority_mux_v1b.vhd
2.  architecture rtl of priority_mux_v1b is
3.  begin
4.     q <= d when sel(3) else
5.          c when sel(2) else
6.          b when sel(1) else
7.          a when sel(0) else
8.          '0';
9.  end architecture;
```

VHDL 代码 3-31 综合后的电路如图 3-31 所示。从 d 端口到 q 端口经过了 1 个 LUT，而从 a 端口、b 端口、c 端口到 q 端口经过了 2 个 LUT。若在所有输入端口和输出端口处添加寄存器，则这 8 条时序路径存在不同的逻辑级数。这个电路给了我们一个启示：可将延迟大的信号（晚到的信号）放置到 d 端口，以实现延迟补偿。

图 3-31

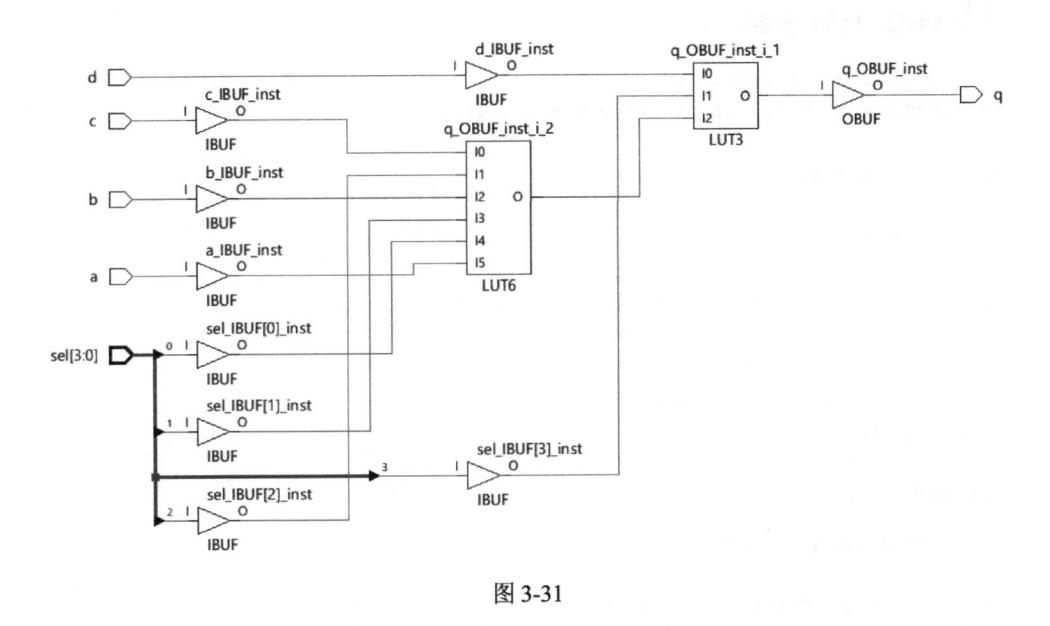

> **设计规则 6**：对于在一个 process（进程）中的多个顺序 if 语句，若条件变量不同，则它们是有优先级的，最后一条 if 语句的优先级最高；若条件变量相同，则综合后的电路不再有优先级。

也可以用 if-elsif-else 语句描述 4 选 1 优先 MUX，如 VHDL 代码 3-33 所示，综合后的电路与图 3-31 完全一致。同样可以采用 case?语句描述，如 VHDL 代码 3-34 所示，或采用 with-select?语句描述，如 VHDL 代码 3-35 所示。虽然这两者的 Elaborated Design Schematic 如图 3-32 所示为没有优先级的 4 选 1 MUX，但是实质上仍然是有优先级的，综合后的结果仍与图 3-31 保持一致。

VHDL 代码 3-33

```
1.  --File: priority_mux_v1c.vhd
2.  architecture rtl of priority_mux_v1c is
3.  begin
4.    process(all)
5.    begin
6.      if sel(3) then
7.        q <= d;
8.      elsif sel(2) then
9.        q <= c;
10.     elsif sel(1) then
11.       q <= b;
12.     elsif sel(0) then
13.       q <= a;
14.     else
15.       q <= '0';
16.     end if;
17.   end process;
18. end architecture;
```

VHDL 代码 3-34

```vhdl
1.  --File: priority_mux_v2a.vhd
2.  architecture rtl of priority_mux_v2a is
3.  begin
4.    process(all)
5.    begin
6.      case? sel is
7.        when "1---" => q <= d;
8.        when "01--" => q <= c;
9.        when "001-" => q <= b;
10.       when "0001" => q <= a;
11.       when others => q <= '0';
12.     end case?;
13.   end process;
14. end architecture;
```

VHDL 代码 3-35

```vhdl
1.  --File: priority_mux_v2b.vhd
2.  architecture rtl of priority_mux_v2b is
3.  begin
4.    with sel select?
5.      q <= d when "1---",
6.           c when "01--",
7.           b when "001-",
8.           a when "0001",
9.           '0' when others;
10. end architecture;
```

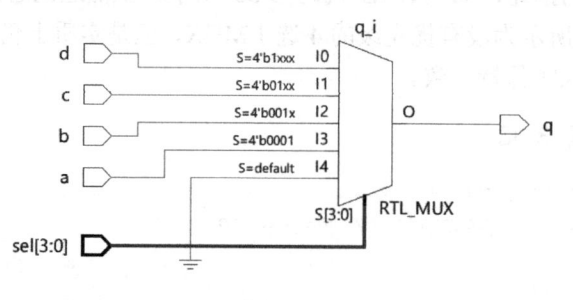

图 3-32

在此基础上，我们可以设计一个参数化的优先 MUX，假定输入数据位宽为 1，共有 W 个输入数据，那么选择信号的宽度也为 W，将该值设置为参数，用 for 循环结合 if 语句描述，如 VHDL 代码 3-36 所示。注意，代码第 18 行是不可缺少的，同时，其位置必须在 for 循环之前。因为代码第 19 行 din'REVERSE_RANGE 的返回值为 0~W-1，所以 din 的最高位优先级也最高。如果期望 din 的最低位优先级最高，只需要用 RANGE 替换 REVERSE_RANGE 即可。

VHDL 代码 3-36

```vhdl
1.  --File: priority_mux.vhd
2.  library ieee;
```

```vhdl
3.  use ieee.std_logic_1164.all;
4.
5.  entity priority_mux is
6.    generic ( W : positive := 4 );
7.    port (
8.      din : in std_logic_vector(W - 1 downto 0);
9.      sel : in std_logic_vector(W - 1 downto 0);
10.     q   : out std_logic
11.   );
12. end entity;
13.
14. architecture rtl of priority_mux is
15. begin
16.   process(all)
17.   begin
18.     q <= '0';
19.     for i in din'REVERSE_RANGE loop
20.       if sel(i) then q <= din(i); end if;
21.     end loop;
22.   end process;
23. end architecture;
```

3.3.2 多路解复用器代码风格

多路解复用器（Demultiplexer，后面统一用 DMUX 表示）的电路功能与多路复用器正好相反。多路解复用器仅有一条输入数据线，有 M 条输出数据线和 K 位选择信号，其中，M 和 K 的关系可表示为

$$K = \lceil \log_2^M \rceil$$

以 1 输入 4 输出 DMUX 为例，其电路结构如图 3-33 所示。可以发现，此电路由 4 个 MUX 构成，这 4 个 MUX 共享选择信号和输入数据信号。相应的 RTL 代码如 VHDL 代码 3-37 所示，仿真结果如图 3-34 所示。

📄 VHDL 代码 3-37

```vhdl
1.  --File: demux_1to4.vhd
2.  library ieee;
3.  use ieee.std_logic_1164.all;
4.
5.  entity demux_1to4 is
6.    generic ( W : positive := 8 );
7.    port (
8.      sel : in std_logic_vector(1 downto 0);
9.      din : in std_logic_vector(W - 1 downto 0);
10.     y0, y1, y2, y3 : out std_logic_vector(W - 1 downto 0)
11.   );
12. end entity;
13.
14. architecture rtl of demux_1to4 is
15. begin
```

```
16.    y3 <= din when sel = "11" else (others => '0');
17.    y2 <= din when sel = "10" else (others => '0');
18.    y1 <= din when sel = "01" else (others => '0');
19.    y0 <= din when sel = "00" else (others => '0');
20. end architecture;
```

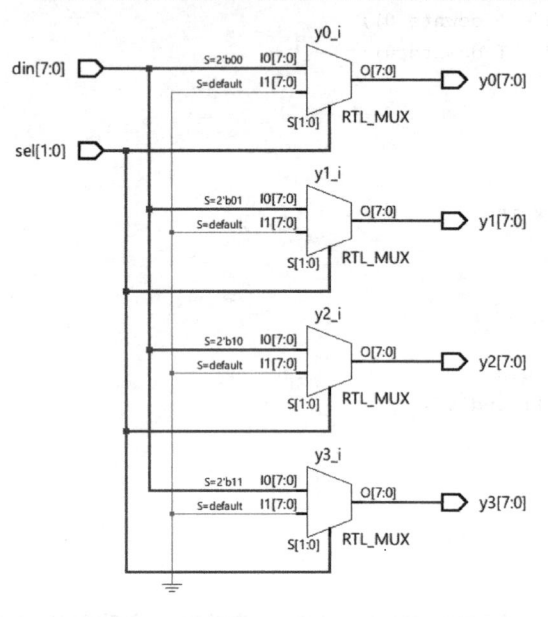

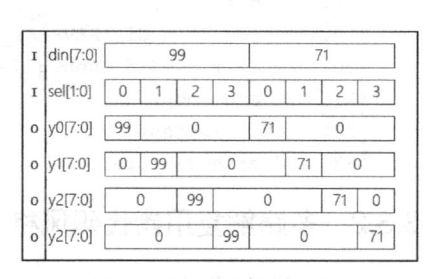

图 3-33

图 3-34

从代码可复用性的角度出发，我们尝试将 DMUX 参数化，如 VHDL 代码 3-38 所示，其中定义了三个参数：W 为输入数据位宽，M 为输出数据个数，K 为选择信号位宽。代码核心部分为第 32 行至第 34 行，通过 for 循环使代码简洁化。

📄 **VHDL 代码 3-38**

```
1.  --File: demux_1toM.vhd
2.  library ieee;
3.  use ieee.std_logic_1164.all;
4.
5.  package bus_mux_pkg is
6.    type bus_array is array(natural range <>) of std_logic_vector;
7.  end package;
8.
9.  library ieee;
10. use ieee.std_logic_1164.all;
11. use ieee.numeric_std.all;
12. use ieee.math_real.all;
13. use work.bus_mux_pkg.all;
14.
15. entity demux_1toM is
16.   generic (
17.     W : positive := 8; -- data width
18.     M : positive := 4; -- #N of output
```

```
19.        K : positive := integer(ceil(log2(real(M)))) -- select width
20.    );
21.    port (
22.      sel : in std_logic_vector(K - 1 downto 0);
23.      din : in std_logic_vector(W - 1 downto 0);
24.      y   : out bus_array(M - 1 downto 0)(W - 1 downto 0)
25.    );
26. end entity;
27.
28. architecture rtl of demux_1toM is
29. begin
30.    process(all)
31.    begin
32.      for i in y'REVERSE_RANGE loop
33.        y(i) <= din when to_integer(unsigned(sel)) = i else (others => '0');
34.      end loop;
35.    end process;
36. end architecture;
```

3.4 加法器与累加器

3.4.1 加法器代码风格

谈到加法器，我们先从一位全加器说起，因为它是实现多位加法器的基础。其输入端是被加数 a、加数 b 及较低位的进位 c_{in}；输出端是本位、s 及较高位的进位 c_{out}。根据二进制加法运算法则可知，其真值表如表 3-7 所示。

表 3-7

a	b	c_{in}	s	c_{out}
0	0	0	0	0
0	0	1	1	0
0	1	0	1	0
0	1	1	0	1
1	0	0	1	0
1	0	1	0	1
1	1	0	0	1
1	1	1	1	1

根据真值表利用卡诺图化简可得输出与输入的逻辑关系式：

$$c_{out} = (a\,\&\,b)|(b\,\&\,c_{in})|(a\,\&\,c_{in}) = (a\,\&\,b)|(a\oplus b)\,\&\,c$$
$$s = (a\oplus b)\oplus c_{in}$$

式中，"$\oplus$"表示异或运算（XOR），"&"表示与运算，"|"表示或运算。根据逻辑关系表达式，可进一步得出一位全加器的硬件电路图，如图 3-35 所示，相应的 RTL 代码如 VHDL 代码 3-39 所示。

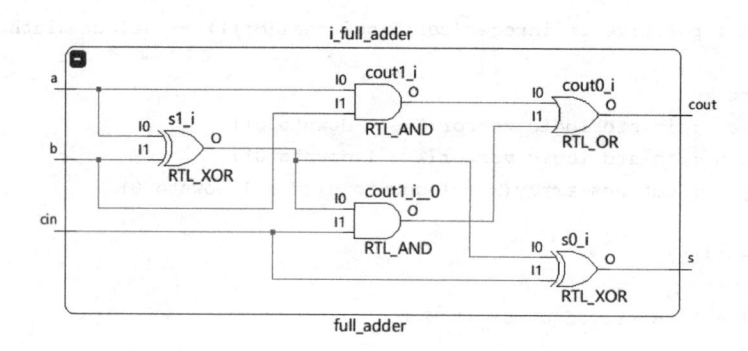

图 3-35

> **VHDL 代码 3-39**

```
1.  --File: full_adder.vhd
2.  library ieee;
3.  use ieee.std_logic_1164.all;
4.
5.  entity full_adder is
6.    port (
7.      a, b, cin : in std_logic;
8.      s, cout   : out std_logic
9.    );
10. end entity;
11.
12. architecture rtl of full_adder is
13. signal temp : std_logic;
14. begin
15.   temp <= a xor b;
16.   s    <= temp xor cin;
17.   cout <= (a and b) or (temp and cin);
18. end architecture;
```

将多个一位全加器级联即可构成一个多位二进制数加法器，图 3-36 所示为一个 4 位二进制数加法器，图中，每个 FA 为如图 3-35 所示的电路。在图 3-36 中，末级（最左侧）全加器的输出端 c_{out} 最终构成和的最高位 s_4。这个电路的特点是后一级的进位输入依赖前一级的进位输出，进位信号从最右侧进入，如波浪般一级一级被传递至末级，所以此加法器被称为行波进位加法器（Ripple Carry Adder, RCA）。不难看出，越靠近后级，延迟越大。图 3-36 相应的 RTL 代码如 VHDL 代码 3-40 所示。

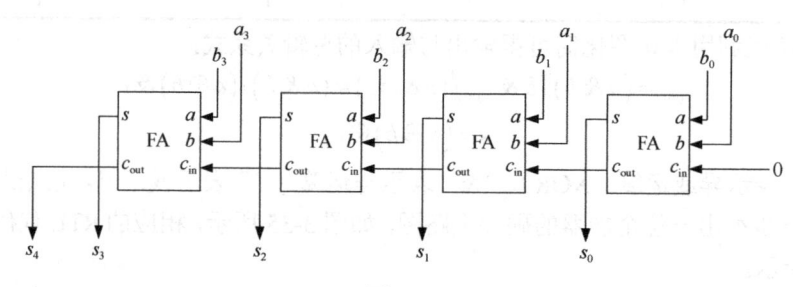

图 3-36

> **VHDL 代码 3-40**

```vhdl
1.  --File: adder_v1.vhd
2.  library ieee;
3.  use ieee.std_logic_1164.all;
4.
5.  entity adder_v1 is
6.    generic ( W : positive := 4 );
7.    port (
8.      a, b : in std_logic_vector(W - 1 downto 0);
9.      cin  : in std_logic;
10.     sum  : out std_logic_vector(W - 1 downto 0);
11.     cout : out std_logic
12.   );
13. end entity;
14.
15. architecture rtl of adder_v1 is
16.   signal cout_i : std_logic_vector(W - 1 downto 0);
17. begin
18.   i_full_adder : entity work.full_adder
19.     port map ( a => a(0), b => b(0), cin => cin, s => sum(0), cout => cout_i(0) );
20.
21.   gen0:
22.   for i in 1 to W - 1 generate
23.     i_full_adder : entity work.full_adder
24.       port map ( a => a(i), b => b(i), cin => cout_i(i-1), s => sum(i), cout => cout_i(i) );
25.   end generate;
26.
27.   cout <= cout_i(W-1);
28. end architecture.
```

VHDL 代码 3-40 的仿真结果如图 3-37 所示。图中，信号 cout_sum 由 cout 和 sum 拼接构成，且 cout 为最高位。这里将 a 和 b 均视为无符号数，仿真结果是正确无误的。若将其视为有符号数，如图 3-38 所示，则第一个结果就是错误的，即 5（对应二进制 4'b0101）与 -4（对应二进制 4'b1100）的和的结果为 -15（对应二进制 5'b10001）。那么错误的原因是什么呢？这里我们就要分析一下二进制加法的原理。

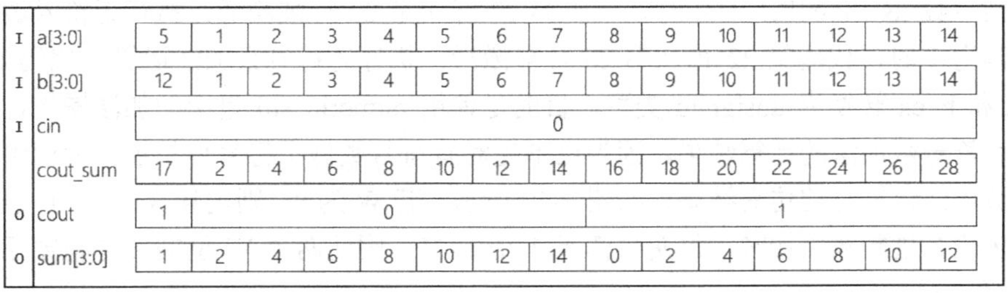

I	a[3:0]	5	1	2	3	4	5	6	7	8	9	10	11	12	13	14
I	b[3:0]	12	1	2	3	4	5	6	7	8	9	10	11	12	13	14
I	cin	0
	cout_sum	17	2	4	6	8	10	12	14	16	18	20	22	24	26	28
O	cout	1	0							1						
O	sum[3:0]	1	2	4	6	8	10	12	14	0	2	4	6	8	10	12

图 3-37

I	a[3:0]	5	1	2	3	4	5	6	7	-8	-7	-6	-5	-4	-3	-2
I	b[3:0]	-4	1	2	3	4	5	6	7	-8	-7	-6	-5	-4	-3	-2
I	cin	0															
	cout_sum	-15	2	4	6	8	10	12	14	-16	-14	-12	-10	-8	-6	-4
o	cout	1	0								1						
o	sum[3:0]	1	2	4	6	-8	-6	-4	-2	0	2	4	6	-8	-6	-4

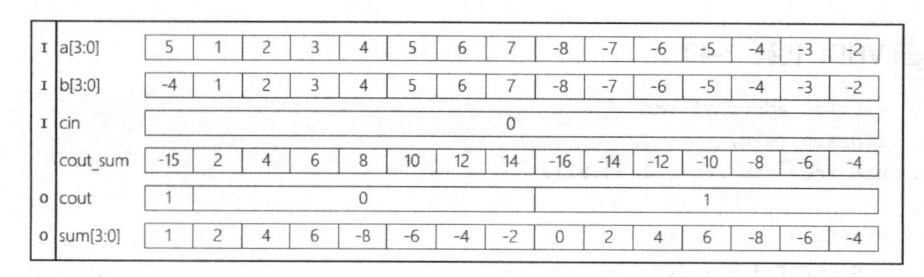

图 3-38

以图 3-39 为例（图中均为二进制数，图中的 1、2、3、4 对应案例 1、案例 2、案例 3、案例 4）。案例 1 中的 2'b011 和 2'b100 相加，若将二者视为无符号数，则分别对应十进制数的 3 和 4，两者之和 2'b111 为十进制数 7。若将其视为有符号数，则二者分别对应十进制数的 3 和-4，两者之和 2'b111 为十进制数-1。从这个角度而言，两个 N 位数相加，如果未溢出，和也为 N 位，那么无符号数加法和有符号数加法并没有区别。案例 2 中的两个 3 位数相加，和为 4 位，发生溢出，若只取低 3 位，结果必然是错误的。因此，两个 N 位数相加，为防止溢出，应将和设置为 N+1 位。若将案例 2 中的数据视为无符号数，可先将其高位补 0，形成 4 位二进制数再相加，和为 4 位，则结果正确，如案例 3 所示。若将案例 2 中的数据视为有符号数，可先将符号位扩展，形成 4 位二进制数再相加，和仍为 4 位，则结果正确，如案例 4 所示。由后 3 个案例可以看出，无符号数加法和有符号数加法的处理方式是不同的。

1	未溢出	2	溢出	3	无符号扩展	4	有符号扩展
	0 1 1		0 1 1		0 0 1 1		0 0 1 1
+	1 0 0	+	1 0 1	+	0 1 0 1	+	1 1 0 1
	1 1 1		1 0 0 0		1 0 0 0		0 0 0 0

图 3-39

💡 **设计规则 7**：两个 N 位二进制数相加，为防止溢出，应将和设置为 N+1 位。

VHDL 是强类型编程语言（Strongly Typed Language），意味着赋值操作左右两侧的数据必须是同一类型且具有相同位宽。因此，对于无符号数加法和有符号数加法，若要将其统一写在一个模块里实现，就要将数据统一为 std_logic_vector 类型，如 VHDL 代码 3-41 所示。当参数 IS_SIGNED 为 0 时，代码第 22 行和第 23 行被激活，执行无符号数加法运算；当其为 1 时，代码第 22 行和第 26 行被激活，执行有符号数加法运算。代码第 29 行将 a_ex 和 b_ex 转换为 unsigned 类型，目的是使用 numeric_std 包中的加法操作，但这并不影响有符号数加法的计算结果，因为已事先对有符号数做了符号位扩展。我们给出执行有符号数加法运算时的仿真结果，如图 3-40 所示。若 W 为 8，则消耗 8 个 LUT。事实上，当两个多位二进制数相加时，每两位消耗 1 个 LUT，因此两个 W 位的二进制数相加，会消耗 W 个 LUT。

📄 **VHDL 代码 3-41**

```
1.  --File: adder_v2.vhd
2.  library ieee;
3.  use ieee.std_logic_1164.all;
```

```
4.  use ieee.numeric_std.all;
5.
6.  entity adder_v2 is
7.    generic (
8.      IS_SIGNED : natural := 0;
9.      W         : positive := 8
10.   );
11.   port (
12.     a, b : in std_logic_vector(W - 1 downto 0);
13.     sum  : out std_logic_vector(W downto 0)
14.   );
15. end entity;
16.
17. architecture rtl of adder_v2 is
18. signal a_ex, b_ex : std_logic_vector(W downto 0);
19. begin
20.   gen0:
21.   if IS_SIGNED = 0 generate
22.     a_ex <= '0' & a;
23.     b_ex <= '0' & b;
24.   else generate
25.     a_ex <= a(a'HIGH) & a;
26.     b_ex <= b(b'HIGH) & b;
27.   end generate;
28.
29.   sum <= std_logic_vector(unsigned(a_ex) + unsigned(b_ex));
30. end architecture;
```

VHDL-2008 新特性

VHDL-2008 新增了 if-elsif-else generate 语句，这样对于同一变量的不同条件，可以使用该语句描述，如 VHDL 代码 3-41 中的第 21 行至第 27 行所示。而在 VHDL-93 版本中，则需要多个 if generate 语句分开描述，从而使代码变得冗长，并降低代码的可读性。

I	a[3:0]	5	1	2	3	4	5	6	7	-8	-7	-6	-5	-4	-3	-2	-1
I	b[3:0]	-4	1	2	3	4	5	6	7	-8	-7	-6	-5	-4	-3	-2	-1
O	sum[4:0]	1	2	4	6	8	10	12	14	-16	-14	-12	-10	-8	-6	-4	-2

图 3-40

当然，有符号数加法也可以直接使用 signed 类型描述，定义好输入数据与输出数据之间的位宽关系。以 3 个有符号整数相加为例，如 VHDL 代码 3-42 所示。若输入数据位宽为 W，那么输出数据位宽应设置为 W+2，才可防止溢出（N 个有符号整数相加，位宽增长 $\lceil \log_2^N \rceil$）。代码第 17 行使用了 VHDL 的聚合操作符 "=>"，将 a_ex 的高两位赋值为 a1 的最高位。相应的仿真结果如图 3-41 所示。若 W 为 8，则该加法器会消耗 10 个 LUT。事实上，若 3 个数相加，每 3 位会消耗 1 个 LUT，但此时总位宽要按输出位宽计算。因此 3 个 W 位有符号整数相加，输出位宽为 W+2，则共消耗 3(W+2)/3 也就是 W+2 个 LUT。

VHDL 代码 3-42

```vhdl
1.  --File: adder_v3.vhd
2.  library ieee;
3.  use ieee.std_logic_1164.all;
4.  use ieee.numeric_std.all;
5.
6.  entity adder_v3 is
7.    generic ( W : positive := 8 );
8.    port (
9.      a1, a2, a3 : in  signed(W - 1 downto 0);
10.     sum        : out signed(W + 1 downto 0)
11.   );
12. end entity;
13.
14. architecture rtl of adder_v3 is
15.   signal a1_ex, a2_ex, a3_ex : signed(W + 1 downto 0);
16. begin
17.   a1_ex <= (W + 1 downto W => a1(W - 1)) & a1;
18.   a2_ex <= (W + 1 downto W => a2(W - 1)) & a2;
19.   a3_ex <= (W + 1 downto W => a3(W - 1)) & a3;
20.   sum   <= a1_ex + a2_ex + a3_ex;
21. end architecture;
```

I	a0[3:0]	0	1	2	3	4	5	6	7	-8	-7	-6	-5	-4	-3	-2
I	a1[3:0]	1	2	3	4	5	6	7	-8	-7	-6	-5	-4	-3	-2	-1
I	a2[3:0]	2	3	4	5	6	7	-8	-7	-6	-5	-4	-3	-2	-1	0
O	sum[4:0]	3	6	9	12	15	18	5	-8	-21	-18	-15	-12	-9	-6	-3

图 3-41

无论是 unsigned 类型还是 signed 类型，执行加法运算时都要手工进行补齐位宽操作。好在 VHDL-2008 提供了定点数据类型，可以自动补齐。这里，我们先介绍一下 VHDL-2008 中定义的无符号定点数和有符号定点数。无符号定点数用 ufixed(M-1 downto -N) 来表示，有符号定点数用 sfixed(M-1 downto -N) 来表示。其中，M 表示整数部分位宽，N 表示小数部分位宽。当 M 为 0 时，表明该定点数只有小数部分（纯小数）；当 N 为 0 时，表示该定点数只有整数部分。表 3-8 给出了无符号定点数 9.75 的二进制形式，表 3-9 给出了有符号定点数-9.75 的二进制形式。

表 3-8

数 据 类 型	ufixed(3 downto -4)								
位 索 引 号	3	2	1	0	-1	-2	-3	-4
二进制数据	1	0	0	1	.	1	1	0	0
权 值	8	4	2	1		0.5	0.25	0.125	0.0625
十进制数据	1×8+0×4+0×2+1×1+1×0.5+1×0.25=9.75								

表 3-9

数 据 类 型	sfixed(4 downto -4)								
位 索 引 号	4	3	2	1	0	−1	−2	−3	−4
二进制数据	1	0	1	1	0 .	0	1	0	0
权 值	16	8	4	2	1	0.5	0.25	0.125	0.0625
十进制数据	−1×16+1×4+1×2+0×1+0×0.5+1×0.25=−9.75								

实数（real）、无符号整数（unsigned）、std_logic_vector 类型的数据都可以转换为无符号定点数，而无符号定点数也可以转换为实数（to_real）、无符号整数（to_unsigned）、整数（to_integer）或 std_logic_vector 类型（to_slv），具体案例与使用方法如表 3-10 所示。相应地，有符号定点数也可以执行类似的数据类型转换，如表 3-11 所示。此外，还可以借助函数 resize 实现不同位宽之间的转换，如表 3-12 所示。

不同于 unsigned 和 signed 两种数据类型，两个 ufixed 或 sfixed 数据相加或相减，其结果的位宽会自动发生变化，如表 3-13 所示，表中，max 表示选取两者中的较大者，而 min 表示选取两者中的较小者。这样，我们就不需要手工进行位宽扩展了，从而防止溢出。

表 3-10

uf1: ufixed(3 downto -4) uint1: unsigned(7 downto 0) := "00001010" uint2: unsigned(8 downto 0) := "101100110" uint3: unsigned(7 downto 0) int1 : integer slv1 : std_logic_vector(7 downto 0) := "10011100" slv2 : std_logic_vector(7 downto 0)	真实值 （十进制）	真实值 （二进制）	备 注
uf1 <= to_ufixed(9.75, 3, −4)	9.75	"1001.1100"	3 为 uf1 的最高位，−4 为最低位
uf1 <= to_ufixed(9.75, uf1'HIGH, uf1'LOW)	9.75	"1001.1100"	uf1'HIGH 表示 uf1 的最高位，uf1'LOW 为最低位
uf1 <= to_ufixed(9.75, uf1)	9.75	"1001.1100"	9.75 是实数
uf1 <= to_ufixed(2#1001.1100#, uf1)	9.75	"1001.1100"	2#1001.1100#是实数
slv2 <= to_slv(uf1)		"10011100"	slv2 与 uf1 的位宽必须一致
uint3 <= to_unsigned(uf1, uint3'LENGTH)	10	"00001010"	四舍五入
int1 <= to_integer(uf1)	10	"1010"	四舍五入
uf1 <= to_ufixed(uint1, uf1)	10	"1010.0000"	
uf1 <= to_ufixed(uint2, uf1)	15.9375	"1111.1111"	发生溢出
uf1 <= to_ufixed(slv1, uf1)	9.75	"1001.1100"	slv1 与 uf1 的位宽必须一致

表 3-11

	真实值（十进制）	真实值（二进制）	备注
sf1: sfixed(4 downto −4)			
slv1: std_logic_vector(8 downto 0)			
sint1 : signed(4 downto 0)			
slv2 : std_logic_vector(8 downto 0) := "101100100"			
sint2 : signed(4 downto 0) := "10110"			
sint3 : signed(5 downto 0) := "011111"			
sf1 <= to_sfixed(−9.75, 4, −4)	−9.75	"10110.0100"	4 为 sf1 的最高位，−4 为最低位
sf1 <= to_sfixed(−9.75, sf1'HIGH, sf1'LOW)	−9.75	"10110.0100"	sf1'HIGH 表示 uf1 的最高位，sf1'LOW 为最低位
sf1 <= to_sfixed(−9.75, sf1)	−9.75	"10110.0100"	
slv1 <= to_slv(sf1)		"101100100"	sf1 与 slv1 的位宽必须一致
sint1 <= to_signed(sf1, sint1'LENGTH)	−10	"10110"	四舍五入
int1 <= to_integer(sf1)	−10	"10110"	四舍五入
sf1 <= to_sfixed(slv2, sf1);	−9.75	"10110.0100"	slv2 与 sf1 的位宽必须一致
sf1 <= to_sfixed(sint2, sf1)	−10	"101100000"	
sf1 <= to_sfixed(sint3, sf1)	15.9375	"011111111"	发生溢出

表 3-12

	真实值（十进制）	真实值（二进制）	备注
uf1: ufixed(3 downto −4)			
uf2: ufixed(4 downto −5)			
sf1: sfixed(4 downto −4);			
sf2: sfixed(4 downto −1)			
sf3: sfixed(2 downto −4)			
uf1 <= to_ufixed(9.75, 3, −4)	9.75	"1001.1100"	
uf2 <= resize(uf1, uf2)	9.75	"01001.11000"	
uf2 <= resize(uf1, uf2'HIGH, uf2'LOW)	9.75	"01001.11000"	
sf1 <= to_sfixed(−9.75, sf1)	−9.75	"101100100"	
sf2 <= resize(sf1, sf2)	−10	"101100"	发生截尾
sf3 <= resize(sf1, sf3)	−4	"1000000"	发生溢出

表 3-13

操　作	输出数据位宽
A+B	max(A'LEFT, B'LEFT) + 1 down to min(A'RIGHT, B'RIGHT)
A−B	max(A'LEFT, B'LEFT) + 1 down to min(A'RIGHT, B'RIGHT)

　　如 VHDL 代码 3-43 所示。代码第 3 行表明需要使用定点包 fixed_pkg，这样才能使用 ufixed（无符号定点数）和 sfixed（有符号定点数）。代码第 7 行定义了输入数据整数部分的位宽为 4，第 8 行定义了其小数部分的位宽 4，代码第 12 行确定了输出和的整数部分的位宽为 5，小数部分位宽为 4。

📄 **VHDL 代码 3-43**

```
1.  --File: adder_fixed_v1.vhd
2.  library ieee;
3.  use ieee.fixed_pkg.all;
4.
5.  entity adder_fixed_v1 is
6.    generic (
7.      IW : integer := 4; -- width of integer part
8.      FW : integer := 4  -- width of fraction part
9.    );
10.   port (
11.     a1, a2 : in sfixed(IW - 1 downto -FW);
12.     sum    : out sfixed(IW   downto -FW)
13.   );
14. end entity;
15.
16. architecture rtl of adder_fixed_v1 is
17. begin
18.   sum <= a1 + a2;
19. end architecture;
```

进行电路设计时，我们可以充分利用 ufixed 和 sfixed 数据类型在执行加/减运算时的位宽自动扩展以防止溢出的功能。以 VHDL 代码 3-44 为例，将 4 个定点数相加。这 4 个定点数均为 ufixed(3 downto -2)。若采用代码第 4 行的描述方式，则是加法器级联，从输入到输出位宽会增加 3 位，如图 3-42 所示。若采用代码第 5 行的描述方式，则是加法树结构，从输入到输出位宽会增加 2 位，如图 3-43 所示。但两者相比，加法器级联结构会消耗 18 个 LUT，而加法树结构会消耗 19 个 LUT。

📄 **VHDL 代码 3-44**

```
1.  --File: fixed_add_v1.vhd
2.  architecture rtl of fixed_add_v1 is
3.  begin
4.    sum <= a1 + a2 + a3 + a4;
5.    --sum <= (a1 + a2) + (a3 + a4);
6.  end architecture;
```

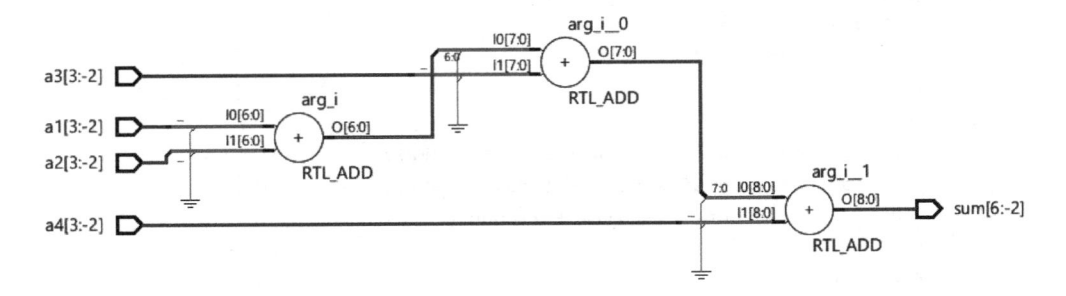

图 3-42

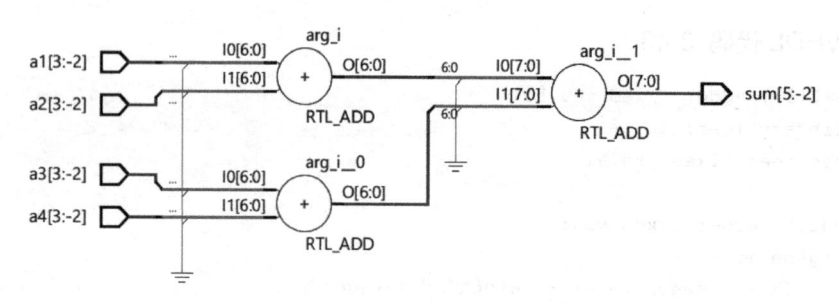

图 3-43

应用案例 2：并行 6 个数相加求和

并行的 6 个数（数据类型一致，位宽相等）相加，可以每两个数一组用一个加法器，如图 3-44 左侧所示。也可以每三个数一组用一个加法器，如图 3-44 右侧所示。但它们的资源消耗量是不一样的，尽管两者从输入到输出位宽都增加了 3 位，如 VHDL 代码 3-45 所示，代码第 20 行对应图 3-44 左侧结构，代码第 19 行对应图 3-44 右侧结构。使用 Vivado 2022.1 综合，目标芯片为 7 系列 FPGA，最终，左侧结构消耗 43 个 LUT，而右侧结构消耗 44 个 LUT。

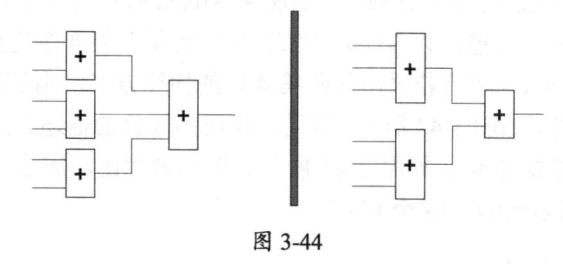

图 3-44

VHDL 代码 3-45

```vhdl
1.  --File: adder_6_data_v2.vhd
2.  library ieee;
3.  use ieee.fixed_pkg.all;
4.
5.  entity adder_6_data_v2 is
6.    generic (
7.      IW : integer := 4;
8.      FW : integer := 4
9.    );
10.   port (
11.     a1, a2, a3 : in sfixed(IW - 1 downto -FW);
12.     a4, a5, a6 : in sfixed(IW - 1 downto -FW);
13.     sum        : out sfixed(IW + 2 downto -FW)
14.   );
15. end entity;
16.
17. architecture rtl of adder_6_data_v2 is
18. begin
19.   -- sum <= (a1 + a2 + a3) + (a4 + a5 + a6);
```

```
20.     sum <= (a1 + a2) + (a3 + a4) + (a5 + a6);
21. end architecture;
```

应用案例 3：求两个有符号整数的平均数

求两个数的平均数只需要把这两个数相加求和，再除以 2 即可，除以 2 等效于右移一位。而实际上，移位操作也是可以避免的。为便于说明，我们以两个 4 位有符号整数为例进行说明，如表 3-14 所示。表中，第一行为操作数 a0 的每一位，这里对 a0 进行了符号位扩展；第二行为操作数 a1 的每一位，同样对 a1 进行了符号位扩展。第三行为 a0 和 a1 的和 sum，第四行为将 sum 右移一位的结果，可以看到，第四行的灰色部分就是 a0 与 a1 之和的平均数。根据此原理，可形成 VHDL 代码 3-46，相应的仿真结果如图 3-45 所示。

表 3-14

操作数 a0	a0(3)	a0(3)	a0(2)	a0(1)	a0(0)
操作数 a1	a1(3)	a1(3)	a1(2)	a1(1)	a1(0)
和	sum(4)	sum(3)	sum(2)	sum(1)	sum(0)
右　移	sum(4)	sum(4)	sum(3)	sum(2)	sum(1)

VHDL 代码 3-46

```
1.  --File: average_v1.vhd
2.  library ieee;
3.  use ieee.numeric_std.all;
4.
5.  entity average_v1 is
6.    generic ( W : positive := 8 );
7.    port (
8.       a0, a1 : in signed(W - 1 downto 0);
9.       avg    : out signed(W - 1 downto 0)
10.    );
11. end entity;
12.
13. architecture rtl of average_v1 is
14. signal sum : signed(W downto 0);
15. begin
16.    sum <= (a0(W - 1) & a0) + (a1(W - 1) & a1);
17.    avg <= sum(W downto 1);
18. end architecture;
```

I	a0[3:0]	0	1	2	3	4	5	6	7	-8	-7	-6	-5	-4	-3	-2
I	a1[3:0]	1	2	3	4	5	6	7	-8	-7	-6	-5	-4	-3	-2	-1
I	sum[3:0]	1	3	5	7	9	11	13	-1	-15	-13	-11	-9	-7	-5	-3
O	avg[4:0]	0	1	2	3	4	5	6	-1	-8	-7	-6	-5	-4	-3	-2

图 3-45

下列公式可以计算两个有符号数的平均值，其中，前者为向上取整，后者为向下取整，式中，sra 表示算术右移。

$$(x+y)/2 = (x|y) - ((x \oplus y)\text{sra1})$$
$$(x+y)/2 = (x\&y) + ((x \oplus y)\text{sra1})$$

VHDL 代码 3-47 的描述方式可以根据参数 IS_CEIL 来控制是向上取整还是向下取整。当该参数为 1 时，表示向上取整，当该参数为 0 时表示向下取整。对于无符号数，只需要将代码中的算术右移替换为逻辑右移，同时去掉 signed 关键字即可。该代码的仿真结果如图 3-46 所示，对应位宽 W 为 4。其中，上部为向上取整的仿真结果，下部为向下取整的仿真结果。

VHDL 代码 3-47

```vhdl
1.  --File: average_v2.vhd
2.  library ieee;
3.  use ieee.numeric_std.all;
4.
5.  entity average_v2 is
6.    generic (
7.      W       : positive := 8;
8.      IS_CEIL : boolean := true
9.    );
10.   port (
11.     a0, a1 : in signed(W - 1 downto 0);
12.     avg    : out signed(W - 1 downto 0)
13.   );
14. end entity;
15.
16. architecture rtl of average_v2 is
17. begin
18.   gen0:
19.   if IS_CEIL generate
20.     avg <= (a0 or a1) - ((a0 xor a1) sra 1);
21.   else generate
22.     avg <= (a0 and a1) + ((a0 xor a1) sra 1);
23.   end generate;
24. end architecture;
```

I	a0[3:0]	0	1	2	3	4	5	6	7	-8	-7	-6	-5	-4	-3	-2
I	a1[3:0]	1	2	3	4	5	6	7	-8	-7	-6	-5	-4	-3	-2	-1
O	avg[4:0]	1	2	3	4	5	6	7	0	-7	-6	-5	-4	-3	-2	-1

I	a0[3:0]	0	1	2	3	4	5	6	7	-8	-7	-6	-5	-4	-3	-2
I	a1[3:0]	1	2	3	4	5	6	7	-8	-7	-6	-5	-4	-3	-2	-1
O	avg[4:0]	0	1	2	3	4	5	6	-1	-8	-7	-6	-5	-4	-3	-2

图 3-46

如果加法器除了两个输入数据，还有进位输入端，同时，输出也包含进位输出端，那么可以采用如 VHDL 代码 3-48 所示的描述方式。可以看到 unsigned 的每一位都是 std_logic

类型（此结论对 signed 类型也成立）。这里是无符号整数，对于有符号整数，需要将 unsigned 替换为 signed，同时将代码第 19 行的高位补零操作替换为符号位扩展操作。

VHDL 代码 3-48

```vhdl
1.  --File: adder_v4.vhd
2.  library ieee;
3.  use ieee.std_logic_1164.all;
4.  use ieee.numeric_std.all;
5.
6.  entity adder_v4 is
7.    generic ( W : positive := 4 );
8.    port (
9.      a0, a1 : in unsigned(W - 1 downto 0);
10.     ci     : in std_logic;
11.     sum    : out unsigned(W - 1 downto 0);
12.     co     : out std_logic
13.   );
14. end entity;
15.
16. architecture rtl of adder_v4 is
17. signal temp: unsigned(W downto 0);
18. begin
19.   temp <= ('0' & a0) + ('0' & a1) + ci;
20.   sum  <= temp(W - 1 downto 0);
21.   co   <= temp(W);
22. end architecture;
```

VHDL-2008 新特性

VHDL-2008 对"+"和"−"操作进行了重定义，使得其中一个操作数可以是向量（如 unsigned 或 signed 类型），而另一个操作数可以是 std_logic 类型（见 VHDL 代码 3-48 第 19 行）。若未将 VHDL 代码 3-48 对应文件声明为 VHDL-2008 类型，那么 Vivado 会将其作为 VHDL-93 类型进行编译，在综合阶段会显示如图 3-47 所示的错误信息。

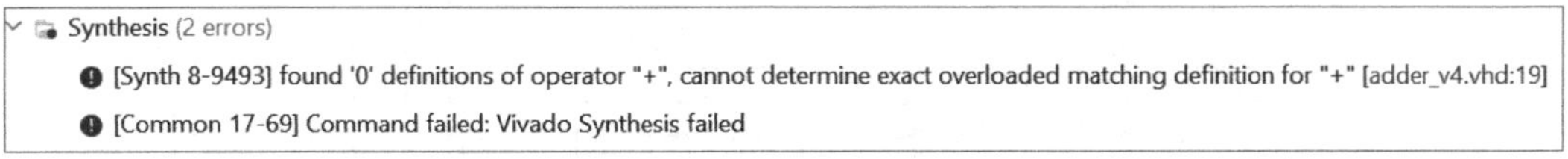

```
Synthesis (2 errors)
  [Synth 8-9493] found '0' definitions of operator "+", cannot determine exact overloaded matching definition for "+" [adder_v4.vhd:19]
  [Common 17-69] Command failed: Vivado Synthesis failed
```

图 3-47

既然加法器既可以用一位全加器的形式（见 VHDL 代码 3-40）搭建出来，又可以采用加法运算操作符"+"直接描述，那么二者有什么区别呢？这里我们以输入操作数均为 16 位无符号整数为例。采用 VHDL 代码 3-40 的形式（一位全加器）会消耗 29 个 LUT，采用 VHDL 代码 3-48 的形式（直接用"+"描述）会消耗 16 个 LUT 和 3 个 CARRY8（进位链）。显然，后者消耗的资源更少，同时，后者使用了进位链，这样可以降低逻辑延迟。

在加法运算的基础上，我们再看减法运算。减法运算本质上可以看作加法运算，因为 $a-b$ 等效于 $a+(-b)$，从二进制的角度而言，$-b$ 对应的二进制补码可先对 b 对应的二进制补码取反，再加 1 获取。图 3-48 显示了这种方法的计算过程。这里每个数据均以二进制补码的形式表示，为了防止溢出，进行了符号位扩展（由 4 位扩展为 5 位）。

①						②							③						
	0	0	1	1	1		取反	0	0	1	0	1			0	0	1	1	1
−	0	0	1	0	1			1	1	0	1	0		+	1	1	0	1	1
	0	0	0	1	0		+					1			0	0	0	1	0
								1	1	0	1	1							

①						②							③						
	1	1	0	0	1		取反	1	1	1	0	0			1	1	0	0	1
−	1	1	1	0	0			0	0	0	1	1		+	0	0	1	0	0
	1	1	1	0	1		+					1			1	1	1	0	1
								0	0	1	0	0							

图 3-48

取反加 1 的方法很容易映射到基于一位全加器的多位加法器上，只需要增加取反逻辑，同时将进位输入端改为 1，如图 3-49 所示。此外，还可以将加法器和减法器合二为一，引入控制端 control，通过它来切换加法操作与减法操作，这样就形成了如图 3-50 所示的电路结构。图中，当 control 为 0 时执行 $a+b$，当 control 为 1 时执行 $a-b$。该电路利用了两个布尔等式，式中，$-x$ 表示对 x 取反。

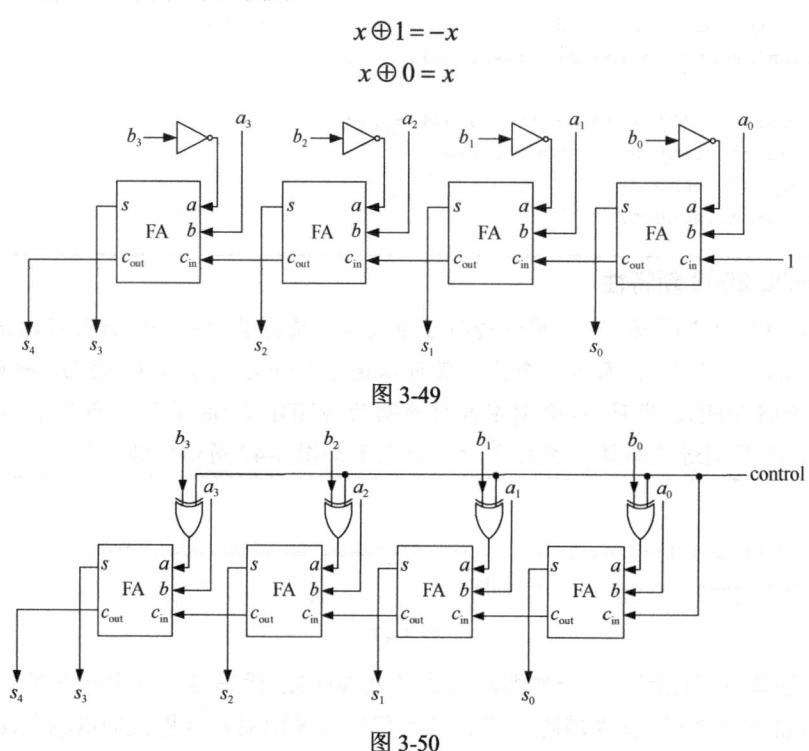

$$x \oplus 1 = -x$$
$$x \oplus 0 = x$$

图 3-49

图 3-50

从行为级的角度看，动态可切换的加法器/减法器的电路结构如图 3-51 所示。两个输入数据同时执行加法运算和减法运算，将结果传输至 2 选 1 数据选择器的输入端，而数据选择器的控制端 add_sub 用于选择输出是取两者之差还是取两者之和。相应的 RTL 代码如 VHDL 代码 3-49 所示。

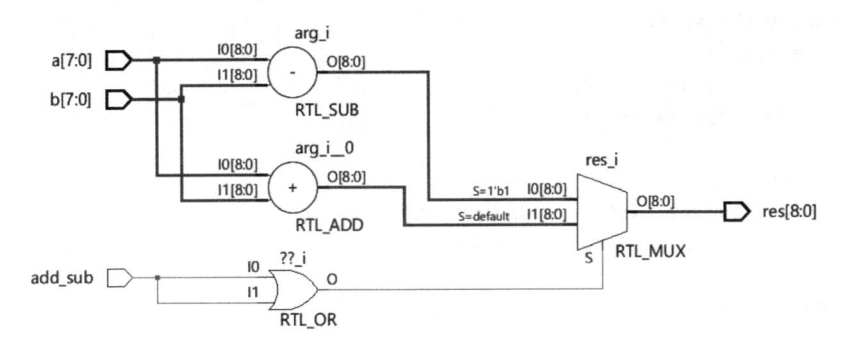

图 3-51

VHDL 代码 3-49

```
1.  --File: dynamic_adder.vhd
2.  library ieee;
3.  use ieee.std_logic_1164.all;
4.  use ieee.fixed_pkg.all;
5.
6.  entity dynamic_adder is
7.    generic ( W : positive := 8 );
8.    port (
9.      add_sub : in std_logic; --0: +; 1 -
10.     a, b    : in sfixed(W - 1 downto 0);
11.     res     : out sfixed(W downto 0)
12.   );
13. end entity;
14.
15. architecture rtl of dynamic_adder is
16. begin
17.   res <= a - b when add_sub else a + b;
18. end architecture;
```

应用案例 4：求有符号数的绝对值

对于正数，其二进制原码与二进制补码相同，因此其绝对值就是其原码对应的十进制数。对于负数，从十进制的角度而言，其绝对值是其相反数。这个相反数其实执行的是减法操作，被减数为 0。从二进制的角度而言，求绝对值执行的是取反加 1 的操作。因为是减法，所以要考虑字长增长效应。若输入数据位宽为 W，那么输出数据位宽应为 W+1 才能防止溢出。VHDL-2008 提供了函数 abs，可直接用于绝对值运算，其输入为 sfixed 类型。假定输入数据为 a，那么输出数据的位宽范围为 a'LEFT+1 downto a'RIGHT。显然，输出比输入增加了一位。VHDL 代码 3-50 给出了采用 abs 函数的实现方案。

VHDL 代码 3-50

```
1.  --File: abs_v1.vhd
2.  library ieee;
3.  use ieee.std_logic_1164.all;
4.  use ieee.fixed_pkg.all;
5.
```

```
6.  entity abs_v1 is
7.    generic (
8.      IW : natural := 4;
9.      FW : natural := 4
10.   );
11.   port (
12.     a : in sfixed(IW - 1 downto -FW);
13.     y : out sfixed(IW downto -FW)
14.   );
15. end entity;
16.
17. architecture rtl of abs_v1 is
18. begin
19.   y <= abs(a);
20. end architecture;
```

3.4.2 累加器代码风格

累加器也需要加法运算完成，但并不是纯粹的组合逻辑，而是既包含组合逻辑单元，又包含时序逻辑单元。之所以在此处介绍，是因为其核心是加法器。累加器是对输入数据流累加求和，因此必然存在字长增长效应。假定输入数据位宽为 W，累加次数为 N，输出数据位宽为 ACCW，那么三者的关系可表示为

$$\text{ACCW} = W + \left\lceil \log_2^N \right\rceil$$

在这种情况下，才可以保证数据不溢出。

第一种累加器的电路结构如图 3-52 所示。该电路共包含两部分：加法器和触发器。复位和时钟使能都直接与触发器的对应端口相连，而加法器的两个输入数据分别来自外部输入数据（acc_in）和触发器输出端（acc_out）。相应的 RTL 代码如 VHDL 代码 3-51 所示，如果希望采用 DSP48 实现，可添加综合属性 use_dsp，并将其值设置为"yes"，如代码第 18 行和第 19 行所示，相应的仿真结果如图 3-53 所示，其中，输入数据位宽为 4。从仿真结果可以看出，当复位信号 rst 有效时，累加器输出为 0，此时，对应的输入数据不会被累加。

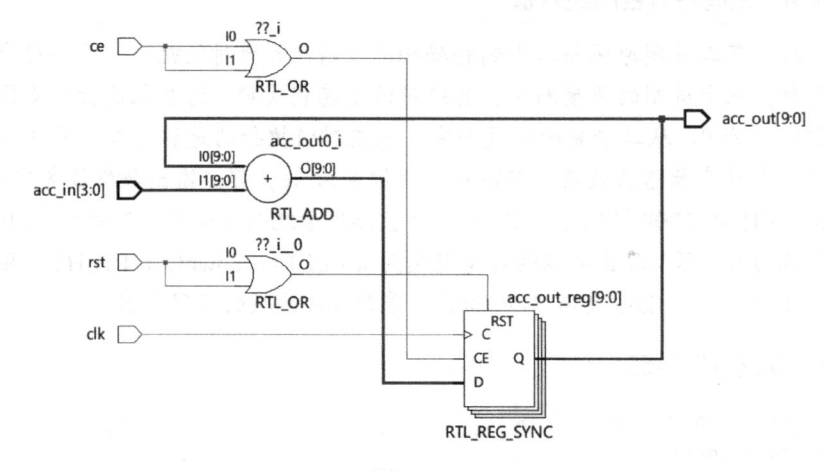

图 3-52

> **VHDL 代码 3-51**

```vhdl
1.  --File: acc_v1.vhd
2.  library ieee;
3.  use ieee.std_logic_1164.all;
4.  use ieee.numeric_std.all;
5.
6.  entity acc_v1 is
7.    generic (
8.      DW   : positive := 4;
9.      ACCW : positive := 10
10.   );
11.   port (
12.     clk     : in std_logic;
13.     rst     : in std_logic;
14.     ce      : in std_logic;
15.     acc_in  : in signed(DW - 1 downto 0);
16.     acc_out : out signed(ACCW - 1 downto 0)
17.   );
18.   attribute use_dsp : string;
19.   attribute use_dsp of acc_v1 : entity is "yes";
20. end entity;
21.
22. architecture rtl of acc_v1 is
23. signal acc_in_ex : signed(ACCW - 1 downto 0);
24. begin
25.   acc_in_ex <= (ACCW - 1 downto DW => acc_in(DW - 1)) & acc_in;
26.   process(clk)
27.   begin
28.     if rising_edge(clk) then
29.       if rst then
30.         acc_out <= (others => '0');
31.       elsif ce then
32.         acc_out <= acc_out + acc_in_ex;
33.       end if;
34.     end if;
35.   end process;
36. end architecture;
```

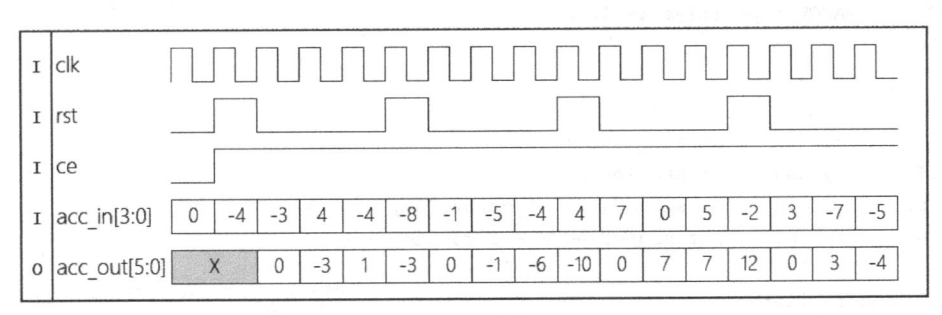

I	clk															
I	rst															
I	ce															
I	acc_in[3:0]	0	-4	-3	4	-4	-8	-1	-5	-4	4	7	0	5	-2	3	-7	-5
O	acc_out[5:0]	X	0	-3	1	-3	0	-1	-6	-10	0	7	7	12	0	3	-4

图 3-53

> 🔍 **VHDL-2008 新特性**
> VHDL 提供了两种输出类型 out 和 buffer。当需要读取输出信号供 architecture 内部使用时，需要将该输出声明为 buffer 类型。VHDL-2008 取消了这一限制。这意味着，对于 out 类型，architecture 内部是可以直接读取的，如 VHDL 代码 3-51 第 32 行所示。

第二种累加器不需要复位信号，取而代之的是 bypass 信号，当其为高时，累加器直接输出当前输入端口对应的数据；当其为低时，执行累加操作，电路结构如图 3-54 所示。与第一种累加器相比，第二种累加器增加了一个 2 选 1 数据选择器。相应的 HDL 代码如 VHDL 代码 3-52 所示，仿真结果如图 3-55 所示。从图 3-55 中可以看出，当 ce 有效时，bypass 周期就是一个累加周期。同样可以使用综合属性 use_dsp 使其映射为 DSP48。

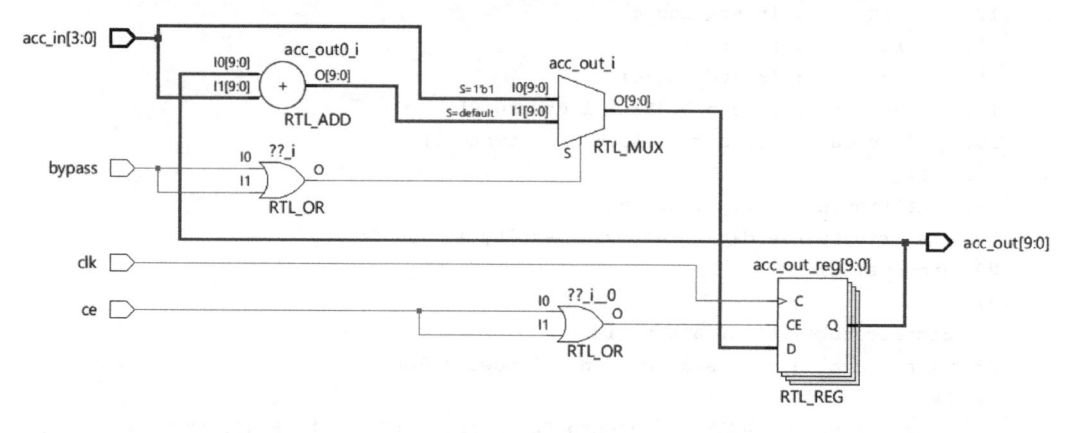

图 3-54

📄 **VHDL 代码 3-52**

```vhdl
1.  --File: acc_v2.vhd
2.  library ieee;
3.  use ieee.std_logic_1164.all;
4.  use ieee.numeric_std.all;
5.
6.  entity acc_v2 is
7.    generic (
8.      DW   : positive := 4;
9.      ACCW : positive := 10
10.   );
11.   port (
12.     clk     : in std_logic;
13.     ce      : in std_logic;
14.     bypass  : in std_logic;
15.     acc_in  : in signed(DW - 1 downto 0);
16.     acc_out : out signed(ACCW - 1 downto 0)
17.   );
18. end entity;
19.
20. architecture rtl of acc_v2 is
21. signal acc_in_ex : signed(ACCW - 1 downto 0);
```

```
22. begin
23.     acc_in_ex <= (ACCW - 1 downto DW => acc_in(DW - 1)) & acc_in;
24.     process(clk)
25.     begin
26.       if rising_edge(clk) then
27.         if ce then
28.           if bypass then
29.             acc_out <= acc_in_ex;
30.           else
31.             acc_out <= acc_out + acc_in_ex;
32.           end if;
33.         end if;
34.       end if;
35.     end process;
36. end architecture;
```

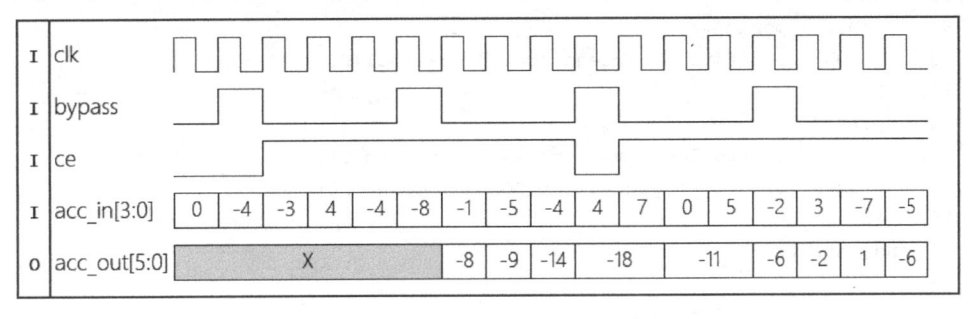

图 3-55

应用案例 5：获取指定位宽的二进制数据中位元为 1 的个数

以 4 位二进制数为例，若其为 4'b0001，则只有一个位元为 1；若其为 4'b1001，则有 2 个位元为 1。不难看出，只需要把 4 个位元相加求和即可得到位元为 1 的个数，其数学表达式为

$$\text{num}_{\text{ones}} = \sum_{i=0}^{\text{DIW}-1} \text{din}[i] = \text{din}[0] + \text{din}[1] + \text{din}[2] + \text{din}[3]$$

式中，din 为指定二进制数，其位宽为 DIW，这里，DIW 为 4。对应的硬件电路结构如图 3-56 所示。

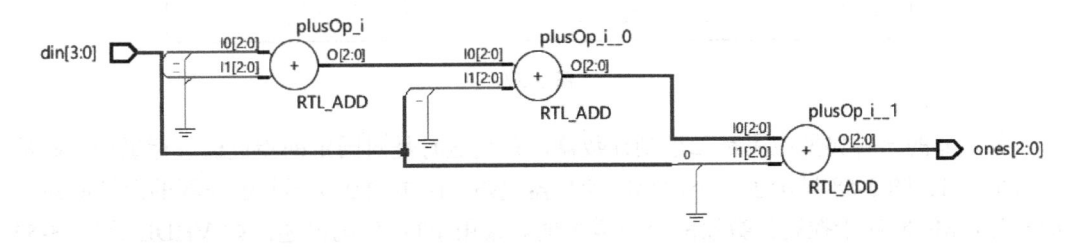

图 3-56

在如图 3-56 所示的电路中，如果采用 RTL 代码描述，更为高效的方式是使用 for 循环，而不是用加号将每个位元连接起来写在一个表达式里，如 VHDL 代码 3-53 所示。注意，代码第 21 行是不可或缺的。此外，还要注意输入数据和输出数据之间的位宽关系，如代码第 10 行所示，其中，DOW 为输出数据位宽。相应的仿真结果如图 3-57 所示。

📄 **VHDL 代码 3-53**

```vhdl
1.  --File: num_ones.vhd
2.  library ieee;
3.  use ieee.std_logic_1164.all;
4.  use ieee.numeric_std.all;
5.  use ieee.math_real.all;
6.
7.  entity num_ones is
8.    generic (
9.      DIW : positive := 4;
10.     DOW : positive := integer(ceil(log2(real(DIW)))) + 1
11.   );
12.   port (
13.     din : in std_logic_vector(DIW - 1 downto 0);
14.     ones: out std_logic_vector(DOW - 1 downto 0)
15.   );
16. end entity;
17.
18. architecture rtl of num_ones is
19. begin
20.   process(all)
21.     variable ones_i : unsigned(DOW - 1 downto 0);
22.   begin
23.     ones_i := (others => '0');
24.     for i in din'RANGE loop
25.       ones_i := unsigned(ones_i) + din(i);
26.     end loop;
27.     ones <= std_logic_vector(ones_i);
28.   end process;
29. end architecture;
```

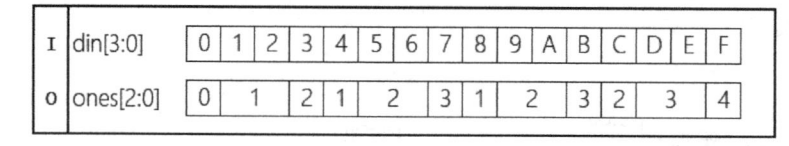

I	din[3:0]	0	1	2	3	4	5	6	7	8	9	A	B	C	D	E	F
O	ones[2:0]	0	1	1	2	1	2	2	3	1	2	2	3	2	3	3	4

图 3-57

累加器的一种典型应用是二进制计数器，只是这时是对固定的步进值进行累加，通常步进值为 1。图 3-58 显示了一个简单的计数器电路，图中的加法器与触发器的反馈回路形成累加，MUX 用于控制计数器的计数最大值。采用 RTL 代码描述，如 VHDL 代码 3-54 所示，相应的仿真结果如图 3-59 所示。

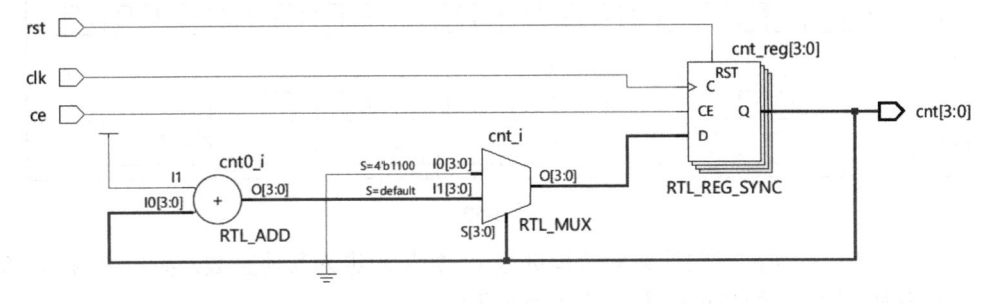

图 3-58

📄 **VHDL 代码 3-54**

```vhdl
1.  --File: counter_v1.vhd
2.  library ieee;
3.  use ieee.std_logic_1164.all;
4.  use ieee.numeric_std.all;
5.
6.  entity counter_v1 is
7.    generic (
8.      W       : positive := 4;
9.      CNT_MAX : positive := 12
10.   );
11.   port (
12.     clk : in std_logic;
13.     rst : in std_logic;
14.     ce  : in std_logic;
15.     cnt : out std_logic_vector(W - 1 downto 0)
16.   );
17. end entity;
18.
19. architecture rtl of counter_v1 is
20. constant CNT_MAX_U : unsigned(W - 1 downto 0) := to_unsigned(CNT_MAX, W);
21. signal cnt_u       : unsigned(W - 1 downto 0);
22. begin
23.   cnt <= std_logic_vector(cnt_u);
24.   process(clk)
25.   begin
26.     if rising_edge(clk) then
27.       if rst then
28.         cnt_u <= (others => '0');
29.       elsif ce then
30.         if cnt_u = CNT_MAX_U then
31.           cnt_u <= (others => '0');
32.         else
33.           cnt_u <= cnt_u + 1;
34.         end if;
35.       end if;
36.     end if;
37.   end process;
38. end architecture;
```

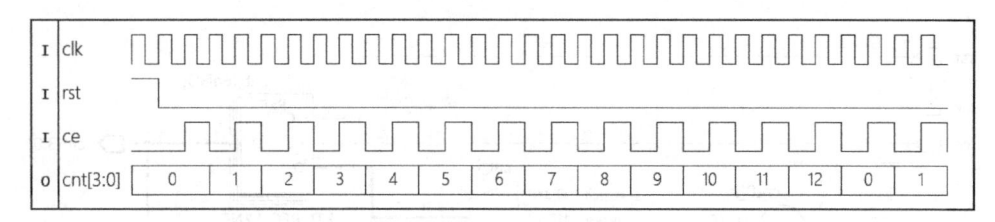

图 3-59

VHDL 代码 3-54 描述的计数器的步进为 1，如果步进为其他整数，如步进为 2，那么只需要将代码第 33 行的步进值由 1 替换为 2 即可。

计数器还可以通过外部输入信号加载新的计数值，使计数器在此值的基础上计数，其电路结构如图 3-60 所示。此时，图 3-60 中的 MUX 在外部信号 load 的控制下用于选择加载外部计数值（load 为 1 时）或当前计数值（load 为 0 时）。采用 RTL 代码描述，如 VHDL 代码 3-55 所示，相应的仿真结果如图 3-61 所示。

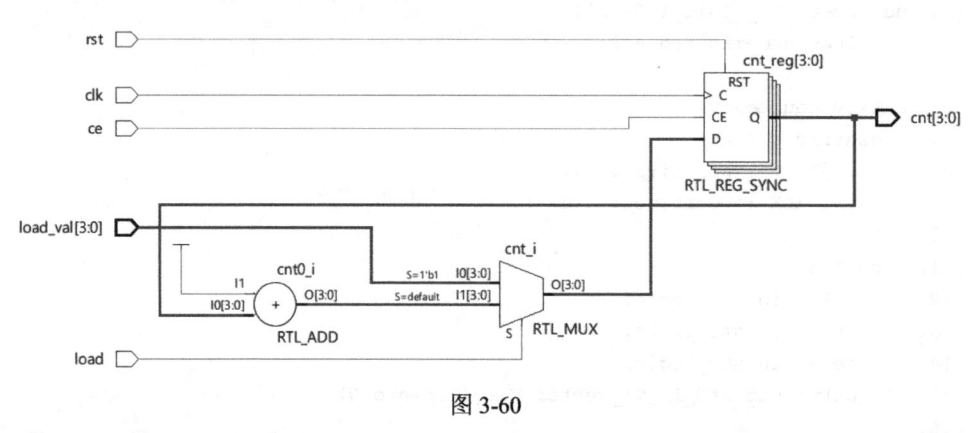

图 3-60

📄 VHDL 代码 3-55

```vhdl
1.  --File: counter_v2.vhd
2.  library ieee;
3.  use ieee.std_logic_1164.all;
4.  use ieee.numeric_std.all;
5.
6.  entity counter_v2 is
7.    generic ( W : positive := 4 );
8.    port (
9.      clk      : in std_logic;
10.     rst      : in std_logic;
11.     ce       : in std_logic;
12.     load     : in std_logic;
13.     load_val : in std_logic_vector(W - 1 downto 0);
14.     cnt      : out std_logic_vector(W - 1 downto 0)
15.   );
16. end entity;
17.
18. architecture rtl of counter_v2 is
19.   signal cnt_u : unsigned(W - 1 downto 0);
20. begin
```

```
21.    cnt <= std_logic_vector(cnt_u);
22.    process(clk)
23.    begin
24.      if rising_edge(clk) then
25.        if rst then
26.          cnt_u <= (others => '0');
27.        elsif ce then
28.          if load then
29.            cnt_u <= unsigned(load_val);
30.          else
31.            cnt_u <= cnt_u + 1;
32.          end if;
33.        end if;
34.      end if;
35.    end process;
36. end architecture;
```

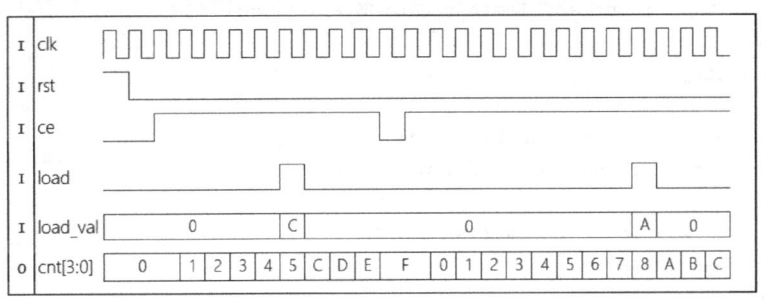

图 3-61

除了递增计数，还可以递减计数，可增加一个外部信号，当其为高电平时，计数器递增计数，否则计数器递减计数，其电路结构如图 3-62 所示。可以看出，这个计数器既可以通过 load 信号加载新值，也可以通过 up_down 信号控制计数方向。该电路包含级联 MUX，意味着存在优先级，可以推测，使用 RTL 代码描述时需要用到 if-elsif-else 语句。但实际上，case?语句也有优先级，会让代码更简洁，因此这里我们用 case?语句描述，相应的 HDL 代码如 VHDL 代码 3-56 所示。

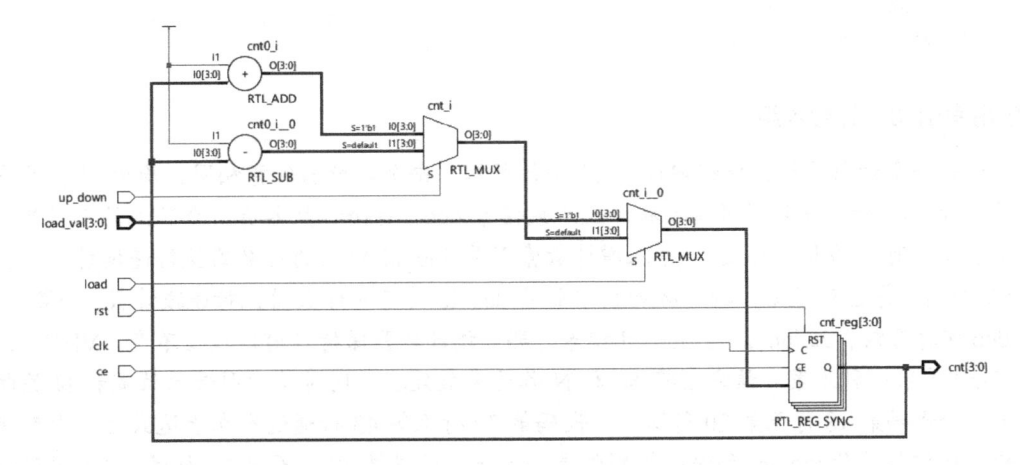

图 3-62

VHDL 代码 3-56

```vhdl
1.  --File: counter_v3.vhd
2.  library ieee;
3.  use ieee.std_logic_1164.all;
4.  use ieee.numeric_std.all;
5.
6.  entity counter_v3 is
7.    generic ( W : positive := 16 );
8.    port (
9.      clk     : in std_logic;
10.     rst     : in std_logic;
11.     ce      : in std_logic;
12.     load    : in std_logic;
13.     load_val: in std_logic_vector(W - 1 downto 0);
14.     up_down : in std_logic;
15.     cnt     : out std_logic_vector(W - 1 downto 0)
16.   );
17. end entity;
18.
19. architecture rtl of counter_v3 is
20. signal cnt_u : unsigned(W - 1 downto 0);
21. begin
22.   cnt <= std_logic_vector(cnt_u);
23.   process(clk)
24.   begin
25.     if rising_edge(clk) then
26.       case? std_logic_vector'(rst, ce, load, up_down) is
27.         when "1---" => cnt_u <= (others => '0');
28.         when "011-" => cnt_u <= unsigned(load_val);
29.         when "0101" => cnt_u <= cnt_u + 1;
30.         when "0100" => cnt_u <= cnt_u - 1;
31.         when others => cnt_u <= cnt_u;
32.       end case?;
33.     end if;
34.   end process;
35. end architecture;
```

应用案例 6：计数器组

这是一个计数器组，该组内包含 N 个计数器，每个计数器位宽相同，均为 W，共享复位信号 rst、计数递增信号 inc 和计数递减信号 dec。每个计数器有一个独立的索引号。当 inc 有效时（高电平有效），输入增计数索引号 inc_id 对应的计数器执行递增计数。当 dec 有效时（高电平有效），输入减计数索引号 dec_id 对应的计数器执行递减计数。如果 inc 和 dec 同时有效，且 inc_id 与 dec_id 相等，那么该计数器保持当前计数值不变。VHDL 代码 3-57 是第一种方案。该方案将 W 和 N 作为参数处理，同时将索引号的位宽和 N 关联起来，如代码第 18 行至第 20 行所示。代码第 39 行至第 43 行通过布尔表达式生成内部使用的递增控制信号 inc_int 和递减控制信号 dec_int。代码第 49 行至第 51 行通过 for 循环对计数器进行复位。代码第 53 行和第 54 行对计数器进行计数操作。但该方案的问题也在于此，若 inc_id 和 dec_id 相等，则第 54 行代码的优先级将高于第 53 行代码的优先级。最终会导致当 inc 为高电平而 dec 为低电平且 inc_id 和 dec_id 相等时，对应的计数器不是递增计数，而是保持不变，这违背了设计要求。

VHDL 代码 3-57

```vhdl
1.  --File: array_counter_v1.vhd
2.  library ieee;
3.  use ieee.std_logic_1164.all;
4.  use ieee.numeric_std.all;
5.
6.  package mypkg is
7.    type myarray is array(natural range <>) of unsigned;
8.  end package;
9.
10. library ieee;
11. use ieee.std_logic_1164.all;
12. use ieee.numeric_std.all;
13. use ieee.math_real.all;
14. use work.mypkg.all;
15.
16. entity array_counter_v1 is
17.   generic (
18.     W   : positive := 6; -- counter width
19.     N   : positive := 8; -- #N counter
20.     IDW : integer  := integer(ceil(log2(real(N)))) -- ID width
21.   );
22.   port (
23.     clk    : in std_logic;
24.     rst    : in std_logic;
25.     inc    : in std_logic;
26.     dec    : in std_logic;
27.     inc_id : in std_logic_vector(IDW - 1 downto 0);
28.     dec_id : in std_logic_vector(IDW - 1 downto 0);
29.     cnt    : out myarray(0 to N - 1)(W - 1 downto 0)
30.   );
31. end entity;
32.
33. architecture rtl of array_counter_v1 is
34. signal inhibit, inc_int, dec_int : std_logic;
35. signal inc_id_i : integer;
36. signal dec_id_i : integer;
37. begin
38.
39.   inc_id_i <= to_integer(unsigned(inc_id));
40.   dec_id_i <= to_integer(unsigned(dec_id));
41.   inhibit  <= inc and dec and (inc_id ?= dec_id);
42.   inc_int  <= inc and (not inhibit);
43.   dec_int  <= dec and (not inhibit);
44.
45.   process(clk)
```

```
46.   begin
47.     if rising_edge(clk) then
48.       if rst then
49.         for i in cnt'RANGE loop
50.           cnt(i) <= (others => '0');
51.         end loop;
52.       else
53.         cnt(inc_id_i) <= cnt(inc_id_i) + inc_int;
54.         cnt(dec_id_i) <= cnt(dec_id_i) - dec_int;
55.       end if;
56.     end if;
57.   end process;
58. end architecture;
```

对第一种方案的代码的第 53 行和第 54 行进行微小改动，即可实现设计要求，改动后的结果如 VHDL 代码 3-58 所示（这里只给出结构体部分），代码的其余部分保持不变。这是第一种方案的优化版本。可以看到，当 inc 为 1 而 dec 为 0 且 inc_id 和 dec_id 均为 0 时，0 号计数器执行递增计数；当 inc 为 0 而 dec 为 1 且 inc_id 和 dec_id 均为 1 时，1 号计数器执行递减计数。从代码风格的角度看，代码第 23 行 cnt(inc_id)和第 26 行 cnt(dec_id)的写法实际上对应的是二进制码 MUX，如图 3-63 所示。这也意味着 MUX 之后的加法器/减法器是共享的。

📄 VHDL 代码 3-58

```
1.  --File: array_counter_v1_opt.vhd
2.  architecture rtl of array_counter_v1_opt is
3.  signal inhibit, inc_int, dec_int : std_logic;
4.  signal inc_id_i : integer;
5.  signal dec_id_i : integer;
6.  begin
7.
8.    inc_id_i <= to_integer(unsigned(inc_id));
9.    dec_id_i <= to_integer(unsigned(dec_id));
10.   inhibit <= inc and dec and (inc_id ?= dec_id);
11.   inc_int <= inc and (not inhibit);
12.   dec_int <= dec and (not inhibit);
13.
14.   process(clk)
15.   begin
16.     if rising_edge(clk) then
17.       if rst then
18.         for i in cnt'RANGE loop
19.           cnt(i) <= (others => '0');
20.         end loop;
21.       else
22.         if inc_int then
23.           cnt(inc_id_i) <= cnt(inc_id_i) + 1;
24.         end if;
25.         if dec_int then
26.           cnt(dec_id_i) <= cnt(dec_id_i) - 1;
27.         end if;
```

```
28.        end if;
29.      end if;
30.    end process;
31. end architecture;
```

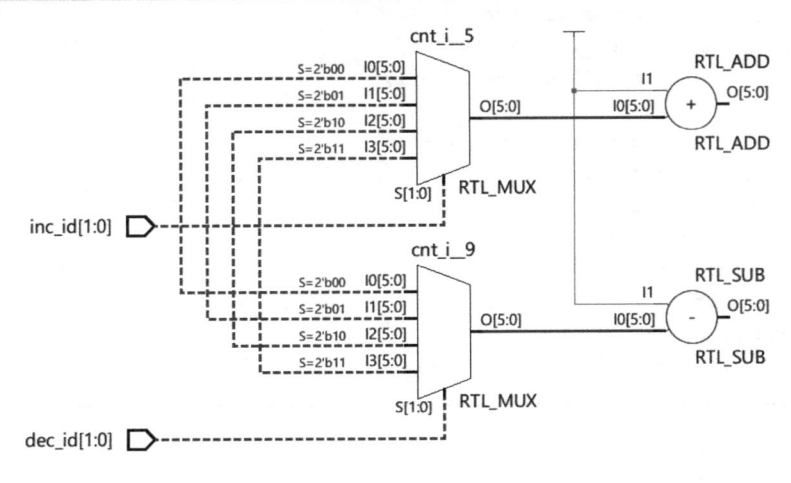

图 3-63

对第一种方案的优化版本进一步改进，形成第二种方案，如 VHDL 代码 3-59 所示。代码第 7 行和第 8 行完成了从二进制码到独热码的转换，同时生成 inc_dec 信号，用于指示需要执行递增计数或递减计数的计数器（相应位为 1）。代码第 19 行至第 27 行通过 for 循环完成计数器组的计数功能。相比第一种方案的优化版本，该方案采用了独热码 MUX，但其不足之处是会生成 N 个加法器和 N 个减法器，如图 3-64 所示（将 N 设置为 4），这源于代码第 19 行至第 27 行的描述方式。该电路结构非常清晰，独热码 MUX 用于从加法器和减法器输出结果中进行选择，从而实现递增计数和递减计数。异或门的输出结果作为触发器的时钟使能信号，这样当使能信号无效时，触发器的输出值保持不变，这其实对应的就是计数器在 inc 无效或 dec 无效或 inc_id 与 dec_id 相等且 inc 与 dec 相等时，计数器输出值保持不变。

📄 VHDL 代码 3-59

```vhdl
1.  --File: array_counter_v2.vhd
2.  architecture rtl of array_counter_v2 is
3.  signal inc_int : std_logic_vector(N - 1 downto 0);
4.  signal dec_int : std_logic_vector(N - 1 downto 0);
5.  signal inc_dec : std_logic_vector(N - 1 downto 0);
6.  begin
7.    inc_int <= ((N - 1 downto 1 => '0') & inc) sll (to_integer(unsigned(inc_id)));
8.    dec_int <= ((N - 1 downto 1 => '0') & dec) sll (to_integer(unsigned(dec_id)));
9.    inc_dec <= inc_int xor dec_int;
10.
11.   process(clk)
12.   begin
13.     if rising_edge(clk) then
14.       if rst then
15.         for i in cnt'RANGE loop
```

```
16.            cnt(i) <= (others => '0');
17.          end loop;
18.        else
19.          for i in cnt'RANGE loop
20.            if inc_dec(i) then
21.              if inc_int(i) then
22.                cnt(i) <= cnt(i) + 1;
23.              else
24.                cnt(i) <= cnt(i) - 1;
25.              end if;
26.            end if;
27.          end loop;
28.        end if;
29.      end if;
30.    end process;
31. end architecture;
```

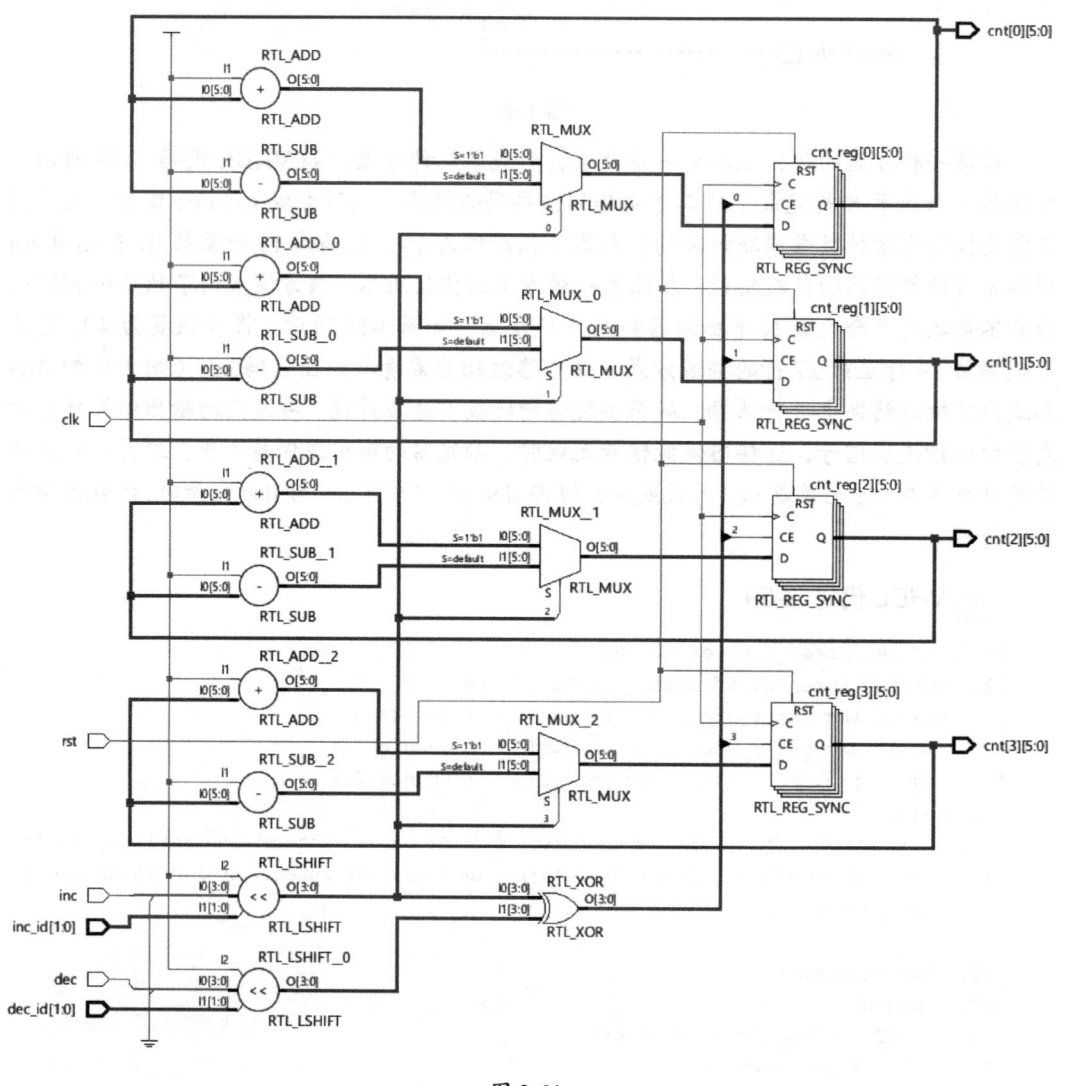

图 3-64

对第二种方案进一步改进，以降低加法器和减法器的个数，形成第三种方案，如 VHDL 代码 3-60 所示，在第二种方案的基础上新增了代码第 17 行和第 18 行的内容，同时对 for 循环进行调整，如代码第 28 行至第 33 行所示。但该方案没有独热码 MUX 的优势，这是因为代码第 17 行和第 18 行的描述方式，该方式将会映射为二进制码 MUX，如图 3-65 所示。代码第 30 行仍然为独热码 MUX。

VHDL 代码 3-60

```
1.  --File: array_counter_v3.vhd
2.  architecture rtl of array_counter_v3 is
3.    signal inc_id_int : integer range 0 to N - 1;
4.    signal dec_id_int : integer range 0 to N - 1;
5.    signal inc_int    : unsigned(N - 1 downto 0);
6.    signal dec_int    : unsigned(N - 1 downto 0);
7.    signal inc_dec    : unsigned(N - 1 downto 0);
8.    signal cnt_plus   : unsigned(W - 1 downto 0);
9.    signal cnt_minus  : unsigned(W - 1 downto 0);
10. begin
11.
12.   inc_id_int <= to_integer(unsigned(inc_id));
13.   dec_id_int <= to_integer(unsigned(dec_id));
14.   inc_int    <= ((N - 1 downto 1 => '0') & inc) sll inc_id_int;
15.   dec_int    <= ((N - 1 downto 1 => '0') & dec) sll dec_id_int;
16.   inc_dec    <= inc_int xor dec_int;
17.   cnt_plus   <= cnt(inc_id_int) + 1;
18.   cnt_minus  <= cnt(dec_id_int) + 1;
19.
20.   process(clk)
21.   begin
22.     if rising_edge(clk) then
23.       if rst then
24.         for i in cnt'RANGE loop
25.           cnt(i) <= (others => '0');
26.         end loop;
27.       else
28.         for i in cnt'RANGE loop
29.           if inc_dec(i) then
30.             cnt(i) <= cnt_plus when inc_int(i) else
31.                       cnt_minus;
32.           end if;
33.         end loop;
34.       end if;
35.     end if;
36.   end process;
37. end architecture;
```

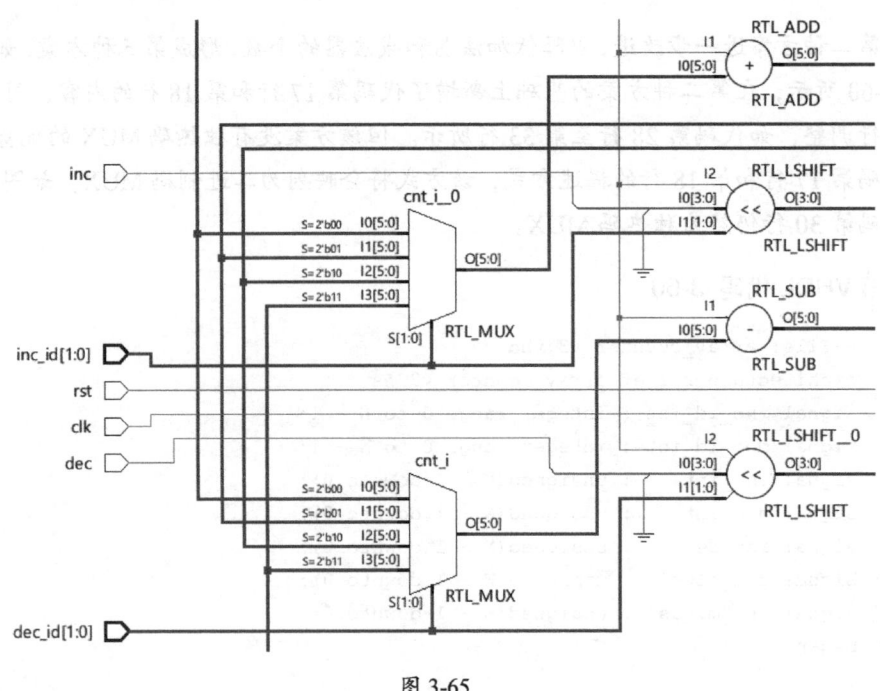

图 3-65

> 💡 **设计规则 8**：在可综合的 VHDL 代码中使用 integer 时，如 VHDL 代码 3-60 第 3 行和第 4 行所示，需要声明 integer 的取值范围（通过 range 实现）。本质上，integer 是 32 位 signed 类型。如果没有声明其取值范围，那么在综合时，工具可能会按 32 位处理，从而影响综合质量。在仿真时，工具不会自动推断出其取值范围，从而会导致仿真无法进行。

对第一种方案的优化版本、第二种方案和第三种方案进行比较，如表 3-15 所示，其中，目标芯片为 xcvu3p-ffvc1517-2-i，时钟频率为 500MHz，Vivado 版本为 2021.2。不难发现，对于 W=6 且 N=4 的情形，第二种方案无论是在资源上还是在 Fmax 上都最具有优势，紧随其后的是第一种方案的优化版本，而第三种方案在三者中最为糟糕。尽管第二种方案使用了 N 个加法器和 N 个减法器，但独热码的优势仍然更为明显。

表 3-15

位宽 W	计数器个数 N	版 本	特 征	WNS	WHS	最大逻辑级数	LUT	FF	F7MUX	CARRY8
6	4	array_counter_v1_opt	二进制码 MUX 1 个加法器 1 个减法器	1.235	0.097	3	45	24	0	0
		array_counter_v2	独热码 MUX N 个加法器 N 个减法器	1.262	0.044	2	28	24	0	0
		array_counter_v3	独热码 MUX 二进制码 MUX 1 个加法器 1 个减法器	1.186	0.073	3	48	24	0	0

续表

位宽 W	计数器个数 N	版本	特征	WNS	WHS	最大逻辑级数	LUT	FF	F7MUX	CARRY8
6	8	array_counter_v1_opt	二进制码 MUX 1 个加法器 1 个减法器	0.768	0.108	4	77	64	12	0
		array_counter_v2	独热码 MUX N 个加法器 N 个减法器	1.210	0.043	2	60	48	0	0
		array_counter_v3	独热码 MUX 二进制码 MUX 1 个加法器 1 个减法器	0.792	0.160	4	90	80	12	0
12	8	array_counter_v1_opt	二进制码 MUX 1 个加法器 1 个减法器	0.498	0.104	5	154	112	24	0
		array_counter_v2	独热码 MUX N 个加法器 N 个减法器	1.743	0.061	3	108	96	0	16
		array_counter_v3	独热码 MUX 二进制码 MUX 1 个加法器 1 个减法器	0.567	0.102	5	171	168	24	0

3.5 其他组合逻辑电路

3.5.1 移位器代码风格

移位器需要用到移位操作符，VHDL-2008 提供了 6 种移位操作符，如表 3-16 所示。总体来说，分为两大类：逻辑移位操作符和算术移位操作符。对于左移而言，逻辑左移和算术左移是一致的。对于右移而言，当数据为有符号数时，算术右移需要对空余位填补数据的符号位，除此之外，与逻辑右移保持一致。

表 3-16

操作符	使用方法	含义	左操作数	右操作数	结果	案例	备注
sll	y = a sll b	将 a 逻辑左移 b 位	std_logic_vector	整数	和 a 同类型	a sll 1	a(2 downto 0) & '0'
srl	y = a srl b	将 a 逻辑右移 b 位	std_logic_vector	整数	和 a 同类型	a srl 1	'0' & a(3 downto 1)
sla	y = a sla b	将 a 算术左移 b 位	unsigned / signed	整数	和 a 同类型	a sla 1(a 为 unsigned)	a(2 downto 0) & '0'
						a sla 1(a 为 signed)	a(2 downto 0) & '0'

续表

操作符	使用方法	含义	左操作数	右操作数	结果	案例	备注
sra	y = a sra b	将a算术右移b位	unsigned / signed	整数	和a同类型	a sra 1(a 为 unsigned)	'0' & a(3 downto 1)
						a sra 1(a 为 signed)	a(3) & a(3 downto 1)
rol	y = a rol b	将a循环左移b位	std_logic_vector	整数	和a同类型	a rol 1	a(2 downto 0) & a(3)
ror	y = a ror b	将a循环右移b位	std_logic_vector	整数	和a同类型	a ror 1	a(0) & a(3 downto 1)

常规移位器可借助移位操作符实现。若移位量是固定常数，则相应的移位器就是静态移位器，这种移位器不会消耗任何逻辑资源。若移位量是变量，则相应的移位器就是动态移位器，如 VHDL 代码 3-61 所示。如果输入数据位宽为 8，那么移位量最大值为 7，输出数据位宽为 15，三者的关系如代码第 9 行至第 11 行所示。当 n 为 7 时，输入位字段与输出 y 左对齐，当 n 为 0 时，输入位字段与输出 y 右对齐。仿真结果如图 3-66 所示。不同于静态移位器，该电路会消耗 18 个 LUT。

📄 VHDL 代码 3-61

```vhdl
1.  --File: myshift_left.vhd
2.  library ieee;
3.  use ieee.std_logic_1164.all;
4.  use ieee.numeric_std.all;
5.  use ieee.math_real.all;
6.
7.  entity myshift_left is
8.    generic (
9.      DIW : positive := 8;
10.     SW  : positive := integer(ceil(log2(real(DIW))));
11.     DOW : positive := 2 * DIW - 1
12.   );
13.   port (
14.     a : in  std_logic_vector(DIW - 1 downto 0);
15.     n : in  std_logic_vector(SW  - 1 downto 0);
16.     y : out std_logic_vector(DOW - 1 downto 0)
17.   );
18. end entity;
19.
20. architecture rtl of myshift_left is
21. signal a_ex  : unsigned(DOW - 1 downto 0);
22. signal shift : integer;
23. begin
24.   shift <= to_integer(unsigned(n));
25.   a_ex  <= unsigned((DOW - 1 downto DIW => '0') & a);
26.   y     <= std_logic_vector(a_ex sll shift);
27. end architecture;
```

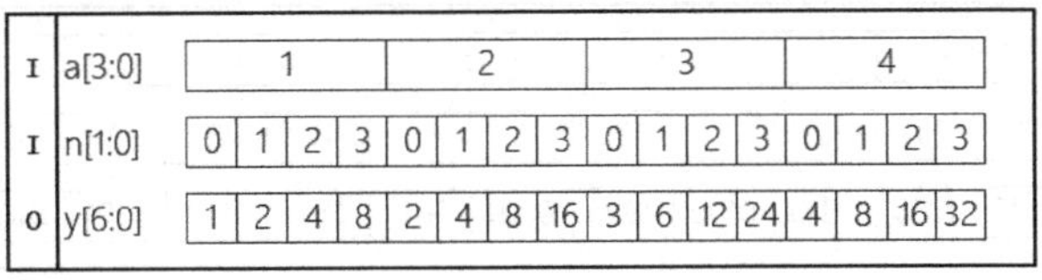

I	a[3:0]			1				2				3				4	
I	n[1:0]	0	1	2	3	0	1	2	3	0	1	2	3	0	1	2	3
o	y[6:0]	1	2	4	8	2	4	8	16	3	6	12	24	4	8	16	32

图 3-66

除此之外，还有一种所谓的桶形移位器，其工作原理是将数据左移，然后将移出的数据位填补到最低位，就像所有的数据位在一个圆上循环。以 4'b1000 为例，将其输入桶形移位器，其输出结果如表 3-17 所示。

表 3-17

原 始 数 据	桶形移位器输出结果		
	1 位	2 位	3 位
4'b1000	0001	0010	0100

采用 RTL 代码描述桶形移位器，如 VHDL 代码 3-62 所示，其仿真结果如图 3-67 所示，图中，a 和 y 以二进制形式显示，n 以十进制形式显示。

📄 VHDL 代码 3-62

```vhdl
1.  --File: barrel_shift.vhd
2.  library ieee;
3.  use ieee.std_logic_1164.all;
4.  use ieee.numeric_std.all;
5.  use ieee.math_real.all;
6.
7.  entity barrel_shift is
8.    generic (
9.      DW : positive := 8; -- data width
10.     SW : positive := integer(ceil(log2(real(DW))))
11.   );
12.   port (
13.     a : in std_logic_vector(DW - 1 downto 0);
14.     n : in std_logic_vector(SW - 1 downto 0);
15.     y : out std_logic_vector(DW - 1 downto 0)
16.   );
17. end entity;
18.
19. architecture rtl of barrel_shift is
20. signal apad : unsigned(2 * DW - 2 downto 0);
21. signal shift_amt : integer;
22. signal shift_out : std_logic_vector(2 * DW - 2 downto 0);
23.
24. begin
25.   apad <= unsigned( (2 * DW - 2 downto DW => '0') & a);
26.   shift_amt <= to_integer(unsigned(n));
27.   shift_out <= std_logic_vector(apad sll shift_amt);
28.   y <= shift_out(DW - 1 downto 0) or ('0' & shift_out(2 * DW - 2 downto DW));
29. end architecture;
```

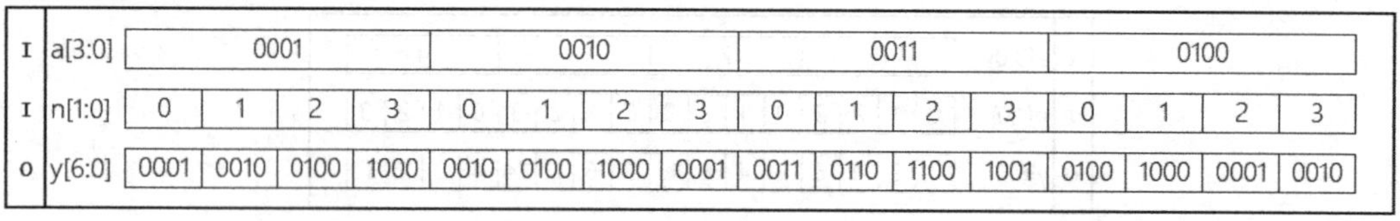

I	a[3:0]	0001				0010				0011				0100			
I	n[1:0]	0	1	2	3	0	1	2	3	0	1	2	3	0	1	2	3
O	y[6:0]	0001	0010	0100	1000	0010	0100	1000	0001	0011	0110	1100	1001	0100	1000	0001	0010

图 3-67

移位操作符还可用于实现移位寄存器，移位寄存器的具体功能可查看第 4 章。这里给出另一种描述方式。先实现一个输入为 1 位的移位寄存器，如 VHDL 代码 3-63 所示。然后，在此基础上通过 for generate 语句描述一个位宽可用参数设定的移位寄存器，如 VHDL 代码 3-64 所示。

📄 **VHDL 代码 3-63**

```vhdl
1.  --File: single_bit_sreg.vhd
2.  library ieee;
3.  use ieee.std_logic_1164.all;
4.  use ieee.numeric_std.all;
5.
6.  entity single_bit_sreg is
7.    generic ( DEPTH : positive := 4 );
8.    port (
9.      clk : in std_logic;
10.     ce  : in std_logic;
11.     sin : in std_logic;
12.     sout: out std_logic
13.   );
14. end entity;
15.
16. architecture rtl of single_bit_sreg is
17.   signal sin_dly : std_logic_vector(DEPTH - 1 downto 0);
18. begin
19.   sout <= sin_dly(DEPTH - 1);
20.   process(clk)
21.   begin
22.     if rising_edge(clk) then
23.       if ce then
24.         sin_dly <= (sin_dly sll 1);
25.         sin_dly(0) <= sin;
26.       end if;
27.     end if;
28.   end process;
29. end architecture;
```

VHDL 代码 3-63 第 24 行和第 25 行都会给 sin_dly 的最低位也就是 sin_dly(0) 赋值，这意味着在同一个 process 中给同一信号多次赋值，那么到底哪个赋值会最终生效呢？答案是 process 最后一条赋值语句的优先级最高，最终生效。

📄 **VHDL 代码 3-64**

```vhdl
1.  --File: multi_bit_sreg.vhd
2.  library ieee;
```

```vhdl
3.  use ieee.std_logic_1164.all;
4.
5.  entity multi_bit_sreg is
6.    generic (
7.      DEPTH : natural  := 32;
8.      DW    : positive := 8
9.    );
10.   port (
11.     clk : in std_logic;
12.     ce  : in std_logic;
13.     din : in std_logic_vector(DW - 1 downto 0);
14.     dout : out std_logic_vector(DW - 1 downto 0)
15.   );
16. end entity;
17.
18. architecture rtl of multi_bit_sreg is
19. begin
20.   gen0:
21.   for i in 0 to DW - 1 generate
22.     i_single_bit_sreg : entity work.single_bit_sreg
23.       generic map ( DEPTH => DEPTH )
24.       port map (
25.         clk  => clk,
26.         ce   => ce,
27.         sin  => din(i),
28.         sout => dout(i)
29.       );
30.   end generate;
31. end architecture;
```

3.5.2 比较器代码风格

比较器可借助关系运算符完成。表 3-18 列出了 Vivado 支持的可综合的 VHDL 关系运算表达式，其中，灰色部分是 VHDL-2008 新引入的。对于可综合的关系运算表达式，要求两个操作数均为同一数据类型。

表 3-18

操作符	含义	操作数数据类型	结果数据类型
=	相等	除文件类型和保护类型（Protected Type）外的其他类型	布尔类型
/=	不相等		布尔类型
<	小于	任意标量类型或数组类型	布尔类型
<=	小于或等于	任意标量类型或数组类型	布尔类型
>	大于	任意标量类型或数组类型	布尔类型
>=	大于或等于	任意标量类型或数组类型	布尔类型
?=	匹配相等	bit 或 std_ulogic，也可以是 bit_vector 或 std_ulogic_vector	若为标量，则与标量类型相同；若为向量，则与向量元素类型相同，即若操作数为 bit_vector，那么结果的数据类型为 bit；若操作数为 std_ulogic_vector，那么结果的数据类型为 std_ulogic
?/=	匹配不相等		

续表

操作符	含义	操作数数据类型	结果数据类型
?<	匹配小于	bit 或 std_ulogic	与操作数同类型
?<=	匹配小于或等于	bit 或 std_ulogic	与操作数同类型
?>	匹配大于	bit 或 std_ulogic	与操作数同类型
?>=	匹配大于或等于	bit 或 std_ulogic	与操作数同类型

这里需要注意的是对于匹配关系运算，只有?=和?/=支持操作数数据类型为向量类型，其余的都必须为 bit 或 std_ulogic（位宽为 1）。std_logic 是 std_ulogic 的子类型（subtype），所以表 3-18 中的所有操作符对于 std_logic 类型都是支持的。

本质上，比较器是逐位进行比较的。下面以两个 4 位无符号数为例，比较两个数是否相等，需要 4 个同或门和 1 个与门，如图 3-68 所示。图中，xnor 表示同或运算，and 表示与运算。

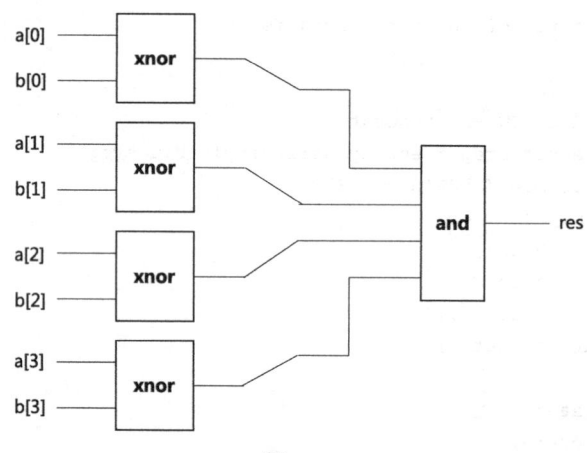

图 3-68

采用 RTL 代码描述时，可直接从行为级的角度描述，如 VHDL 代码 3-65 所示。若 a>b，则 cgt 为 1，否则 cgt 为 0；若 a<b，则 clt 为 1，否则 cgt 为 0；若 a=b，则 ceq 为 1，否则 cgt 为 0。

📄 VHDL 代码 3-65

```
1.  --File: cmp_v1.vhd
2.  library ieee;
3.  use ieee.std_logic_1164.all;
4.
5.  entity cmp_v1 is
6.    generic ( W : positive := 4 );
7.    port (
8.      a, b : in std_logic_vector(W - 1 downto 0);
9.      cgt, clt, ceq : out std_logic
10.   );
11. end entity;
12.
13. architecture rtl of cmp_v1 is
14. begin
15.   cgt <= '1' when a > b else '0';
```

```
16.    clt <= '1' when a < b else '0';
17.    ceq <= '1' when a = b else '0';
18. end architecture;
```

VHDL 代码 3-65 综合后的电路如图 3-69 所示。共消耗了 6 个 LUT，其中，每个输出对应 2 个 LUT。还可发现每个 LUT6 的 6 个输入端口中有 3 个来自 a、有 3 个来自 b，这在本质上体现了逐位比较的特征。但当数据位宽较大时，综合工具会将比较运算转换为减法运算，如 a>b 等效于 b-a<0，以 sub 表示 b-a 的结果，那么当 sub 最高位（符号位）为 1 时，表明 sub 为负数，也就是 b-a<0，因此只需要观察 sub 的最高位即可判定 a>b 是否成立。类似地，a>=b 等效于 a-b>=0，则只需要看 a 与 b 的差的最高位即可判定 a 是否大于或等于 b。因此，并不需要手工进行这种转换。这也就不难解释对于两个 18 位无符号数 a 和 b，判定 a>b 是否成立，除了需要消耗 9 个 LUT，还会消耗 2 个进位链（CARRY8，目标芯片为 UltraScale+ FPGA）。

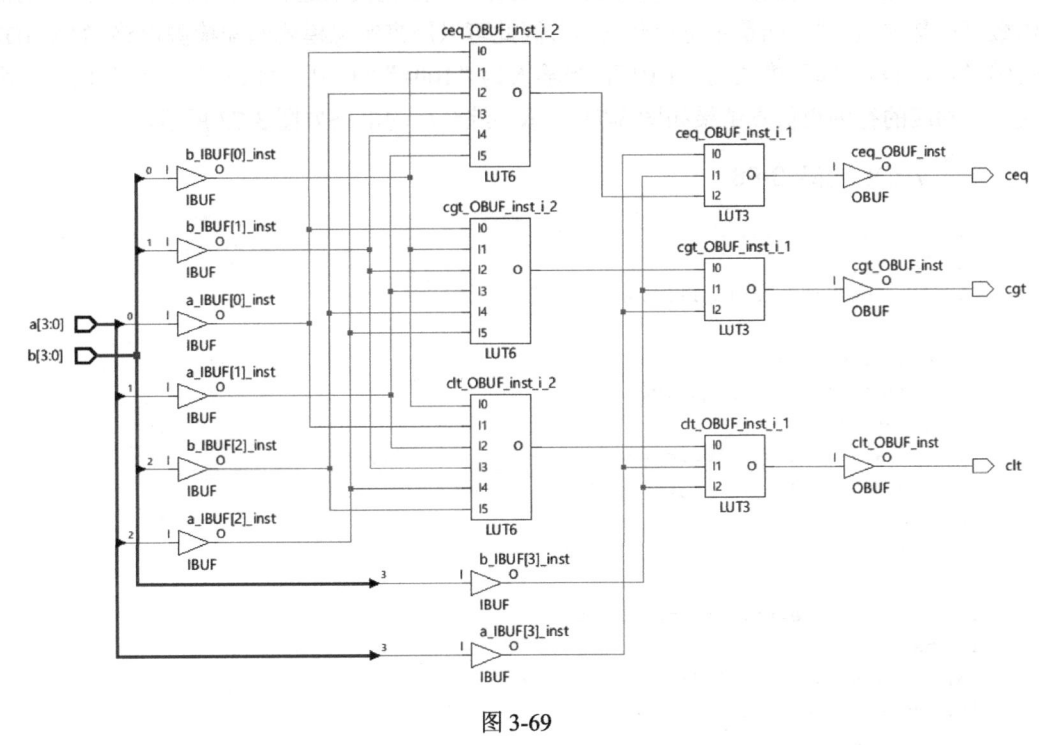

图 3-69

VHDL 代码 3-65 中的两个操作数的位宽相等，根据表 3-18 可知，对于关系运算符（不含匹配关系运算符），并不要求操作数的位宽相等。那么当两个操作数的位宽不相等时，会出现什么情形呢？若两个操作数均为 unsigned 类型，则工具会自动对位宽较小者进行高位补零，以满足两个操作数的位宽相等。若两个操作数均为 signed 类型，则工具会自动对位宽较小者进行符号位扩展，即复制最高位填补到高位字段，以满足两个操作数的位宽相等。若两个数均为 std_logic_vector 类型，则要小心，工具可能会给出错误结果。例如，a 为 4 位，b 为 2 位，Elaborated Design 视图如图 3-70 所示，可以看到，ceq 恒接地，这显然不合理。

图 3-70

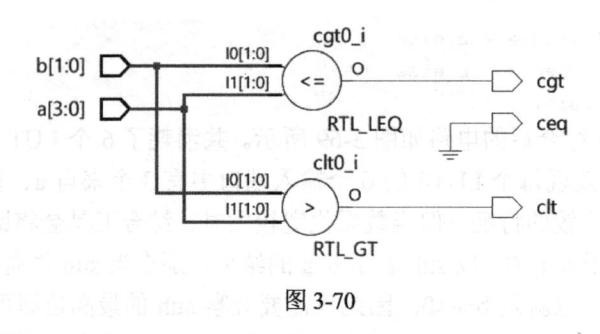

> **设计规则 9**：使用关系运算符时，应尽可能保证两个操作数的位宽相等，或者使用 unsigned 或 signed 数据类型。

除了常见的关系运算符，VHDL-2008 还引入了匹配关系运算符，其中，?=和?/=的操作数可以是向量，同时向量中可以包含"-"，因此可以用来实现模式匹配检测电路，如 VHDL 代码 3-66 所示。匹配模式为"1-01"，当输入为"1001""1101""1x01""1z01"时，匹配成功，相应的行为级仿真结果如图 3-71 所示，综合后的电路如图 3-72 所示。

VHDL 代码 3-66

```
1.  --File: cmp_v2.vhd
2.  library ieee;
3.  use ieee.std_logic_1164.all;
4.
5.  entity cmp_v2 is
6.    generic ( W : positive := 4 );
7.    port (
8.      a   : in std_logic_vector(W - 1 downto 0);
9.      res : out std_logic
10.   );
11. end entity;
12.
13. architecture rtl of cmp_v2 is
14. begin
15.   res <= a ?= "1-01";
16. end architecture;
```

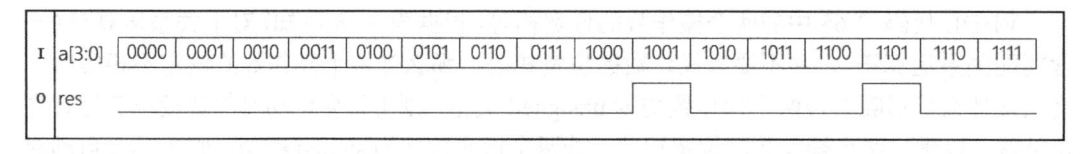

图 3-71

观察如图 3-72 所示的电路，从 LUT3 对应的布尔表达式不难看出，这里，所谓模式匹配，就是要检测最高位和最低位均为 1 且次低位为 0，当三者同时成立时，模式匹配成功。

图 3-72

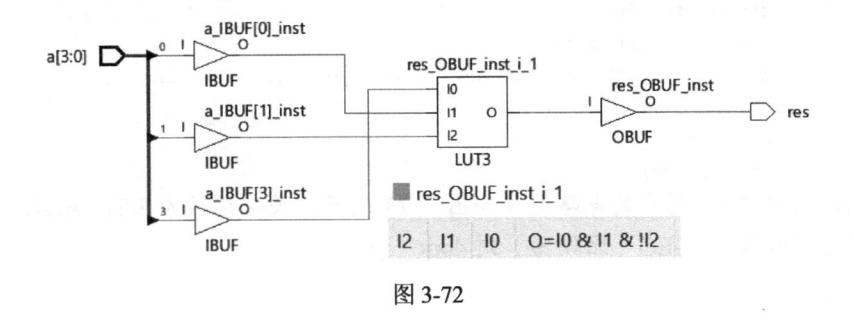

应用案例 7：优化代码中的比较器

目标芯片：xcvc1902-vsva2197-1LP-i-S，时钟频率：500MHz。

VHDL 代码 3-67 描述的是一个包含比较器的电路。该代码的核心部分由两部分构成：代码第 22 行至第 25 行是一个 with-select 语句描述的 MUX，实际上是通过 sel_d1 控制计数器的行为的。代码第 26 至第 34 行是对输入信号、中间生成信号和输出信号进行寄存，其中，输出由比较器判定。

📄 VHDL 代码 3-67

```vhdl
1.  --File: cmp_opt_before.vhd
2.  library ieee;
3.  use ieee.std_logic_1164.all;
4.  use ieee.numeric_std.all;
5.
6.  entity cmp_opt_before is
7.    generic ( W : positive := 10 );
8.    port (
9.      clk  : in std_logic;
10.     sel  : in std_logic_vector(1 downto 0);
11.     din  : in std_logic_vector(W - 1 downto 0);
12.     dout : out std_logic
13.   );
14. end entity;
15.
16. architecture rtl of cmp_opt_before is
17. signal sel_d1 : std_logic_vector(1 downto 0) := "00";
18. signal din_d1 : std_logic_vector(W - 1 downto 0) := (others => '0');
19. signal cnt     : unsigned(w - 1 downto 0) := (others => '0');
20. signal cnt_nxt : unsigned(W - 1 downto 0) := (others => '0');
21. begin
22.   with sel_d1 select
23.     cnt_nxt <= cnt + 1 when "10",
24.                cnt - 1 when "01",
25.                cnt     when others;
26.   process(clk)
27.   begin
28.     if rising_edge(clk) then
29.       sel_d1 <= sel;
```

```
30.      din_d1 <= din;
31.      cnt    <= cnt_nxt;
32.      dout   <= '1' when cnt_nxt >= unsigned(din_d1) else '0';
33.    end if;
34.  end process;
35. end architecture;
```

VHDL 代码 3-67 对应的电路结构如图 3-73 所示。关键路径为 RTL_ADD/RTL_SUB 经 RTL_MUX 和 RTL_GEQ（比较器）到 dout_reg。

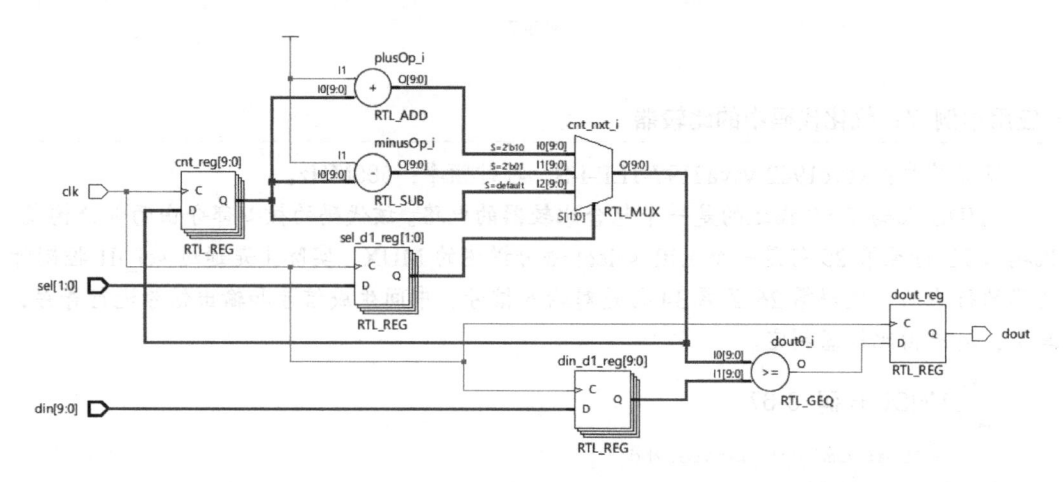

图 3-73

在此基础上，我们对 VHDL 代码 3-67 进行优化，优化后的代码如 VHDL 代码 3-68 所示。与原始代码相比，增加了第 16 至第 22 行，同时，将比较器变换为减法器并放置在 MUX 之前完成，如代码第 33 至第 36 行所示。原始代码要判定 cnt_nxt 是否大于或等于 din_d1，若成立，则输出 1，否则输出 0。换为减法运算，即判断 cnt_nxt 与 din_d1 的差，而 cnt_nxt 的值由 sel_d1 确定，分三种情形，因此其与 din_d1 的差也分三种情形，从而构成了 3 个并行的减法器。为了确保减法运算不溢出，事先对操作数 sum/sub 和 din_d1 进行位扩展，由 10 位变为 11 位。但从减法运算的角度看，当 cnt_nxt 大于或等于 din_d1 时，其差值的最高位也就是符号位为 0，而不是期望的 1，这就需要对差值的最高位再进行取反。这个取反操作可以放在 sum/sub 的位扩展时完成，即对其最高位补 1，而不是补 0。这从理论上解释了为什么这种优化并没有改变电路的整体功能。优化后电路结构的核心部分如图 3-74 所示。

VHDL 代码 3-68

```
1.  --File: cmp_opt_after.vhd
2.  library ieee;
3.  use ieee.std_logic_1164.all;
4.  use ieee.numeric_std.all;
5.
6.  entity cmp_opt_after is
7.    generic (W : positive := 10 );
8.    port (
9.      clk  : in std_logic;
10.     sel  : in std_logic_vector(1 downto 0);
11.     din  : in std_logic_vector(W - 1 downto 0);
```

```vhdl
12.         dout : out std_logic
13.     );
14. end entity;
15.
16. architecture rtl of cmp_opt_after is
17. signal sel_d1    : std_logic_vector(1 downto 0) := "00";
18. signal din_d1    : unsigned(W - 1 downto 0) := (others => '0');
19. signal cnt       : unsigned(W - 1 downto 0) := (others => '0');
20. signal cnt_nxt   : unsigned(W - 1 downto 0) := (others => '0');
21. signal sum       : unsigned(W - 1 downto 0);
22. signal sub       : unsigned(W - 1 downto 0);
23. signal dout_nxt  : unsigned(W downto 0);
24. begin
25.     with sel_d1 select
26.         cnt_nxt <= cnt + 1 when "10",
27.                    cnt - 1 when "01",
28.                    cnt     when others;
29.
30.     sum <= cnt + 1;
31.     sub <= cnt - 1;
32.
33.     with sel_d1 select
34.         dout_nxt <= ('1' & sum) - ('0' & din_d1) when "10",
35.                     ('1' & sub) - ('0' & din_d1) when "01",
36.                     ('1' & cnt) - ('0' & din_d1) when others;
37.
38.     process(clk)
39.     begin
40.         if rising_edge(clk) then
41.             sel_d1 <= sel;
42.             din_d1 <= unsigned(din);
43.             cnt    <= cnt_nxt;
44.             dout   <= dout_nxt(W);
45.         end if;
46.     end process;
47. end architecture;
```

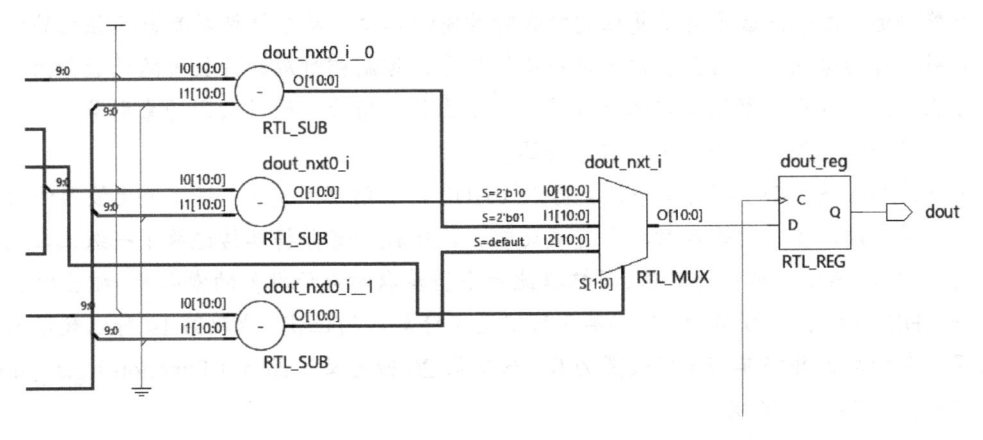

图 3-74

对比优化前后的性能（使用 Vivado 2021.1），从时序的角度看，WNS 由 0.333 提升到 0.667，如图 3-75 所示。这得益于逻辑级数的降低，如图 3-76 所示，尽管最高逻辑级数均为 4，但优化后的逻辑级数整体低于优化前的逻辑级数。从资源利用率的角度看，优化前后没有明显变化，如图 3-77 所示。

Name	Constraints	WNS	TNS	WHS	THS	TPWS	Total Power	LUT	FF
✓ synth_2	constrs_1							34	23
✓ impl_2	constrs_1	0.333	0.000	0.055	0.000	0.000	0.839	32	23

Name	Constraints	WNS	TNS	WHS	THS	TPWS	Total Power	LUT	FF
✓ synth_2	constrs_1							39	23
✓ impl_2	constrs_1	0.667	0.000	0.052	0.000	0.000	0.837	34	23

图 3-75

Logic Level Distribution

General Information
Logic Level Distribution

End Point Clock	Requirement	1	2	3	4
clk	2.000ns	1	8	1	1

General Information
Logic Level Distribution

End Point Clock	Requirement	1	2	3	4
clk	2.000ns	16	3	1	1

图 3-76

Name	CLB LUTs (394080)	CLB Registers (788160)	CARRY8 (49260)	CLB (49260)	LUT as Logic (394080)
cmp_opt_before	32	23	3	9	32

Name	CLB LUTs (394080)	CLB Registers (788160)	CARRY8 (49260)	CLB (49260)	LUT as Logic (394080)
cmp_opt_after	34	23	2	7	34

图 3-77

应用案例 8：获取 4 个计数器中计数值最小的计数器的索引号

功能描述：4 个计数器共享复位端口和时钟使能端口，每个计数器有自己独立的索引号。索引号作为输入信号结合复位和时钟使能信号来控制相对应的计数器的计数行为。在此基础上，比较 4 个计数器当前的计数值，并给出计数值最小的计数器的索引号。

（1）第一种方案：采用 3 个串行比较器。

电路结构如图 3-78 所示，4 个计数器以 VHDL 的数组形式定义。将 3 个比较器依次串联起来，每个比较器接收输入数据并完成比较，输出较小者并将其传递给下一级比较器，最终，末级比较器输出最小计数值，再根据此最小值获取对应计数器的索引号。相应的 RTL 代码如 VHDL 代码 3-69 所示。代码第 7 行定义了计数器的位宽，代码第 19 行以数组形式定义了 4 个计数器并将其初始值设置为 0。代码第 20 行定义了函数（Function）get_min，描述了如图 3-78 所示的功能。

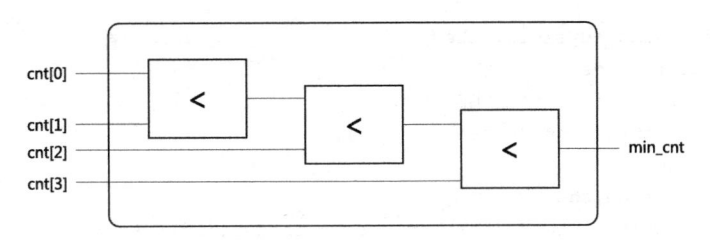

图 3-78

VHDL 代码 3-69

```vhdl
1.  --File: array_cmp_v1.vhd
2.  library ieee;
3.  use ieee.std_logic_1164.all;
4.  use ieee.numeric_std.all;
5.
6.  entity array_cmp_v1 is
7.    generic ( W : positive := 12 );
8.    port (
9.      clk    : in std_logic;
10.     rst    : in std_logic;
11.     ce     : in std_logic;
12.     id     : in unsigned(1 downto 0);
13.     min_id : out unsigned(1 downto 0)
14.   );
15. end entity;
16.
17. architecture rtl of array_cmp_v1 is
18. type lane is array(natural range <>) of unsigned;
19. signal cnt : lane(0 to 3)(W - 1 downto 0) := (others => (others => '0'));
20. function get_min (cnt : lane(0 to 3)(W - 1 downto 0)) return unsigned is
21.   variable sel : unsigned(1 downto 0) := "00";
22.   variable min_cnt : unsigned(W - 1 downto 0) := cnt(0);
23. begin
24.   for i in 1 to 3 loop
25.     if cnt(i) < min_cnt then
26.       min_cnt := cnt(i);
27.       sel := to_unsigned(i, 2);
28.     end if;
29.   end loop;
30.   return sel;
31. end function;
32. begin
33.   process(clk)
34.   begin
35.     if rising_edge(clk) then
36.       min_id <= get_min(cnt);
37.     end if;
38.   end process;
39.
40.   process(clk)
41.   begin
```

```
42.     if rising_edge(clk) then
43.       if rst then
44.         for i in cnt'RANGE loop
45.           cnt(i) <= (others => '0');
46.         end loop;
47.       elsif ce then
48.         cnt(to_integer(id)) <= cnt(to_integer(id)) + 1;
49.       end if;
50.     end if;
51.   end process;
52. end architecture;
```

（2）第二种方案：采用半并行结构。

依然是 3 个比较器，如图 3-79 所示。第一级（最左侧）的两个比较器分别接收两个输入数据并完成比较，输出较小者给后级比较器，最终，末级比较器输出最小计数值。相应的 RTL 代码如 VHDL 代码 3-70 所示。代码第 20 行定义了 c32 和 c10，当 cnt[3] 小于 cnt[2] 时，c32 为 1，否则 c32 为 0。当 cnt[1] 小于 cnt[0] 时，c10 为 1，否则 c10 为 0。代码第 33 行通过逻辑表达式获取最小计数值对应的计数器的索引号。

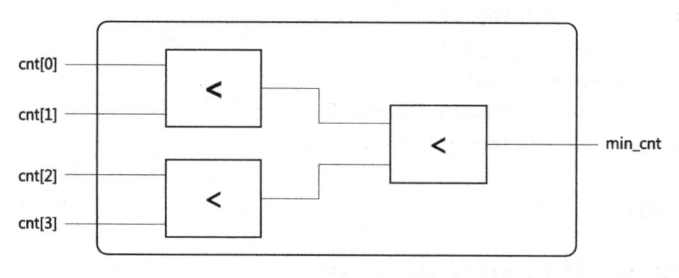

图 3-79

VHDL 代码 3-70

```
1.  --File: array_cmp_v2.vhd
2.  library ieee;
3.  use ieee.std_logic_1164.all;
4.  use ieee.numeric_std.all;
5.
6.  entity array_cmp_v2 is
7.    generic ( W : positive := 12 );
8.    port (
9.      clk    : in std_logic;
10.     rst    : in std_logic;
11.     ce     : in std_logic;
12.     id     : in unsigned(1 downto 0);
13.     min_id : out unsigned(1 downto 0)
14.   );
15. end entity;
16.
17. architecture rtl of array_cmp_v2 is
18.   type lane is array(natural range <>) of unsigned;
19.   signal cnt : lane(0 to 3)(W - 1 downto 0) := (others => (others => '0'));
20.   signal c32, c10, c32_vs_c10, cx : std_logic;
21.   signal cnt32, cnt10 : unsigned(W - 1 downto 0);
```

```vhdl
22.  begin
23.    c32 <= '1' when cnt(3) < cnt(2) else '0';
24.    c10 <= '1' when cnt(1) < cnt(0) else '0';
25.    cnt32 <= minimum(cnt(3), cnt(2));
26.    cnt10 <= minimum(cnt(1), cnt(0));
27.    c32_vs_c10 <= '1' when cnt32 < cnt10 else '0';
28.    cx   <= c32 when c32_vs_c10 else c10;
29.
30.    process(clk)
31.    begin
32.      if rising_edge(clk) then
33.        min_id <= c32_vs_c10 & cx;
34.      end if;
35.    end process;
36.
37.    process(clk)
38.    begin
39.      if rising_edge(clk) then
40.        if rst then
41.          for i in cnt'RANGE loop
42.            cnt(i) <= (others => '0');
43.          end loop;
44.        elsif ce then
45.          cnt(to_integer(id)) <= cnt(to_integer(id)) + 1;
46.        end if;
47.      end if;
48.    end process;
49. end architecture;
```

🔍 VHDL-2008 新特性

VHDL-2008 提供了函数 minimum，其输入可以是两个标量操作数，数据类型可以是 std_logic_vector、signed、unsiged、ufixed 或 sfixed 等，输出为其中的较小者；也可以是一个数组，如 integer_vector：

constant VECT : integer_vector := (16, 5, 32);
constant VECT_MIN : integer := minimum(VECT)。

此时，返回的是数组元素的最小值。此外，还提供了函数 maximum，其用法与 minimum 相同，只是返回的是最大值。无论是 minimum 还是 maximum，本质上都是执行 ">"。从可综合的角度而言，minimum(a, b) 和 maximum(a, b)（这里，a 和 b 均为标量，且是可综合的数据类型）是可综合的。

（3）第三种方案：采用全并行结构。

此方案的电路结构如图 3-80 所示，相应的 RTL 代码如 VHDL 代码 3-71 所示。本方案共 6 个比较器，对应代码第 23 行至第 28 行。代码第 20 行和第 21 行定义的 cmn（m=3,2,1, n=2,1,0）是 cnt(m) 和 cnt(n) 相比较的结果，当 cnt(m) 小于 cnt(n) 时，cmn 为 1，否则为 0。目标索引号 min_id 与 cmn 的对应关系如表 3-19 所示。代码第 33 行和第 34 行中的逻辑表达式正是依据此表生成的。

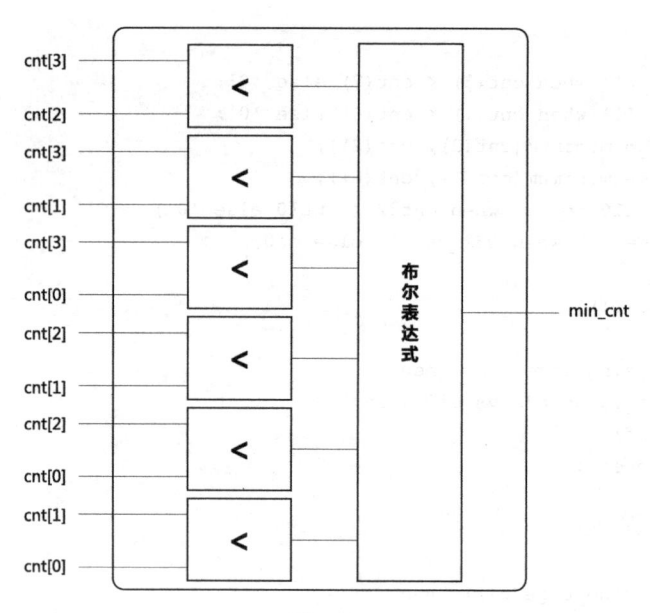

图 3-80

表 3-19

c32	c31	c30	c21	c20	c10	min_id[1]	min_id[0]
1	1	1	x	x	x	1	1
0	x	x	1	1	x	1	0
x	x	0	0	x	1	0	1
x	x	0	x	0	0	0	0

VHDL 代码 3-71

```
1.  --File: array_cmp_v3.vhd
2.  library ieee;
3.  use ieee.std_logic_1164.all;
4.  use ieee.numeric_std.all;
5.
6.  entity array_cmp_v3 is
7.    generic ( W : positive := 12 );
8.    port (
9.      clk    : in std_logic;
10.     rst    : in std_logic;
11.     ce     : in std_logic;
12.     id     : in unsigned(1 downto 0);
13.     min_id : out unsigned(1 downto 0)
14.   );
15. end entity;
16.
17. architecture rtl of array_cmp_v3 is
18. type lane is array(natural range <>) of unsigned;
19. signal cnt : lane(0 to 3)(W - 1 downto 0) := (others => (others => '0'));
20. signal c32, c31, c30 : std_logic;
21. signal c21, c20, c10 : std_logic;
22. begin
```

```vhdl
23.    c32 <= '1' when cnt(3) < cnt(2) else '0';
24.    c31 <= '1' when cnt(3) < cnt(1) else '0';
25.    c30 <= '1' when cnt(3) < cnt(0) else '0';
26.    c21 <= '1' when cnt(2) < cnt(1) else '0';
27.    c20 <= '1' when cnt(2) < cnt(0) else '0';
28.    c10 <= '1' when cnt(1) < cnt(0) else '0';
29.
30.    process(clk)
31.    begin
32.      if rising_edge(clk) then
33.        min_id(1) <= (c31 and c30) or (c21 and c20);
34.        min_id(0) <= (c32 and c30) or (c10 and not c21);
35.      end if;
36.    end process;
37.
38.    process(clk)
39.    begin
40.      if rising_edge(clk) then
41.        if rst then
42.          for i in cnt'RANGE loop
43.            cnt(i) <= (others => '0');
44.          end loop;
45.        elsif ce then
46.          cnt(to_integer(id)) <= cnt(to_integer(id)) + 1;
47.        end if;
48.      end if;
49.    end process;
50. end architecture;
```

对比这三种电路结构，从串行到半并行再到全并行，不难推断出，全并行方案所能达到的 F_{max}（设计最高频率）最大，而串行结构的 F_{max} 最小。从资源利用率的角度而言，串行方案和半并行方案都使用 3 个比较器，因此 LUT 的消耗量应不相上下，而全并行方案使用 6 个比较器，理论上，全并行方案消耗的 LUT 最多。实际布线后的结果如表 3-20 所示。其中，目标芯片为 xcvu3p-ffvc1517-2-i，时钟频率为 500MHz，Vivado 版本为 2021.2。无论计数器位宽是 6 还是 12，全并行方案所获得的 WNS 都最大，这意味着其 F_{max} 最高。例如，当 W 为 12 时，全并行方案的 WNS 为 0.807，因此

$$F_{max} = 1000 \times \frac{1}{(2 - 0.807)} = 838.223 \text{MHz}$$

式中，2 为 500MHz 时钟对应的时钟周期，单位为 ns（纳秒）。还可以看到，全并行方案消耗的 LUT 也最少。读者可以根据所提供的代码在 Vivado 下执行综合和布局布线，分析其中的原因。当位宽较小时，比较器不会消耗 CARRY8（进位链），但当位宽较大时，比较器会消耗 CARRY8。

表 3-20

位宽 W	版 本	特 征	WNS	WHS	最大逻辑级数	LUT	FF	CARRY8
6	array_cmp_v1	串行	0.652	0.129	9	32	26	0
6	array_cmp_v2	半并行	0.895	0.060	5	32	26	0
6	array_cmp_v3	全并行	1.217	0.086	4	28	26	0

续表

位宽 W	版本	特征	WNS	WHS	最大逻辑级数	LUT	FF	CARRY8
12	array_cmp_v1	串行	0.142	0.146	7	77	50	3
	array_cmp_v2	半并行	0.482	0.110	6	67	50	3
	array_cmp_v3	全并行	0.807	0.094	4	62	50	6

3.5.3 奇偶校验电路代码风格

奇偶校验是一种简单、实现代价小的检错方式，常用于芯片内部数据传输或外部数据总线上的数据传输，传统的 PCI 总线中就使用了奇偶校验。其原理是在发送端对一组并行传输的数据（通常为 8 位二进制数）先计算出其奇偶校验位，然后将该奇偶校验位与原始数据一起传输。接收端根据接收到的数据重新计算其奇偶校验位并与接收到的奇偶校验值比较，若二者不相等，则可以确定数据传输过程中出现了错误；若二者相等，则可以确定传输过程中没有出错或出现了偶数个错误（出现这种情况的概率极低）。

奇偶校验包括奇校验和偶校验两种类型。仍以 8 位二进制数为例，这样发送端最终发送的数据为 9 位。对于偶校验，在这 9 位数据中，1 的个数是偶数。对于奇校验，在这 9 位数据中，1 的个数是奇数。例如，8 位原始数据为 8'b1001_0101，其中，共 4 个 1，偶校验时，校验结果为 0，这样原始数据和校验位共包含的 1 的个数是偶数（仍为 4）。奇校验时，校验结果为 1，这样原始数据和校验位共包含的 1 的个数是奇数 5。

奇偶校验值的生成也很简单。对于偶校验，只需要将输入数据按位异或即可得到偶校验值，这可通过 VHDL 中的位缩减运算符实现。对偶校验值取反即可得到奇校验值。相应的电路如图 3-81 所示，图中，even_parity 为偶校验值，odd_parity 为奇校验值。采用 RTL 代码描述，如 VHDL 代码 3-72 所示。

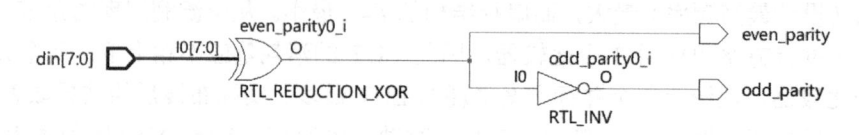

图 3-81

📄 **VHDL 代码 3-72**

```vhdl
1.  --File: parity_checker.vhd
2.  library ieee;
3.  use ieee.std_logic_1164.all;
4.
5.  entity parity_checker is
6.    generic ( W : positive := 8 );
7.    port (
8.      din         : in  std_logic_vector(W - 1 downto 0);
9.      even_parity : out std_logic;
10.     odd_parity  : out std_logic
11.   );
12. end entity;
```

```
13.
14. architecture rtl of parity_checker is
15. begin
16.     even_parity <= xor din;
17.     odd_parity  <= not even_parity;
18. end architecture
```

对于 8 位输入数据，要计算 8 输入异或运算以获取偶校验值，而 LUT 为 6 输入 LUT，因此需要 2 个 LUT。VHDL 代码 3-72 综合后的电路如图 3-82 所示，可以看到，共用了 3 个 LUT。只计算偶校验值或只计算奇校验值都只用 2 个 LUT。实际上，同时计算奇校验值和偶校验值也只用 2 个 LUT。图 3-82 中的两个 LUT5 因为共享输入端，最终 Vivado 通过 LUT 整合将其放在 1 个 LUT6 中实现，even_parity 和 odd_parity 分别通过 LUT6 的 O5 和 O6 输出。

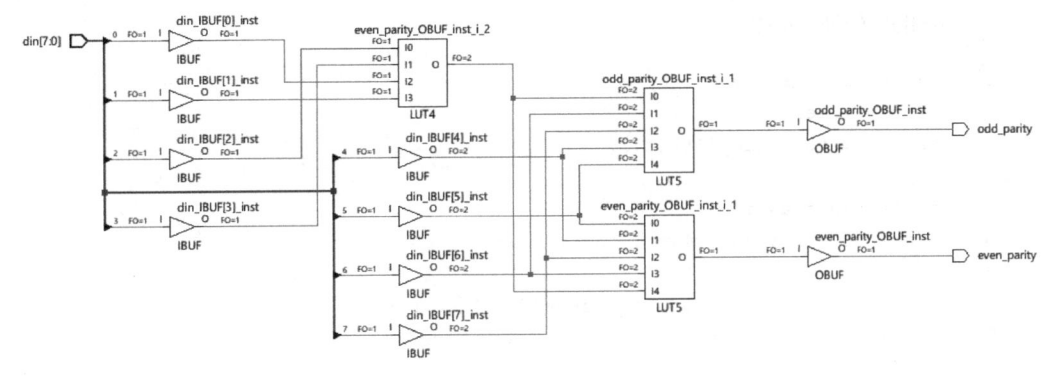

图 3-82

3.5.4 二进制码与格雷码互转电路代码风格

格雷码是由弗兰克·格雷于 1953 年发明的。其主要特点是相邻编码值中只有一位发生变化，表 3-21 显示了三位二进制码及与之对应的格雷码，从中可以看出，这一特点很明显。

表 3-21

数 值	二 进 制 码	格 雷 码
0	000	000
1	001	001
2	010	011
3	011	010
4	100	110
5	101	111
6	110	101
7	111	100

格雷码还有一个特点，即镜像对称。对于两位格雷码，如图 3-83 左侧所示，图中，灰色部分的数字以图中的粗线为镜呈现对称特征。对于三位格雷码，如图 3-83 右侧所示，也呈现此特征。利用此特征可以很方便地手工写出格雷码。

若已知二进制码，如何将其转换为对应的格雷码呢？下面以三位二进制码为例，相应的电路结构如图 3-84 所示。图中，bin_val[2:0]为三位二进制码，gray_val[2:0]为转换后的格雷码。bin_val[0]与 bin_val[1]异或可得 gray_val[0]，bin_val[1]与 bin_val[2]异或可得 gray_val[1]，而 gray_val[2]与 bin_val[2]是相等的。此电路的 RTL 代码如 VHDL 代码 3-73 所示。

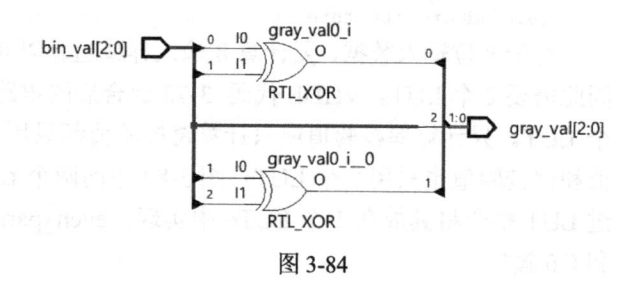

图 3-83

图 3-84

VHDL 代码 3-73

```vhdl
1.  --File: bin2gray.vhd
2.  library ieee;
3.  use ieee.std_logic_1164.all;
4.
5.  entity bin2gray is
6.    generic ( W : positive := 3 );
7.    port (
8.      bin_val : in std_logic_vector(W - 1 downto 0);
9.      gray_val: out std_logic_vector(W - 1 downto 0)
10.   );
11. end entity;
12.
13. architecture rtl of bin2gray is
14. begin
15.   gray_val <= (bin_val srl 1) xor bin_val;
16. end architecture;
```

对于三位二进制码转换为格雷码的电路，由于格雷码的最高位与二进制码的最高位相等，因此此位不会消耗 LUT，这样只有低两位会消耗 LUT，理论上会消耗两个 LUT，如图 3-85 所示，但由于这两个 LUT 有共同的输入信号，因此 Vivado 会通过 LUT 整合，使其位于同一个 LUT6 中，从而最终只消耗一个 LUT。

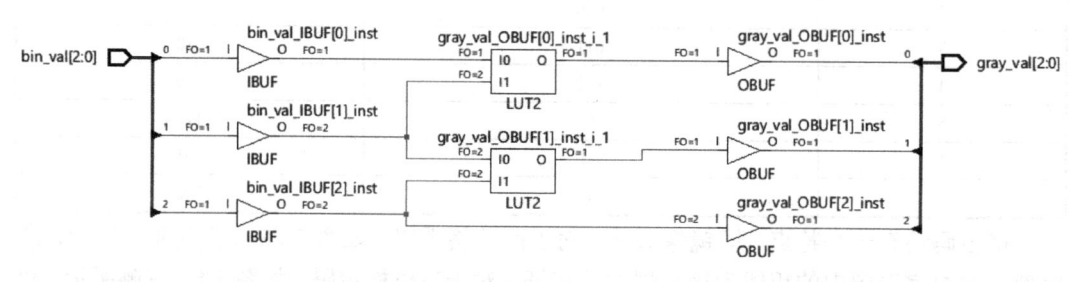

图 3-85

若已知格雷码，如何将其转换为二进制码呢？这个过程也很简单，这里直接给出 RTL 代码，如 VHDL 代码 3-74 所示。

> **VHDL 代码 3-74**

```vhdl
1.  --File: gray2bin.vhd
2.  library ieee;
3.  use ieee.std_logic_1164.all;
4.
5.  entity gray2bin is
6.    generic ( W : positive := 3 );
7.    port (
8.      gray_val : in  std_logic_vector(W - 1 downto 0);
9.      bin_val  : out std_logic_vector(W - 1 downto 0)
10.   );
11. end entity;
12.
13. architecture rtl of gray2bin is
14. begin
15.   bin_val(W - 1) <= gray_val(W - 1);
16.   xor_gates :
17.   for i in W - 2 downto 0 generate
18.     bin_val(i) <= bin_val(i + 1) xor gray_val(i);
19.   end generate;
20. end architecture;
```

相邻两个格雷码只有一位不同的特点使得格雷码广泛应用于异步 FIFO（First In First Out）中。在异步 FIFO 中，写地址和读地址根据读/写操作发生连续改变，其地址是用二进制计数器表示的。以三位二进制计数器为例，地址由 0 至 7，到达 7 之后再归 0。由于读/写时钟异步，因此需要进行跨时钟域处理，如图 3-86 所示。在 clka 时钟域，二进制计数器输出结果先通过二进制码转格雷码模块（bin2gray）转为格雷码，此格雷码再通过寄存器输出传送给 clkb 时钟域。在 clkb 时钟域有两级同步器，同步器输出结果再经格雷码转二进制码模块（gray2bin）转为二进制码。

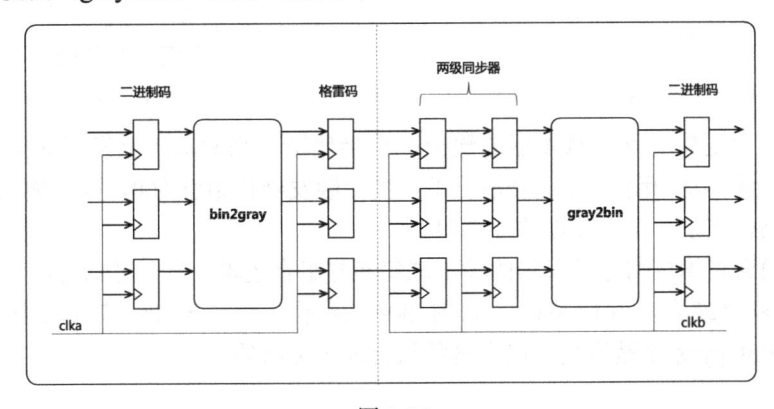

图 3-86

那么，这种码制之间的转换会带来什么好处呢？以二进制计数器输出 5（3'b101）为例，其对应的格雷码为 3'b111，两级同步器将接收到 3'b111，下一个 clka 周期，计数器将输出 6（3'b110），对应的格雷码为 3'b101，经同步器之后，接收端时钟域的值变为 3'b101，或者暂时保持为原来的 3'b111，待下一个时钟周期之后才变为 3'b101。可以看到，无论是 3'b101 还是 3'b111，最终传递的结果都是按序出现的合法值，如表 3-22 所示。如果不使用二进制码和格雷码互转电路会出现什么情况呢？仍以 clka 时钟域计数器由 5 跳变到 6 为例，同步器的旧值为 3'b101，而新值可能是 3'b101、3'b110、3'b100 或 3'b111。这是因为 clka 和 clkb 相互独立，同步器的 3 个输入值有两个发生改变，而这两个改变的值可能会出现在不同的时钟周期上。尽管同步器最终会输出正确的值，但这中间可能会产生非法值。对于 FIFO 而言，这些非法值可能会导致 FIFO 产生错误的空、满状态标志，从而使外部电路对其内部存储数据量产生错误判断。例如，将一个存有数据的 FIFO 当成是空的，或者将一个没有写满的 FIFO 当成是满的，从而造成系统工作错误，甚至导致系统崩溃。

表 3-22

十进制数值	二进制码值	格 雷 码 值	两级同步器输出值
5	101	111	111
6	110	101	111/101

3.6 避免组合逻辑环路

在设计组合逻辑电路时，要避免出现组合逻辑环路（Combinational Loop）。组合逻辑环路是起始于某个组合逻辑单元，之后经过其他组合逻辑单元又回到起始组合逻辑单元的逻辑环路。

有三种典型的组合逻辑环路。第一种是将组合逻辑输出端通过其他组合逻辑单元输出后又反馈到该组合逻辑输入端。图 3-87 就是这种情形。与门的输出经过非门之后又回到与门的输入端。这种电路采用 Vivado 综合之后，打开综合后的网表，执行命令 report_methodology，会给出如图 3-88 所示的违例信息。

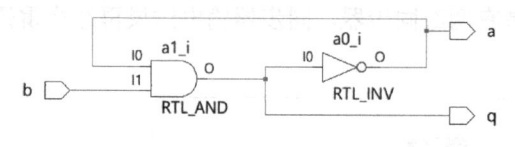

图 3-87

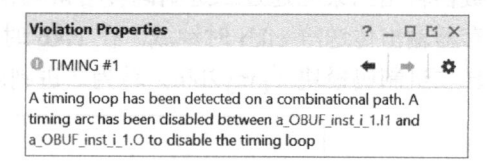

图 3-88

第二种组合逻辑环路是触发器的输出端直接反馈回来驱动该触发器的异步复位或异步置位端，如图 3-89 所示。在这种情况下，由于触发器初始值为 0，因此输出 q 恒为 0，综合后的电路仅剩余一个恒接地的 q 端口。

第三种组合逻辑环路是触发器的输出端口参与组合逻辑，组合逻辑的输出又驱动该触发器的异步复位或异步置位端口，如图 3-90 所示。在 Vivado 综合后的网表中运行 report_methodology 命令是检查不出来这种组合逻辑环路的。

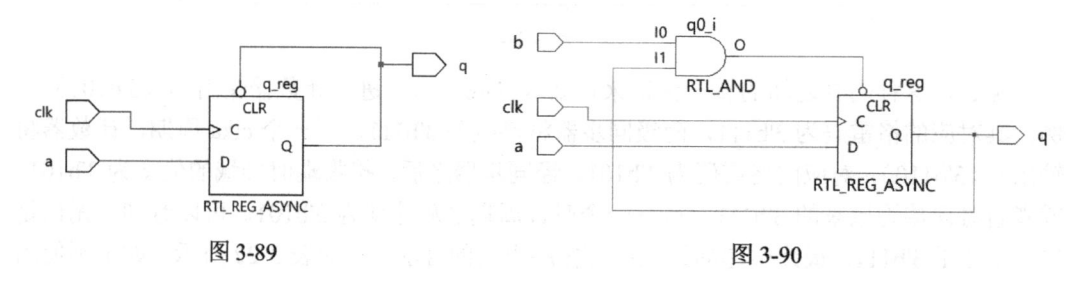

图 3-89

图 3-90

无论是哪种组合逻辑环路，在 FPGA 设计中都应避免。因为组合逻辑环路违反了同步设计原则，会极大地影响系统的稳定性和可靠性。

3.7 思考空间

1. 试用 VHDL-2008 描述一个二进制码转独热码的模块，其中，二进制码位宽为 3。
2. 试用 VHDL-2008 描述一个 64 选 1 二进制码 MUX，并将其分别在 7 系列 FPGA、UltraScale FPGA 和 Versal ACAP 上实现，然后给出三者在资源利用率上的差别。
3. 试解释 VHDL 代码 3-13 为什么会出现综合前后的仿真结果不匹配的情形。
4. 试解释在什么情况下 if-elsif-else 语句没有优先级，在什么情况下有优先级。
5. 试分析在什么情况下 if-elsif-else 语句与 case 语句等效。
6. 试描述 VHDL-2008 中 case 与 case?的差异。
7. 试解释动态移位器为什么会消耗 LUT。
8. 试用 MUX 和静态移位器实现一个动态左移的移位器，移位量最大为 15。
9. 若 a 为 W 位有符号数，对其累加 N 次，试计算累加器的位宽，以保证累加结果不溢出。
10. 若 a 是位宽为 4 的无符号整数（unsigned），b 是位宽为 6 的无符号整数（unsigned），则在 VHDL-2008 中：

（1）当描述 a+b 时，是否需要对其进行高位补 0 使位宽相等？

（2）若 a 和 b 是有符号整数（signed），则描述 a+b 时，需要对位宽如何操作？

（3）若 a 和 b 是 4 位有符号定点数（sfixed），则 VHDL-2008 会认为 a+b 的位宽为多少？

第 4 章

优化触发器

4.1 触发器资源

4.1.1 7 系列 FPGA 中的触发器资源

7 系列 FPGA 中的每个 SLICE 包含 8 个触发器。这 8 个触发器共享时钟使能端口 CE、时钟端口 CLK 和复位/置位端口 SR（复位端口和置位端口是同一个端口，意味着触发器只可以单独实现复位功能或置位功能），如图 4-1 所示。时钟使能信号、时钟信号和复位/置位信号是触发器的控制信号，这三者一起构成触发器的控制集（Control Set）。CE 和 SR 均为高电平有效，且 SR 的优先级高于 CE。

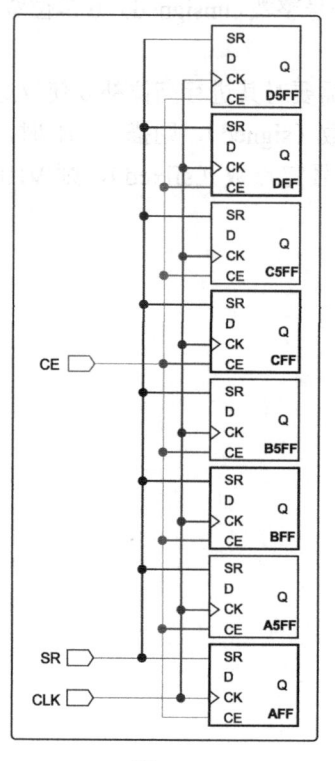

图 4-1

> 设计规则 1：用 RTL 代码描述 D 触发器时，始终遵循 SR 的优先级高于 CE，且应避免既复位又置位。

这 8 个触发器并非完全相同。其中，D5FF/C5FF/B5FF/A5FF 构成一组，仅可以配置为边沿敏感型的 D 触发器。DFF/CFF/BFF/AFF 不仅可以配置为边沿敏感型的 D 触发器，还可以配置为电平敏感型的锁存器，但是，一旦这 4 个触发器中有一个被当作锁存器使用，那么 D5FF～A5FF 就报废了。

> 💡 **设计规则 2**：在设计中应尽量避免使用锁存器，可将其转换为带时钟使能信号的 D 触发器。

当将这些触发器用作 D 触发器时，既支持上升沿采样，又支持下降沿采样。进行下降沿采样时，意味着需要对输入时钟取反，用取反后的时钟上升沿采样。此时，这个时钟路径上的反相器会被自动吸收到触发器内部。因此，如 VHDL 代码 4-1 所示的描述方式只会消耗一个触发器，并不会消耗额外的查找表，如图 4-2 所示。触发器的属性 IS_C_INVERTED 值为 1，表明当前时钟为下降沿采样。

📄 **VHDL 代码 4-1**

```
1.  --File: neg_edge.vhd
2.  architecture rtl of neg_edge is
3.  begin
4.    process(clk)
5.    begin
6.      if falling_edge(clk) then
7.        q <= '0' when rst else d;
8.      end if;
9.    end process;
10. end architecture
```

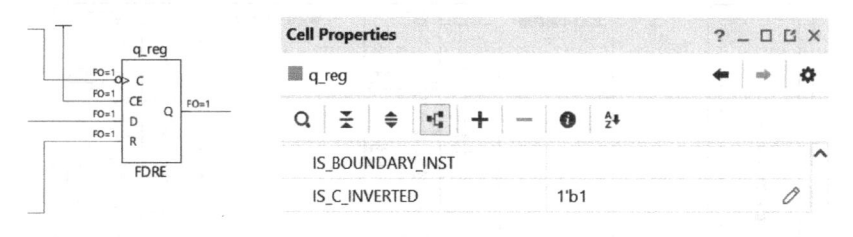

图 4-2

🔍 **VHDL-2008 新特性**

VHDL-2008 允许在 process 中使用 when-else 语句，这样可以借助其描述同步复位，如 VHDL 代码 4-1 第 7 行所示。尽管其功能与 if-else 语句一致，但比 if-else 语句更为简洁。

> 💡 **设计规则 3**：在设计中应始终保持触发器时钟采样的极性一致。

使用 Tcl 代码 4-1 可以查看 7 系列 FPGA 一个 SLICE 内触发器的分布状况。在代码第 6 行中，TYPE 为 FF_INIT，说明该触发器只可以用作 D 触发器；若 TYPE 为 REG_INIT，则说明该触发器既可以用作 D 触发器，也可以用作锁存器。

Tcl 代码 4-1

```
1.  #File: check_ff_7fpga.tcl
2.  set mypart [get_parts xc7k70tfbg484-3]
3.  link_design -part $mypart
4.  set slice SLICE_X0Y0
5.  => SLICE_X0Y0
6.  set ff_in_slice [get_bels -filter "TYPE==FF_INIT" \
7.      -of [get_sites $slice]]
8.  => SLICE_X0Y0/D5FF SLICE_X0Y0/C5FF SLICE_X0Y0/B5FF SLICE_X0Y0/A5FF
9.  set ffL_in_slice [get_bels -filter "TYPE==REG_INIT" \
10.     -of [get_sites $slice]]
11. => SLICE_X0Y0/DFF SLICE_X0Y0/CFF SLICE_X0Y0/BFF SLICE_X0Y0/AFF
12. puts "#N FF in one slice: \
13.     [expr {[llength $ff_in_slice] + [llength $ffL_in_slice]}]"
14. => #N FF in one slice:  8
```

4.1.2 UltraScale/UltraScale+ FPGA 中的触发器资源

在 UltraScale/UltraScale+ FPGA 中，每个 SLICE 包含 16 个触发器。这 16 个触发器完全相同，既可配置为边沿敏感型的 D 触发器，也可配置为电平敏感型的锁存器。这 16 个触发器在同一列，每 8 个为一组，分为上下两组，如图 4-3 所示。每组中若有一个触发器被当作锁存器使用，那么该组的其余触发器要么也被当作锁存器使用，要么不能被使用，总之不能再被当作 D 触发器使用。同时，每组的 8 个触发器共享时钟端口和复位/置位端口，每 4 个触发器有一个时钟使能端口，每组有两个时钟使能端口。因此，这 16 个触发器最多可以有 4 个时钟使能信号、2 个时钟信号和 2 个复位/置位信号。相比 7 系列 FPGA，其控制信号更丰富了一些。

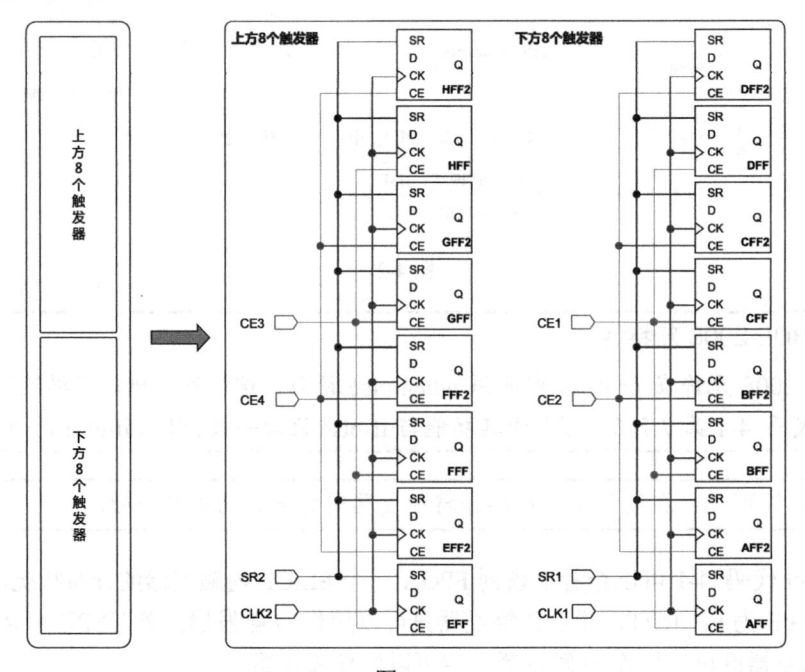

图 4-3

与 7 系列 FPGA 一样，当触发器用时钟的下降沿采样时，反相器会被自动吸收到触发器内部。SR 的优先级高于 CE。同时，SR 既支持高电平有效也支持低电平有效，但 CE 只支持高电平有效。如 VHDL 代码 4-2 所示的描述方式，在 7 系列 FPGA 中会消耗一个查找表和一个触发器，如图 4-4 左侧所示，而在 UltraScale/UltraScale+ FPGA 中，只会消耗一个触发器，如图 4-4 右侧所示，两者均需要打开布线后的设计查看资源利用率。

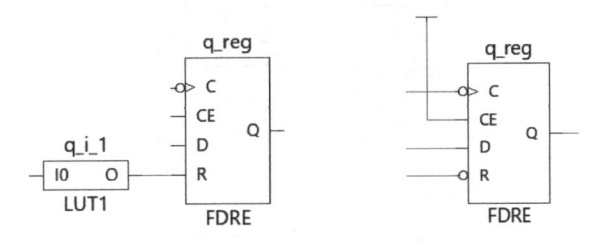

图 4-4

VHDL 代码 4-2

```
1.  --File: neg_edge_rst.vhd
2.  architecture rtl of neg_edge_rst is
3.  begin
4.    process(clk)
5.    begin
6.      if falling_edge(clk) then
7.        q <= '0' when (not rst) else d;
8.      end if;
9.    end process;
10. end architecture
```

借助 Tcl 代码 4-2 可以查看 UltraScale/UltraScale+ FPGA 中每个 SLICE 内触发器的分布状况。由于返回值较多，代码第 7 行只给出前 3 个返回值，其余返回值以省略号代替。

Tcl 代码 4-2

```
1.  #File: check_ff_us.tcl
2.  set mypart [get_parts xcvu5p-flva2104-2-i]
3.  link_design -part $mypart
4.  set slice SLICE_X0Y0
5.  => SLICE_X0Y0
6.  set ff_in_slice [get_bels "*FF*" -of [get_sites $slice]]
7.  => SLICE_X0Y0/AFF SLICE_X0Y0/AFF2 SLICE_X0Y0/BFF ...
8.  puts "#N FF in one slice: [llength $ff_in_slice]"
9.  => #N FF in one slice: 16
```

4.1.3　Versal ACAP 中的触发器资源

与 UltraScale/UltraScale+ FPGA 一样，Versal ACAP 中的一个 SLICE 内有 16 个触发器，这 16 个触发器既可配置为边沿敏感型的 D 触发器，也可配置为电平敏感型的锁存器。但是，只要有一个触发器被配置为锁存器，其他触发器就不能再被配置为 D 触发器了。当触发器被配置为 D 触发器时，可支持上升沿采样，也可支持下降沿采样；复位/置位既可以是高电平有效，也可以是低电平有效，而不会消耗额外的查找表。就控制集而言，这 16

个触发器共享时钟端口、复位/置位端口。由上至下,相邻的每4个触发器为一组(不再是交织分组),共享一个时钟使能信号,如图4-5所示。

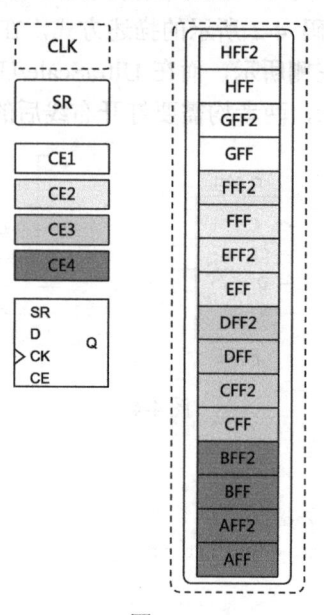

图 4-5

借助 Tcl 代码 4-3 可以查看 Versal ACAP 中每个 SLICE 内触发器的分布状况。

Tcl 代码 4-3

```
1.  #File: check_ff_versal.tcl
2.  set mypart [get_parts xcvc1902-vsva2197-2MP-e-S]
3.  link_design -part $mypart
4.  set slice SLICE_X40Y0
5.  => SLICE_X40Y0
6.  set ff_in_slice [get_bels -filter "TYPE==FF" -of [get_sites $slice]]
7.  => SLICE_X40Y0/AFF SLICE_X40Y0/AFF2 SLICE_X40Y0/BFF ...
8.  puts "#N FF in one slice: [llength $ff_in_slice]"
9.  => #N FF in one slice: 16
```

由 Tcl 代码 4-3 第 6 行的返回值可以看到,触发器的名称由两部分构成:所在 SLICE 和处于 16 个触发器中的位置。因此,无论使用哪个系列的 FPGA,如果要将设计中的某个触发器单元固定在指定位置上,就需要两条约束:第一条约束为 BEL,第二条约束为 LOC,如 Tcl 代码 4-4 所示。

Tcl 代码 4-4

```
1.  #File: fix_ff_loc.sv
2.  set_property BEL DFF2 [get_cells LDCE_inst]
3.  set_property LOC SLICE_X190Y205 [get_cells LDCE_inst]
4.  set_property BEL CFF [get_cells FDRE_inst_ce2]
5.  set_property LOC SLICE_X191Y205 [get_cells FDRE_inst_ce2]
```

对 7 系列 FPGA、UltraScale/UltraScale+ FPGA 和 Versal ACAP 中的触发器进行比较,如表 4-1 所示。不难看出,无论是哪个系列的 FPGA,当将触发器配置为 D 触发器时,都可以既无复位又无置位（对应 VHDL 代码 4-3，最终映射为 FDRE，复位端接地，时钟使能端恒为高电平），可以是同步复位（对应 VHDL 代码 4-4），可以是同步置位（对应 VHDL 代码 4-5），可以是异步复位（对应 VHDL 代码 4-6），可以是异步置位（对应 VHDL 代码 4-7）。从描述方式的角度而言，同步和异步的差别在于同步时敏感变量列表里只有时钟 clk，异步时则既有时钟又有复位/置位信号。

表 4-1

指　　标	7 系列 FPGA	UltraScale/UltraScale+ FPGA	Versal ACAP
每个 SLICE 内包含的触发器的个数	8	16	16
用作 D 触发器时的具体类型			
既无复位又无置位	√	√	√
同步复位（对应原语 FDRE）	√	√	√
同步置位（对应原语 FDSE）	√	√	√
异步复位（对应原语 FDCE）	√	√	√
异步置位（对应原语 FDPE）	√	√	√
控制集{CE, CLK, SR}	{1, 1, 1}	{4, 2, 2}	{4, 1, 1}
时钟采样极性	↑↓	↑↓	↑↓
SR 有效极性	高电平	高/低电平	高/低电平

VHDL 代码 4-3

```
1.  --File: myff.vhd
2.  architecture rtl of myff is
3.  begin
4.    process(clk)
5.    begin
6.      if rising_edge(clk) then
7.        q <= d;
8.      end if;
9.    end process;
10. end architecture;
```

VHDL 代码 4-4

```
1.  --File: myfdre.vhd
2.  architecture rtl of myfdre is
3.  begin
4.    process(clk)
5.    begin
6.      if rising_edge(clk) then
7.        q <= '0' when rst else
8.             d   when ce  else
9.             q;
10.     end if;
11.   end process;
12. end architecture;
```

VHDL 代码 4-5

```
1.  --File: myfdse.vhd
```

```
2.  architecture rtl of myfdse is
3.  begin
4.    process(clk)
5.    begin
6.      if rising_edge(clk) then
7.        q <= '1' when set else
8.             d   when ce  else
9.             q;
10.     end if;
11.   end process;
12. end architecture;
```

📄 **VHDL 代码 4-6**

```
1.  --File: myfdce.vhd
2.  architecture rtl of myfdce is
3.  begin
4.    process(clk, clr)
5.    begin
6.      if clr then
7.        q <= '0';
8.      elsif rising_edge(clk) then
9.        q <= d when ce else q;
10.     end if;
11.   end process;
12. end architecture;
```

📄 **VHDL 代码 4-7**

```
1.  --File: myfdpe.vhd
2.  architecture rtl of myfdpe is
3.  begin
4.    process(clk, pre)
5.    begin
6.      if pre then
7.        q <= '1';
8.      elsif rising_edge(clk) then
9.        q <= d when ce else q;
10.     end if;
11.   end process;
12. end architecture;
```

由于触发器只有 4 种原语，因此很容易根据原语找到设计中指定的触发器。例如，要找到设计中所有的异步复位触发器，就可以用如 Tcl 代码 4-5 所示的描述方式。

🪶 **Tcl 代码 4-5**

```
1.  #File: get_specified_ff.tcl
2.  set ref_name FDCE
3.  set myff [get_cells -hier -filter "REF_NAME==$ref_name"]
4.  if {[llength $myff] > 0} {
5.    show_objects $myff -name "my$ref_name"
6.  }
```

4.2 建立时间和保持时间

建立时间和保持时间是触发器的两个重要指标，是触发器特性的直接体现，也是时序收敛必须考虑的因素。这里我们所讨论的触发器均为边沿敏感型触发器。

如图 4-6 所示，建立时间是指在触发器时钟上升沿到来之前，数据稳定存在于触发器 D 端口保持不变的时间。如果建立时间不够，那么数据将无法在这个上升沿被稳定地打入触发器，T_{su} 是指时钟上升沿到来之前数据最小的稳定时间。保持时间是指在触发器时钟信号上升沿到来之后，数据还应稳定存在于触发器 D 端口一段时间保持不变。如果保持时间不够，那么数据同样无法在这个上升沿被稳定地打入触发器，T_h 是指时钟上升沿到来之后数据最小的稳定时间。T_{su} 和 T_h 共同决定了最小数据有效窗。T_{co} 是输出响应时间，也被称为时钟到输出时间。描述的是触发器输出端在时钟上升沿到来之后的多长时间内发生变化，也就是触发器输出延时。T_{co} 必然大于 0，这反映了自然界的变化规律：是"渐变"而非"突变"，即时钟上升沿到来之后，输出端不会立即响应，而是逐渐发生变化。

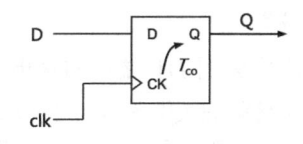

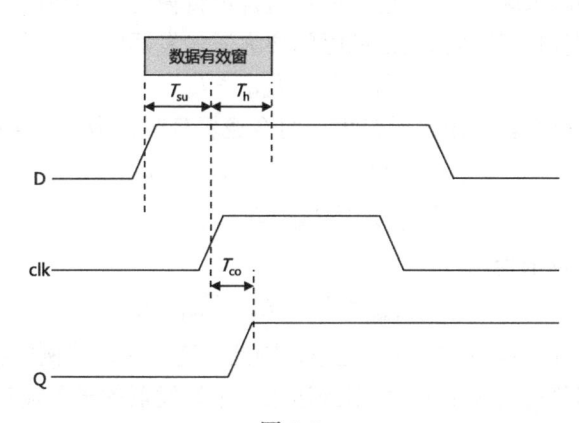

图 4-6

那么 T_{su} 和 T_h 是如何影响时序收敛的呢？我们以一个经典的时序路径为例，如图 4-7 所示。这是一个典型的单周期时序路径，也是最容易发生时序违例的路径。时序收敛的目标是保证数据能够在这两个触发器上正确传输，从而确定中间组合逻辑的传输延时范围。

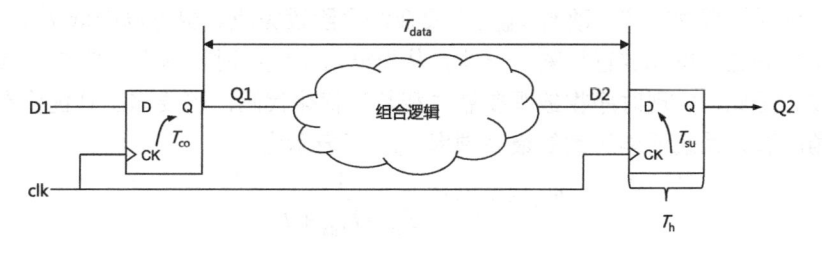

图 4-7

先看建立时间裕量，这里暂不考虑时钟偏移的影响（关于时钟偏移对时序收敛的影响请阅读第 2.2 节），如图 4-8 所示。

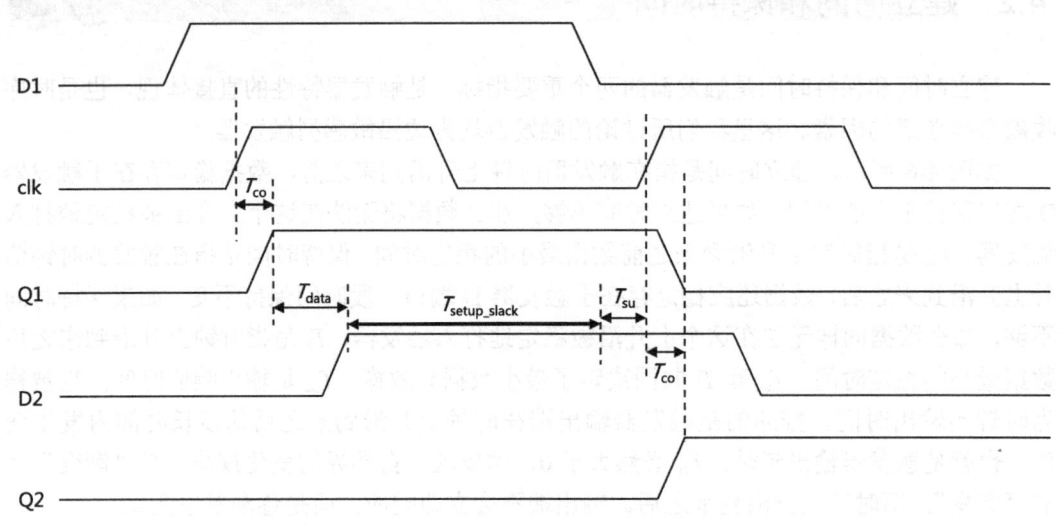

图 4-8

在时钟信号 clk 的第一个上升沿，触发器采集到 D1 端口的高电平，并将其打入触发器，经过 T_{co} 的触发器输出延时到达第一个触发器的 Q1 端口。再经过组合逻辑延时 T_{data} 到达第二个触发器的输入端口 D2（这里假定组合逻辑输出仍为高电平）。在第二个时钟上升沿到来之前，D2 端口上的数据稳定的时间要大于触发器的建立时间 T_{su}，用公式表示为

$$T_{clk} - T_{co} - T_{data} \geqslant T_{su}$$

考虑最坏情况，触发器输出延时最大，组合逻辑延时也最大，可得

$$T_{clk} - T_{co\_max} - T_{data\_max} \geqslant T_{su}$$

$$T_{data\_max} \leqslant T_{clk} - T_{co\_max} - T_{su}$$

建立时间裕量可表示为

$$T_{setup\_slack} = T_{clk} - T_{co} - T_{data} - T_{su}$$

对于已选定的芯片，T_{co} 和 T_{su} 为固定值，因此，建立时间违例的根本原因就是 T_{data} 过大（在不考虑时钟偏移的情况下），数据到达第二个触发器的输入端口"太晚了"。T_{data} 可表示为

$$T_{data} = T_{logic} + T_{net}$$

式中，T_{logic} 为逻辑延时，T_{net} 为布线延时。为了降低 T_{data}，可以降低 T_{logic}，本质上就是降低路径的逻辑级数，这需要修改 RTL 代码，插入流水寄存器，或使用重定时（Retiming）技术。但无论采用哪种方式，改善 T_{logic} 主要在综合阶段完成。也可以降低 T_{net}，其本质是在布局时将相关逻辑单元放置得紧凑一些，从而缩短布线延时，这主要依靠工具在布局布线阶段完成。因此，实现时序收敛既要靠"智慧"优化代码，也要靠工具优化布局布线。从另一个角度看，系统所能达到的最高频率 F_{max} 可表示为

$$F_{max} = \frac{1}{T_{clk}} \leqslant \frac{1}{T_{co} + T_{data} + T_{su}}$$

再看保持时间裕量，如图 4-9 所示。在图 4-8 的基础上，第二个时钟上升沿采集到第

一个触发器端口 D1 上的低电平，经过 T_{co} 到达 Q1 端口，再经过 T_{data} 到达第二个触发器的 D2 端口。

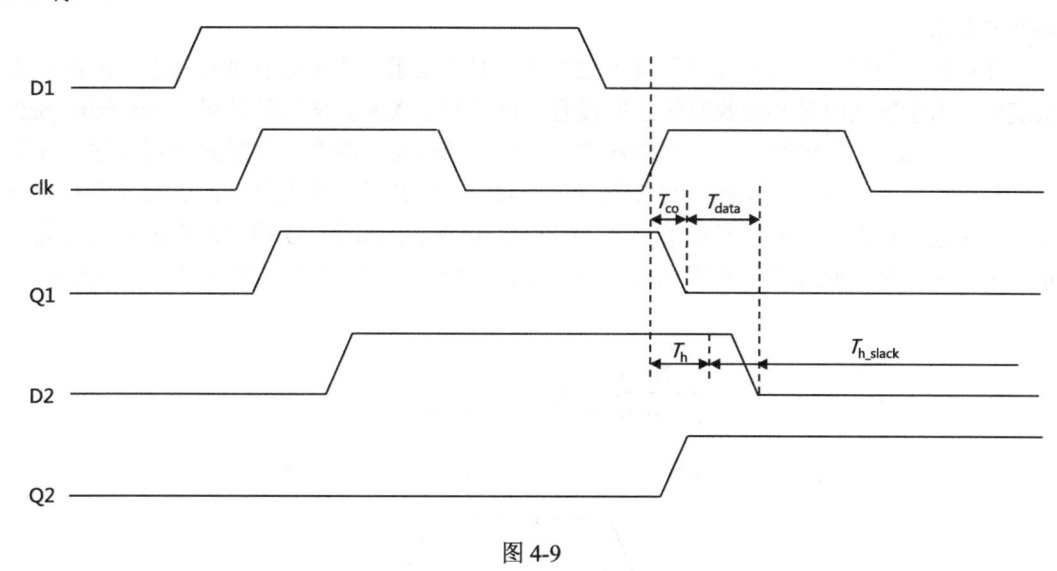

图 4-9

经过 $T_{co}+T_{data}$ 的延时，D2 上原本的高电平在第二个时钟上升沿到来之后稳定的时间必须大于或等于第二个触发器的保持时间。满足此条件，第二个触发器才能稳定地接收到最初由 D1 传过来的高电平，即

$$T_{co} + T_{data} \geqslant T_h$$

考虑最坏的情况，触发器输出延时最小，组合逻辑延时也最小，可得

$$T_{co\_min} + T_{data\_min} \geqslant T_h$$

$$T_{data\_min} \geqslant T_h - T_{co\_min}$$

保持时间裕量可表示为

$$T_{h\_slack} = T_{co} + T_{data} - T_h$$

因此，保持时间违例的根本原因是 T_{data} 太小了，数据到达第二个触发器的输入端口"太早了"。结合建立时间裕量和保持时间裕量，可得 T_{data} 的范围

$$T_{clk} - T_{co\_max} - T_{su} \geqslant T_{data} \geqslant T_h - T_{co\_min}$$

由此可见，建立时间 T_{su} 决定了组合逻辑的最大延时，保持时间 T_h 决定了组合逻辑的最小延时。这也是为什么 Vivado 中跟一些时序相关的 Tcl 命令选项 -max 表示的就是建立时间，选项 -min 表示的就是保持时间。

4.3 亚稳态

如 4.2 节所述，为了保证触发器稳定、可靠地工作，触发器输入端的数据必须在时钟有效沿到来之前稳定地存在于输入端口的最小时间为 T_{su}，且在时钟有效沿到来之后仍然在输入端口保持稳定的最小时间为 T_h，之后，经 T_{co} 在触发器 Q 端口输出。如果数据传输过程中不满足 T_{su} 和 T_h 的要求，就可能产生亚稳态，此时触发器输出端口 Q 在时钟有效沿之后比较长的一段时间内处于不确定的状态，在这段时间内，Q 端在 0 和 1 之间处于振荡状态，而不是等于数据输入端 D 的值。这段时间称为决断时间（Resolution Time），用 T_{met} 表示。经过 T_{met} 之后，Q 端将稳定到 0 或 1 上，但是稳定到 0 或 1 是随机的，与输入没有必然的关系。

在同步系统中，输入信号必须满足触发器的时序要求，才不会发生亚稳态。但在异步系统中，由于发送时钟与接收时钟之间没有明确的相位关系，同时往往采用 set_false_path 或 set_max_delay 等进行约束，这样发送时钟发出的数据可能在任何时间点到达与之异步的接收寄存器，从而出现无法满足接收寄存器对 T_{su} 和 T_h 的要求的情况，进而导致亚稳态发生，这是亚稳态发生的典型场景之一。如图 4-10 所示，图中，D 端口的数据来自与之异步的另一时钟域的触发器的输出端口。经过 T_{met} 之后，Q 端可能稳定到 0 上，也可能稳定到 1 上。

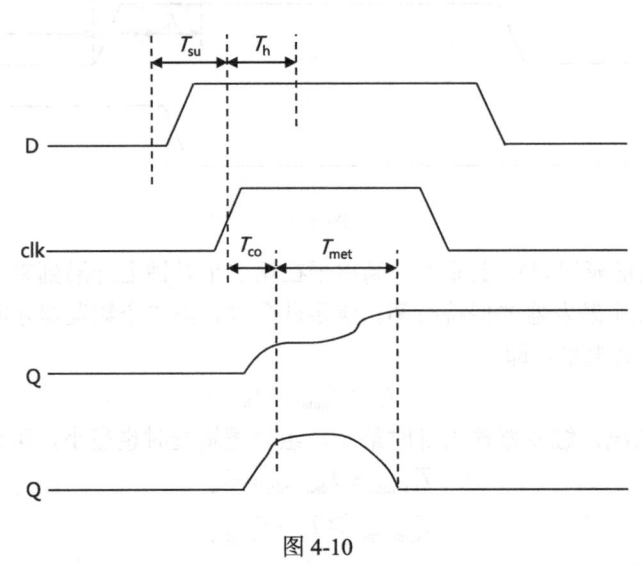

图 4-10

为了降低发生亚稳态的概率，对于控制信号（通常为 1 位），可采用如图 4-11 所示的由级联触发器构成的同步链，也称为双触发器同步器。该电路有 5 个特点：①同步链中的触发器由同一个时钟驱动；②同步链中的第一个触发器的 Q 端扇出必须为 1；③对同步链中的两个触发器要施加综合属性 ASYNC_REG，并将其值设置为 "TRUE"，其目的是阻止工具优化，并在布局时将两者放置在同一个 SLICE 内；④同步链中第一个触发器的输入数据来自与之异步的另一时钟域的触发器的输出端，而不能是组合逻辑的输出端；⑤clk2 时钟域最终使用的是同步链末级触发器的输出结果，换言之，中间结果不能使用。

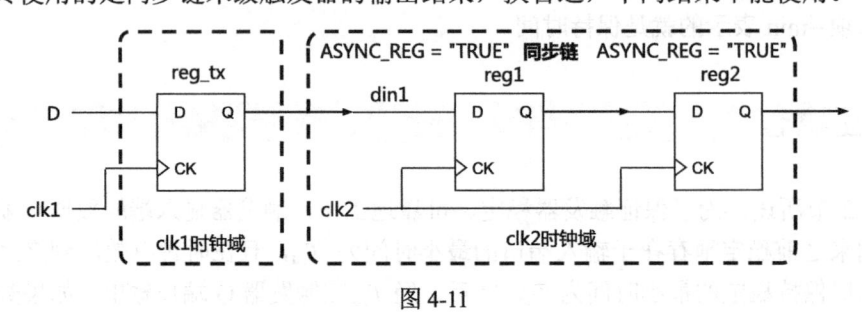

图 4-11

从约束的角度而言，需要对 reg_tx 和 reg1 之间的路径使用 set_max_delay 进行约束，如 Tcl 代码 4-6 所示。set_max_delay 的值为 clk1 和 clk2 两者中时钟周期的最小值，这可由代码第 4 行确定。

📝 **Tcl 代码 4-6**

```
1.  #File: cdc_constraint.tcl
2.  set clk_tx_period [get_property PERIOD [get_clocks clk1]]
3.  set clk_rx_period [get_property PERIOD [get_clocks clk2]]
4.  set delay [expr {min($clk_tx_period, $clk_rx_period)}]
5.  set_max_delay -from [get_pins reg_tx/C] -to [get_pins reg1/D] \
6.  -datapath_only $delay
```

同步链电路并不能百分之百地保证消除亚稳态，如图 4-12 所示。此时，同步链第一级触发器发生亚稳态，但第二级触发器稳定地捕获到输入信号，避免了亚稳态的传递。但也可能发生如图 4-13 所示的情形，亚稳态从同步链的第一级触发器传递到第二级触发器，这意味着第二级触发器最终可能稳定到 0，也可能稳定到 1。

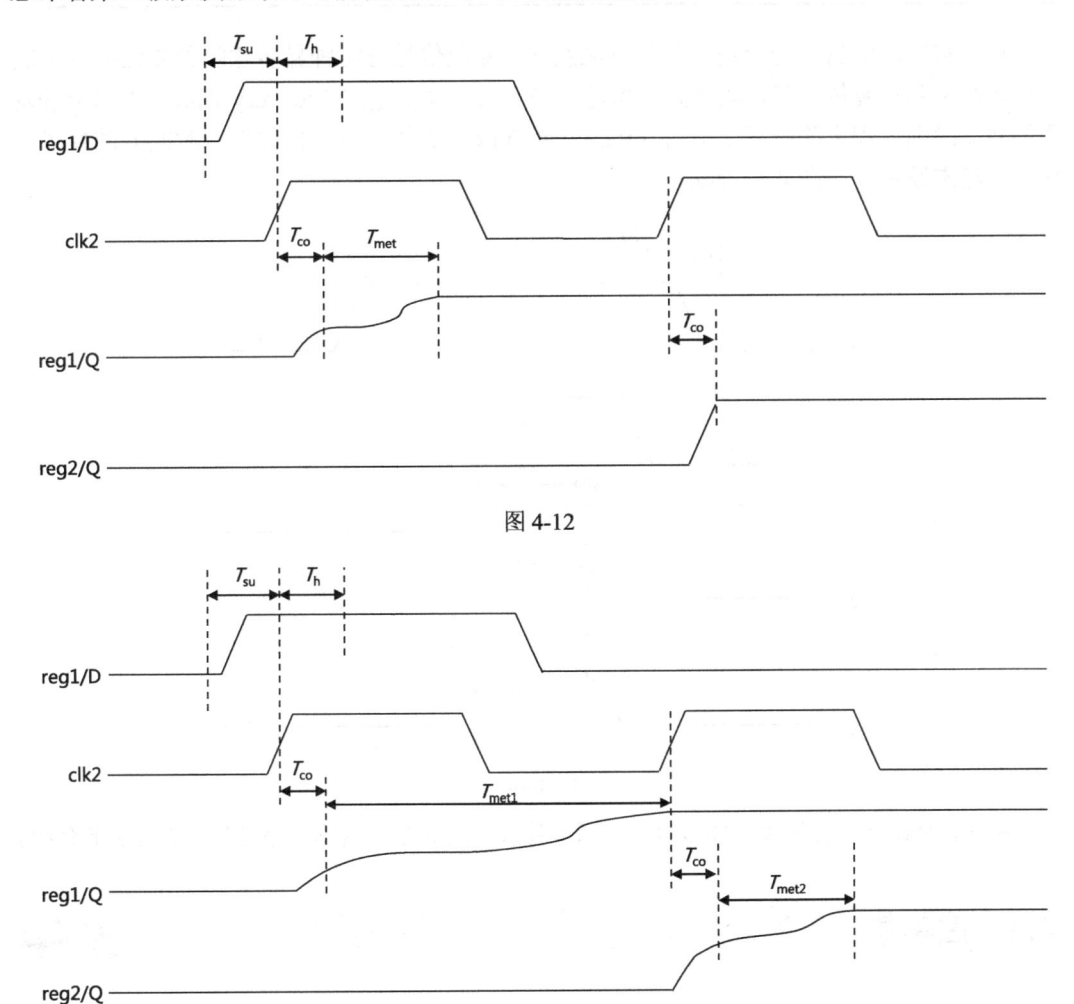

图 4-12

图 4-13

对于如图 4-11 所示的电路，通常采用平均无故障时间 MTBF（Mean Time Between Failures）衡量其稳健性。它表征的是相邻两次亚稳态发生的时间间隔的数学期望值。显然，MTBF 越大，说明该电路越稳健。MTBF 可表示为

$$\text{MTBF} = \frac{e^{\frac{T_{met}}{C_2}}}{C_1 \cdot f_{clk2} \cdot f_{data}}$$

式中，C_1 和 C_2 是由芯片工艺和工作条件决定的固定常数，f_{clk2} 为接收时钟的时钟频率，f_{data} 为同步链中第一个触发器接收到的异步信号的翻转率。由此式可知，clk2 的频率越高或异步信号的翻转率越高，MTBF 越低。

对于多位宽的数据信号，就不能采用如图 4-11 所示的电路实现跨时钟域操作。这是因为数据信号的每一位可能会被不同的 clk2 上升沿捕获到，从而导致接收数据错误。因此，这种情形应该使用 FIFO 实现跨时钟域操作。

> 💡 设计规则 4：对于多位宽数据信号的跨时钟域，应使用 FIFO 进行隔离。

亚稳态出现的另一场合是异步复位触发器。复位信号与时钟异步意味着复位信号可能在任意时刻到达复位端口，从而无法满足对恢复时间 $T_{recovery}$（Recovery Time，类似于触发器的建立时间）和去除时间 $T_{removal}$（Removal Time，类似于触发器的保持时间）的要求，导致亚稳态发生，如图 4-14 所示。

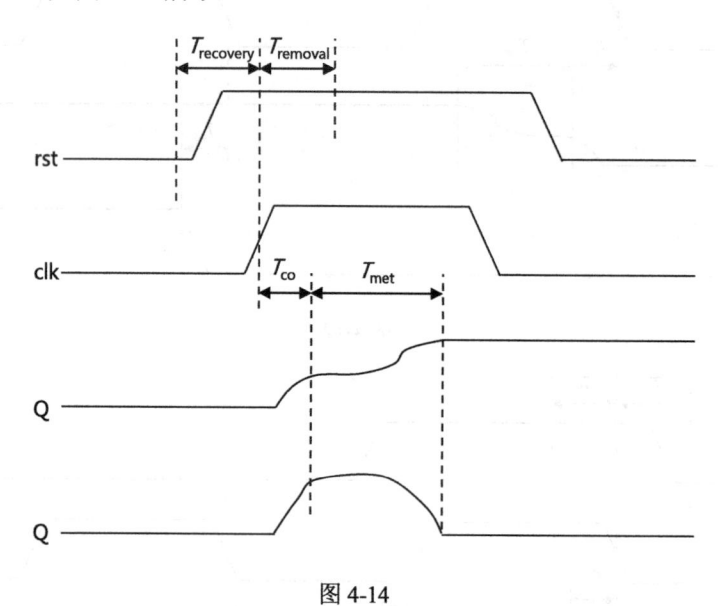

图 4-14

在这种情况下，应遵循"异步复位，同步释放"的原则，具体电路结构将在 4.5 节介绍。

4.4 控制集

当触发器被用作时序逻辑时，也就是被配置为 D 触发器时，其时钟、时钟使能和复位/置位构成了触发器的控制集。例如，两个触发器共享时钟端口，但一个复位信号由其他逻辑

提供，另一个复位信号恒接地，那么这两个触发器的控制集是不同的。不同芯片一个SLICE内所支持的触发器的控制集的个数也是不同的。例如，7系列FPGA每个SLICE内的8个触发器共享时钟端口、时钟使能端口和复位/置位端口。而UltraScale/UltraScale+ FPGA一个SLICE内的16个触发器可以支持2个时钟信号、4个时钟使能信号和2个复位/置位信号。控制集的限制会导致工具在布局阶段移动一些触发器及其输入端连接的查找表的位置。在某些情况下，这些位置并不是最佳位置，从而会造成线延迟过大，这不仅会对资源利用率造成负面影响，而且会对布局质量和功耗造成负面影响。因此，应尽量减少设计中控制集的个数。

> 💡**设计规则 5**：Xilinx给出了控制集的指导建议：当设计中的控制集个数占总控制集的7.5%时是可以接受的；介于7.5%和15%之间时，建议减少控制集个数；大于15%时，则必须减少控制集个数。

这个指导建议给出的建议值并不需要手工计算，只需要查看相应报告即可，这就需要用到如Tcl代码4-7所示的两条命令。第2行代码生成的报告由四部分构成，这里给出前三部分。报告的第一部分是总结，如图4-15所示。在这部分可以看到设计中控制集的个数，在这里为37。同时，总结部分还指出执行report_qor_suggestions命令可自动生成控制集优化建议。报告的第二部分如图4-16所示，显示了控制信号的扇出分布状况。这对设置综合选项-control_set_opt_threshold有帮助。报告的第三部分如图4-17所示，给出了触发器的分布状况。从此报告中可以看出设计中是否存在异步复位/置位触发器。第3行代码生成的报告如图4-18所示。在这个报告中可以直接看到控制集的状态，Status列为OK，若为REVIEW，则需要减少控制集个数。

Tcl 代码 4-7

```
1.  #File: control_set.tcl
2.  report_control_sets -verbose
3.  xilinx::designutils::report_failfast
```

```
1. Summary
   ---------

   +------------------------------------------------------------+-------+
   |                         Status                             | Count |
   +------------------------------------------------------------+-------+
   | Total control sets                                         |  37   |
   |     Minimum number of control sets                         |  37   |
   |     Addition due to synthesis replication                  |   0   |
   |     Addition due to physical synthesis replication         |   0   |
   | Unused register locations in slices containing registers   |  102  |
   +------------------------------------------------------------+-------+
   * Control sets can be merged at opt_design using control_set_merge or merge_equivalent_drivers
   ** Run report_qor_suggestions for automated merging and remapping suggestions
```

图 4-15

```
2. Histogram
----------

+----------------------+--------+
| Fanout               | Count  |
+----------------------+--------+
| Total control sets   |   37   |
| >= 0 to < 4          |    4   |
| >= 4 to < 6          |    5   |
| >= 6 to < 8          |    1   |
| >= 8 to < 10         |    2   |
| >= 10 to < 12        |    6   |
| >= 12 to < 14        |    2   |
| >= 14 to < 16        |    0   |
| >= 16                |   17   |
+----------------------+--------+
* Control sets can be remapped at either synth_design or opt_design
```

图 4-16

```
3. Flip-Flop Distribution
-------------------------

+--------------+---------------------+----------------------+-----------------+---------------+
| Clock Enable | Synchronous Set/Reset | Asynchronous Set/Reset | Total Registers | Total Slices |
+--------------+---------------------+----------------------+-----------------+---------------+
| No           | No                  | No                   | 143             | 45            |
| No           | No                  | Yes                  | 6               | 3             |
| No           | Yes                 | No                   | 147             | 55            |
| Yes          | No                  | No                   | 43              | 14            |
| Yes          | No                  | Yes                  | 0               | 0             |
| Yes          | Yes                 | No                   | 287             | 86            |
+--------------+---------------------+----------------------+-----------------+---------------+
```

图 4-17

```
# +----------------------------------------------------------------+
# | Design Summary                                                 |
# | checkpoint_wave_gen_routed                                     |
# | xc7k70tfbg676-1                                                |
# +----------------------------------------------------------------+
# | Criteria                              | Guideline | Actual  | Status |
# +---------------------------------------+-----------+---------+--------+
# | LUT                                   | 70%       | 1.87%   | OK     |
# | FD                                    | 50%       | 0.76%   | OK     |
# | LUTRAM+SRL                            | 25%       | 0.01%   | OK     |
# | MUXF7                                 | 15%       | 0.01%   | OK     |
# | DSP                                   | 80%       | 0.00%   | OK     |
# | RAMB/FIFO                             | 80%       | 0.74%   | OK     |
# | DSP+RAMB+URAM (Avg)                   | 70%       | 0.74%   | OK     |
# | BUFGCE* + BUFGCTRL                    | 24        | 3       | OK     |
# | DONT_TOUCH (cells/nets)               | 0         | 21      | REVIEW |
# | MARK_DEBUG (nets)                     | 0         | 0       | OK     |
# | Control Sets                          | 769       | 37      | OK     |
# | Average Fanout for modules > 100k cells | 4       | 0       | OK     |
# | Non-FD high fanout nets > 10k loads   | 0         | 0       | OK     |
# +---------------------------------------+-----------+---------+--------+
```

图 4-18

有多种方法可以减少控制集个数。从代码风格的角度看，只在必要时才使用时钟使能或复位/置位的方法。例如，在数据路径上经常会使用很多流水寄存器（只要提到寄存器，就表明该触发器被配置为 D 触发器），如图 4-19 所示。在流水线上，老的数据总会被新的数据"冲掉"，因此，只用在首级和末级使用复位即可。应保持控制信号的极性一致，要么都是高电平有效，要么都是低电平有效。此外，应避免使用低扇出的异步复位/置位，因为综合工具无法将其搬移到数据路径上。

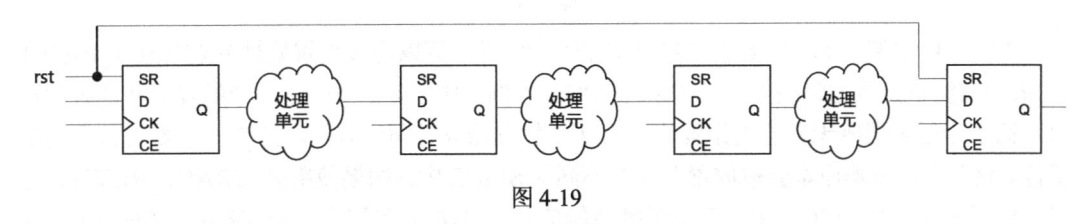

图 4-19

应用案例 1：将同步复位信号搬移到触发器的数据路径上

对于 VHDL 代码 4-8 所描述的同步复位触发器，默认情况下，Vivado 会将其映射为 FDRE，其中，SR 端口连接 rst。可通过如 Tcl 代码 4-8 所示的约束将其搬移到数据路径上（注意代码第 2 行 get_cells 后的对象的名字，对于寄存器，综合后的名字为寄存器输出端的名字加上"_reg"）。EXTRACT_RESET 设置为 no 意味着不允许工具提取出复位信号，如图 4-20 所示。图 4-20 右侧显示了 LUT2 对应的真值表，其实现功能用逻辑表达式可表示为(!rst) & d。对于同步置位，使用同样的约束也可以将其搬移到数据路径上，此时 LUT2 的功能为 set | d。由于异步复位/置位信号不受时钟的控制，因此无法将异步复位/置位信号搬移到数据路径上。由此可见，异步复位/置位必然会增加控制集个数。

VHDL 代码 4-8

```
1.  --File: move_sync_rst.vhd
2.  architecture rtl of move_sync_rst is
3.  begin
4.    process(clk)
5.    begin
6.      if rising_edge(clk) then
7.        q <= '0' when rst else d;
8.      end if;
9.    end process;
10. end architecture;
```

Tcl 代码 4-8

```
1.  #File: extract_rst.tcl
2.  set_property EXTRACT_RESET no [get_cells q_reg]
```

图 4-20

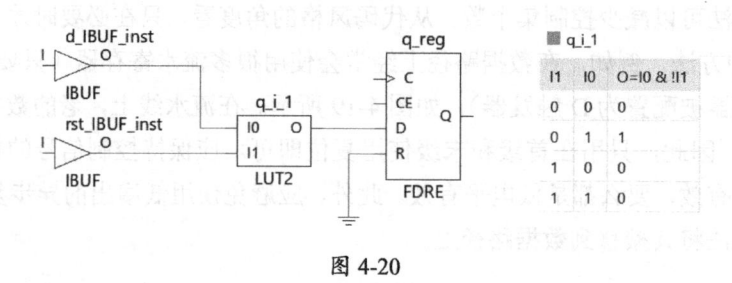

根据应用案例 1 可知，对于同步复位/置位触发器，可以通过设置属性 EXTRACT_RESET 为 no 将其搬移到数据路径上。以 7 系列 FPGA 为例，如图 4-21 所示，3 个触发器共享时钟端口，第 1 个触发器既无复位也无置位，第 2 个触发器只有同步置位，第 3 个触发器只有同步复位。这 3 个触发器原本会被放置在 3 个不同的 SLICE 中，如果使用 EXTRACT_RESET，可将其放置在同一个 SLICE 中。对于时钟使能信号，可通过设置属性 EXTRACT_ENABLE 为 no 将其搬移到数据路径上，如图 4-22 所示。

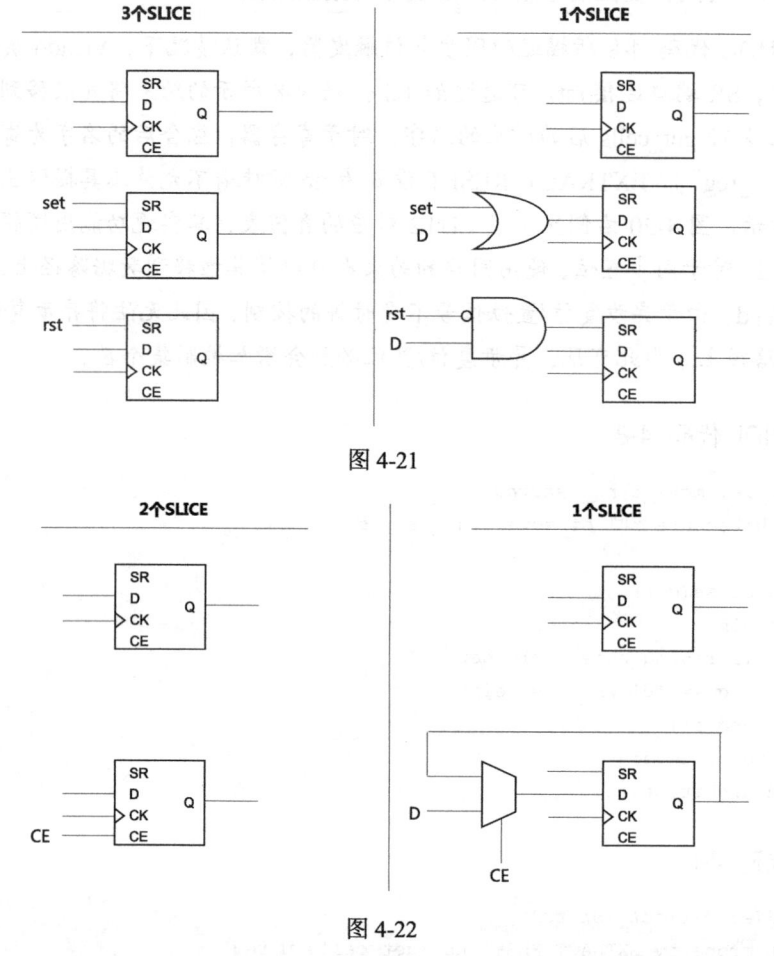

图 4-21

图 4-22

此外，从工具层面，Vivado 也提供了一些方法来减少控制集个数。
（1）慎用 MAX_FANOUT 降低扇出。因为 MAX_FANOUT 会增加控制集。
（2）根据图 4-16 中控制信号的扇出分布状况，增大综合设置选项-control_set_opt_

threshold 的值，如图 4-23 所示。

图 4-23

（3）-control_set_opt_threshold 是一个全局选项，更精细化的方式是采用模块化综合技术，对控制集过多的模块设置合理的 CONTROL_SET_THRESHOLD 值，如 Tcl 代码 4-9 所示。

Tcl 代码 4-9

```
1.  #File: ctrl_set_threshold.tcl
2.  set_property BLOCK_SYNTH.CONTROL_SET_THRESHOLD 16 [get_cells uart/uart_tx]
```

（4）在 opt_design 阶段，添加选项-control_set_merge 或-merge_equivalent_drivers，合并等效控制信号。

（5）设置 CONTROL_SET_REMAP 属性。该属性的值可以是 RESET、ENABLE、ALL 或 NONE，对象必须是触发器，如 Tcl 代码 4-10 所示。可将名字以 ff 开头的触发器的时钟使能信号搬移到数据路径上。

Tcl 代码 4-10

```
1.  #File: ctrl_set_remap.tcl
2.  set_property CONTROL_SET_REMAP ENABLE [get_cells ff*]
```

通常情况下，工具支持的优化方式可先通过命令 report_qor_suggestions 查看，该命令可自动生成相应的 Tcl 脚本或.rqs 文件，其中包含相应的优化建议。

4.5 复位信号的代码风格

4.5.1 异步复位还是同步复位

如果复位是必须的，那么 Xilinx 建议使用同步复位而不是异步复位。这是因为同步复位能更好地匹配硬件架构，如 7 系列 FPGA 及 UltraScale/UltraScale+ FPGA 的 DSP48、BRAM 和 UltraRAM 中的寄存器只支持同步复位，使用异步复位将会导致工具无法将这些寄存器放置在 DSP48、BRAM 或 UltraRAM 的内部，从而影响性能，也增加了触发器的资源利用率。同时，相比异步复位，同步复位更加灵活，这是因为在综合时工具可根据控制集的要求将其搬移到数据路径上，这对于后期改善布局质量或对布局进行微调是很有帮助的。此外，由于同步复位受时钟的控制，因此，时钟可以过滤复位信号上的毛刺，从而防止触发器误动作。异步触发则不受时钟控制，因此复位信号上的毛刺极易导致触发器误动作，如图 4-24 所示。在图 4-24 中，上半部分为同步复位触发器的仿真结果，下半部分为异步复位触发器的仿真结果。尽管如此，如果同步复位信号上的毛刺出现在时钟有效沿附近，也会导致触发器处于亚稳态。使用同步复位时还要保证复位信号有足够的宽度可以被有效时钟沿稳定地捕获到。

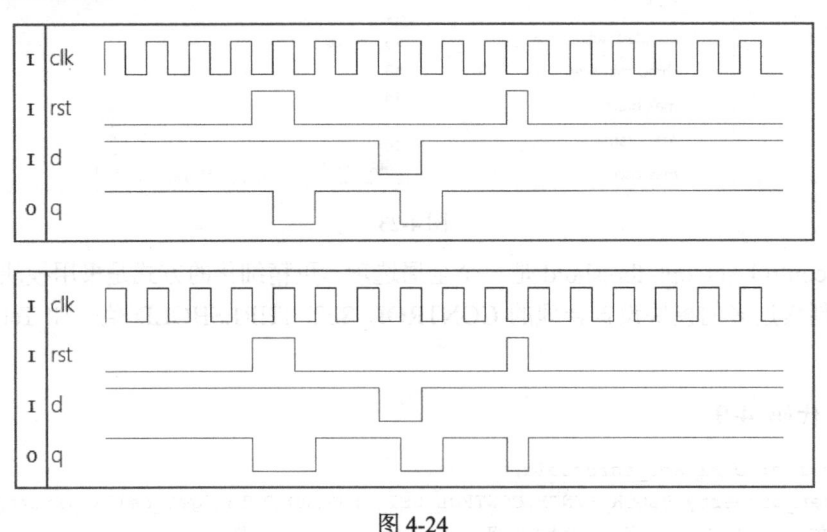

图 4-24

> **设计规则 6**：如果设计中确实需要复位信号，应优先选择使用同步复位。

如果复位信号确实是异步的，可将其采用如图 4-25 所示的方式同步化（同步到时钟域 clk 下），以保证该异步复位信号最终能在时钟 clk 下同步释放。该电路称为"复位桥"，可采用如 VHDL 代码 4-9 所示的方式描述。代码第 6 行的参数 N 应控制在 8 以内。注意，代码第 17 行至第 19 行添加了综合属性 ASYNC_REG，并将其值设置为 TRUE，以保证相应的寄存器在布局时被放置在同一个 SLICE 内。图 4-25 中的电路输出信号 srst 就是同步化复位信号，给需要复位的模块提供复位信号，而这些模块必须是同步复位。

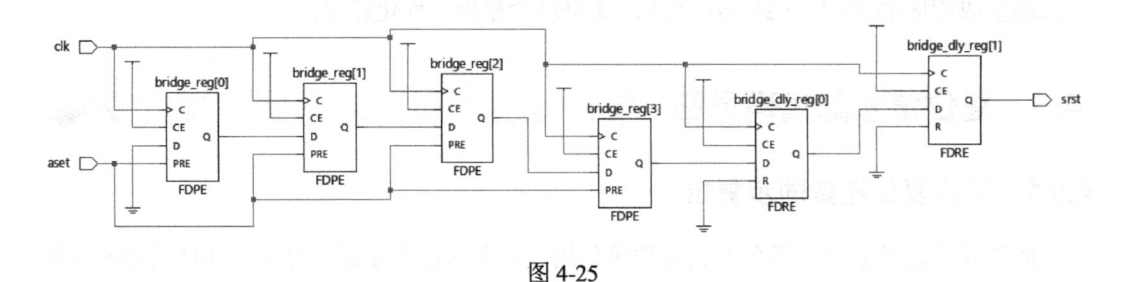

图 4-25

📄 **VHDL 代码 4-9**

```
1.  --File: rst_bridge.vhd
2.  library ieee;
3.  use ieee.std_logic_1164.all;
4.
5.  entity rst_bridge is
```

```vhdl
6.    generic ( N : positive := 4 );
7.    port (
8.      clk  : in std_logic;
9.      aset : in std_logic;
10.     srst : out std_logic
11.   );
12. end entity;
13.
14. architecture rtl of rst_bridge is
15. signal bridge     : std_logic_vector(N - 1 downto 0) := (others => '0');
16. signal bridge_dly : std_logic_vector(    1 downto 0) := (others => '0');
17. attribute ASYNC_REG : string;
18. attribute ASYNC_REG of bridge     : signal is "TRUE";
19. attribute ASYNC_REG of bridge_dly : signal is "TRUE";
20.
21. begin
22.   srst <= bridge_dly(1);
23.
24.   process(clk, aset)
25.   begin
26.     if aset then
27.       bridge <= (others => '1');
28.     elsif rising_edge(clk) then
29.       bridge <= bridge(N - 2 downto 0) & '0';
30.     end if;
31.   end process;
32.
33.   process(clk)
34.   begin
35.     if rising_edge(clk) then
36.       bridge_dly(0) <= bridge(N - 1);
37.       bridge_dly(1) <= bridge_dly(0);
38.     end if;
39.   end process;
40. end architecture;
```

图 4-25 所示的电路的仿真结果如图 4-26（对应输入复位信号为窄脉冲）和图 4-27（对应输入复位信号为宽脉冲）所示。可以看到该电路具备两个功能：①将输入脉冲展宽；②将输入脉冲同步到目标时钟域下。

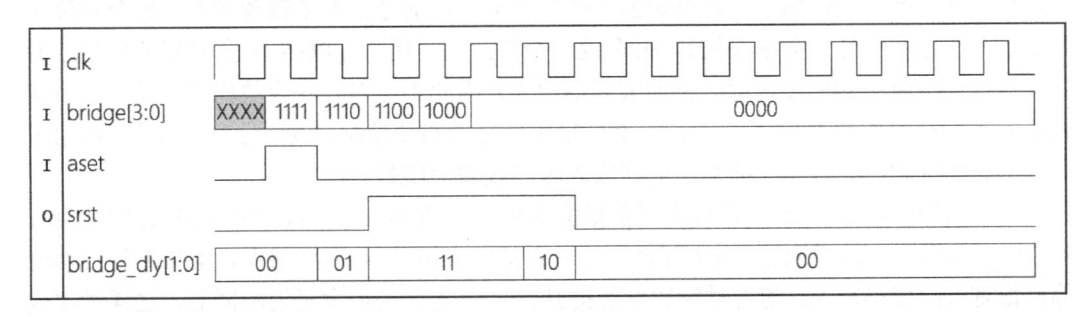

图 4-26

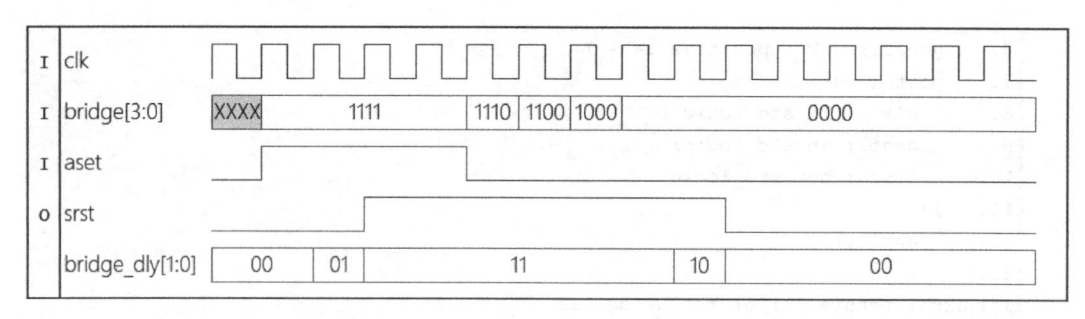

```
  I  clk      ┌┐┌┐┌┐┌┐┌┐┌┐┌┐┌┐┌┐┌┐┌┐┌┐┌┐┌┐
  I  bridge[3:0]  XXXX │ 1111 │1110│1100│1000│     0000
  I  aset         ───┘                    └─────────
  O  srst        ──────┘                  └────────
     bridge_dly[1:0] 00 │ 01 │    11    │ 10 │   00
```

图 4-27

如果设计中还有其他时钟域需要用到此异步复位信号，应将其使用复位桥展宽并同步到相应的时钟域上，如图 4-28 所示。

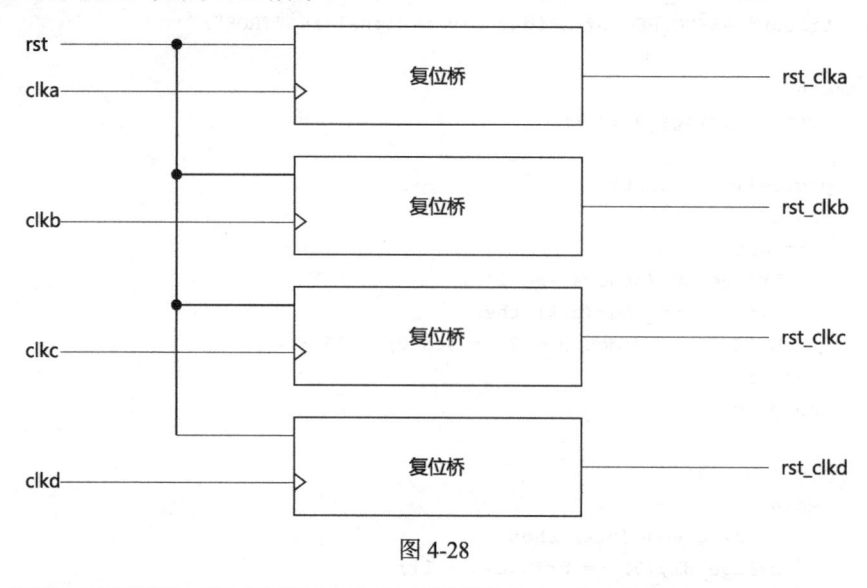

图 4-28

> 💡 **设计规则 7**：对于异步复位，应将其通过复位桥展宽并同步到相应的时钟域上，以保证异步复位、同步释放。

4.5.2 全局复位还是局部复位

从系统的角度考虑，电路板上除了 FPGA，可能还有其他芯片。此时，会在电路板上通过按压开关产生复位信号，也可能由处理器产生复位信号。大多数情况下，这个复位信号是一个宽脉冲信号（频率很低）。很多设计会将此信号用作 FPGA 的全局复位信号，并认为该信号相对于 FPGA 内部时钟而言周期很大，因此不会对时序收敛构成威胁。但事实果真如此吗？图 4-29 展示了某设计中全局复位信号的布线情况，该信号扇出很高，占用了大量的布线资源，导致布线拥塞，最终成为时序收敛的瓶颈。

对于全局复位信号 rst，若其为同步复位信号，如图 4-30 所示，相应的时钟对其会有建立时间的要求，从而限定了 rst 可允许的最大延迟。如果 rst 扇出很大，意味着复位网络偏移很有可能变得很大，随着时钟频率的提升，rst 可允许的最大延迟将降低，这样时序收敛就会面临很大的挑战。

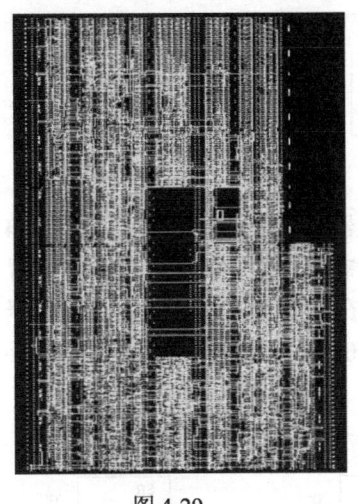

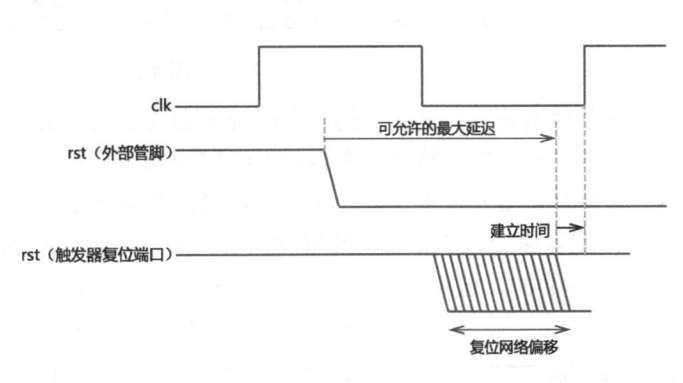

图 4-29   图 4-30

若 rst 为异步复位信号，较大的复位网络偏移将导致受 rst 控制的触发器很难在同一个时钟周期内被激活。如图 4-31 所示，若 rst 在 A 区域内被释放（由高变低），则相应的触发器将在紧随其后的下一个时钟沿被激活；若 rst 在 B 区域内被释放，恰好落在建立时间区间内，则会导致触发器处于亚稳态；若 rst 在 C 区域内被释放，则相应的触发器将在第二个时钟沿被激活。

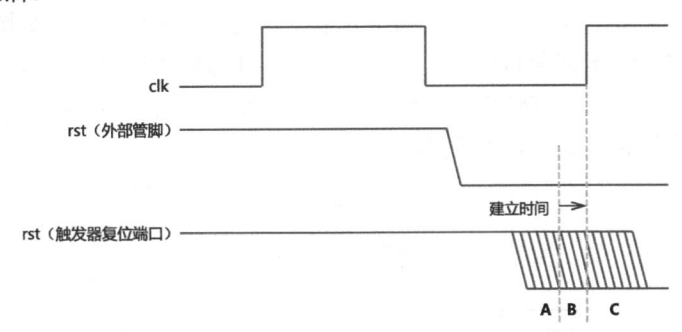

图 4-31

随着时钟频率的提升，考虑到复位网络偏移，受 rst 控制的触发器在同一个时钟周期内被激活将变得越来越困难，如图 4-32 所示。

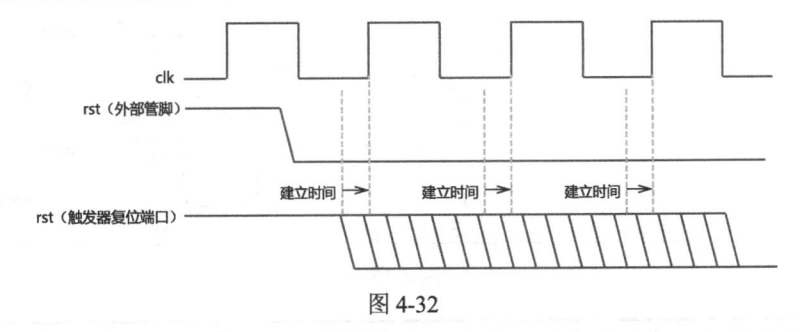

图 4-32

在某些情形下，虽然受 rst 控制的触发器不能在同一个时钟周期内被激活，但是不会影响系统功能，如图 4-33 所示。这是一个典型的数据路径流水线设计，老数据总会被新数据冲刷掉，因此，经过若干个时钟周期之后，该流水线将正常运转。

图 4-33

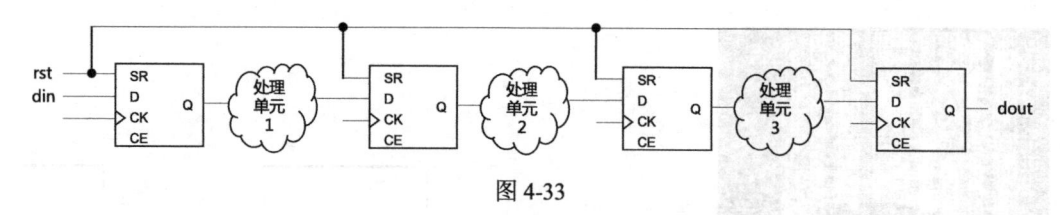

在某些情形下，无法在同一个时钟周期内激活相关触发器将导致功能错误，如图 4-34 所示，这是一条控制路径，产生状态编码方式为独热码的状态机。第一级触发器为异步置位，其余触发器为异步复位。结合图 4-31，若第一级触发器在 A 区域被激活，其余触发器在 C 区域被激活，那么在 A 区域对应的时刻，state 为 4'b1000；在 C 区域对应的时刻，state 将变为 4'b0000，这将导致状态机进入一种无效的状态。

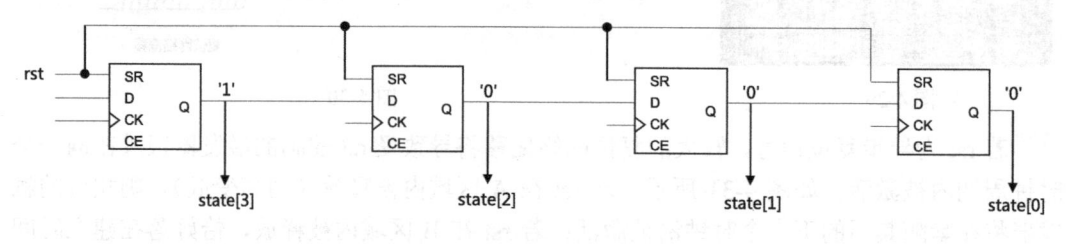

图 4-34

由此可见，全局复位信号潜在的风险在于复位网络偏移较大，而造成此结果的根本原因在于全局复位信号的扇出很大，为此，应通过复位树的方式降低扇出，如图 4-35 所示。如果此复位信号还是一个异步复位信号，那么应先将其通过复位桥同步。对于图 4-35 中的等效寄存器，应添加综合属性 DONT_TOUCH 或使用模块化综合方式 KEEP_EQUIVALENT_REGISTER，防止综合工具将其优化掉。

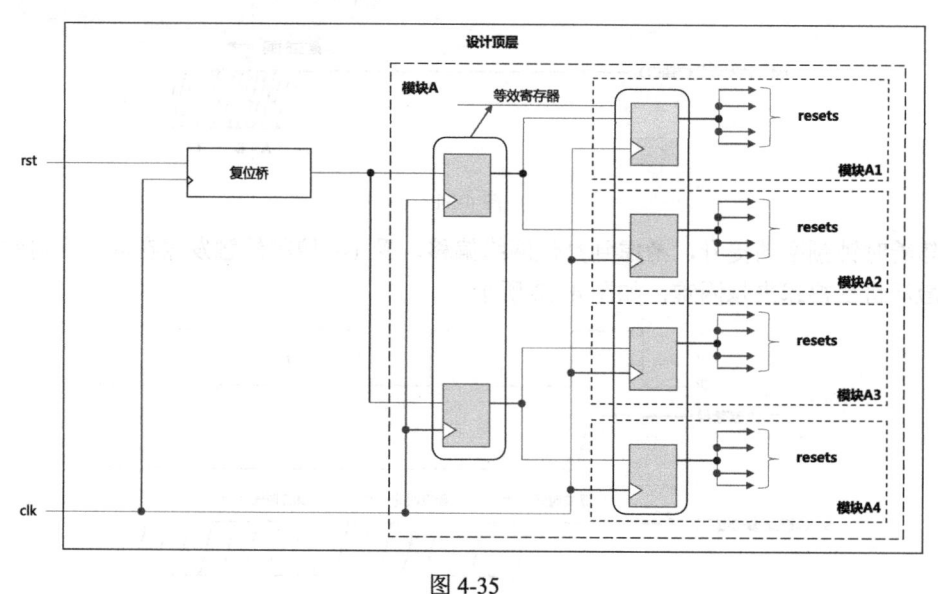

图 4-35

💡 **设计规则 8**：对于全局复位信号，应采用复位树的方式降低扇出，使其变为局部复位信号。

4.5.3 是否需要上电复位

就 FPGA 而言，上电复位是指在上电瞬间执行复位操作，完成对 FPGA 的初始化。初始化的主要对象是设计中的触发器。事实上，对于异步/同步置位触发器，其初始值默认均为 1'b1，对于异步/同步复位触发器，其初始值默认均为 1'b0。另一方面，我们也可以在 HDL 代码里设置触发器的初始值。以 VHDL 代码 4-10 为例，代码第 15 行对信号 cnt_i 完成初始化，实际上，cnt_i 就是一个 4 位触发器。从仿真结果来看，如图 4-36 所示，初始值是有效的。进一步，从最终布线后的网表中查看这 4 个触发器的初始值，如 Tcl 代码 4-11 所示，可以看到，初始值也是生效的。这样看来，如果某触发器的复位信号仅仅是上电复位，那么这个复位信号并没有存在的价值。

VHDL 代码 4-10

```vhdl
1.  --File: counter4.vhd
2.  library ieee;
3.  use ieee.std_logic_1164.all;
4.  use ieee.numeric_std.all;
5.
6.  entity counter4 is
7.    port (
8.      clk : in std_logic;
9.      rst : in std_logic;
10.     cnt : out std_logic_vector(3 downto 0)
11.   );
12. end entity;
13.
14. architecture rtl of counter4 is
15. signal cnt_i : unsigned(3 downto 0) := (others => '0');
16. begin
17.   process(clk)
18.   begin
19.     if rising_edge(clk) then
20.       cnt_i <= (others => '0') when rst else
21.                cnt_i + 1;
22.     end if;
23.   end process;
24.   cnt <= std_logic_vector(cnt_i);
25. end architecture;
```

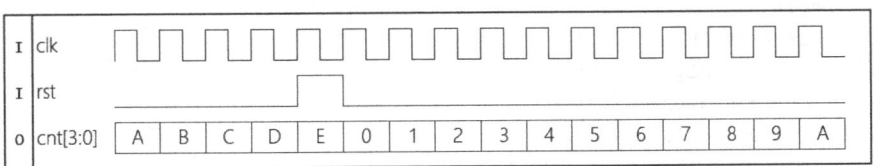

图 4-36

> 💡 **设计规则 9**：如果复位信号仅仅扮演的是上电复位的角色，那么应从设计中移除此复位信号。

Tcl 代码 4-11

```
1.  #File: get_init.tcl
2.  set cnt_i_reg [get_cells cnt_i_reg*]
3.  => cnt_i_reg[0] cnt_i_reg[1] cnt_i_reg[2] cnt_i_reg[3]
4.  get_property INIT $cnt_i_reg
5.  => 1'b0 1'b1 1'b0 1'b1
```

应用案例 2：正确移除设计中不必要的复位信号

如 VHDL 代码 4-11 所示，如果未注释掉代码第 20 行和第 21 行，那么其综合结果将是 3 个级联的异步复位触发器。现在要将前两级的复位信号移除，简单的注释是否有效呢？对比注释前后的综合结果，如图 4-37 所示。图中，上半部分为注释前的综合结果，下半部分为注释后的综合结果。可以看到，注释后，尽管移除了复位信号，但引入了时钟使能信号，且增加了一个查找表，实现取反功能，这是因为时钟使能信号是高电平有效。本质上，if 和 else 分支是等价的，注释后，在 else 分支给 din_dly1 和 din_dly2 赋值，但 if 分支没有对这两个信号进行任何操作，这意味着对这两个信号进行"保持"操作，因此形成了时钟使能信号。

VHDL 代码 4-11

```vhdl
1.  --File: comment_out_rst.vhd
2.  library ieee;
3.  use ieee.std_logic_1164.all;
4.
5.  entity comment_out_rst is
6.    port (
7.      clk  : in std_logic;
8.      arst : in std_logic;
9.      din  : in std_logic;
10.     dout : out std_logic
11.   );
12. end entity;
13.
14. architecture rtl of comment_out_rst is
15. signal din_dly1, din_dly2 : std_logic;
16. begin
17.   process(clk, arst)
18.   begin
19.     if arst then
20.       -- din_dly1 <= '0';
21.       -- din_dly2 <= '0';
22.       dout <= '0';
23.     elsif rising_edge(clk) then
24.       din_dly1 <= din;
25.       din_dly2 <= din_dly1;
26.       dout     <= din_dly2;
27.     end if;
```

```
28.    end process;
29. end architecture;
```

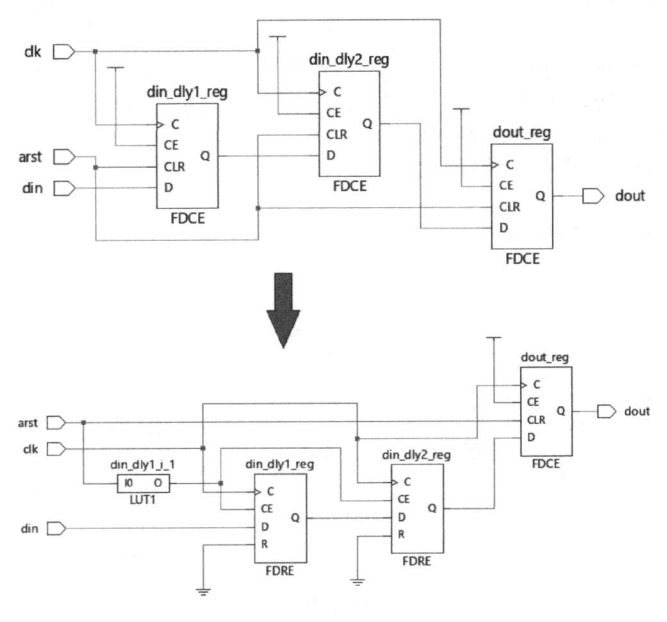

图 4-37

有效移除异步复位信号的方式如 VHDL 代码 4-12 所示,将需要复位的信号和不需要复位的信号写在不同的进程里,此时综合结果如图 4-38 所示。

📄 VHDL 代码 4-12

```
1.  --File: remove_rst.vhd
2.  library ieee;
3.  use ieee.std_logic_1164.all;
4.
5.  entity remove_rst is
6.    port (
7.      arst : in std_logic;
8.      clk  : in std_logic;
9.      din  : in std_logic;
10.     dout : out std_logic
11.   );
12. end entity;
13.
14. architecture rtl of remove_rst is
15. signal din_dly1, din_dly2 : std_logic;
16. begin
17.   process(clk, arst)
18.   begin
19.     if arst then
20.       dout <= '0';
21.     elsif rising_edge(clk) then
22.       dout <= din_dly2;
23.     end if;
```

```
24.    end process;
25.
26.    process(clk)
27.    begin
28.      if rising_edge(clk) then
29.        din_dly1 <= din;
30.        din_dly2 <= din_dly1;
31.      end if;
32.    end process;
33. end architecture;
```

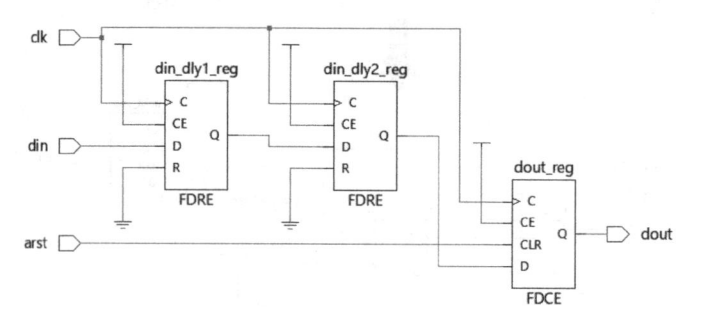

图 4-38

> **设计规则 10**：将需要复位的信号和不需要复位的信号分开写在不同的进程里。

如果复位信号由组合逻辑生成，应将该组合逻辑单独处理。VHDL 代码 4-13 将复位信号 srst 和 hrst 均列在了敏感变量列表里，如代码第 17 行所示。代码第 19 行表明 srst 和 hrst 相或运算之后构成复位信号。这种代码风格是不可取的。VHDL 代码 4-14 则体现了良好的代码风格，srst 和 hrst 相或的逻辑被单独放在第 5 行，在 process 之外。这增强了代码的可读性和可维护性，尽管 VHDL 代码 4-13 和 VHDL 代码 4-14 的综合结果相同，如图 4-39 所示。

📄 **VHDL 代码 4-13**

```
1.  --File: separate_logic_rst_poor.vhd
2.  library ieee;
3.  use ieee.std_logic_1164.all;
4.
5.  entity separate_logic_rst_poor is
6.    port (
7.      clk  : in std_logic;
8.      srst : in std_logic;
9.      hrst : in std_logic;
10.     d    : in std_logic;
11.     q    : out std_logic
12.   );
13. end entity;
14.
15. architecture rtl of separate_logic_rst_poor is
16. begin
```

```
17.    process(clk, srst, hrst)
18.    begin
19.      if (srst or hrst) then
20.        q <= '0';
21.      elsif rising_edge(clk) then
22.        q <= d;
23.      end if;
24.    end process;
25. end architecture
```

📄 **VHDL 代码 4-14**

```
1.  --File: separate_logic_rst_good.vhd
2.  architecture rtl of separate_logic_rst_good is
3.    signal rst : std_logic;
4.  begin
5.    rst <= srst or hrst;
6.    process(clk, rst)
7.    begin
8.      if rst then
9.        q <= '0';
10.     elsif rising_edge(clk) then
11.       q <= d;
12.     end if;
13.   end process;
14. end architecture;
```

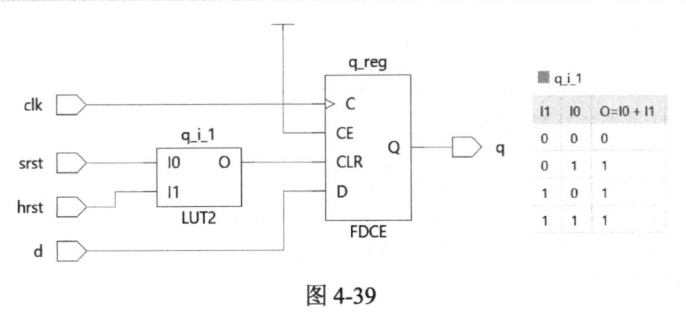

图 4-39

4.6 同步边沿检测电路代码风格

在 FPGA 设计中，我们会经常遇到边沿检测。当信号发生变化时，我们需要检测这种变化，以触发相应的电路操作。有时需要检测信号的上升沿，有时需要检测信号的下降沿，有时则需要同时检测这两个边沿。若输入信号在接收触发器的时钟域内，那么该信号与触发器时钟是同步的。此时可通过简单的组合逻辑检测信号边沿。

在如图 4-40 所示的电路中，输入信号 siga 与时钟 clk 同步，经触发器输出 siga_d1，在此基础上通过组合逻辑生成 siga_fall_edge（用于检测 siga 的下降沿）、siga_rise_edge（用于检测 siga 的上升沿）和 siga_both_edge（用于检测 siga 的上升沿和下降沿）。相应的 RTL 代码如 VHDL 代码 4-15 所示，仿真结果如图 4-41 所示。

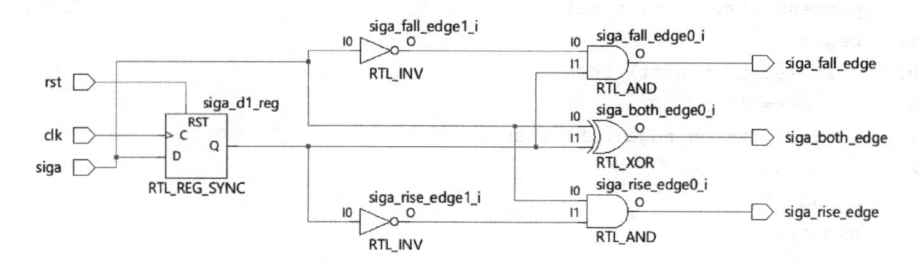

图 4-40

📄 VHDL 代码 4-15

```vhdl
1.  --File: sync_edge_checker.vhd
2.  library ieee;
3.  use ieee.std_logic_1164.all;
4.
5.  entity sync_edge_checker is
6.    port (
7.      clk            : in std_logic;
8.      rst            : in std_logic;
9.      siga           : in std_logic;
10.     siga_rise_edge : out std_logic;
11.     siga_fall_edge : out std_logic;
12.     siga_both_edge : out std_logic
13.   );
14. end entity;
15.
16. architecture rtl of sync_edge_checker is
17. signal siga_d1 : std_logic;
18. begin
19.   siga_rise_edge <= siga and (not siga_d1);
20.   siga_fall_edge <= (not siga) and siga_d1;
21.   siga_both_edge <= siga xor siga_d1;
22.   process(clk)
23.   begin
24.     if rising_edge(clk) then
25.       siga_d1 <= '0' when rst else siga;
26.     end if;
27.   end process;
28. end architecture;
```

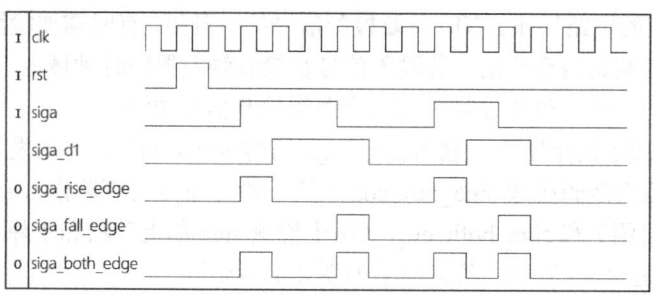

图 4-41

4.7 串并互转电路代码风格

若输入信号与接收触发器时钟不在同一个时钟域，那么需要先对输入信号进行跨时钟域处理，再用如图 4-40 所示的电路完成边沿检测。

在 FPGA 设计中，我们会经常碰到串并转换或并串转换。串并转换是指输入为一位数据，输出为多位并行数据。以 4 位并行数据为例，输入与输出之间的时序关系如图 4-42 所示。串行输入数据 sin 依次输入 0110，对应十进制数 6，接着输入 0111，对应十进制数 7。当 combine 信号为高电平时，说明 4 位数据已到齐，可以捕获输出。

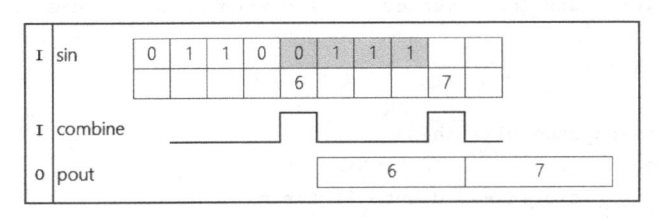

图 4-42

在此时序图的基础上可得相应的电路结构，如图 4-43 所示，图中，左侧的 4 个寄存器串联起来构成一个移位寄存器，右侧的 4 个寄存器执行捕获功能，一旦时钟使能信号有效，就捕获输入数据并将其输出。采用 RTL 代码描述，如 VHDL 代码 4-16 所示。代码第 19 行至第 25 行对应左侧寄存器，代码第 27 行至第 32 行对应右侧寄存器。

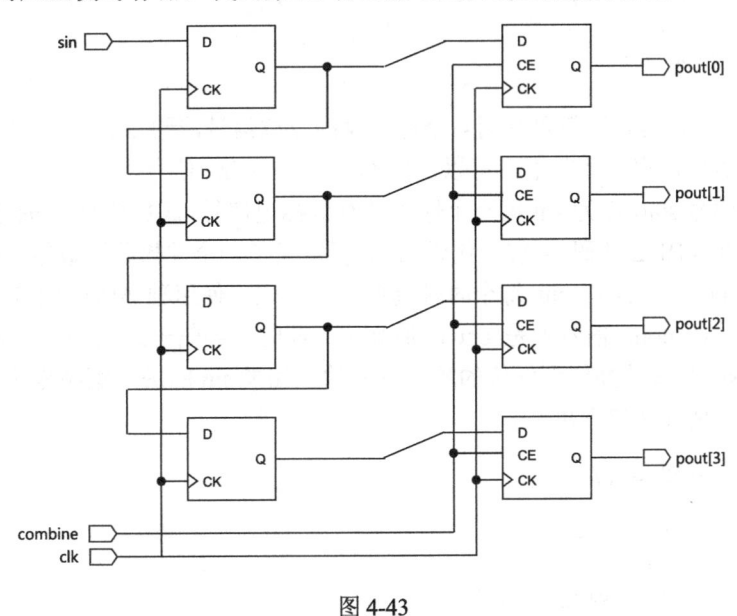

图 4-43

VHDL 代码 4-16

```
1. --File: s2p_v1.vhd
2. library ieee;
3. use ieee.std_logic_1164.all;
4.
```

```vhdl
5.  entity s2p_v1 is
6.    generic ( W : positive := 4 );
7.    port (
8.      clk     : in std_logic;
9.      rst     : in std_logic;
10.     combine : in std_logic;
11.     sin     : in std_logic;
12.     pout    : out std_logic_vector(W - 1 downto 0)
13.   );
14. end entity;
15.
16. architecture rtl of s2p_v1 is
17. signal sreg : std_logic_vector(W - 1 downto 0) := (others => '0');
18. begin
19.   process(clk)
20.   begin
21.     if rising_edge(clk) then
22.       sreg <= (others => '0') when rst else
23.               sreg(W - 2 downto 0) & sin;
24.     end if;
25.   end process;
26.
27.   process(clk)
28.   begin
29.     if rising_edge(clk) then
30.       pout <= sreg when combine else pout;
31.     end if;
32.   end process;
33. end architecture;
```

进一步分析，其实并不需要左侧 4 个寄存器的复位信号，因此可以将此复位信号移除。此外，在已知数据位宽的情况下，可以由内部产生 combine 信号，无须外部提供。同时，可以添加一个信号 start 作为 sin 第一个输入的有效标记信号，添加信号 done 表明串并转换结束。最终形成 VHDL 代码 4-17。代码第 18 行定义了一个独热码计数器 cnt，当 start 为 1 时，cnt 为"0001"，否则，cnt 每个时钟周期左移一位。使用独热码的优势在于可简化后续的译码操作，如 done 信号就是 cnt 的最高位。此时，各信号之间的时序关系如图 4-44 所示。这里，sin 是按从高位到低位的顺序进入的，如果 sin 的进入顺序是从低位到高位，该如何处理呢？请读者自行思考。

📄 **VHDL 代码 4-17**

```vhdl
1.  --File: s2p_v2.vhd
2.  library ieee;
3.  use ieee.std_logic_1164.all;
4.
5.  entity s2p_v2 is
6.    generic ( W : positive := 4 );
7.    port (
8.      clk   : in std_logic;
9.      start : in std_logic;
10.     sin   : in std_logic;
```

```vhdl
11.      pout  : out std_logic_vector(W - 1 downto 0);
12.      done  : out std_logic
13.   );
14. end entity;
15.
16. architecture rtl of s2p_v2 is
17.   signal sreg : std_logic_vector(W - 1 downto 0) := (others => '0');
18.   signal cnt  : std_logic_vector(W - 1 downto 0) := (others => '0');
19. begin
20.   done <= cnt(W - 1);
21.   process(clk)
22.   begin
23.     if rising_edge(clk) then
24.       sreg <= sreg(W - 2 downto 0) & sin;
25.     end if;
26.   end process;
27.
28.   process(clk)
29.   begin
30.     if rising_edge(clk) then
31.       cnt <= (W - 1 downto 1 => '0') & '1' when start else
32.              cnt sll 1;
33.     end if;
34.   end process;
35.
36.   process(clk)
37.   begin
38.     if rising_edge(clk) then
39.       pout <= sreg when cnt(W - 1) else pout;
40.     end if;
41.   end process;
42. end architecture;
```

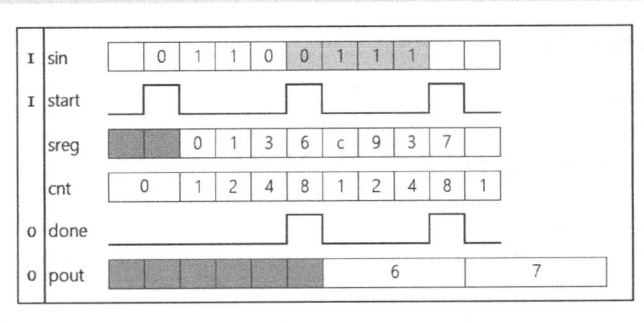

图 4-44

🔍 VHDL-2008 新特性

在 VHDL-2008 中，sll 左操作数可以是 std_logic_vector 类型，如 VHDL 代码 4-17 第 32 行所示。

并入串出正好是串入并出的逆过程。我们期望的时序关系如图 4-45 所示。当 load 信号为高电平时，加载并行数据 pin，之后并串转换电路由高位至低位将 pin 串行输出。基于此时序关系，可得如图 4-46 所示的电路图。这里假定输入并行数据的位宽为 4。

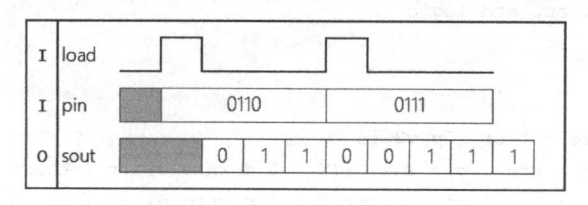

图 4-45

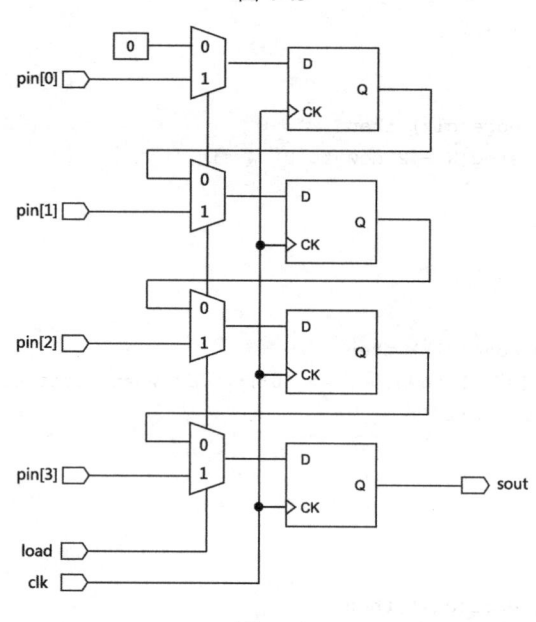

图 4-46

对于如图 4-46 所示的电路，采用 RTL 代码描述，如 VHDL 代码 4-18 所示，这里只是引入了复位信号。

📄 **VHDL 代码 4-18**

```vhdl
1.  --File: p2s_v1.vhd
2.  library ieee;
3.  use ieee.std_logic_1164.all;
4.
5.  entity p2s_v1 is
6.    generic ( W : positive := 4 );
7.    port (
8.      clk  : in std_logic;
9.      rst  : in std_logic;
10.     load : in std_logic;
11.     pin  : in std_logic_vector(W - 1 downto 0);
12.     sout : out std_logic
13.   );
14. end entity;
15.
16. architecture rtl of p2s_v1 is
17.   signal pin_shift : std_logic_vector(W - 1 downto 0);
```

```
18. begin
19.   sout <= pin_shift(W - 1);
20.   process(clk)
21.   begin
22.     if rising_edge(clk) then
23.       case? std_logic_vector'(rst, load) is
24.         when "1-" => pin_shift <= (others => '0');
25.         when "01" => pin_shift <= pin;
26.         when others => pin_shift <= pin_shift sll 1;
27.       end case?;
28.     end if;
29.   end process;
30. end architecture;
```

> **🔍 VHDL-2008 新特性**
>
> 在 VHDL-2008 中新增了 case?语句，对于有优先级的场合，其比嵌套的 if-elsif-else 语句更为直观、简洁，从而有效减少了代码行数，如 VHDL 代码 4-18 第 23 行所示。

进一步分析，这里并不需要复位信号。同时，为了与其他模块互连，我们引入信号 done 用于表示并转串结束；rdy 用于表示可以接收新的数据；busy 用于表示当前为忙碌状态，不能接收新数据，从而形成 VHDL 代码 4-19，各信号之间的时序关系如图 4-47 所示。

📄 **VHDL 代码 4-19**

```
1.  --File: p2s_v2.vhd
2.  library ieee;
3.  use ieee.std_logic_1164.all;
4.
5.  entity p2s_v2 is
6.    generic ( W : positive := 4 );
7.    port (
8.      clk  : in std_logic;
9.      load : in std_logic;
10.     pin  : in std_logic_vector(W - 1 downto 0);
11.     sout : out std_logic;
12.     done : out std_logic;
13.     rdy  : out std_logic;
14.     busy : out std_logic
15.   );
16. end entity;
17.
18. architecture rtl of p2s_v2 is
19. signal pin_shift : std_logic_vector(W - 1 downto 0) := (others => '0');
20. signal cnt       : std_logic_vector(W - 1 downto 0) := (others => '0');
21. begin
22.   sout <= pin_shift(W - 1);
23.   done <= cnt(W - 1);
24.   busy <= or cnt;
25.   rdy  <= not busy;
26.   process(clk)
```

```
27.    begin
28.      if rising_edge(clk) then
29.        if load then
30.          pin_shift <= pin;
31.          cnt       <= (W - 1 downto 1 => '0') & '1';
32.        else
33.          pin_shift <= pin_shift sll 1;
34.          cnt       <= cnt sll 1;
35.        end if;
36.      end if;
37.    end process;
38. end architecture;
```

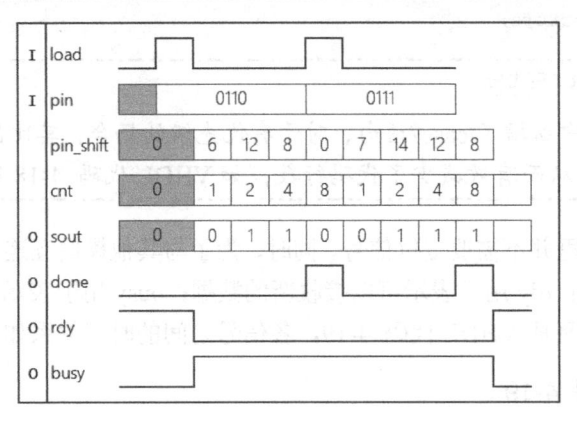

图 4-47

4.8 避免意外生成的锁存器

在 FPGA 设计中，应尽量避免锁存器，一方面，锁存器的存在会浪费一些触发器，另一方面，create_clock 创建的时钟周期约束无法覆盖锁存器的相关路径，如图 4-48 所示（用命令 report_clock_networks 生成时钟网络报告），这将导致静态时序分析不完备，也就无法保证设计能正常工作。

图 4-48

那么在什么情况下会生成锁存器呢？

工具会将不完备的 if 语句推断为锁存器，如 VHDL 代码 4-20 所示，该段代码在 Vivado 下会被综合为 LDCE，VHDL 代码 4-21 会被综合为 LDPE，两者均为锁存器。其根本原因在于 if 语句没有对应的 else 分支。

VHDL 代码 4-20

```vhdl
1.  --File: incomplete_if0.vhd
2.  process(all)
3.  begin
4.    if s then
5.      q <= a;
6.    end if;
7.  end process;
```

VHDL 代码 4-21

```vhdl
1.  --File: incomplete_if1.vhd
2.  process(all)
3.  begin
4.    if s0 then
5.      q <= '1';
6.    elsif s1 then
7.      q <= a;
8.    end if;
9.  end process;
```

对于不完备的设计语句，为了达到同样的功能，可将其转换为带时钟使能端口的触发器，以 VHDL 代码 4-21 为例，优化后的代码如 VHDL 代码 4-22 所示，代码第 5 行对应同步置位功能，代码第 7 行对应同步使能功能。对比两者的综合结果，如图 4-49 所示。由于 Vivado 将同步置位和同步使能信号搬移到了数据路径，所以还会消耗一个查找表。查找表的功能如图 4-49 右侧所示。

VHDL 代码 4-22

```vhdl
1.   --File: incomplete_if1_opt.vhd
2.   process(clk)
3.   begin
4.     if rising_edge(clk) then
5.       if s0 then
6.         q <= '1';
7.       elsif s1 then
8.         q <= a;
9.       end if;
10.    end if;
11.  end process;
```

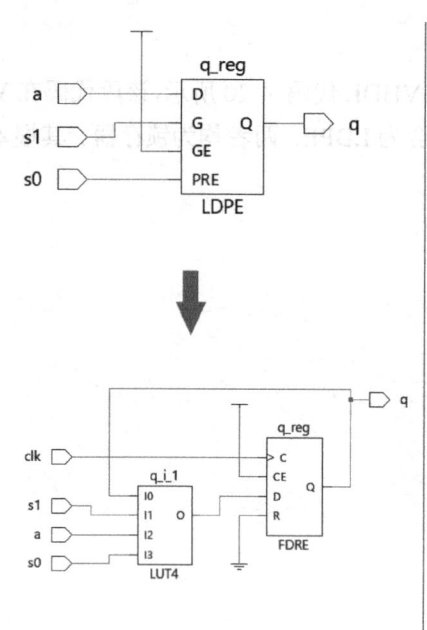

图 4-49

不同于 Verilog 和 System Verilog，VHDL 是强类型语言，使用 case 语句时，要求 case 语句必须完备，否则在综合阶段就会报错。但即使如此，如果在某个条件分支要求输出保持不变，那么依然会导致生成锁存器，如 VHDL 代码 4-23 所示。代码第 7 行表明当 s 为 "10" 或 "11" 时输出值保持不变，综合结果如图 4-50 所示，图中，LDCE 为锁存器。实际上，case 语句中的 when others 分支等效于 if 语句中的 else 分支，因此，在 if 语句中，虽然写明了 else 分支，但是该分支为"输出<=输出"，即输出保持不变，工具也会将其推断为锁存器。

VHDL 代码 4-23

```
1.  --File: case0_latch.vhd
2.  process(all)
3.  begin
4.    case s is
5.      when "00" => q <= a(0);
6.      when "01" => q <= a(1);
7.      when others => q <= q;
8.    end case;
9.  end process;
```

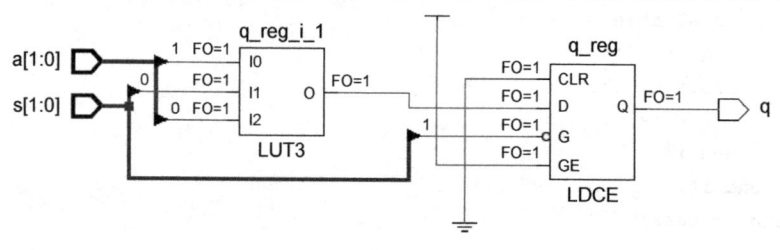

图 4-50

VHDL 代码 4-24 也会被 Vivado 推断为锁存器，这是因为代码第 10 行和第 11 行都出现了输出保持不变的情形。但这段代码也给我们一个好的示范，即当不同的条件分支对应的输出相同时，可以将其放在一行描述，如代码第 5 行和第 7 行所示，这样可以使代码更为简短。

VHDL 代码 4-24

```
1.   --File: case1_latch.vhd
2.   process(all)
3.   begin
4.     case s is
5.       when "00" | "01" =>
6.         q0 <= a(0);
7.       when "10" | "11" =>
8.         q1 <= a(1);
9.       when others =>
10.        q0 <= q0;
11.        q1 <= q1;
12.    end case;
13.  end process;
```

类似地，在使用 when-else 和 with-select 语句时，如果出现输出保持不变的情形，那么工具也会将其推断为锁存器。

4.9 思考空间

1．结合 Tcl 代码 4-1，试用 Tcl 脚本获取指定芯片中一列 SLICE 内所包含的触发器的个数。

2．试写出一个既包含同步置位信号又包含同步复位信号的触发器，在 Vivado 下查看其综合结果，列出其资源利用率，并解释为什么在实际工程中要避免这类触发器。

3．试给出造成建立时间违例的可能的因素，并给出消除建立时间违例的方法。

4．从时序的角度而言，如何评估系统的 F_{max}？

5．试给出亚稳态发生的两种典型场合，并给出相应的降低亚稳态发生概率的方法。

6．试用 VHDL 描述如图 4-11 所示的电路并观察添加 ASYNC_REG 和不添加 ASYNC_REG 时布局布线的差异。

7．分别用 set_false_path 和 set_max_delay 对如图 4-11 所示的电路进行约束，并观察布局布线和时序报告的差异。

8．试评估如图 4-21 所示的电路，若在 UltraScale FPGA 芯片中不将控制信号搬移到数据路径上，会占用几个 SLICE？

9．试评估如图 4-22 所示的电路，若在 UltraScale FPGA 芯片中不将控制信号搬移到数据路径上，会占用几个 SLICE？

10．对于如图 4-25 所示的电路，如何保证前 4 级触发器在布局时可以被放置在同一个 SLICE 内？

11. 对于 VHDL 代码 4-9 中的参数 N，若目标芯片为 Versal ACAP，则 N 的最大值是多少？

12. 对于如图 4-35 所示的电路中的等效寄存器，可以采用哪些方法防止 Vivado 在综合时将其优化掉（提示：有三种方法）？并比较这些方法的优缺点。

13. 试列出全局复位信号可能给设计带来的弊端。

14. 试列出异步复位信号可能给设计带来的弊端。

15. 试将 VHDL 代码 4-20 转换为带时钟使能信号的触发器。

第 5 章

优化移位寄存器

5.1 移位寄存器资源

5.1.1 7 系列 FPGA 中的移位寄存器资源

7 系列 FPGA 中的 SLICE 分为两大类：SLICEL 和 SLICEM。每个 SLICE 内都有 4 个查找表（Look-Up Table，LUT）。其中，SLICEL 中的 LUT 仅可用作逻辑函数发生器，而 SLICEM 中的 LUT 不仅可以用作逻辑函数发生器，还可以用作存储器，这也是 SLICEM 名称的由来（M 表示 Memory）。对比 SLICEL 和 SLICEM 中的 LUT，如图 5-1 所示（左侧为 SLICEL 中的 LUT，右侧为 SLICEM 中的 LUT），不难看出输入/输出端口的差异。

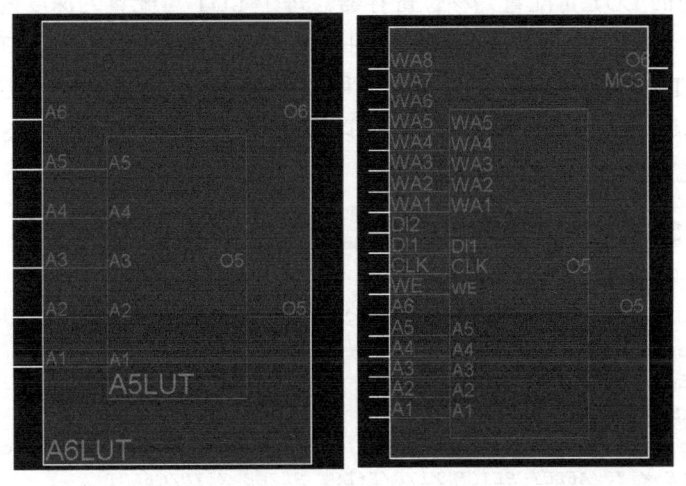

图 5-1

在存储器中，比较典型的一种是移位寄存器。SLICEM 中的每个 LUT 可实现深度为 32 的移位寄存器（等效于 32 个寄存器级联），意味着可将一位数据延迟 32 个时钟周期。而同一个 SLICEM 中的 LUT 又可通过专用级联走线构成深度更大的移位寄存器，最大可构成深度为 128（32×4）的移位寄存器，这意味着可将一位数据延迟 128 个时钟周期。这 4 个 LUT 的级联方向是由上至下。如果需要更大深度（深度大于 128）的移位寄存器，就不得不将不同 SLICEM 中的 LUT 级联起来，这时跨越 SLICEM 之间的走线不是专用走线。

从图 5-1 中还可看出，LUT 并没有置位/复位端口，因此，不管是同步还是异步，将 LUT 用作移位寄存器时都不支持置位/复位。

> **设计规则 1**：用 RTL 代码描述移位寄存器时，如果期望其映射为 SLICEM 中的 LUT，则不能添加置位/复位功能。

借助 Tcl 代码 5-1 可查看 7 系列 FPGA 中指定 SLICEM 内的 LUT 分布状况。

Tcl 代码 5-1

```
1.  #File: check_srl_7fpga.tcl
2.  set mypart [get_parts xc7k70tfbg484-3]
3.  link_design -part $mypart
4.  set slice [get_sites SLICE_X2Y0]
5.  => SLICE_X2Y0
6.  set lutram_in_slice [get_bels -filter "TYPE==LUT_OR_MEM6" -of $slice]
7.  => SLICE_X2Y0/D6LUT SLICE_X2Y0/C6LUT SLICE_X2Y0/B6LUT SLICE_X2Y0/A6LUT
8.  puts "The number of LUTRAM in one SLICEM: [llength $lutram_in_slice]"
9.  => The number of LUTRAM in one SLICEM: 4
```

5.1.2 UltraScale/UltraScale+ FPGA 中的移位寄存器资源

和 7 系列 FPGA 一样，UltraScale/UltraScale+ FPGA 内也有 SLICEL 和 SLICEM 之分，而且 SLICEM 中的 LUT 可配置为移位寄存器。每个 LUT 可配置为深度为 32 的移位寄存器。不同之处在于，在 UltraScale/UltraScale+ FPGA 中，每个 SLICEM 内有 8 个 LUT，因此，这 8 个 LUT 级联可构成深度为 256 的移位寄存器，这意味着对一位数据可实现 256 个时钟周期的延迟，级联方向为由上至下。可以借助 Tcl 代码 5-2 查看 UltraScale/UltraScale+ FPGA 中指定 SLICEM 内的 LUT 分布状况。

Tcl 代码 5-2

```
1.  #File: check_srl_us.tcl
2.  set mypart [get_parts xcvu5p-flva2104-2-i]
3.  link_design -part $mypart
4.  set slice [get_sites SLICE_X1Y0]
5.  => SLICE_X1Y0
6.  set lutram_in_slice [get_bels -filter "NAME=~*6LUT" -of $slice]
7.  => SLICE_X1Y0/A6LUT SLICE_X1Y0/B6LUT SLICE_X1Y0/C6LUT …
8.  puts "The number of LUTRAM in one SLICEM: [llength $lutram_in_slice]"
9.  => The number of LUTRAM in one SLICEM: 8
```

5.1.3 Versal ACAP 中的移位寄存器资源

和 UltraScale/UltraScale+ FPGA 一样，Versal ACAP 的每个 SLICEM 内也有 8 个 LUT。每个 LUT 可配置为深度为 32 的移位寄存器，而且这 8 个 LUT 可级联构成深度为 256 的移位寄存器，级联方向变为由下至上，更为重要的是相邻 SLICEM 中的 LUT 也可级联构成深度更大的移位寄存器，仍然使用专用走线。可以借助 Tcl 代码 5-3 查看 Versal ACAP 中指定 SLICEM 内的 LUT 分布状况。

Tcl 代码 5-3

```
1.  #File: check_srl_versal.tcl
2.  set mypart [get_parts xcvc1902-vsva2197-2MP-e-S]
3.  link_design -part $mypart
4.  set slice [get_sites SLICE_X79Y52]
5.  => SLICE_X79Y52
6.  set lutram_in_slice [get_bels -filter "TYPE==SLICEM_LUT6" -of $slice]
7.  => SLICE_X79Y52/A6LUT SLICE_X79Y52/B6LUT SLICE_X79Y52/C6LUT ...
8.  puts "The number of LUTRAM in one SLICEM: [llength $lutram_in_slice]"
9.  => The number of LUTRAM in one SLICEM: 8
```

以深度为 512 的移位寄存器为例，显然需要消耗 512/32 也就是 16 个 LUT。图 5-2 由左至右分别显示了 7 系列 FPGA、UltraScale/UltraScale+ FPGA 和 Versal ACAP 中这 16 个 LUT 的分布情况和级联方向，其中，箭头方向就是级联方向。不难看出，对 7 系列 FPGA 而言，这 16 个 LUT 分布在 4 个 SLICEM 中。SLICEM 内的 LUT 级联方向为由上至下，跨越 SLICEM 时，级联方向为由下至上。对于 UltraScale/UltraScale FPGA 而言，这 16 个 LUT 分布在 2 个 SLICEM 中，SLICEM 内的 LUT 及 SLICEM 边界的 LUT 的级联方向均为由上至下。对于 Versal ACAP 而言，这 16 个 LUT 也分布在 2 个 SLICEM 中，SLICEM 内的 LUT 与 SLICEM 边界的 LUT 均自下而上级联。

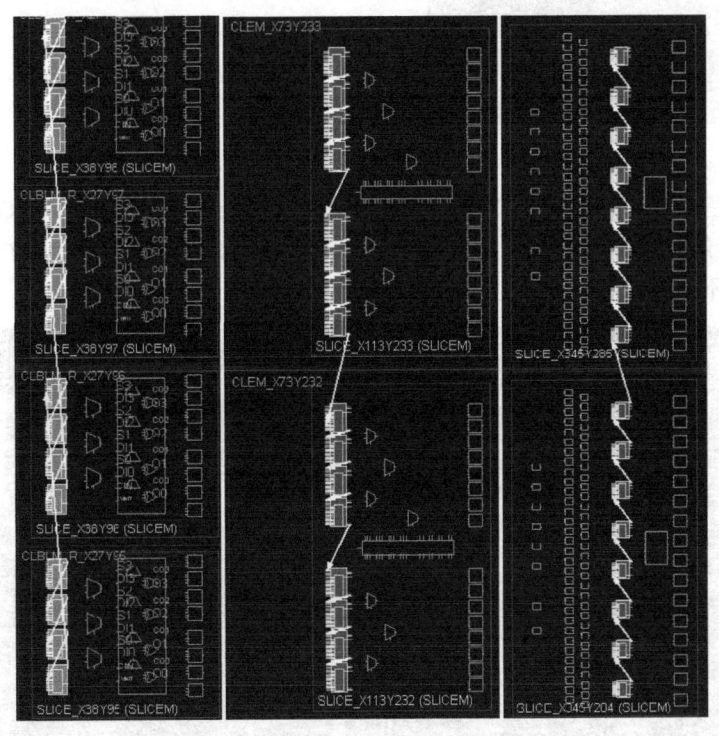

图 5-2

图 5-3 由左至右分别显示了 7 系列 FPGA、UltraScale/UltraScale+ FPGA 和 Versal ACAP 中 SLICEM 内部相邻 LUT 的走线状况。不难看出，7 系列 FPGA 的走线比较长，而 UltraScale/UltraScale+ FPGA 和 Versal ACAP 的走线短了很多，这样可以有效改善布线延迟。

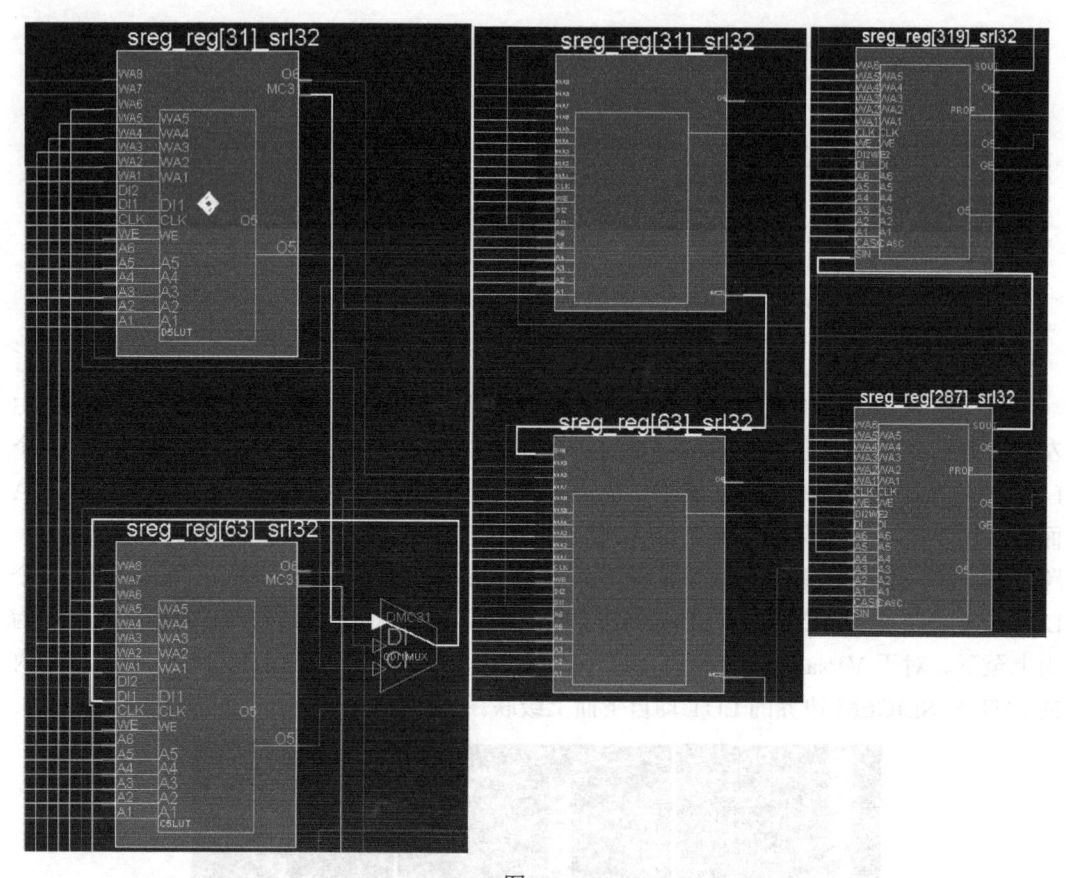

图 5-3

图 5-4 由左至右分别显示了 7 系列 FPGA 和 UltraScale/UltraScale+ FPGA 中 SLICEM 边界 LUT 的走线状况，图 5-5 则显示了 Versal ACAP 中 SLICEM 边界 LUT 的走线状况。将两者对比，不难看出，Versal ACAP 中的走线更短一些。

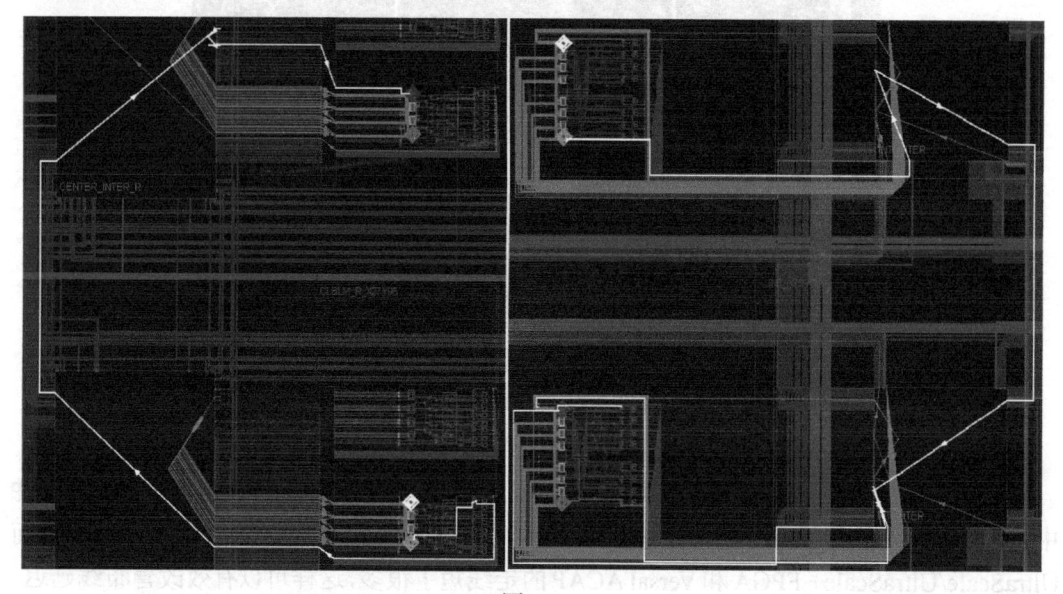

图 5-4

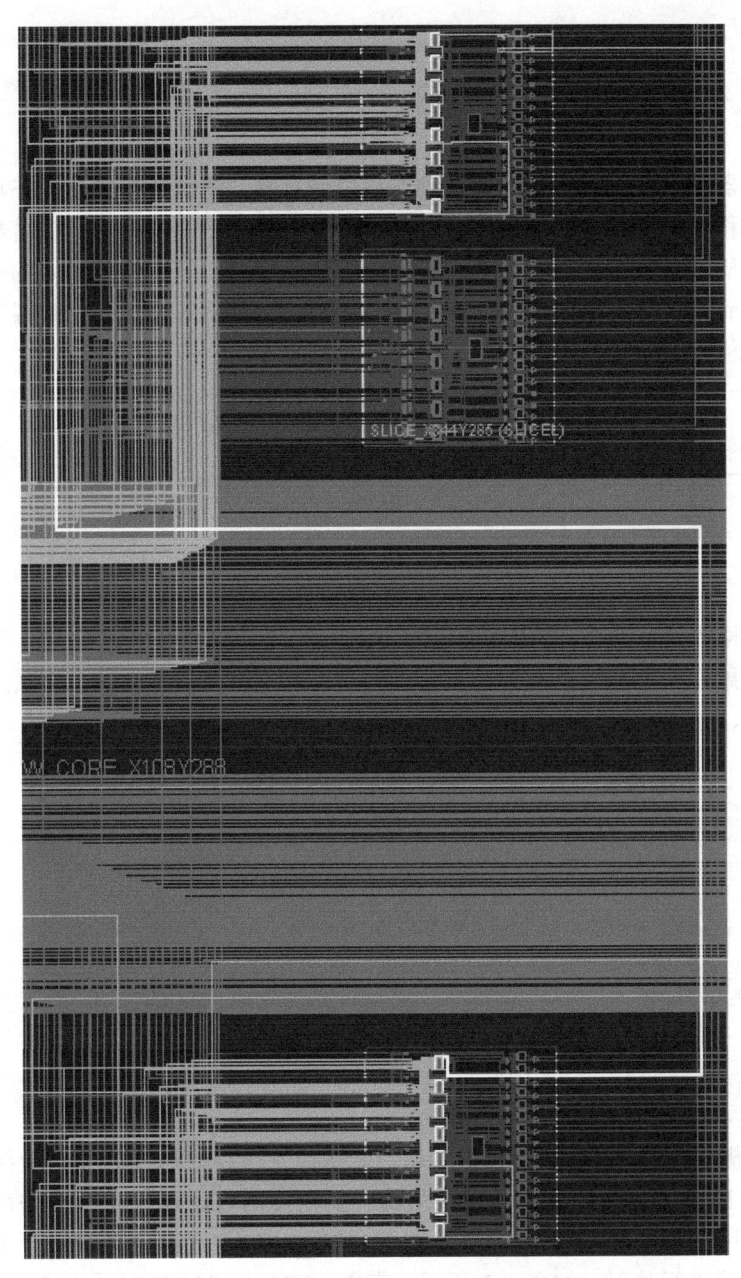

图 5-5

表 5-1 对 7 系列 FPGA、UltraScale/UltraScale+ FPGA 和 Versal ACAP 中 SLICEM 内的 LUT 进行了比较，可以看到其中的共性和差异。

表 5-1

性 能 指 标	7 系列 FPGA	UltraScale/UltraScale+ FPGA	Versal ACAP
每个 SLICEM 中 LUT 的个数	4	8	8
每个 SLICEM 中单个 LUT 可实现的移位寄存器深度	32	32	32
单个 SLICEM 内 LUT 级联可构成的移位寄存器深度	128	256	256
每个 SLICEM 内 LUT 的级联方向	由下至上	由下至上	由上至下

5.2 移位寄存器的代码风格

移位寄存器分为静态移位寄存器和动态移位寄存器。这里我们先从静态移位寄存器谈起。所谓静态移位寄存器,是指移位寄存器的深度是固定不变的,这意味着从输入到输出所需要的时钟周期个数是固定不变的。从功能的角度而言,一个深度为 4 的移位寄存器可用如图 5-6 所示的 4 个级联的寄存器表示。d 为输入数据端口,clk 为时钟端口,ce 为时钟使能端口,q 为输出数据端口。

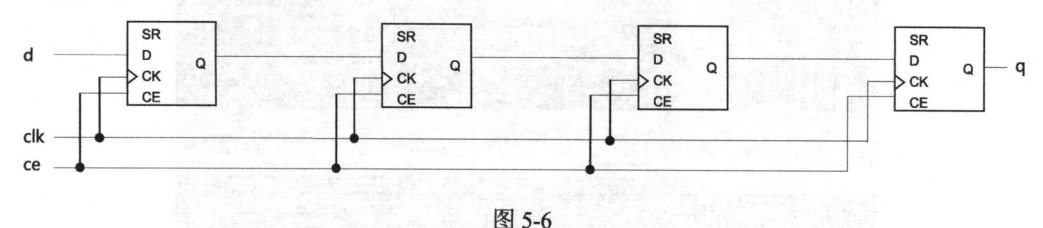

图 5-6

从代码风格的角度而言,图 5-6 所示的电路可以采用如 VHDL 代码 5-1 所示的方式描述。但是,如果输入数据位宽不再是 1 位或移位寄存器深度不再是 4,那么此代码就无法使用了。因此,这样的代码不具有可复用性。

VHDL 代码 5-1

```
1.  --File: static_sreg_poor.vhd
2.  process(clk)
3.  begin
4.    if rising_edge(clk) then
5.      a_dly1 <= a;
6.      a_dly2 <= a_dly1;
7.      a_dly3 <= a_dly2;
8.      q      <= a_dly3;
9.    end if;
10. end process;
```

Xilinx 也提供了两个原语,即 SRL16E 和 SRLC32E,但这样的原语最终只能映射为 SLICEM 中的 LUT,因此缺乏灵活性。从综合的角度而言,移位寄存器是可以由级联寄存器构成的,因此,Vivado 提供了综合属性 SRL_STYLE,用于指导综合工具将移位寄存器按指定形式进行映射,如表 5-2 所示,表中,SRL 为 SLICEM 中的 LUT 实现的移位寄存器。

表 5-2

SRL_STYLE	实 现 方 式
register	寄存器
srl	SRL
srl_reg	SRL + 寄存器
reg_srl	寄存器 + SRL
reg_srl_reg	寄存器 + SRL + 寄存器
block	BRAM

同时，Vivado 综合设置也提供了选项-shreg_min_size，如图 5-7 所示，其默认值为 3，意味着当移位寄存器的深度大于或等于 3 时，Vivado 会将其映射为"SRL+寄存器"的形式。

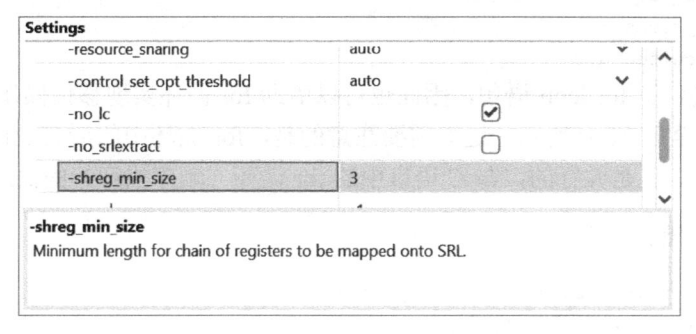

图 5-7

-shreg_min_size 是个全局选项，因此对设计中的移位寄存器都生效，是一种粗粒度的管理形式，相比而言，SRL_STYLE 更精确一些，因为它能具体到指定的移位寄存器。因此，我们首先将移位寄存器的深度和 SRL_STYLE 参数化，数据位宽为固定值 1。结合如图 5-6 所示的级联寄存器，形成 VHDL 代码 5-2。这段代码给我们的启示是可以将移位寄存器链路上的每个寄存器的输出取出来，构成一个向量，如代码第 19 行所示，从而很容易描述"移位"这个动作，如代码第 27 行所示。同时，代码第 19 行显示了如何将移位寄存器的初始值设置为 0。

VHDL 代码 5-2

```vhdl
1.  --File: static_single_bit_sreg_v1.vhd
2.  library ieee;
3.  use ieee.std_logic_1164.all;
4.
5.  entity static_single_bit_sreg_v1 is
6.    generic (
7.      DEPTH         : positive := 8;
8.      SRL_STYLE_VAL : string   := "srl"
9.    );
10.   port (
11.     clk : in  std_logic;
12.     ce  : in  std_logic;
13.     si  : in  std_logic;
14.     so  : out std_logic
15.   );
16. end entity;
17.
18. architecture rtl of static_single_bit_sreg_v1 is
19.   signal sreg : std_logic_vector(DEPTH - 1 downto 0) := (others => '0');
20.   attribute SRL_STYLE : string;
21.   attribute SRL_STYLE of sreg : signal is SRL_STYLE_VAL;
22. begin
23.   so <= sreg(DEPTH - 1);
24.   process(clk)
25.   begin
```

```
26.      if rising_edge(clk) then
27.         sreg <= sreg(DEPTH - 2 downto 0) & si when ce else sreg;
28.      end if;
29.   end process;
30. end architecture;
```

VHDL 还提供了 for-loop 语句,因此也可以借助 for 循环实现移位操作,如 VHDL 代码 5-3 第 29 行至第 31 行所示。这里需要注意的是,for 循环中的所有操作都是在一个时钟周期内完成的,且是并行的,与 C 语言中的 for 循环(顺序执行)的概念是不一样的。

📄 **VHDL 代码 5-3**

```
1.  --File: static_single_bit_sreg_v2.vhd
2.  library ieee;
3.  use ieee.std_logic_1164.all;
4.
5.  entity static_single_bit_sreg_v2 is
6.    generic (
7.      DEPTH         : positive := 8;
8.      SRL_STYLE_VAL : string   := "srl"
9.    );
10.   port (
11.     clk : in  std_logic;
12.     ce  : in  std_logic;
13.     si  : in  std_logic;
14.     so  : out std_logic
15.   );
16. end entity;
17.
18. architecture rtl of static_single_bit_sreg_v2 is
19.   signal sreg : std_logic_vector(DEPTH - 1 downto 0) := (others => '0');
20.   attribute SRL_STYLE : string;
21.   attribute SRL_STYLE of sreg : signal is SRL_STYLE_VAL;
22. begin
23.   so <= sreg(DEPTH - 1);
24.   process(clk)
25.   begin
26.     if rising_edge(clk) then
27.       if ce then
28.         sreg(0) <= si;
29.         for i in 0 to DEPTH - 2 loop
30.           sreg(i + 1) <= sreg(i);
31.         end loop;
32.       end if;
33.     end if;
34.   end process;
35. end architecture;
```

VHDL 代码 5-2 和 VHDL 代码 5-3 都是可行的。在此基础上,很容易实现位宽的参数化,只需要用 for generate 语句即可,如 VHDL 代码 5-4 所示。基于 VHDL 代码 5-3 构成的多位宽移位寄存器的描述方式与 VHDL 代码 5-4 类似,这里不再详述。

VHDL 代码 5-4

```vhdl
1.  --File: static_multi_bit_sreg_v1.vhd
2.  library ieee;
3.  use ieee.std_logic_1164.all;
4.
5.  entity static_multi_bit_sreg_v1 is
6.    generic (
7.      WIDTH         : positive := 4;
8.      DEPTH         : positive := 8;
9.      SRL_STYLE_VAL : string   := "srl_reg"
10.   );
11.   port (
12.     clk : in std_logic;
13.     ce  : in std_logic;
14.     si  : in std_logic_vector(WIDTH - 1 downto 0);
15.     so  : out std_logic_vector(WIDTH - 1 downto 0)
16.   );
17. end entity;
18.
19. architecture rtl of static_multi_bit_sreg_v1 is
20. begin
21.   gen0 :
22.   for i in 0 to WIDTH - 1 generate
23.     i_static_single_bit_sreg_v1 :
24.     entity work.static_single_bit_sreg_v1
25.       generic map (
26.         DEPTH => DEPTH,
27.         SRL_STYLE_VAL => SRL_STYLE_VAL
28.       )
29.       port map (
30.         clk => clk,
31.         ce  => ce,
32.         si  => si(i),
33.         so  => so(i)
34.       );
35.   end generate;
36. end architecture;
```

VHDL 代码 5-4 中输入/输出信号之间的时序关系如图 5-8 所示。当 ce 为周期信号时，输入/输出信号之间的时序关系如图 5-9 所示。结合这两个时序图，读者可以更深入地理解移位寄存器的功能。应注意 ce 在移位寄存器中所起的作用。

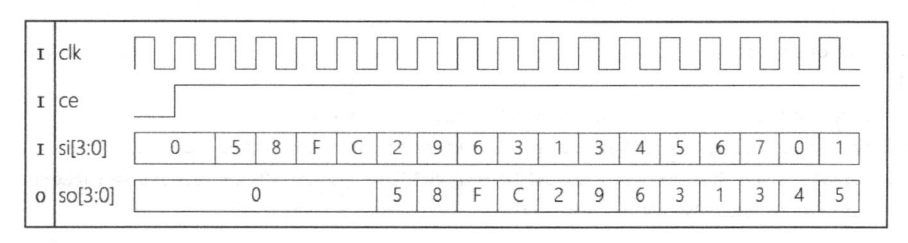

图 5-8

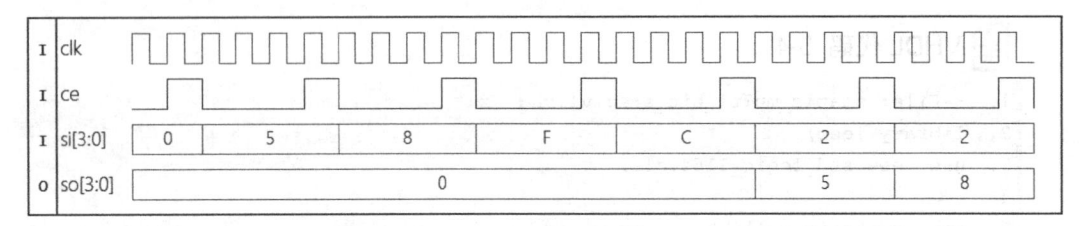

```
 I  clk   ||||||||||||||||||||||||||||||||||
 I  ce    ___|‾‾‾|___|‾‾‾‾‾‾‾|___|‾‾‾‾‾‾‾‾‾|___|‾‾‾‾‾‾‾|___
 I  si[3:0]  0  |  5  |  8   |  F  |  C  |  2  |  2
 O  so[3:0]        0                          |  5  |  8
```

图 5-9

VHDL 代码 5-4 体现的是"搭积木"的电路描述思维，实际上，我们可以借助 VHDL 中的数组直接描述静态移位寄存器。这里将移位寄存器的深度、宽度和综合属性 SRL_STYLE 作为参数，如 VHDL 代码 5-5 所示。代码第 31 行定义了数组 sreg，深度为 DEPTH，宽度为 WIDTH，初始值全为 0。代码第 41 行至第 43 行表明移位操作是通过 for 循环实现的。

VHDL 代码 5-5

```vhdl
1.  /*
2.  //File: static_multi_bit_sreg_v3.sv
3.  //Author: Gao Yajun
4.  //Parameters:
5.  //DEPTH: define the clock cycle from input to output. Namely Latency
6.  //DEPTH must be equal to or greater than 1
7.  //WIDTH: data width
8.  //SRL_STYLE_VAL: register, srl, reg_srl, srl_reg, reg_srl_reg
9.  //If DEPTH < 5, then register is optimal
10. //Otherwise, you can set to srl_reg or reg_srl_reg or reg_srl
11. */
12. library ieee;
13. use ieee.std_logic_1164.all;
14.
15. entity static_multi_bit_sreg_v3 is
16.   generic (
17.     DEPTH         : positive := 64;
18.     WIDTH         : positive := 4;
19.     SRL_STYLE_VAL : string := "srl_reg"
20.   );
21.   port (
22.     clk : in std_logic;
23.     ce  : in std_logic;
24.     si  : in std_logic_vector(WIDTH - 1 downto 0);
25.     so  : out std_logic_vector(WIDTH - 1 downto 0)
26.   );
27. end entity;
28.
29. architecture rtl of static_multi_bit_sreg_v3 is
30. type bus_array is array(natural range<>) of std_logic_vector;
31. signal sreg : bus_array(0 to DEPTH - 1)(WIDTH-1 downto 0) := (others => (others => '0'));
32. attribute SRL_STYLE : string;
```

```
33.   attribute SRL_STYLE of sreg : signal is SRL_STYLE_VAL;
34. begin
35.   so <= sreg(DEPTH - 1);
36.   process(clk)
37.   begin
38.     if rising_edge(clk) then
39.       if ce then
40.         sreg(0) <= si;
41.         for i in 1 to DEPTH - 1 loop
42.           sreg(i) <= sreg(i - 1);
43.         end loop;
44.       end if;
45.     end if;
46.   end process;
47. end architecture;
```

🔍 VHDL-2008 新特性

VHDL-2008 新增了区域注释符，也称为块注释符（Block Comments），即"/* 被注释内容 */"，如 VHDL 代码 5-5 第 1 行至第 11 行所示。此外，VHDL-2008 还允许在定义数组类型时只定义元素数据类型，如 VHDL 代码 5-5 第 30 行所示，这为定义不同规格的数组变量提供了便利，如在此基础上可方便地定义以下数组变量：

signal sreg1 : bus_array(0 to 3)(7 downto 0);
signal sreg2 : bus_array(0 to 15)(31 downto 0);

此外，也可通过 for generate 语句实现多位宽移位寄存器，如 VHDL 代码 5-6 所示。这里只给出结构体部分，其综合结果与 VHDL 代码 5-5 完全一致。

📄 VHDL 代码 5-6

```
1.  --File: static_multi_bit_sreg_v4.vhd
2.  architecture rtl of static_multi_bit_sreg_v4 is
3.    type bus_array is array(natural range<>) of std_logic_vector;
4.    signal sreg : bus_array(0 to DEPTH - 1)(WIDTH - 1 downto 0) := (others => (others => '0'));
5.    attribute SRL_STYLE : string;
6.    attribute SRL_STYLE of sreg : signal is SRL_STYLE_VAL;
7.  begin
8.    so <= sreg(DEPTH - 1);
9.    process(clk)
10.   begin
11.     if rising_edge(clk) then
12.       sreg(0) <= si when ce else sreg(0);
13.     end if;
14.   end process;
15.
16.   gen0:
17.   for i in 1 to DEPTH - 1 generate
18.     process(clk)
19.     begin
```

```
20.         if rising_edge(clk) then
21.            sreg(i) <= sreg(i - 1) when ce else sreg(i);
22.         end if;
23.      end process;
24.   end generate;
25. end architecture;
```

对于深度为 18、宽度为 2 的移位寄存器，目标芯片为 XCVU5P，需要消耗 2 个 SLICEM 中的 LUT，如图 5-10 所示。SRLC32E 的地址端口 A[4:0] 为 5'b10001，对应十进制数 17，因此移位寄存器深度为 18。换言之，对于 SRLC32E，深度为 A + 1。若深度为 36，宽度为 2，则需要消耗 4 个 SLICEM 中的 LUT，如图 5-11 所示。可以看到，每个 SLICEM 中的 LUT 提供的数据宽度为 1 位，深度为 32，借助此特点可评估设计中的移位寄存器所消耗的 LUT 个数。

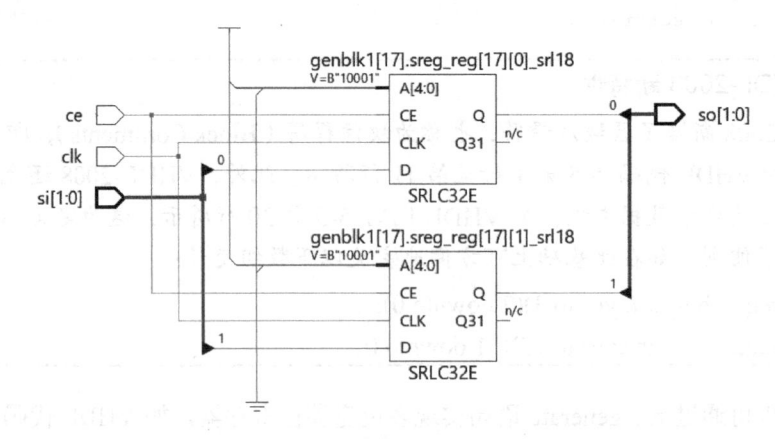

图 5-10

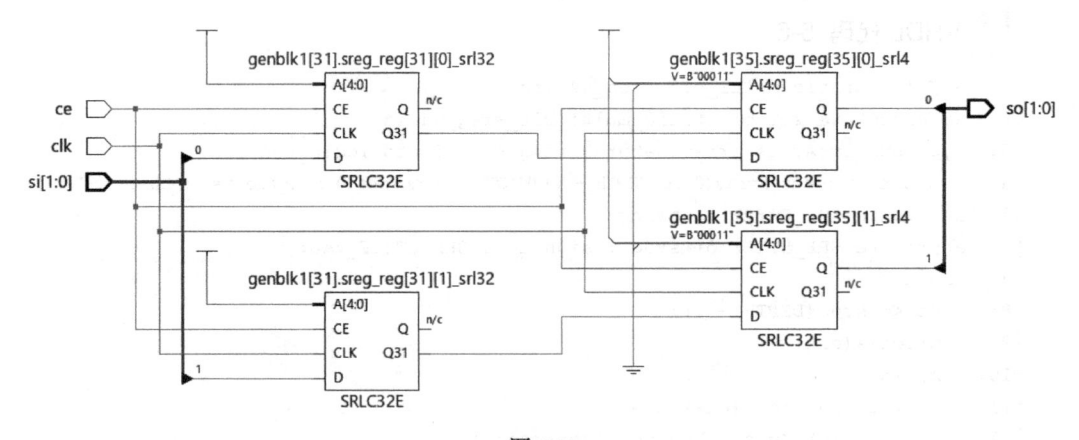

图 5-11

从图 5-10 和图 5-11 中可以看出，此时 SRLC32E 的 A 端口的输入值是固定的，因此相应的移位寄存器称为静态移位寄存器，也叫作固定长度移位寄存器。如果其输入值是可变的，那么相应的移位寄存器就是动态移位寄存器，也叫作可变长度移位寄存器。这里的"可变"体现在移位深度位于 1 和最大深度之间，是通过地址端口实现的，其工作原理如图 5-12 所示。这是一个深度为 4 的移位寄存器，端口 q 最终输出的数据取决于地址端 addr，也就是数据选择器的选择端。例如，若 addr 为 0，则端口 q 输出 1 号寄存器 Q 端口数据，相当于移位深度为 1。若 addr 为 2，则端口 q 输出 3 号寄存器 Q 端口数据，相当于移位深度为 3，这就是"动态"的原因。从图 5-12 中还可以看出，端口 addr 和 q 不受时钟的控制，因此其输出是异步的，有时为改善时序，会在 q 端口添加寄存器。端口 addr 和 q 也与时钟使能端口无关。

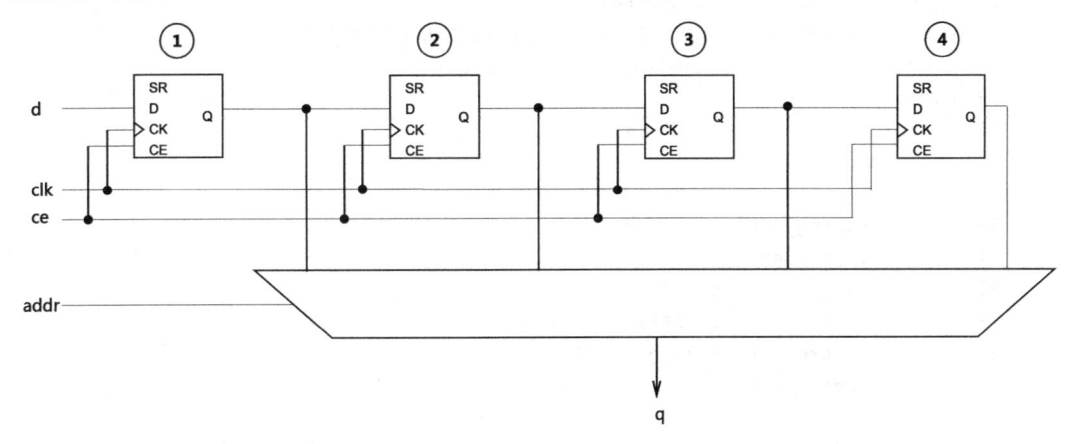

图 5-12

采用 VHDL 代码描述时，需要考虑数据位宽、地址位宽和综合属性 SRL_STYLE。其中，地址位宽决定了移位寄存器的最大深度。同时，考虑到时序因素，输出是否添加寄存器可由参数控制，最终如 VHDL 代码 5-7 所示。从代码第 12 行可以看出移位寄存器最大深度与地址位宽之间的关系。代码第 31 行至第 41 行与静态移位寄存器的描述方式一致。代码第 43 行至第 53 行通过 if generate 语句控制输出端口是否添加寄存器。

📄 **VHDL 代码 5-7**

```
1.   --File: dynamic_sreg.vhd
2.   library ieee;
3.   use ieee.std_logic_1164.all;
4.   use ieee.numeric_std.all;
5.
6.   entity dynamic_sreg is
7.     generic (
8.       AW            : positive := 2;
9.       DW            : positive := 1;
10.      IS_SYNC       : boolean  := true;
11.      SRL_STYLE_VAL : string   := "srl";
12.      DEPTH         : positive := 2 ** AW
13.    );
14.    port (
15.      clk : in std_logic;
16.      ce  : in std_logic;
17.      si  : in std_logic_vector(DW - 1 downto 0);
18.      addr: in std_logic_vector(AW - 1 downto 0);
19.      so  : out std_logic_vector(DW - 1 downto 0)
```

```
20.     );
21. end entity;
22.
23. architecture rtl of dynamic_sreg is
24.   type bus_array is array(natural range<>) of std_logic_vector;
25.   signal sreg : bus_array(0 to DEPTH - 1)(DW - 1 downto 0) := (others => (others => '0'));
26.   attribute SRL_STYLE : string;
27.   attribute SRL_STYLE of sreg : signal is SRL_STYLE_VAL;
28.   signal addr_u : unsigned(AW - 1 downto 0);
29. begin
30.   addr_u <= unsigned(addr);
31.   process(clk)
32.   begin
33.     if rising_edge(clk) then
34.       if ce then
35.         sreg(0) <= si;
36.         for i in 1 to DEPTH - 1 loop
37.           sreg(i) <= sreg(i - 1);
38.         end loop;
39.       end if;
40.     end if;
41.   end process;
42.
43.   gen0:
44.   if IS_SYNC generate
45.     process(clk)
46.     begin
47.       if rising_edge(clk) then
48.         so <= sreg(to_integer(addr_u));
49.       end if;
50.     end process;
51.   else generate
52.     so <= sreg(to_integer(addr_u));
53.   end generate;
54. end architecture;
```

> **VHDL-2008 新特性**
>
> VHDL-2008 的条件生成语句 if-generate 支持 else 分支,这样当条件互斥时,就可以直接用 if-generate 加上 else generate 分支描述,从而使代码更为简洁,如 VHDL 代码 5-7 第 44 行和第 51 行所示。

图 5-13 和图 5-14 反映的是 VHDL 代码 5-7 的输入/输出信号之间的时序关系。其中,前者对应 IS_SYNC 为 true(输出添加寄存器),后者对应 IS_SYNC 为 false(输出未添加寄存器)。

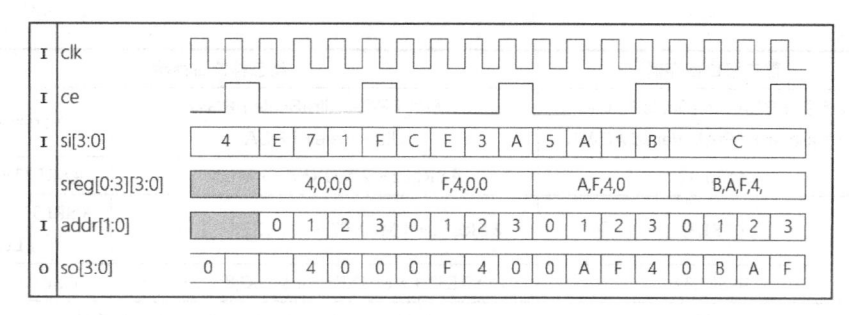

图 5-13

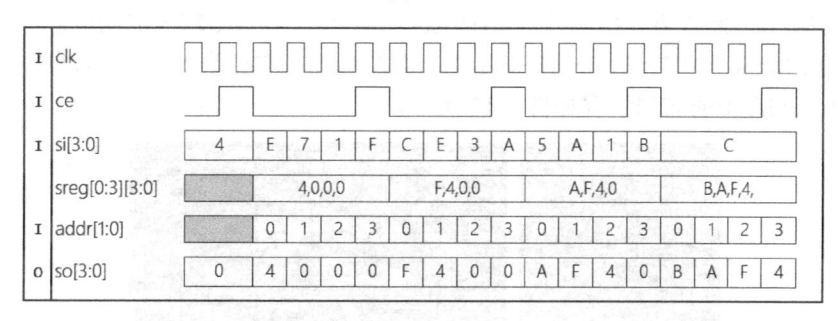

图 5-14

数据位宽为 1、地址位宽为 5 的动态移位寄存器，目标芯片为 XCVU5P，综合结果如图 5-15 所示。可见，最终还是会使用 SRLC32E，也就是 SLICEM 中的 LUT。因为，输出添加寄存器，所以这里会有一个 FDRE。从这个角度而言，对于动态移位寄存器，若输出端口添加了寄存器，那么也可以有相应的同步复位/置位端口。

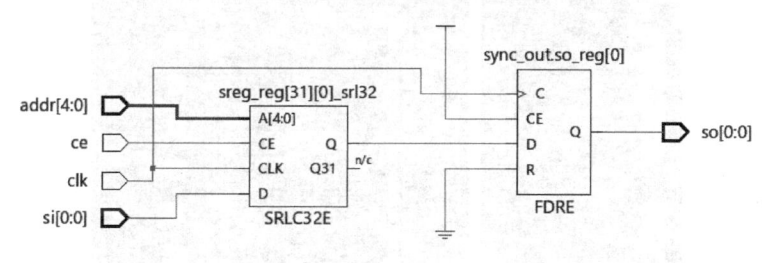

图 5-15

表 5-3 对比了宽度为 1 的不同深度的移位寄存器在 7 系列 FPGA、UltraScale/UltraScale+ FPGA 和 Versal ACAP 中的资源利用率。若为静态移位寄存器，则资源利用率是一致的。若为动态移位寄存器，由于 7 系列 FPGA 和 UltraScale/UltraScale+SLICE 中有 MUXF7 和 MUXF8，而 Versal ACAP 中没有，因此 Versal ACAP 中消耗的 LUT 会更多，但这并不意味着 Versal ACAP 的性能会更差。

表 5-3

深度	静态移位寄存器	动态移位寄存器	
	7 系列 FPGA/UltraScale FPGA/ UltraScale+ FPGA /Versal ACAP	7 系列 FPGA/UltraScale FPGA/ UltraScale+ FPGA	Versal ACAP
16	1 SRL16	1 SRL16	1 SRL16
32	1 SRL32	1 SRL32	1 SRL32
64	2 SRL32	2 SRL32 + 1 MUXF7	2 SRL32 + 1 LUT3

续表

深度	静态移位寄存器	动态移位寄存器	
	7 系列 FPGA/UltraScale FPGA/ UltraScale+ FPGA /Versal ACAP	7 系列 FPGA/UltraScale FPGA/ UltraScale+ FPGA	Versal ACAP
128	4 SRL32	4 SRL32 + 2 MUXF7 + 1 MUXF8	4 SRL32 + 1 LUT6
256	8 SRL32	8 SRL32 + 4 MUXF7 + 2 MUXF8 + 1 LUT3	8 SRL32 + 2 LUT6 + 1 LUT3
512	16 SRL32	16 SRL32 + 8 MUXF7 + 4 MUXF8 + 1 LUT6	16 SRL32 + 5 LUT6

以 512 的深度为例，Versal ACAP 中消耗的 16 个 SRL32 会分布在垂直方向上相邻的两个 SLICEM 中，同时需要用额外的 LUT 实现 MUXF7 和 MUXF8 的功能。如图 5-16 所示，左侧为 UltraScale/UltraScale+ FPGA 中的布局情况，右侧为 Versal ACAP 中的布局情况。7 系列 FPGA 中的布局情况如图 5-17 所示。

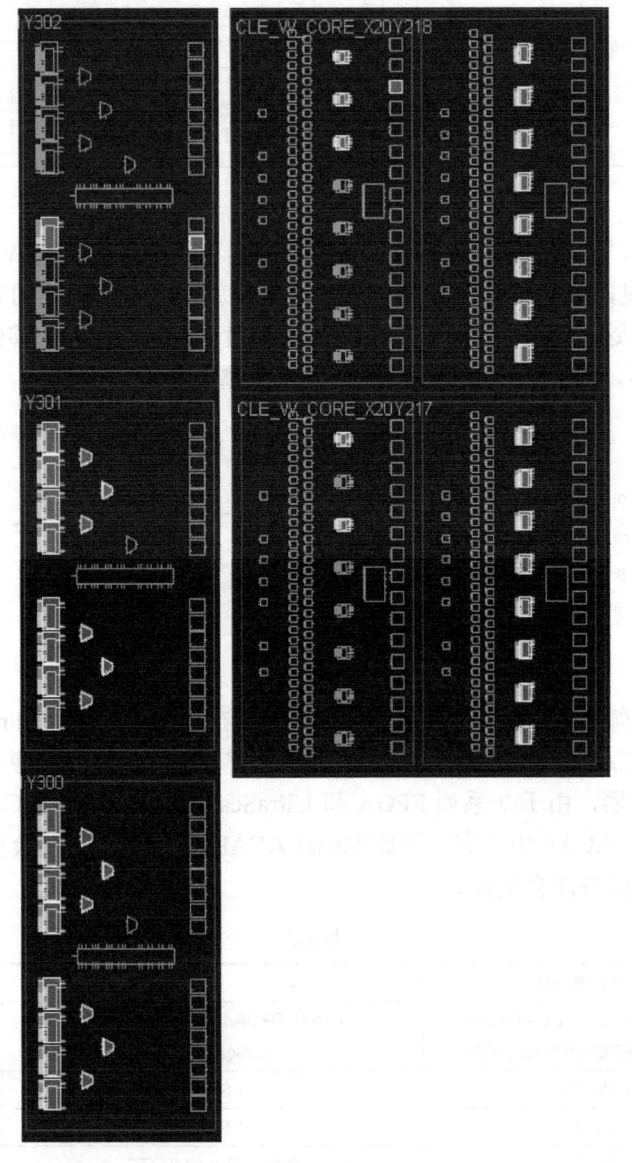

图 5-16

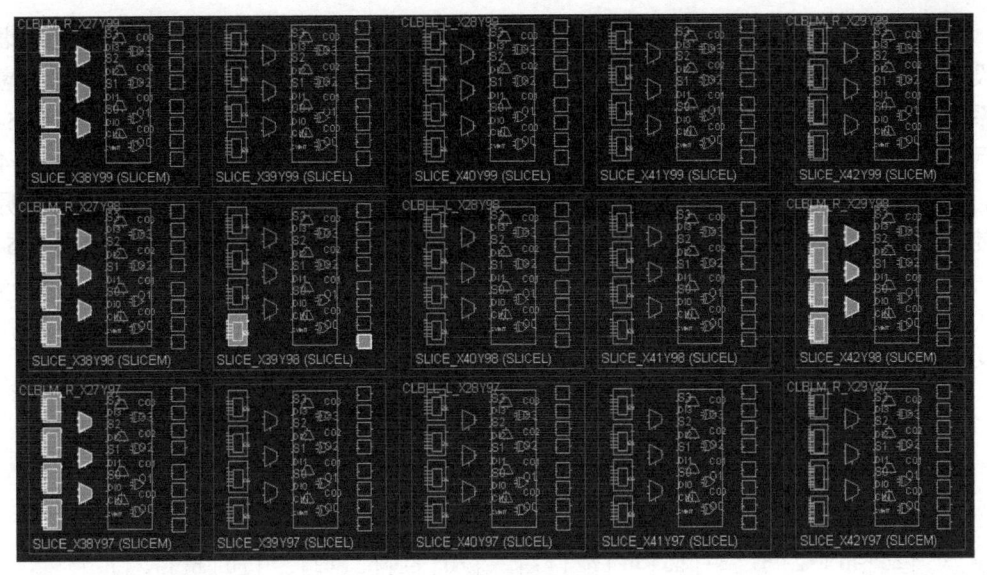

图 5-17

5.3 移位寄存器的应用场景

移位寄存器最典型的应用场景是延迟补偿，以实现数据对齐。

应用案例 1：用移位寄存器实现数据对齐

如图 5-18 所示，I 支路为 64 位数据，要经过操作 A 和操作 B，分别需要 8 个时钟周期和 12 个时钟周期。Q 支路也为 64 位数据，仅有操作 C，需要 3 个时钟周期。I/Q 数据需要在对齐后执行操作 E，因此需要对 Q 支路延迟补偿 17 个时钟周期。

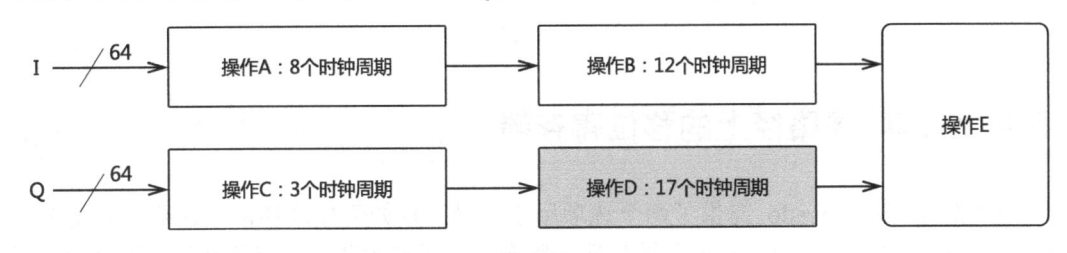

图 5-18

延迟补偿需要用到移位寄存器。将一位数据延迟 17 个时钟周期需要 1 个 SRL，延迟 64 位数据则需要 64 个 SRL，但如果使用触发器实现移位操作，就需要用到 64×17 也就是 1088 个触发器。表 5-4 对比了在不同芯片下对移位寄存器采用两种实现方式的资源利用率，可以看到查找表实现移位寄存器的优势，同时可以看到芯片架构对设计的影响。

表 5-4

实 现 方 式	7 系列 FPGA	UltraScale FPGA /UltraScale+ FPGA/ Versal ACAP
移位寄存器用触发器实现	1088 个触发器，占用 136 个 SLICE	1088 个触发器，占用 68 个 SLICE
移位寄存器用查找表实现	64 个 SRL，占用 16 个 SLICE	64 个 SRL，占用 8 个 SLICE

此外，移位寄存器还可应用于 FIR（Finite Impulse Response，有限脉冲响应）滤波器设计中。FIR 滤波器在本质上执行的是卷积运算。以长度为 4 的 FIR 滤波器为例，滤波器系数为 h0~h3，输入数据为 xn，卷积运算所需要的输入数据的动态变化过程如图 5-19 所示。在图 5-19 中，第一行从左至右依次计算出 y0~y3；第 2 行从左至右依次计算出 y4~y7。每个方格代表一个时钟周期，可以看到每个输入数据会存在 4 个时钟周期，这表明时钟使能信号周期是时钟周期的 4 倍。这种动态变化过程符合动态移位寄存器的规律。图 5-20 是一个深度为 4 的动态移位寄存器的仿真结果。时钟使能信号 ce 是时钟信号 clk 的 4 分频，输入数据 si 为 xn，输出数据 so 为卷积运算所需要的输入数据。so 的速率是 si 的 4 倍。结合图 5-19 可以判断，该仿真结果是符合预期要求的。

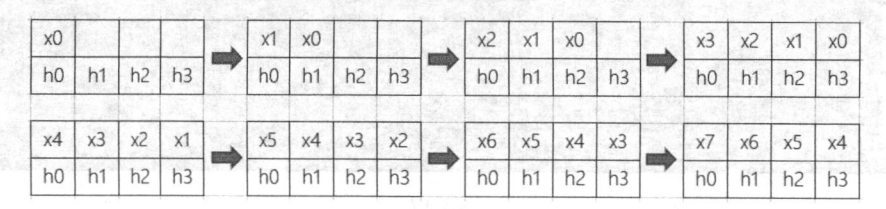

图 5-19

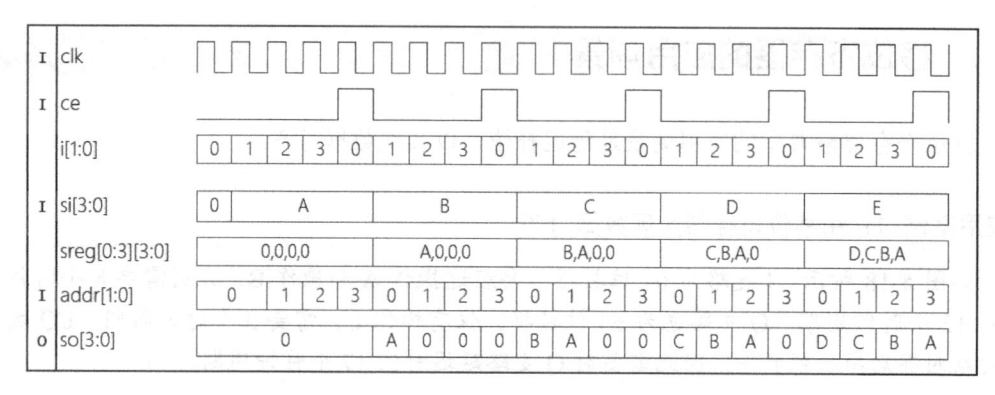

图 5-20

5.4 管理时序路径上的移位寄存器

在综合阶段，Vivado 提供了两个选项用于指导工具如何处理移位寄存器，如图 5-21 所示。其中，选项-no_srlextract 若被勾选，则意味着所有的移位寄存器都将被综合为级联的寄存器，也就是以寄存器链的形式实现移位寄存器。选项-shreg_min_size 则决定了移位寄存器被映射为 SRL 的最小深度。

图 5-21

-no_srlextract 和-shreg_min_size 是全局选项，意味着对设计中的所有模块均适用。因此，通常情况下不会手工修改这两项，而是由综合策略决定其具体设置情况，如表 5-5 所示。可以看到，较大的-shreg_min_size 可以改善布线拥塞。

表 5-5

综 合 策 略	-shreg_min_size	-no_srlextract
Flow_AlternateRoutability	10	未勾选
Flow_PerfOptimized_high	5	未勾选
其他	3	未勾选

Vivado 还提供了综合属性 SHREG_EXTRACT 和 SRL_STYLE。其中，SRL_STYLE 在 5.2 节已经阐述，这里介绍一下 SHREG_EXTRACT。该综合属性可写在 RTL 代码中，也可写在 XDC 约束文件中，其作用对象可以是某个模块，也可以是某个移位寄存器。以 VHDL 代码 5-7 为例，对信号 sreg 使用综合属性 SHREG_EXTRACT（可以全部大写，也可以全部小写，但不能大小写混用），如 VHDL 代码 5-8 所示。若深度为 4，宽度为 1，则综合结果如图 5-22 所示。

📄 **VHDL 代码 5-8**

```
1. architecture rtl of dynamic_sreg is
2. type bus_array is array(natural range<>) of std_logic_vector;
3. signal sreg : bus_array(0 to DEPTH - 1)(DW - 1 downto 0) = others > others > 0'));
4. attribute SHREG_EXTRACT : string;
5. attribute SHREG_EXTRACT of sreg : signal is "no";
```

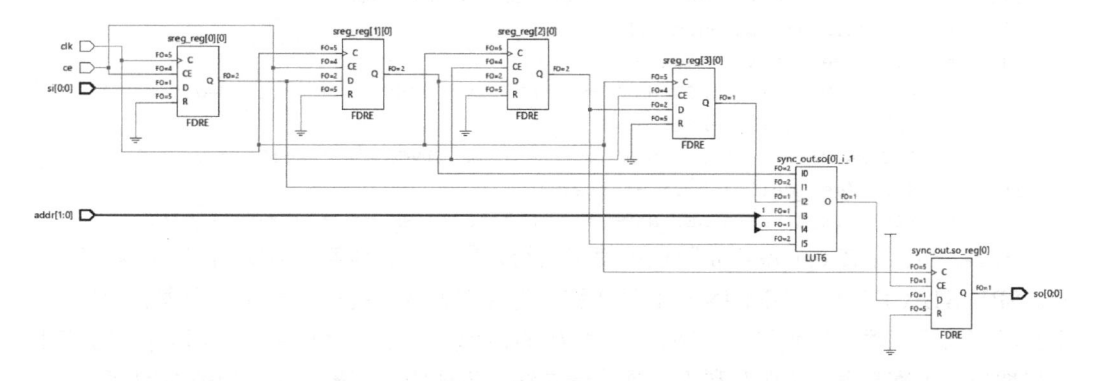

图 5-22

综合属性 SHREG_EXTRACT 的优先级高于 SRL_STYLE。例如，将 SHREG_EXTRACT 设置为"no"，将 SRL_STYLE 设置为"srl"，那么最终移位寄存器会采用寄存器链的形式。综合选项-shreg_min_size 作为全局选项，优先级最低。例如，移位寄存器深度为 10，将 SRL_STYLE 设置为"srl"，将-shreg_min_size 设置为 20，最终还是会将其推断为 SRL。

Vivado 还支持模块化综合技术，这使得对综合的管理粒度更加精细。其中，和移位寄存器相关的属性有两个：SHREG_MIN_SIZE 和 SRL_STYLE，具体使用方法如表 5-6 所示。

表 5-6

名称	可取值	例子
SHREG_MIN_SIZE	3~32	set_property BLOCK_SYNTH.SHREG_MIN_SIZE 8 [get_cells cpuEngine]
SRL_STYLE	REGISTER SRL SRL_REG REG_SRL REG_SRL_REG	set_property BLOCK_SYNTH.SRL_STYLE REG_SRL_REG [get_cells usbEngine0]

> 💡 **设计规则 2**：从对综合的管理粒度的角度考虑，建议通过模块化综合技术设置 SHREG_MIN_SIZE，而不是通过综合全局选项 -shreg_min_size 设置。

无论是综合选项还是综合属性，或者模块化综合技术，使用的前提是 RTL 代码对用户是可见的。若已到了设计实现阶段或用户仅可见综合后的网表文件，则需要使用其他方法对移位寄存器进行优化。此时就需要先找到网表中的移位寄存器。Tcl 代码 5-4 代码第 2 行可以找到设计中所有的移位寄存器。而代码第 4 行、第 6 行和第 8 行可分别找到深度为 1、深度为 2 和深度为 3 的移位寄存器，这三类移位寄存器正是设计中需要格外关注的，尤其是当它们是时序路径终点单元时。因为这三类移位寄存器的深度较浅，所以可考虑是否将其替换为触发器更为合适。

Tcl 代码 5-4

```
1.  #File: get_srl.tcl
2.  set mysrl [get_cells -hier -filter "REF_NAME=~SRL*"]
3.  show_objects $mysrl -name SRL
4.  set srl1 [get_cells -hier -filter \
5.     {IS_PRIMITIVE && REF_NAME =~ SRL* && (NAME =~ *_srl1)} -quiet]
6.  set srl2 [get_cells -hier -filter \
7.     {IS_PRIMITIVE && REF_NAME =~ SRL* && (NAME =~ *_srl2)} -quiet]
8.  set srl3 [get_cells -hier -filter \
9.     {IS_PRIMITIVE && REF_NAME =~ SRL* && (NAME =~ *_srl3)} -quiet]
```

表 5-7 列出了在 opt_design 阶段可使用的约束属性，具体使用方法如 Tcl 代码 5-5 所示。SRL_STAGES_TO_REG_INPUT 可在移位寄存器输入端提取出一个寄存器。例如，对于长时序路径（逻辑级数较高的路径），如果该路径以 SRL 为终点单元，那么借助此属性可提取出一个寄存器，从而有利于工具改善布线延迟或降低逻辑级数，如图 5-23 所示。对于从 BRAM、URAM 或 DSP 出发的路径，如果其终点是 SRL，如图 5-24 所示，那么也可借助此约束属性提取出一个寄存器以改善时序。

表 5-7

属性	属性值	作用对象	有效阶段	例子
REG_TO_SRL	TRUE | FALSE	寄存器	opt_design	若为 TRUE，则将寄存器链转换为移位寄存器
SRL_TO_REG	TRUE | FALSE	静态移位寄存器	opt_design	若为 TRUE，则将移位寄存器转换为寄存器链

续表

属性	属性值	作用对象	有效阶段	例子
SRL_STAGES_TO_REG_INPUT	1|-1	静态移位寄存器	opt_design	若为1,则从移位寄存器输入端提取出一个寄存器链
SRL_STAGES_TO_REG_OUTPUT	1|-1	静态移位寄存器	opt_design	若为1,则从移位寄存器输出端提取出一个寄存器链

Tcl 代码 5-5

```
1.  #File: srl_property.tcl
2.  set_property REG_TO_SRL TRUE [get_cells usbEngine/q_reg]
3.  set_property SRL_TO_REG TRUE [get_cells usbEngine/d_srl3]
4.  set_property SRL_STAGES_TO_REG_INPUT 1 [get_cells usbEngine/d_srl3]
5.  set_property SRL_STAGES_TO_REG_OUTPUT 1 [get_cells usbEngine/d_srl3]
```

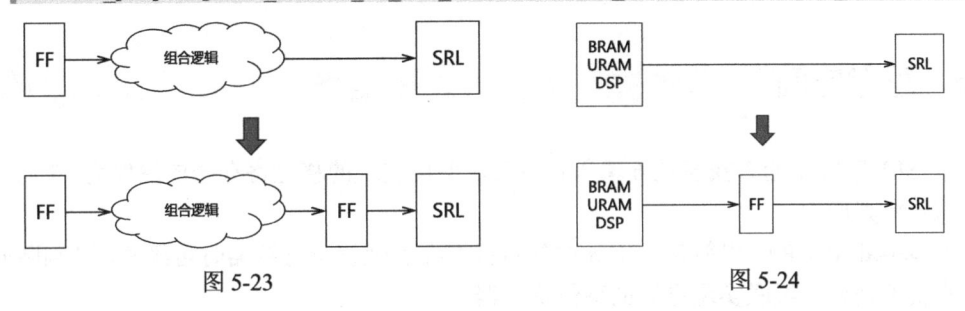

图 5-23

图 5-24

若 SRL 是时序路径的起点单元,则可借助约束属性 SRL_STAGES_TO_REG_OUTPUT 从移位寄存器输出端提取出一个寄存器以改善时序,如图 5-25 和图 5-26 所示。

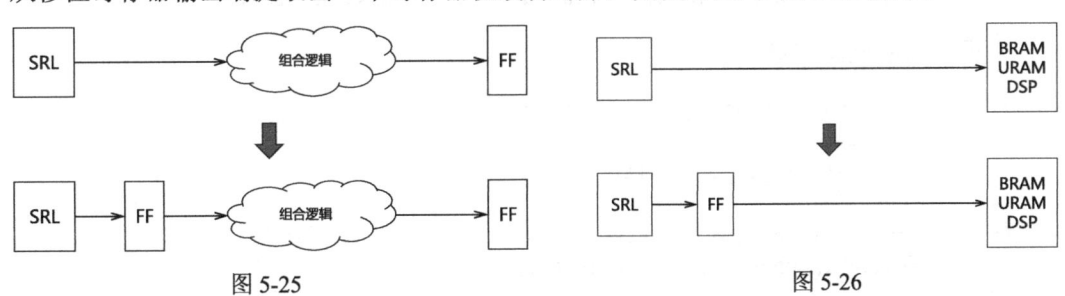

图 5-25

图 5-26

opt_design 还提供了选项-shift_register_opt 和-srl_remap_modes。其中,选项-shift_register_opt 针对驱动高扇出网线的 SRL 进行优化,将其变为 "SRL+寄存器" 的形式,从而使工具在后续流程中可以通过寄存器复制的方式降低扇出。选项-srl_remap_modes 的一种使用方法如 Tcl 代码 5-6 所示。当寄存器链的长度大于参数 min_depth_ffs_to_srl 的指定值时,工具就会把该寄存器链重新映射为 SRL。当移位寄存器的深度小于或等于 max_depth_srl_to_ffs 的指定值时,工具就会把 SRL 重新映射为寄存器链。

Tcl 代码 5-6

```
1.  #File: opt_design_post.tcl
2.  opt_design -srl_remap_modes {{min_depth_ffs_to_srl 3} {max_depth_srl_to_ffs 2}}
3.  puts "-srl_remap_modes DONE"
```

在 Project 模式下,可将 Tcl 代码 5-6 写入文件 opt_design_post.tcl 中,能在如图 5-27 所示的 tcl.post 位置浏览到即可。在 Non-Project 模式下,可直接在 opt_design 之后再次执

行代码第 2 行。

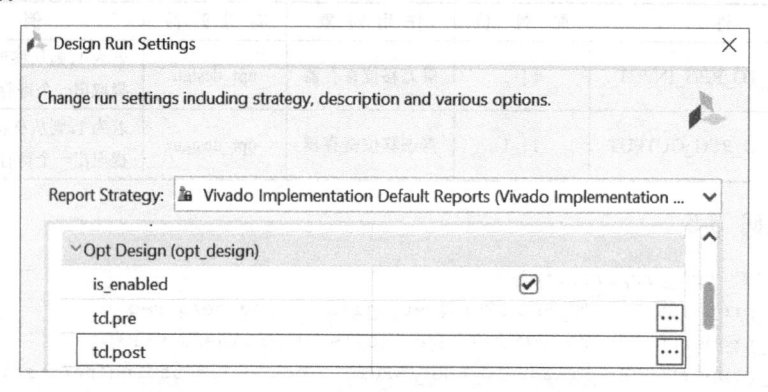

图 5-27

5.5 思考空间

1. SLICEM 中的查找表是 6 输入查找表，但可实现的移位寄存器的深度为 32 而不是 64，试解释其原因。

2. Versal ACAP 中相邻 SLICEM 中的查找表在实现移位寄存器时可级联，试判断同一列的查找表是否可级联实现更深的移位寄存器。

3. 试判断 VHDL 代码 5-9 会使用何种资源实现，并给出资源消耗量（目标芯片可以是 7 系列 FPGA，也可以是 UltraScale/UltraScale+ FPGA 或 Versal ACAP）。

VHDL 代码 5-9

```vhdl
1.  --File: static_single_bit_sreg_ff.vhd
2.  library ieee;
3.  use ieee.std_logic_1164.all;
4.
5.  entity static_single_bit_sreg_ff is
6.    generic ( DEPTH : positive := 8 );
7.    port (
8.      clk : in std_logic;
9.      rst : in std_logic;
10.     ce  : in std_logic;
11.     si  : in std_logic;
12.     so  : out std_logic
13.   );
14. end entity;
15.
16. architecture rtl of static_single_bit_sreg_ff is
17.   signal sreg : std_logic_vector(DEPTH - 1 downto 0);
18. begin
19.   so <= sreg(DEPTH - 1);
20.   process(clk, rst)
21.   begin
22.     if rst then
```

```
23.         sreg <= (others => '0');
24.       elsif rising_edge(clk) then
25.         sreg <= sreg(DEPTH - 2 downto 0) & si when ce else sreg;
26.       end if;
27.     end process;
28. end architecture;
```

4. 对于 VHDL 代码 5-1，如果要使用综合属性 SRL_STYLE，试判断其作用对象应是哪种信号，a_dly1、a_dly2、a_dly3 还是 q？

5. 在某综合后的网表中，发现关键时序路径的终点单元为移位寄存器，现需要将其优化为"寄存器+SRL"的形式，该如何操作？

6. 在某设计中，需要将 8 位数据延迟 512 个时钟周期，目标芯片为 7 系列 FPGA，试给出可行的实现方案。如果使用移位寄存器实现，目标芯片为 UltraScale/UltraScale+ FPGA，试判断会消耗多少个查找表。这些查找表会分布在多少个 SLICEM 中？

7. 如何通过 Tcl 脚本只找到设计中深度为 1 的用查找表实现的移位寄存器？

8. 某设计中的一个模块使用了大量基于查找表实现的移位寄存器，从而造成布线拥塞。从综合策略的角度考虑，可选用什么样的综合策略？

9. 从综合精细化管理的角度看，如何设定一个最小值，使得深度大于此值的移位寄存器映射为查找表？

10. 某移位寄存器使用 SRL 实现，现需要将其重新映射为 SRL+FF，且只能在 opt_design 阶段实施，应使用哪种约束？

第 6 章

优化存储器

6.1 存储器资源

这里的存储器指的是分布式 RAM（Random Access Memory，随机存取存储器）、BRAM 和 UltraRAM。

6.1.1 分布式 RAM

分布式 RAM（Distributed RAM），又称 LUTRAM，是由 SLICEM 中的查找表（Look-Up Table，LUT）实现的。

无论是 7 系列 FPGA，还是 UltraScale/UltraScale+ FPGA 或 Versal ACAP，SLICEM 中的 LUT 都可配置为 RAM 使用。单个 LUT 的存储深度可达 64。对 7 系列 FPGA 而言，由于每个 SLICEM 中有 4 个 LUT，因此可实现 RAM 的最大深度为 256。对 UltraScale/UltraScale+ FPGA 和 Versal ACAP 而言，由于每个 SLICEM 中有 8 个 LUT，因此可实现 RAM 的最大深度为 512。

SLICEM 中的 LUT 可配置为单端口（Single-port）RAM、双端口（Dual-port）RAM 和简单双端口（Simple Dual-port）RAM。当将 LUT 配置为单端口 RAM 时，只有一套地址，读/写地址共享地址端口，写操作同步，读操作异步。当将 LUT 配置为双端口 RAM 时，本质上是将数据写入两份存储空间，一份存储空间共享读/写地址，另一份存储空间的写地址与前一份存储空间共享，同时具备独立的读地址端口。同样地，写操作同步，读操作异步。当将 LUT 配置为简单双端口 RAM 时，仍然是写操作同步，读操作异步。无论是哪种类型的 RAM，LUT 实现时都只有一个时钟端口。

根据这些特征可以很容易判断指定 RAM 的资源消耗量。表 6-1 列出了几种 RAM 的资源利用情况。表中，S 表示单端口 RAM，D 表示双端口 RAM。例如，128×1S 表示深度为 128、位宽为 1 的单端口 RAM；128×1D 表示深度为 128、位宽为 1 的双端口 RAM。LUTRAM 代表 SLCIEM 中的查找表，LUT 代表 SLICEL 中的查找表。这里需要注意的是 UltraScale/UltraScale+ FPGA 的 SLICE 相比 7 系列 FPGA 新增了 F9MUX。从表中可以看出，对于同样规格的 RAM，双端口 RAM 消耗的 LUTRAM 是单端口 RAM 的 2 倍。对 7 系列 FPGA 而言，单个 SLICM 可实现的单端口 RAM 的最大深度为 256、双端口 RAM 的最大深度为 128。对 UltraScale/UltraScale+ FPGA 而言，单个 SLICEM 可实现的单端口 RAM 的最大深度为 512、双端口 RAM 的最大深度为 256。由于 Versal ACAP 中移除了 F7MUX、F8MUX 和 F9MUX，因此需要额外的查找表实现对应功能，这里不再给出具体资源消耗量。

表 6-1

RAM 类型	7 系列 FPGA	UltraScale/UltraScale+ FPGA
128×1S	2 LUTRAM + 1 F7MUX	2 LUTRAM + 1 F7MUX
128×1D	4 LUTRAM + 2 F7MUX	4 LUTRAM + 2 F7MUX
256×1S	4 LUTRAM + 2 F7MUX + 1 F8MUX	4 LUTRAM + 2 F7MUX + 1 F8MUX
256×1D	8 LUTRAM + 4 F7MUX + 4 LUT	8 LUTRAM + 4 F7MUX + 2 F8MUX
512×1S	8 LUTRAM + 4 F7MUX + 2 F8MUX + 3 LUT	8 LUTRAM + 4 F7MUX + 2 F8MUX + 1 F9MUX
512×1D	16 LUTRAM + 8 F7MUX + 6 LUT	16 LUTRAM + 8 F7MUX + 4 F8MUX + 4 LUT

7 系列 FPGA、UltraScale/UltraScale+ FPGA 和 Versal ACAP 还有一个共同点，即对于同一个 SLICEM 内的 LUT，当有一个 LUT 配置为 RAM 时，其余 LUT 就不能配置为 SRL 了，反之亦然。

应用案例 1：评估设计中某个存储器 LUTRAM 的用量

某设计中要用到一个单端口 RAM，深度为 128、宽度为 8，目标芯片为 UltraScale+ FPGA，采用分布式 RAM 实现。由于 SLICEM 中的每个 LUT 可配置为 64×1 的单端口 RAM，因此，128×8 的单端口 RAM 需要 128/64×8 也就是 16 个 LUTRAM。这 16 个 LUTRAM 至少分布在 2 个 SLICEM 中。

6.1.2 BRAM

从 7 系列 FPGA 到 UltraScale/UltraScale+ FPGA 再到 Versal ACAP，都有 BRAM 的身影。每个 BRAM 均为 36Kb（1Kb = 1024bit），由两个独立的 18Kb BRAM 构成（每个 18Kb BRAM 有两个独立的端口，即端口 A 和端口 B，这两个端口共享 18Kb 的存储空间且地位相等），这两个 BRAM 之间是硬核的 FIFO 控制逻辑，但在 Versal ACAP 中，这个硬核的 FIFO 控制逻辑被移除。因此，在 7 系列 FPGA 和 UltraScale/UltraScale+ FPGA 中，一个 36Kb BRAM 可配置为 4 种形式，如图 6-1 所示。但在 Versal ACAP 中，不再有内嵌 FIFO，配置为 FIFO 时需要额外的资源来生成 FIFO 控制逻辑。

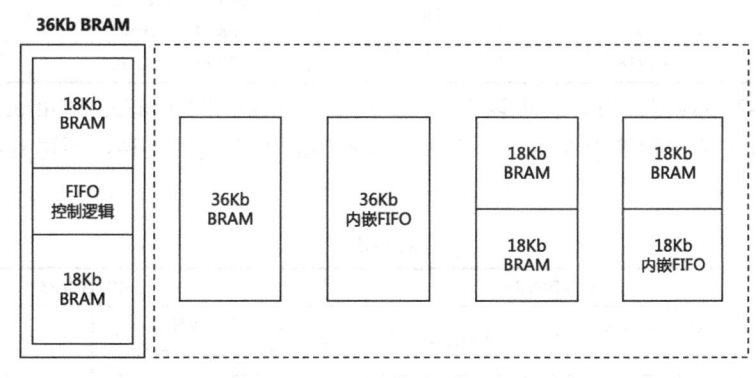

图 6-1

每个 36Kb BRAM 有两个独立端口：端口 A 和端口 B，两者共享存储空间且地位相等，如图 6-2 所示，图中的时钟反相器是可选的。所有的输入信号（控制信号和数据）进入 BRAM 后都会被寄存一拍，即输入端的寄存器是固有的。RAM 输出数据由末级的选择器输出，可以是来自其前端的锁存器或寄存器输出数据。由此可以确定从输入到输出延迟至少为 1 个时钟周期，最多为 2 个时钟周期。

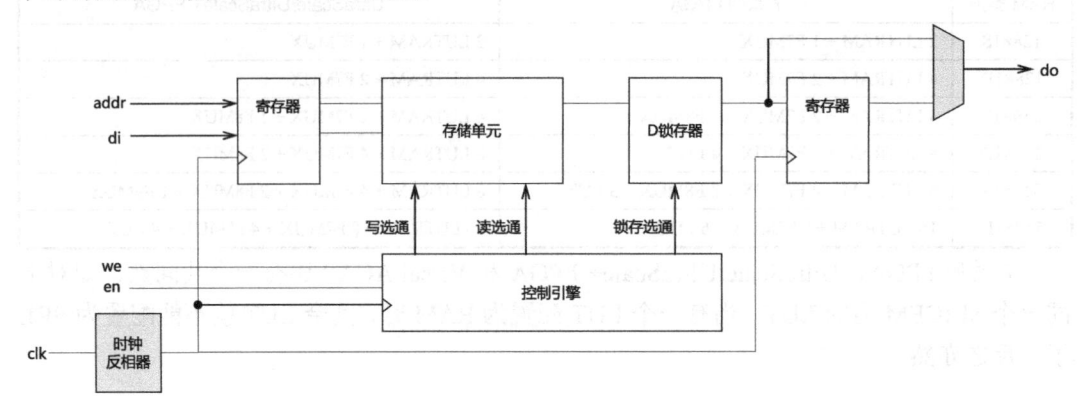

图 6-2

无论是 1 个 18Kb BRAM，还是 1 个 36Kb BRAM，均可配置为单端口 RAM、简单双端口 RAM 或真双端口（True Dual-port）RAM，如表 6-2 所示。表中，RAM 的规格以深度×宽度的形式表示。例如，2K×9 表示 RAM 深度为 2K（2048，1K=1024）、数据位宽为 9。

表 6-2

工作模式	18Kb BRAM		36Kb BRAM	
	7 系列/UltraScale/UltraScale+ FPGA	Versal ACAP	7 系列/UltraScale/UltraScale+ FPGA	Versal ACAP
单端口	16K×1，8K×2，4K×4，2K×9，1K×18，512×36	2K×9，1K×18，512×36	32k×1，16K×2，8K×4，4K×9，2K×18，1K×36，512×72	4K×9，2K×18，1K×36
简单双端口	16K×1，8K×2，4K×4，2K×9，1K×18，512×36	2K×9，1K×18，512×36	32K×1，16K×2，8K×4，4K×9，2K×18，1K×36，512×72	4K×9，2K×18，1K×36，512×72
真双端口	16K×1，8K×2，4K×4，2K×9，1K×18	2K×9，1K×18	32K×1，16K×2，8K×4，4K×9，2K×18，1K×36	4K×9，2K×18，1K×36

结合表 6-2 不难得出表 6-3。可以看到，在 7 系列 FPGA 和 UltraScale/UltraScale+ FPGA 中，BRAM 可支持的位宽有 7 种，而在 Versal ACAP 中降到了 4 种，不再支持小位宽 1、2、4。

表 6-3

工作模式	18Kb BRAM		36Kb BRAM	
	7 系列/UltraScale/UltraScale+ FPGA	Versal ACAP	7 系列/UltraScale/UltraScale+ FPGA	Versal ACAP
单端口和简单双端口	1、2、4、9、18、36	9、18、36	1、2、4、9、18、36、72	9、18、36、72
真双端口	1、2、4、9、18	9、18	1、2、4、9、18、36	9、18、36

应用案例 2：评估设计中某个存储器 BRAM 的用量

某设计中要用到一个简单双端口 RAM，深度为 2048，宽度为 96，采用 BRAM 实现。因为一个 36Kb BRAM 可配置为 2K×18 的简单双端口 RAM，一个 18Kb BRAM 可配置为 2K×9 的简单双端口 RAM。所以对于规格为 2048×96 的简单双端口 RAM，需要消耗 96/18 也就是 5 个 36Kb BRAM（对应位宽为 18×5=90）和 1 个 18Kb BRAM（对应位宽为 96-90=6）。

BRAM 支持三种写模式，以决定同时对同一地址进行读/写操作时读出数据的内容。这三种写模式分别为写优先、读优先和保持，如表 6-4 所示。

表 6-4

写 模 式	读/写顺序	输 出 数 据	备 注
写优先（WRITE_FIRST）	先写后读	新数据（写入该地址上的数据）	
读优先（READ_FIRST）	先读后写	原始数据（该地址上被新数据覆盖之前的数据）	较保持模式功耗高 15%左右
保持（NO_CHANGE）	只写不读	前一次读操作读出的数据	最省功耗

写操作模式可针对每个端口单独设置。当 BRAM 被配置为真双端口 RAM 时，若两个端口同时进行读操作，那么读操作可成功完成。若两个端口同时对同一地址进行写操作，那么最终写入的数据是不确定的。若两个端口一个进行读操作，另一个进行写操作，那么最终结果如表 6-5（两个端口共享时钟，所谓共享时钟是指这两个时钟由同一个 BUFG 驱动）和表 6-6 所示（两个端口有各自的时钟，即由不同的 BUFG 驱动）。其中，RF 代表读优先，WF 代表写优先，NC 代表保持。写使能高电平有效。DIA 代表 A 端口写入数据，DIB 代表 B 端口写入数据。表 6-5 和表 6-6 最大的区别在于当端口 B 为读优先且执行写操作、端口 A 执行读操作时，若两个端口共享时钟，则端口 A 和端口 B 均输出原始数据；否则，端口 A 的输出数据不确定，端口 B 的输出数据为原始数据。实际设计时，一定要避免同时向同一地址进行写操作，尽管这不会损坏 BRAM，但最终写入的数据是不确定的。

表 6-5

时钟类型	端口 A 写模式	端口 B 写模式	端口 A 写使能信号	端口 B 写使能信号	端口 A 输出数据	端口 B 输出数据	写入数据
共享时钟	RF/WF/NC	RF/WF/NC	0	0	原始数据	原始数据	未写入
共享时钟	RF	RF/WF/NC	1（DIA）	0	原始数据	原始数据	DIA
共享时钟	WF	RF/WF/NC	1（DIA）	0	DIA	X	DIA
共享时钟	NC	RF/WF/NC	1（DIA）	0	保持不变	X	DIA
共享时钟	RF/WF/NC	RF	0	1（DIB）	原始数据	原始数据	DIB
共享时钟	RF/WF/NC	WF	0	1（DIB）	X	DIB	DIB
共享时钟	RF/WF/NC	NC	0	1（DIB）	X	保持不变	DIB
共享时钟	RF/WF/NC	RF/WF/NC	1	1	X	X	X

表 6-6

时钟类型	端口 A 写模式	端口 B 写模式	端口 A 写使能信号	端口 B 写使能信号	端口 A 输出数据	端口 B 输出数据	写入数据
独立时钟	RF/WF/NC	RF/WF/NC	0	0	原始数据	原始数据	未写入
独立时钟	RF	RF/WF/NC	1（DIA）	0	原始数据	X	DIA

时钟类型	端口 A 写模式	端口 B 写模式	端口 A 写使能信号	端口 B 写使能信号	端口 A 输出数据	端口 B 输出数据	写入数据
独立时钟	WF	RF/WF/NC	1（DIA）	0	DIA	X	DIA
独立时钟	NC	RF/WF/NC	1（DIA）	0	保持不变	X	DIA
独立时钟	RF/WF/NC	RF	0	1（DIB）	X	原始数据	DIB
独立时钟	RF/WF/NC	WF	0	1（DIB）	X	DIB	DIB
独立时钟	RF/WF/NC	NC	0	1（DIB）	X	保持不变	DIB
独立时钟	RF/WF/NC	RF/WF/NC	1	1	X	X	X

> 💡 **设计规则 1**：当 BRAM 被配置为简单双端口或真双端口时，如果两个端口的时钟是独立时钟（也就是异步时钟），应避免同时对同一地址用不同时钟进行读/写操作。

对于 7 系列 FPGA 同列两个相邻的 BRAM，若每个 BRAM 规格均为 32K×1，则两者可级联构成 64K×1 的规格而不会消耗额外的 SLICE 中的资源。其他情形则无法使用专用级联布线资源。在 UltraScal/UltraScal+ FPGA 和 Versal ACAP 中，BRAM 的级联功能更加强大，如图 6-3 所示，显示了 3 个 BRAM 级联的情形。级联的 BRAM 有一个属性 CASCADE_ORDER，其值表明了对应 BRAM 在级联链路中的位置。在 Vivado 中，若设计中的 BRAM 级联，则可在 Schematic 视图中查看到，如图 6-4 所示。需要注意的是级联仅限于同一个时钟区域（Clock Region）内，不能跨时钟区域。因此，对 UltraScale/UltraScale+ FPGA 而言，最多可允许 12 个 BRAM 级联；对 Versal ACAP 而言，最多可允许 24 个 BRAM 级联。

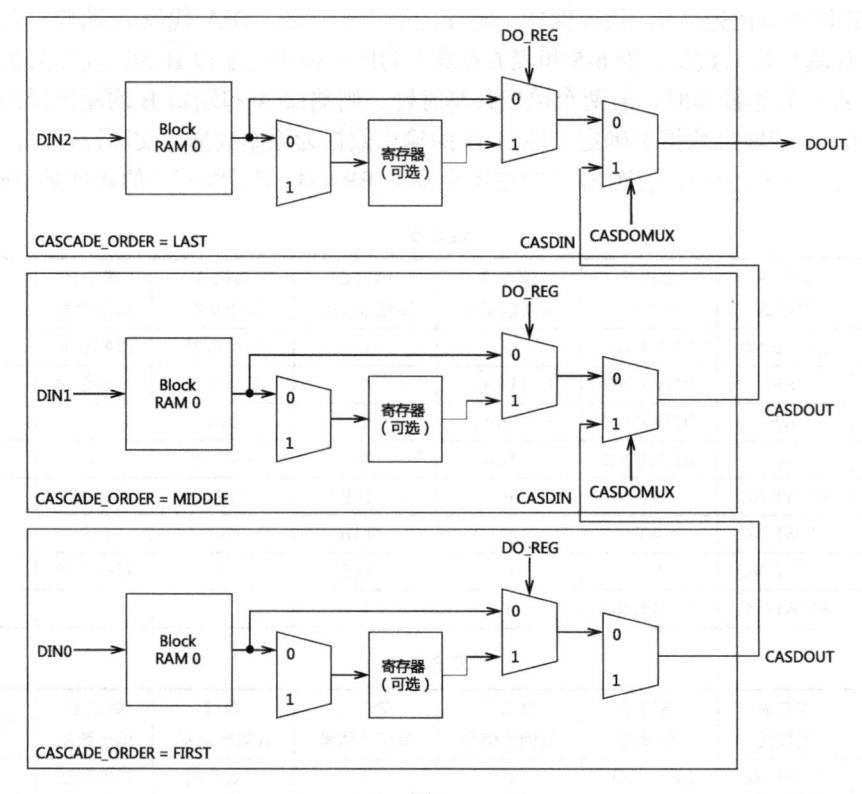

图 6-3

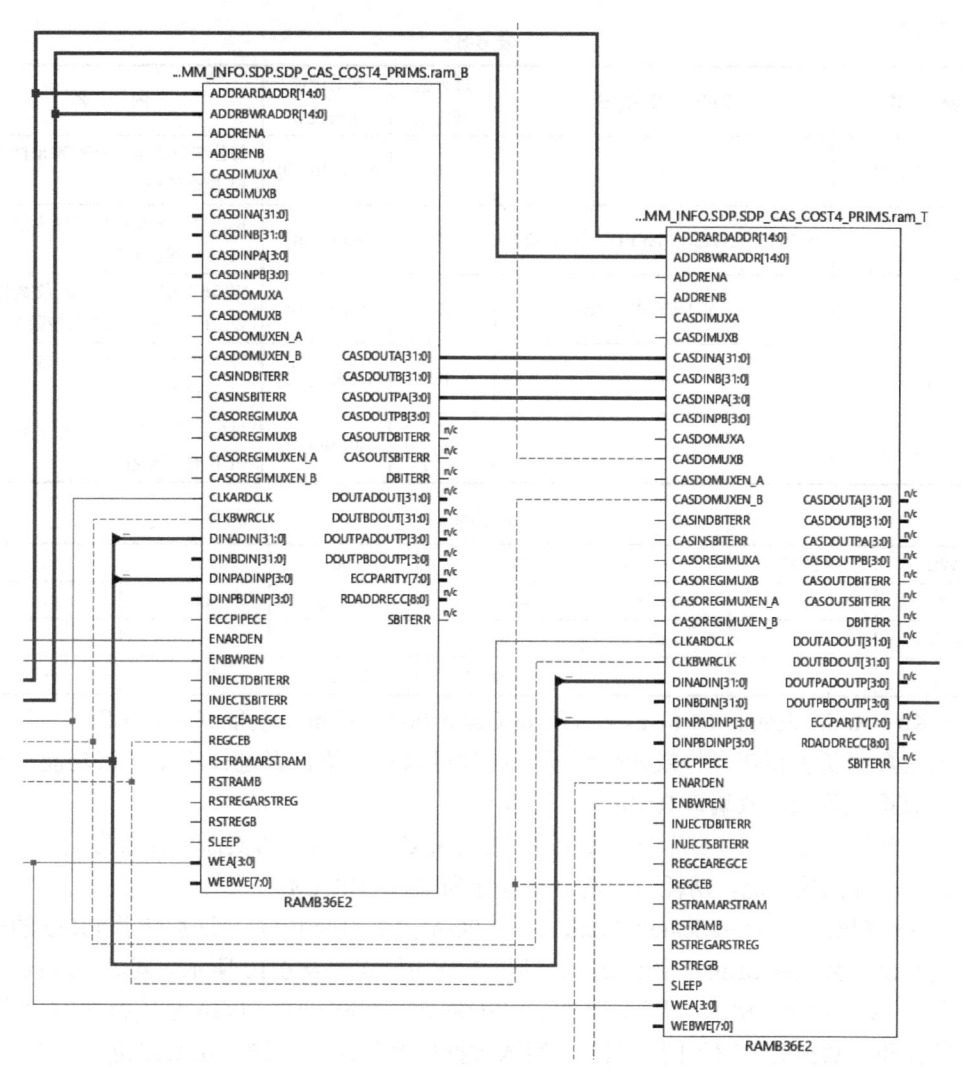

图 6-4

BRAM 支持写操作字节选通功能。只有当对应字节选通信号有效时（高电平有效），该字节才可以被写入 BRAM。7 系列 FPGA、UltraScale/UltraScale+ FPGA 字节选通功能如表 6-7 所示。Versal ACAP 字节选通功能如表 6-8 所示，新增属性 BWE_MODE_B，表中，TDP 表示真双端口，SDP 表示简单双端口，×表示不支持。当 BRAM 与微处理器存在数据交换时，该功能非常有用。当读优先时，BRAM 将输出指定地址上的原始数据。当写优先时，BRAM 将输出字节选通功能对应位与前一时钟周期输出数据的组合，如表 6-9 所示，表中，mem(aa)表示地址 aa 上的原始数据，注意表中的灰色部分。

表 6-7

原 语	最大数据位宽	字节使能信号位宽	内 容
RAMB36E2 TDP	36	4	每位字节使能信号控制 8 位数据+1 位奇偶校验位
RAMB36E2 SDP	72	8	每位字节使能信号控制 8 位数据+1 位奇偶校验位
RAMB18E2 TDP	18	2	每位字节使能信号控制 8 位数据+1 位奇偶校验位
RAMB18E2 SDP	36	4	每位字节使能信号控制 8 位数据+1 位奇偶校验位

表 6-8

原　　语	BWE_MODE_B	最大数据位宽	字节使能信号位宽	内　　容
RAMB36E5 TDP	×	36	WEA/B[3:0]	每位字节使能信号控制 8 位数据 +1 位奇偶校验位
RAMB36E5 SDP	PARITY_INTERLEAVED（默认值）	72	WEB[7:0]	每位字节使能信号控制 8 位数据 +1 位奇偶校验位
RAMB36E5 SDP	PARITY_INDEPENDENT	72	WEB[8:0]	WEB[7:0]每位控制 8 位数据，WEB[8]控制 8 位奇偶校验位
RAMB18E2 TDP	×	18	WEA/B[1:0]	每位字节使能信号控制 8 位数据 +1 位奇偶校验位
RAMB18E2 SDP	×	36	WEB[3:0]	每位字节使能信号控制 8 位数据 +1 位奇偶校验位

表 6-9

WE	4'b0000	4'b0000	4'b1111	4'b0011	
DI	xxxx	xxxx	4'h1111	4'h2222	xxxx
ADDR		aa	bb	bb	cc
DO			mem(aa)	4'h1111	4'h1122	mem(cc)

36Kb BRAM 集成了 64 位 ECC（Error Correction Coding）功能（硬核 ECC），可用于检测输出数据是否存在一位或两位错误，同时可以对一位错误进行纠正。该功能只有当 BRAM 被配置为简单双端口 RAM 时才可使用。

从功耗的角度看，对芯片中最终未使用的 BRAM，工具会将其断电以节省功耗，这就是所谓的功耗门控功能。功耗门控的最小单元为 18Kb BRAM。

7 系列 FPGA 和 UltraScale/UltraScale+ FPGA 提供内嵌 FIFO，这意味着 FIFO 的控制逻辑是硬核不会消耗 SLICE 中的资源，其可支持的位宽如表 6-10 所示。同时，内嵌 FIFO 支持硬核 ECC 功能。相比 7 系列 FPGA，UltraScale/UltraScale+ FPGA 还支持非对称位宽，即写数据和读数据的位宽不同。例如，写入数据位宽若为 18，则读出数据位宽可以是 9、18 或 36。Versal 不再提供内嵌 FIFO，这意味着 FIFO 的控制逻辑需要额外的 SLICE 资源生成，尽管如此，这类 FIFO 仍支持非对称位宽。

表 6-10

	18Kb 内嵌 FIFO	36Kb 内嵌 FIFO
芯片	7 系列/UltraScale/UltraScale+ FPGA	7 系列/UltraScale/UltraScale+ FPGA
规格	4K×4、2K×9、1K×18、512×36	8K×4、4K×9、2K×18、1K×36、512×72
位宽	4、9、18、36	4、9、18、36、72

7 系列 FPGA、UltraScale/UltraScale+ FPGA 和 Versal ACAP 中的 BRAM 都有自带的输出寄存器，该寄存器可支持同步复位或同步置位，由管脚 RSTREG 控制。

💡设计规则 2：BRAM 自带的输出寄存器仅支持同步复位或同步置位，不支持异步复位或异步置位。因此，在用 RTL 代码描述时，为了保证工具将相应寄存器推断为 BRAM 自带的输出寄存器，应使用同步复位或同步置位。

采用 Tcl 代码 6-1 可以查看指定 7 系列 FPGA 芯片中 BRAM 的分布情况。从代码第 7

行的返回值可以看出每列包含的 36Kb BRAM 的个数。从代码第 11 行的返回值可以看出每列包含的 18Kb BRAM 的个数。

Tcl 代码 6-1

```
1.  #File: check_bram_7fpga.tcl
2.  set mypart [get_parts xc7k70tfbg484-3]
3.  => xc7k70tfbg484-3
4.  link_design -part $mypart
5.  set bram36_one_col [get_sites "RAMB36_X0Y*" -of [get_slrs SLR0]]
6.  => RAMB36_X0Y39 RAMB36_X0Y38 RAMB36_X0Y37 ...
7.  puts "The number of RAMB36 in one column: [llength $bram36_one_col]"
8.  => The number of RAMB36 in one column: 40
9.  set bram18_one_col [get_sites "RAMB18_X0Y*" -of [get_slrs SLR0]]
10. => RAMB18_X0Y78 RAMB18_X0Y79 RAMB18_X0Y76 ...
11. puts "The number of RAMB18 in one column: [llength $bram18_one_col]"
12. => The number of RAMB18 in one column: 80
```

采用 Tcl 代码 6-2 可以看出指定 UltraScale/UltraScale+ FPGA 芯片内 BRAM 的分布情况。从代码第 36 行的返回值可以看出 BRAM 列的个数。从代码第 38 行的返回值可以看出一个 SLR 内每列 BRAM 的个数。从代码第 40 行的返回值可以看出指定时钟区域内 BRAM 列的个数。从代码第 42 行可以看出指定时钟区域内一列 BRAM 的个数。这部分 Tcl 脚本也适用于 Versal 器件。

Tcl 代码 6-2

```
1.  #File: check_bram_us.tcl
2.  set mypart [get_parts xcvu5p-flva2104-2-i]
3.  link_design -part $mypart
4.  set slr [get_slrs SLR0]
5.  => SLR0
6.  set bram_one_slr [get_sites RAMB36* -of $slr]
7.  => RAMB36_X0Y59 RAMB36_X1Y59 ...
8.  set x {}
9.  set y {}
10. foreach bram_one_slr_i $bram_one_slr {
11.     set loc [lindex [split $bram_one_slr_i _] 1]
12.     set xy  [split $loc XY]
13.     lappend x [lindex $xy 1]
14.     lappend y [lindex $xy 2]
15. }
16. set col_index_slr [lsort -integer -unique $x]
17. => 0 1 2 3 4 5 6 7 8 9 10 11
18. set row_index_slr [lsort -integer -unique $y]
19. => 0 1 2 3 4 5...55 56 57 58 59
20. set cr [get_clock_regions X0Y0]
21. => X0Y0
22. set brams_one_cr [get_sites RAMB36* -of $cr]
23. => RAMB36_X0Y11 RAMB36_X1Y11 ...
24. set x {}
25. set y {}
```

```
26. foreach brams_one_cr_i $brams_one_cr {
27.     set loc [lindex [split $brams_one_cr_i _] 1]
28.     set xy [split $loc XY]
29.     lappend x [lindex $xy 1]
30.     lappend y [lindex $xy 2]
31. }
32. set col_index [lsort -integer -unique $x]
33. => 0 1 2
34. set row_index [lsort -integer -unique $y]
35. => 0 1 2 3 4 5 6 7 8 9 10 11
36. puts "Number of BRAM columns in SLR $slr: [llength $col_index_slr]"
37. => Number of BRAM columns in SLR SLR0: 12
38. puts "Height of SLR by BRAM: [llength $row_index_slr]"
39. => Height of SLR by BRAM: 60
40. puts "Number of BRAM columns in CR $cr: [llength $col_index]"
41. => Number of BRAM columns in CR X0Y0: 3
42. puts "Height of CR by BRAM: [llength $row_index]"
43. => Height of CR by BRAM: 12
```

6.1.3 UltraRAM

UltraScale+ FPGA 新增了 UltraRAM。每个 UltraRAM 的大小为 288Kb，有两个端口，但仅有一个时钟管脚，由这两个端口共用。输入数据和输出数据均为 72 位，其基本结构如图 6-5 所示。这里仅给出其中一个端口的结构。另一个端口与之相同，两个端口共享存储空间。

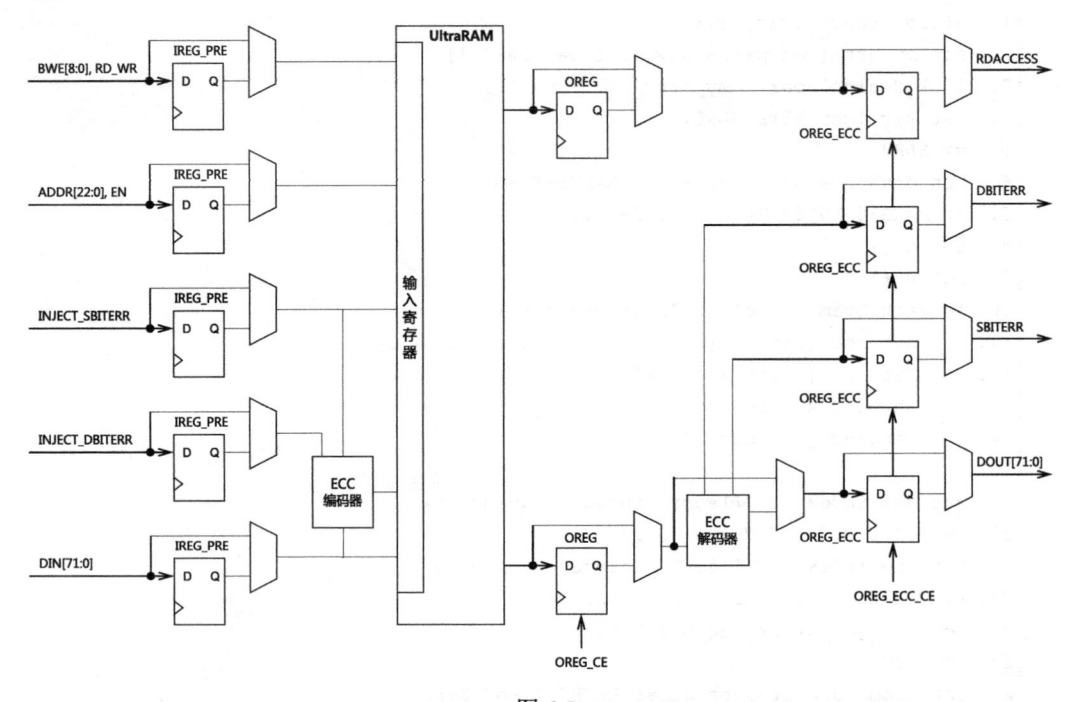

图 6-5

图 6-5 中有 4 类寄存器：输入寄存器（固有的）、位于输入寄存器之前的寄存器（IREG_PRE，可选）、输出寄存器（OREG，可选）、ECC 输出寄存器（OREG_ECC，可选）。

将这 4 类寄存器提取出来，形成如图 6-6 所示的结构。不难看出，从输入到输出延迟最大为 4 个时钟周期，最小为 1 个时钟周期。即使不使用 ECC 功能，OREG_ECC 也是可以使用的。同时，每个端口都有独立的 ECC 功能。图 6-6 中还显示了输入和输出级联寄存器，不同于 BRAM，同一列的 UltraRAM 均可级联，只要其在一个 SLR（Super Logic Region）之内即可。

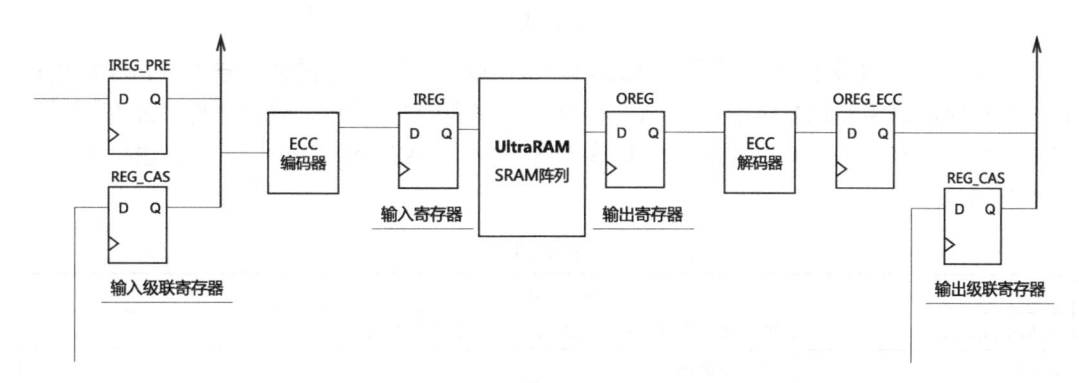

图 6-6

就级联而言，UltraRAM 有专用的级联寄存器。下面以 4 个 UltraRAM 在深度方向的级联为例，如图 6-7 所示。图中，IREG_CAS 为输入级联寄存器，OREG_CAS 为输出级联寄存器。假定 DMUX 的 OREG 和 OREG_ECC 均被使用，则从输入到输出共需要 7 个时钟周期。无论数据在哪个 UltraRAM 中，均是如此。例如，数据在最右侧的 UltraRAM，读地址从输入端到末级 UltraRAM 地址端需要 5 个时钟周期，数据输出需要 2 个时钟周期，总计 7 个时钟周期。因此，可以看到输入到输出的时钟周期个数等于控制信号从输入端到存储数据的 UltraRAM 控制端所需要的时钟周期个数加上数据从 UltraRAM 存储单元到输出端所需要的时钟周期个数。换言之，控制路径会补偿数据路径的延迟。

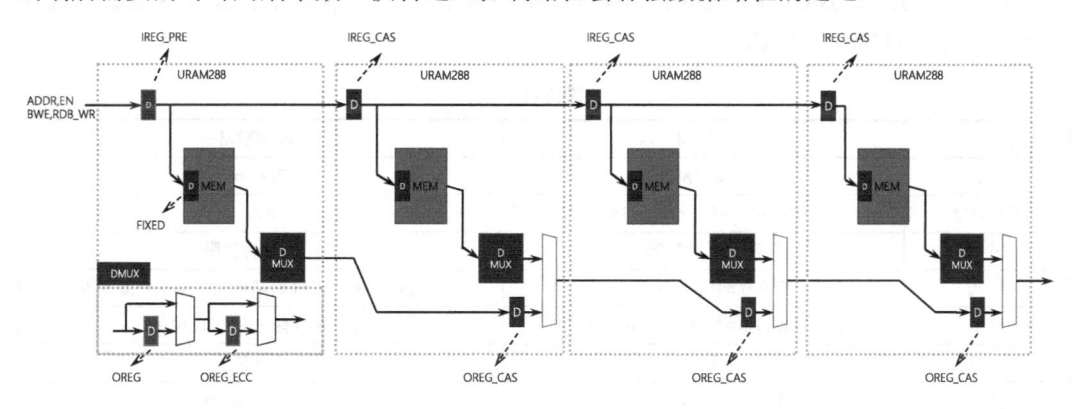

图 6-7

UltraRAM 的读/写行为有固定的模式，即在同一个时钟周期，每个端口仅可执行一次读操作或写操作，这是因为每个端口都有一个控制读/写操作的管脚 RDB_WR_A/B，当其为 0 时，该端口执行读操作，当其为 1 时，该端口执行写操作，这更像是简单双端口的超集。对同一地址，当 A 端口执行写操作、B 端口执行读操作时，B 端口读出的数据为新数据，即 A 端口写入的数据；当 A 端口执行读操作、B 端口执行写操作时，A 端口读出的数据为原始数据，即该地址的原有数据；若 A 端口和 B 端口同时向该地址写入数据，则最终 B 端口的数据会被写入，如图 6-8 所示。UltraRAM 还提供了 BWE_A/B 信号，用于写操作字节选通。

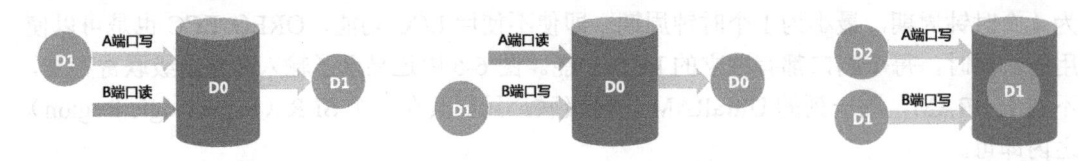

图 6-8

表 6-11 给出了端口 A 的主要输入端口和输出端口。端口 B 的主要输入端口和输出端口与之对应，只需要把名称中的"_A"替换为"_B"，含义相同。除此之外，还有级联端口，均以"CAS_"打头，其中，级联输入端口以"CAS_IN_"打头；级联输出端口以"CAS_OUT_"打头。

表 6-11

输 入 端 口	
ADDR_A[22:0]	读/写地址端口（ADDR_A[22:12]仅在级联时使用）
EN_A	读/写使能端口
RDB_WR_A	读/写选择端口。0 为读操作，1 为写操作
BWE_A[8:0]	字节选通端口
DIN_A[71:0]	输入数据端口
OREG_CE_A	输出寄存器使能端口
OREG_ECC_CE_A	ECC 解码器输出寄存器使能端口
RST_A	输出寄存器复位端口（可支持同步，也可支持异步）
输 出 端 口	
DOUT_A[71:0]	输出数据端口
RDACCESS_A	读输出状态端口，1 表示输出新数据

表 6-12 列出了 BRAM 和 UltraRAM 之间的主要区别。在实际工程设计时根据此区别可决定选用哪种类型的 RAM。

表 6-12

特 征	BRAM	UltraRAM
大小	36Kb/18Kb	288Kb
端口位宽	×1、×2、×4、×9、×18、×36、×72	×72
时钟	两个时钟管脚	一个时钟管脚
模式	简单双端口/真双端口	两个端口，同一时钟周期可执行读操作或写操作
ECC	每个 36Kb BRAM 有一个 ECC 模块	每个端口有一个 ECC 模块
内嵌 FIFO	支持	不支持
初始化	支持用户定义初始值	所有地址上的数据均为 0
写操作字节选通	支持	支持
级联	同列同一个时钟区域内可级联	同列同一 SLR 内可级联
写操作模式	读优先/写优先/保持	固定模式

Versal ACAP 中也有 UltraRAM，但比 UltraScale+ FPGA 中的更灵活，两者的主要差别如表 6-13 所示。可以看到，Versal ACAP 中的 UltraRAM 支持用户定义初始值，同时可支持 4 种不同的位宽。

表 6-13

差 异 点	UltraScale+ UltraRAM	Versal UltraRAM
初 始 值	所有地址上的数据均为 0	支持用户定义初始值
位 宽	×72	×9、×18、×36、×72

UltraScale+UltraRAM 和 Versal UltraRAM 都自带输出寄存器，与 BRAM 不同，该寄存器既可支持同步复位，也可支持异步复位，但不支持置位操作。

采用 Tcl 代码 6-3 可以查看指定 UltraScale+ UltraRAM 的分布情况。通常 UltraRAM 列的个数要比 BRAM 列的个数少，同时，并不是每个时钟区域内都有 UltraRAM。这里的芯片为 XCVU5P。从代码第 39 行的返回值可以看出有 4 列 UltraRAM，在一个 SLR 内，每列 UltraRAM 的个数为 80（对应代码第 41 行的返回值）。在指定的时钟区域内有一列 UltraRAM（对应代码第 43 行的返回值），时钟区域的高度以 UltraRAM 衡量为 16（时钟区域内 UltraRAM 列的高度，对应代码第 45 行的返回值）。这部分脚本也适用于 Versal 器件。

Tcl 代码 6-3

```
1.  #File: check_uram_usp.tcl
2.  set mypart [get_parts xcvu5p-flva2104-2-i]
3.  link_design -part $mypart
4.  set slr [get_slrs SLR0]
5.  => SLR0
6.  set uram_one_slr [get_sites URAM288* -of $slr]
7.  => URAM288_X0Y76 URAM288_X0Y77
8.  set x {}
9.  set y {}
10. foreach uram_one_slr_i $uram_one_slr {
11.     set loc [lindex [split $uram_one_slr_i _] 1]
12.     set xy  [split $loc XY]
13.     lappend x [lindex $xy 1]
14.     lappend y [lindex $xy 2]
15. }
16. set col_index_slr [lsort -integer -unique $x]
17. => 0 1 2 3
18. set row_index_slr [lsort -integer -unique $y]
19. => 0 1 2 3 ... 79
20. highlight_objects $uram_one_slr -color red
21. set cr_with_uram [lsort -unique [get_property CLOCK_REGION $uram_one_slr]]
22. => X1Y0 X1Y1 X1Y2 X1Y3 ...
23. set cr [get_clock_regions [lindex $cr_with_uram 0]]
24. => X1Y0
25. set urams_one_cr [get_sites URAM288* -of $cr]
26. => URAM288_X0Y12 URAM288_X0Y13 ...
27. set x {}
28. set y {}
29. foreach urams_one_cr_i $urams_one_cr {
30.     set loc [lindex [split $urams_one_cr_i _] 1]
31.     set xy [split $loc XY]
32.     lappend x [lindex $xy 1]
33.     lappend y [lindex $xy 2]
34. }
```

```
35. set col_index [lsort -integer -unique $x]
36. => 0
37. set row_index [lsort -integer -unique $y]
38. => 0 1 2 3 4 5 6 7 8 9 10 11 12 13 14 15
39. puts "Number of URAM columns in SLR $slr: [llength $col_index_slr]"
40. => Number of URAM columns in SLR SLR0: 4
41. puts "Height of SLR by URAM: [llength $row_index_slr]"
42. => Height of SLR by URAM: 80
43. puts "Number of URAM columns in CR $cr: [llength $col_index]"
44. => Number of URAM columns in CR X1Y0: 1
45. puts "Height of CR by URAM: [llength $row_index]"
46. => Height of CR by URAM: 16
```

6.2 单端口 RAM 代码风格

单端口 RAM，顾名思义，从数据的角度而言只有一个输入数据端口和一个输出数据端口。只有一个时钟端口，共享读/写地址，写操作受写使能控制。

采用原语的方式，如设计一个 32×4 的单端口 RAM，使用原语 RAM32x1S，如 VHDL 代码 6-1 所示。注意需要声明 UNISIM 库，如代码第 3 行和第 4 行所示。可以很明确地判断出最终会消耗 4 个 LUTRAM，电路形式如图 6-9 所示。

📄 VHDL 代码 6-1

```vhdl
1.  --File: sp_32xn_pmt.vhd
2.  --Single port RAM using primitive RAM32x1S
3.  Library UNISIM;
4.  use UNISIM.vcomponents.all;
5.
6.  library ieee;
7.  use ieee.std_logic_1164.all;
8.
9.  entity sp_32xn_pmt is
10.   generic ( DW : positive := 4 );
11.   port (
12.     wclk : in std_logic;
13.     we   : in std_logic;
14.     addr : in std_logic_vector(4 downto 0);
15.     din  : in std_logic_vector(DW - 1 downto 0);
16.     dout : out std_logic_vector(DW - 1 downto 0)
17.   );
18. end entity;
19.
20. architecture rtl of sp_32xn_pmt is
21. begin
22.   gen0:
23.   for i in 0 to DW - 1 generate
24.     i_ram32x1s : RAM32X1S
25.       generic map (INIT => X"00000000")
26.       port map (
```

```
27.         O    => dout(i),
28.         A0   => addr(0),
29.         A1   => addr(1),
30.         A2   => addr(2),
31.         A3   => addr(3),
32.         A4   => addr(4),
33.         D    => din(i),
34.         WCLK => wclk,
35.         WE   => we
36.      );
37. end generate;
38. end architecture;
```

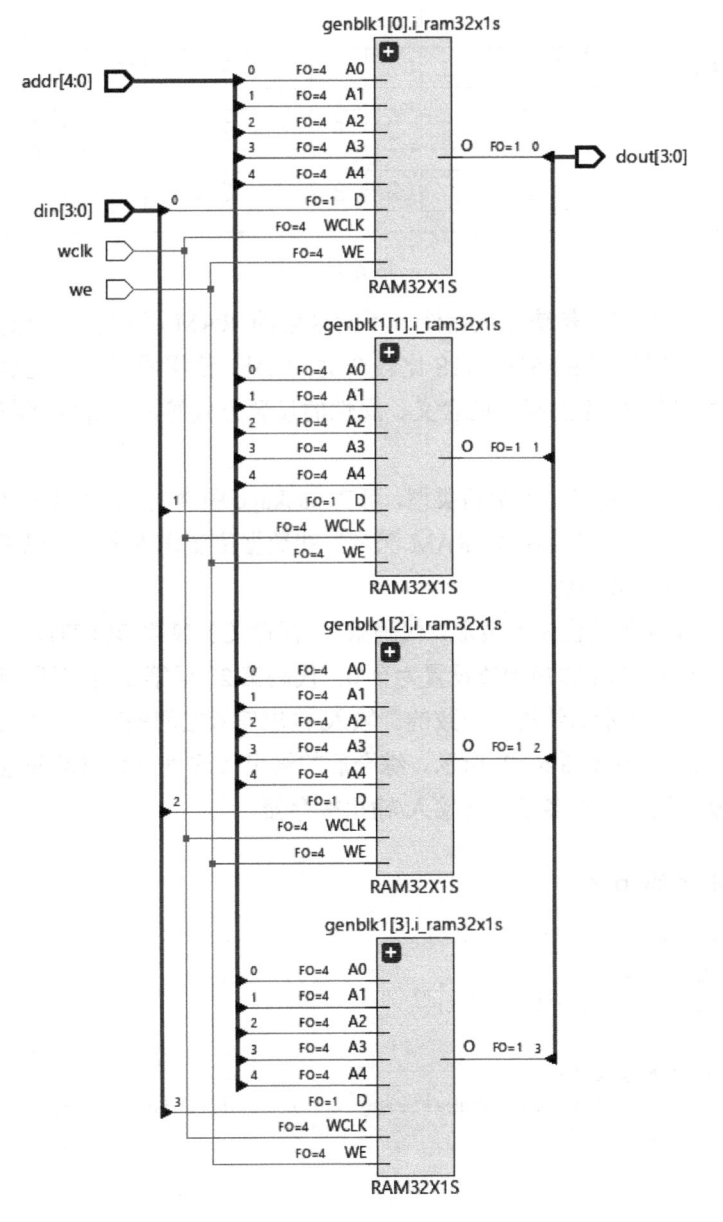

图 6-9

VHDL 代码 6-1 的输入/输出信号之间的时序关系如图 6-10 所示。可以看到，写操作是同步的。当 we 为高电平时，向 0 号地址写入 9，向 1 号地址写入 B。这期间 dout 输出为 0，其实就是 RAM 的初始值。当 we 为低电平时，从 0 号地址读出 9，从 1 号地址读出 B。注意地址和输出数据之间的关系，这表明读操作是异步的。通常为了提高系统性能，会在 LUTRAM 的输出端添加寄存器。不难看出，这其实就是读优先模式，而分布式 RAM 也只有这一种模式，如果要实现其他写操作模式，就需要额外的逻辑。

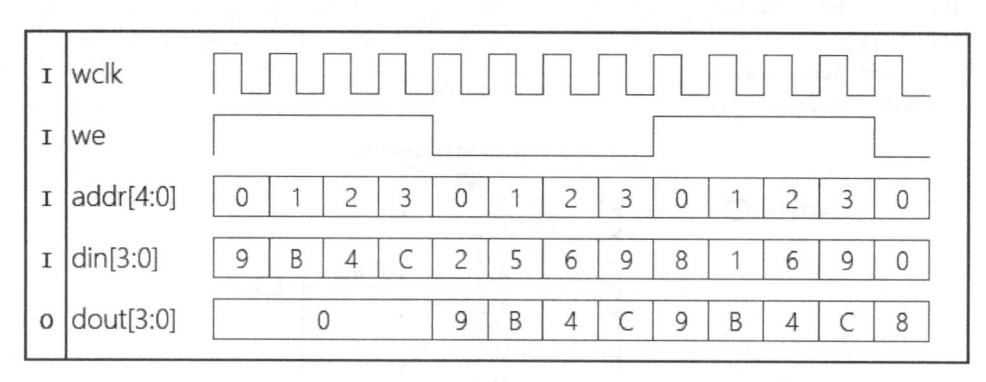

图 6-10

原语方式最大的弊端是缺乏灵活性，不同规格的 RAM 可能需要用到不同的原语，如 128×4 的 RAM 用原语 RAM128x1S 比较合适。此外，使用原语的前提是对原语有深入的理解，要清楚其输入/输出端口的含义、属性的设置方式等，这都需要花费一定的时间和精力。

采用 VHDL，需要定义二维数组类型，可在 package 中声明，如 VHDL 代码 6-2 所示。同时，可以将地址位宽（最终决定 RAM 深度）和数据位宽作为参数，以提高代码的可复用性，这可通过 generic 声明。

VHDL 代码 6-3 是读优先的情形。代码第 11 行定义了参数 DEPTH，该参数由 AW 决定。代码第 23 行将 RAM 的初始值设置为全 0。代码第 28 至第 35 行对所有输入信号寄存一拍，这和如图 6-2 所示的结构是一致的。输入/输出信号之间的时序关系如图 6-11 所示。Elaborated Design 视图如图 6-12 所示。综合后的结果（这里只给出部分电路）如图 6-13 所示，与图 6-9 相比，只是多了一些输入/输出寄存器。

📄 **VHDL 代码 6-2**

```
1.  --File: mem_pkg.vhd
2.  library ieee;
3.  use ieee.std_logic_1164.all;
4.
5.  package mem_pkg is
6.    type mem_t is array(natural range<>) of std_logic_vector;
7.  end package;
```

VHDL 代码 6-3

```vhdl
1.  --File: sp_ram_rf.vhd
2.  library ieee;
3.  use ieee.std_logic_1164.all;
4.  use ieee.numeric_std.all;
5.  use work.mem_pkg.all;
6.
7.  entity sp_ram_rf is
8.    generic (
9.      AW    : positive := 5;
10.     DW    : positive := 4;
11.     DEPTH : positive := 2 ** AW
12.   );
13.   port (
14.     clk  : in std_logic;
15.     we   : in std_logic;
16.     addr : in std_logic_vector(AW - 1 downto 0);
17.     din  : in std_logic_vector(DW - 1 downto 0);
18.     dout : out std_logic_vector(DW - 1 downto 0)
19.   );
20. end entity;
21.
22. architecture rtl of sp_ram_rf is
23.   signal mem: mem_t(0 to DEPTH - 1)(DW - 1 downto 0) := (others => (others => '0'));
24.   signal we_d1   : std_logic;
25.   signal addr_d1 : unsigned(AW - 1 downto 0);
26.   signal din_d1  : std_logic_vector(DW - 1 downto 0);
27. begin
28.   process(clk)
29.   begin
30.     if rising_edge(clk) then
31.       we_d1   <= we;
32.       addr_d1 <= unsigned(addr);
33.       din_d1  <= din;
34.     end if;
35.   end process;
36.
37.   process(clk)
38.   begin
39.     if rising_edge(clk) then
40.       if we_d1 then
41.         mem(to_integer(addr_d1)) <= din_d1;
42.       end if;
43.       dout <= mem(to_integer(addr_d1));
44.     end if;
45.   end process;
46. end architecture;
```

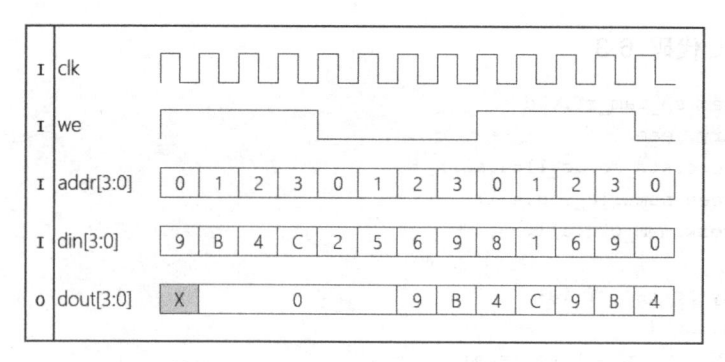

图 6-11

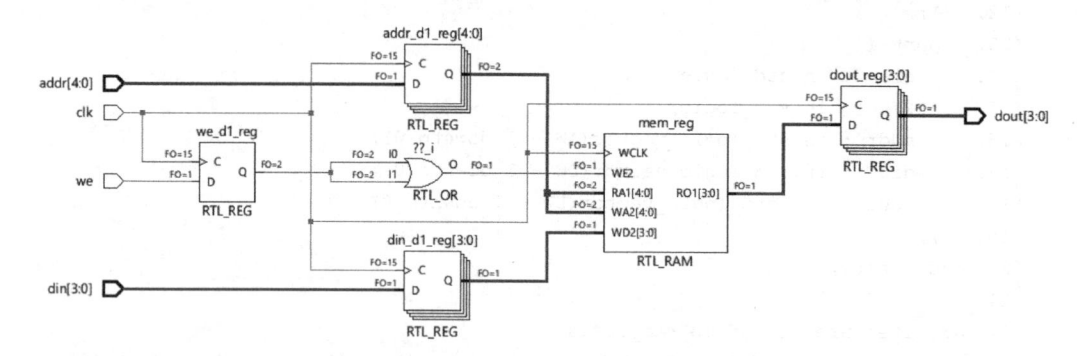

图 6-12

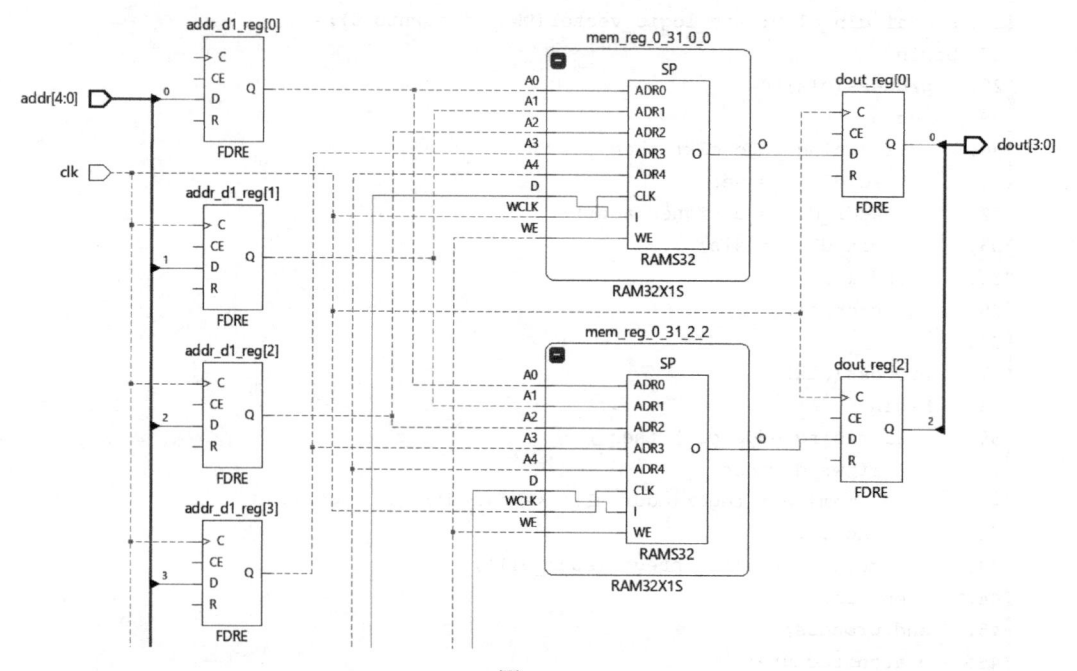

图 6-13

🔍 VHDL-2008 新特性

VHDL-2008 提供了两类数组：约束数组（Constrained Array）和未约束数组（Unconstrained Array），其中，约束数组又分为全约束数组和部分约束数组。例如：

　　type mem_t is array(natural range<>) of std_logic_vector;
　　type cmem_t is array(0 to 31) of std_logic_vector(3 downto 0);
　　type pcmem_t1 is array(natural range<>) of std_logic_vector(3 downto 0);
　　type pcmem_t2 is array(0 to 31) of std_logic_vector。

cmem_t 是全约束数组，数组行、列的个数均已明确指定（32 行 8 列）。pcmem_t1 和 pcmem_t2 为部分约束数组（数组的行或列未明确指定）。mem_t 为未约束数组，数组的行和列均未明确指定。可通过如下方式定义信号为指定的数组类型。当行或列是明确指定的部分约束数组类型时，用关键字 open 表征，如 mem3 和 mem4 所示：

　　signal mem1: mem_t(0 to DEPTH - 1)(DW - 1 downto 0);
　　signal mem2: cmem_t;
　　signal mem3: pcmem_t2(open)(DW - 1 downto 0);
　　signal mem4: pcmem_t1(0 to 31)(open)。

💡 **设计规则 3**：采用 VHDL-2008 描述数组时，可将数组类型设置为未约束数组，在定义数组信号时，再指定具体的行列范围。同时，若该数组类型需要在多个 entity 中使用，则可将其定义在 package 中，这样可最大化地提高代码的可复用性和简洁性。

💡 **设计规则 4**：VHDL-2008 支持在 generic 中引用已定义的参数，尽管对新参数和被引用参数（如 VHDL 代码 6-3 第 11 行所示 DEPTH : positive := 2 ** AW）的位置未提出明确要求，只需要保证新参数在被引用参数之后即可，但从代码风格的角度而言，应将新参数放在 generic 列表的最后一个位置。

对于写优先模式，可以采用如 VHDL 代码 6-4 所示的方式描述。从代码第 40 行和第 43 行构成的 if-else 语句来看，当写使能为高电平时，输出与输入相同，否则输出 RAM 对应地址上的数据，这正是写优先的本质。输入/输出信号之间的时序关系如图 6-14 所示。Elaborated Design 视图如图 6-15 所示，图中的 RTL_MUX 对应的就是代码中的 if-else 语句。综合后的结果（这里只给出部分电路图）如图 6-16 所示，图中，查找表的功能可由对应的真值表判断，实质上实现的就是 if-else 语句的功能，与图 6-9 相比，多了一些输入/输出寄存器，也多了一些查找表。

VHDL 代码 6-4

```vhdl
1.  --File: sp_ram_wf.vhd
2.  library ieee;
3.  use ieee.std_logic_1164.all;
4.  use ieee.numeric_std.all;
5.  use work.mem_pkg.all;
6.
7.  entity sp_ram_wf is
8.    generic (
9.      AW    : positive := 5;
10.     DW    : positive := 4;
11.     DEPTH : positive := 2 ** AW
12.   );
13.   port (
14.     clk  : in std_logic;
15.     we   : in std_logic;
16.     addr : in std_logic_vector(AW - 1 downto 0);
17.     din  : in std_logic_vector(DW - 1 downto 0);
18.     dout : out std_logic_vector(DW - 1 downto 0)
19.   );
20. end entity;
21.
22. architecture rtl of sp_ram_wf is
23. signal mem: mem_t(0 to DEPTH - 1)(DW - 1 downto 0) := (others => (others => '0'));
24. signal we_d1   : std_logic;
25. signal addr_d1 : unsigned(AW - 1 downto 0);
26. signal din_d1  : std_logic_vector(DW - 1 downto 0);
27. begin
28.   process(clk)
29.   begin
30.     if rising_edge(clk) then
31.       we_d1   <= we;
32.       addr_d1 <= unsigned(addr);
33.       din_d1  <= din;
34.     end if;
35.   end process;
36.
37.   process(clk)
38.   begin
39.     if rising_edge(clk) then
40.       if we_d1 then
41.         mem(to_integer(addr_d1)) <= din_d1;
42.         dout                     <= din_d1;
43.       else
44.         dout <= mem(to_integer(addr_d1));
45.       end if;
46.     end if;
47.   end process;
48. end architecture;
```

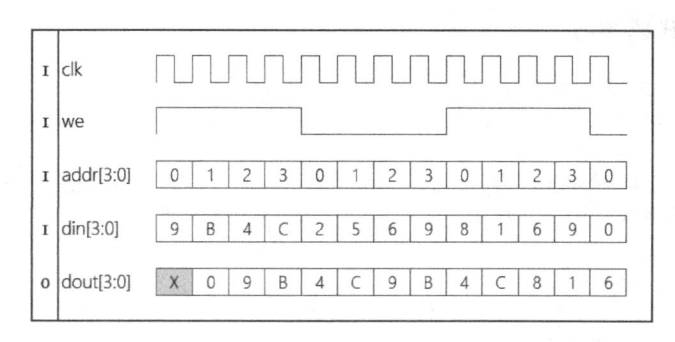

图 6-14

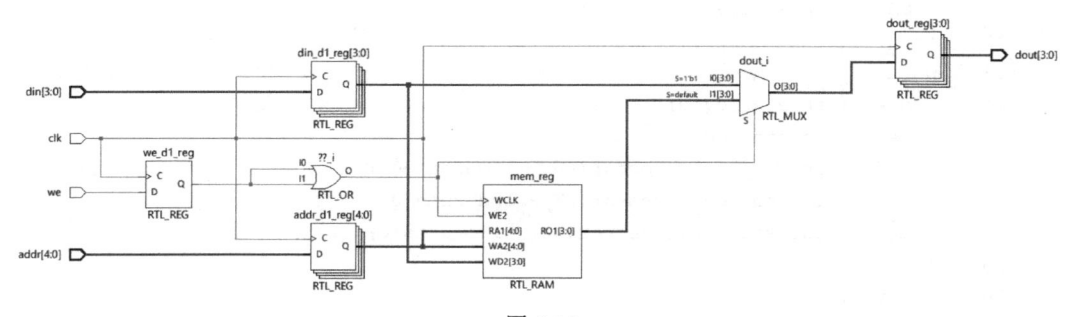

图 6-15

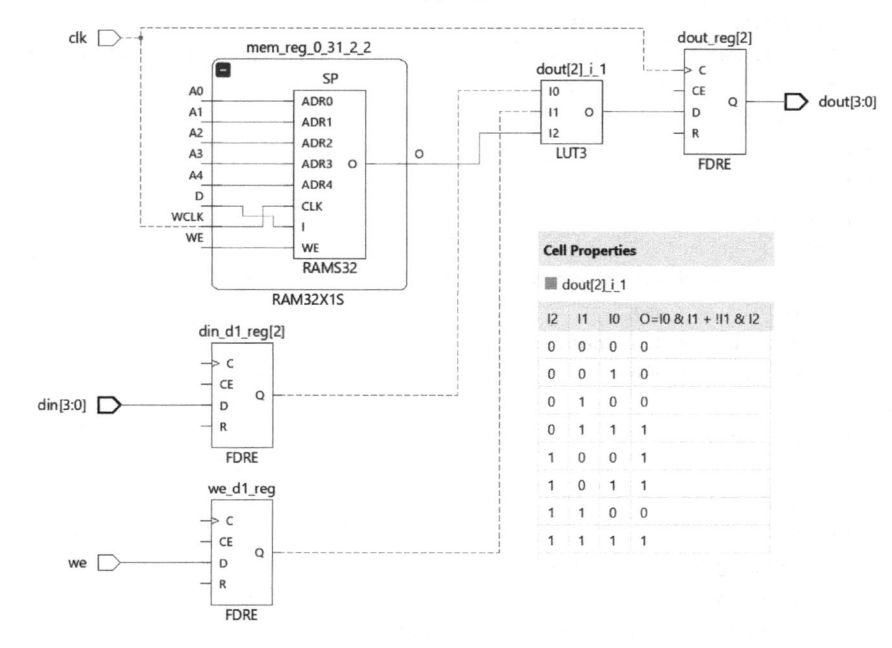

图 6-16

对于保持模式，可采用如 VHDL 代码 6-5 所示的方式描述。从代码第 40 行和第 42 行的 if-else 语句来看，由于 dout 在 if 分支没有被赋值，因此可推断这将综合为一个以 we_d1 为时钟使能信号的寄存器，且 we_d1 为低电平有效。输入/输出信号之间的时序关系如图 6-17 所示。Elaborated Design 视图如图 6-18 所示，图中，RTL_MUX 实际上实现的是取反功能（对 we_d1 取反）。综合后的结果（这里只给出部分电路图）如图 6-19 所示，与图 6-9 相比，多了一些输入/输出寄存器和一个实现取反功能的查找表。

VHDL 代码 6-5

```vhdl
1.  --File: sp_ram_nc.vhd
2.  library ieee;
3.  use ieee.std_logic_1164.all;
4.  use ieee.numeric_std.all;
5.  use work.mem_pkg.all;
6.  
7.  entity sp_ram_nc is
8.    generic (
9.      AW    : positive := 5;
10.     DW    : positive := 4;
11.     DEPTH : positive := 2 ** AW
12.   );
13.   port (
14.     clk  : in std_logic;
15.     we   : in std_logic;
16.     addr : in std_logic_vector(AW - 1 downto 0);
17.     din  : in std_logic_vector(DW - 1 downto 0);
18.     dout : out std_logic_vector(DW - 1 downto 0)
19.   );
20. end entity;
21. 
22. architecture rtl of sp_ram_nc is
23. signal mem: mem_t(0 to DEPTH - 1)(DW - 1 downto 0) := (others => (others => '0'));
24. signal we_d1   : std_logic;
25. signal addr_d1 : unsigned(AW - 1 downto 0);
26. signal din_d1  : std_logic_vector(DW - 1 downto 0);
27. begin
28.   process(clk)
29.   begin
30.     if rising_edge(clk) then
31.       we_d1   <= we;
32.       addr_d1 <= unsigned(addr);
33.       din_d1  <= din;
34.     end if;
35.   end process;
36. 
37.   process(clk)
38.   begin
39.     if rising_edge(clk) then
40.       if we_d1 then
41.         mem(to_integer(addr_d1)) <= din_d1;
42.       else
43.         dout <= mem(to_integer(addr_d1));
44.       end if;
45.     end if;
46.   end process;
47. end architecture;
```

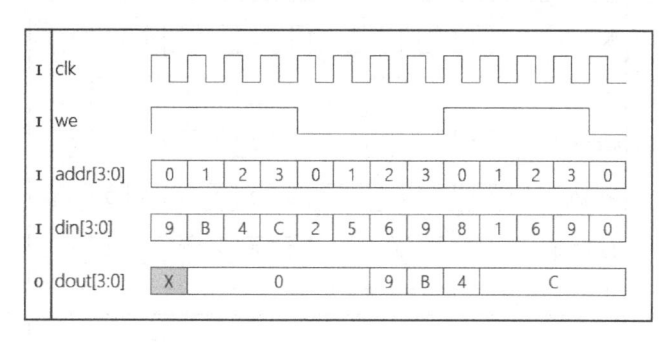

图 6-17

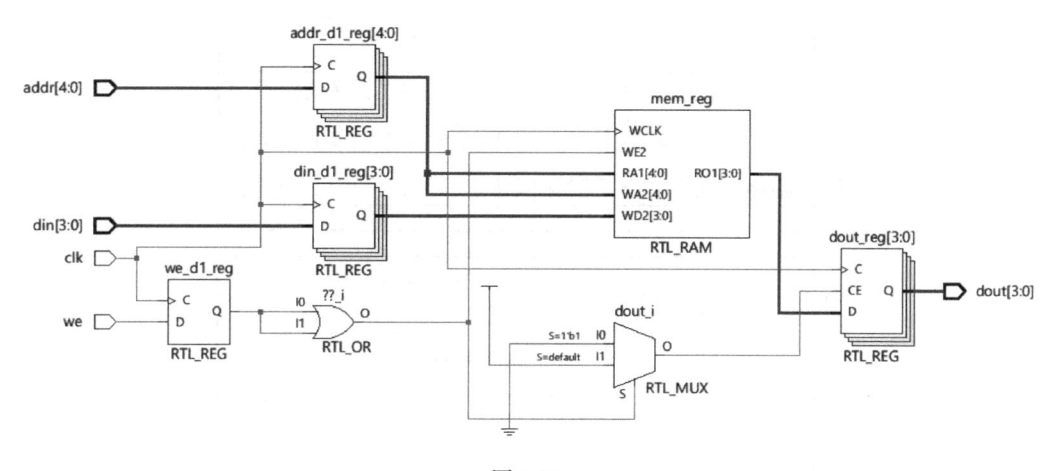

图 6-18

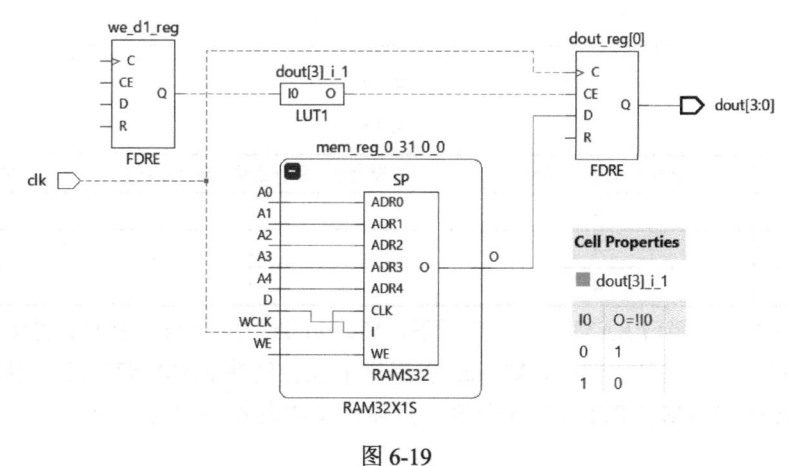

图 6-19

> 💡 **设计规则 5**：对于分布式 RAM，在高速设计中应将其输出寄存一拍（在输出端添加流水寄存器），以改善时序性能。

将读优先、写优先和保持模式的输入/输出信号之间的时序关系放在一张图里，如图 6-20 所示（由上至下依次为读优先、写优先和保持模式），可以加深对这三者的理解。同时，对于 32×4 的单端口 RAM，从资源利用率的角度看，这三者的差别如表 6-14 所示。可以看出，写优先模式消耗的资源最多。

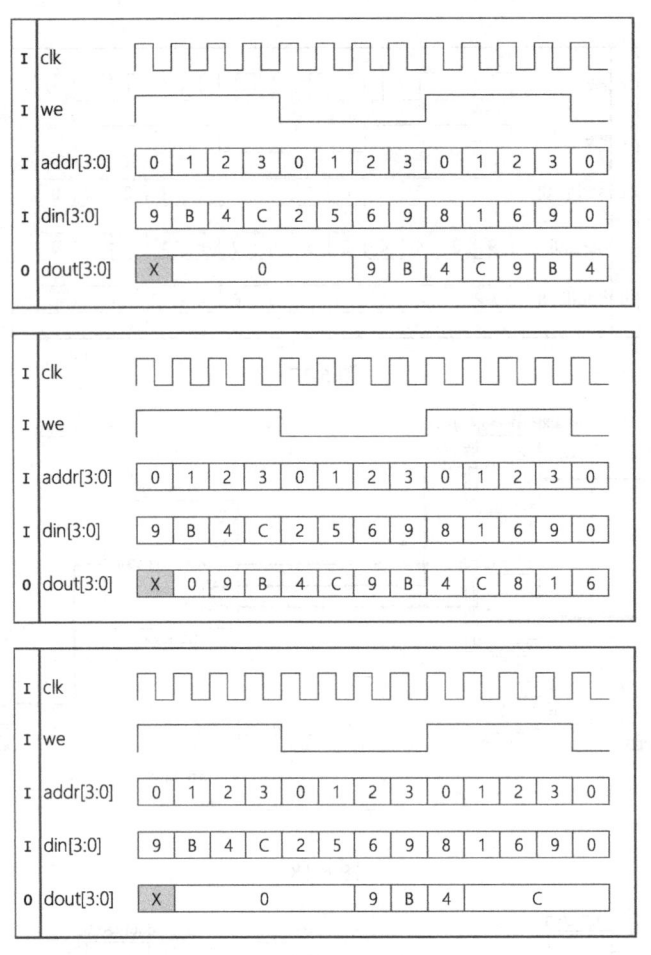

图 6-20

表 6-14

写操作模式	查找表（用作 RAM）	查找表（用作逻辑函数发生器）	触 发 器
读优先	4	0	14
写优先	4	2	14
保持	4	1	14

对于 32×4 的单端口 RAM，默认情形下，Vivado 会采用分布式 RAM 实现，因为这种实现方式最经济、最高效。但是，Vivado 也提供了综合属性 RAM_STYLE，用来指导工具如何推断 RAM。RAM_STYLE 有 5 个值，具体含义如表 6-15 所示。

表 6-15

RAM_STYLE	RAM 实现方式	备　　注
block	BRAM	
distributed	分布式 RAM（LUTRAM）	
registers	寄存器	
ultra	UltraRAM	仅适用于 UltraScale+ FPGA /Versal ACAP
mixed	BRAM/LUTRAM/UltraRAM 三者的组合	节省资源

在高速设计中，如果使用 BRAM 实现 RAM，无论是单端口还是双端口，都要使用 BRAM 自带的输出寄存器，即图 6-2 中的末级寄存器，这是因为输出寄存器对时钟到输出延迟有很大影响，从而对时序收敛也会造成很大影响，如表 6-16 所示。一旦使用输出寄存器，结合图 6-2，从输入到输出就需要 2 个时钟周期。

表 6-16

芯　片	是否使用输出寄存器	速度等级		
		-3	-2	-1
Virtex-7	时钟到输出延迟（未使用输出寄存器）	1.57	1.8	2.08
	时钟到输出延迟（使用输出寄存器）	0.54	0.63	0.75
Virtex-UltraScale FPGA	时钟到输出延迟（未使用输出寄存器）	1.13	1.44	1.64
	时钟到输出延迟（使用输出寄存器）	0.37	0.44	0.49
Virtex-UltraScale+ FPGA	时钟到输出延迟（未使用输出寄存器）	0.91	1.02	1.11
	时钟到输出延迟（使用输出寄存器）	0.27	0.29	0.3

💡**设计规则 6**：在高速设计中使用 BRAM 时，应确保使用了 BRAM 自带的寄存器，以改善时序性能。

考虑到高速设计的需求及代码的可复用性，将地址位宽 AW、数据位宽 DW、写操作模式 WRITE_MODE 和 RAM_STYLE 的值作为参数。这里我们首先创建一个 package，如 VHDL 代码 6-6 所示；在 package 中声明数组数据类型，如代码第 6 行所示；声明枚举类型 ram_wmode_t，用于确定 RAM 的三种写操作模式，如代码第 7 行所示；声明枚举类型 ram_style_t，用于确定 RAM 的实现方式。ram_style_t 的每个值的含义如代码第 9 行至第 12 行所示；声明函数 ram_style_str，如代码第 13 行所示。函数的具体内容如代码第 17 行至第 26 行所示，其目的是完成枚举类型到字符串的转换。枚举类型的好处是限定了取值范围，从而避免非法值造成的干扰。对于代码第 7 行和第 8 行，尽管都使用了枚举类型，但代码第 8 行没有直接给出相应的真实值，而是通过函数完成枚举类型到字符串的转换，这是因为真实值中含有"block"，而"block"是 VHDL 中的保留关键字，直接像代码第 7 行那样描述会引发编译错误。函数 ram_style_str 没有定义返回变量，而是直接通过 return 给出返回值。这里需要了解 VHDL 中字符串 string 的定义。本质上，string 是一系列字符构成的数组，且数组的索引必须为正数，不能是 0，如 VHDL 代码 6-7 所示。因此，若函数的返回值为 string，则要求 string 的长度一致。

在此基础上可形成 VHDL 代码 6-8，使用 case generate 语句将三种写操作模式放在一个文件中，如代码第 43 行至第 77 行所示。代码第 29 行通过 subtype 表明 we_d1 和 we 是同一种数据类型。

📄 **VHDL 代码 6-6**

```
1.  --File: mem_pkg.vhd
2.  library ieee;
3.  use ieee.std_logic_1164.all;
4.
```

```
5.  package mem_pkg is
6.    type mem_t is array(natural range<>) of std_logic_vector;
7.    type ram_wmode_t is (read_first, write_first, no_change);
8.    type ram_style_t is (B, D, R, U, M);
9.    /* B : BRAM, D : Distributed RAM (LUTRAM)
10.      R : Registers, U : UltraRAM (UltraScale/UltraScale+/Versal)
11.      M : Mixed, a mix of BRAM, URAM and LUTRAM
12.   */
13.   function ram_style_str (ram_type : ram_style_t) return string;
14. end package;
15.
16. package body mem_pkg is
17.   function ram_style_str (ram_type : ram_style_t) return string is
18.   begin
19.     case ram_type is
20.       when B => return "block";
21.       when D => return "distributed";
22.       when R => return "registers";
23.       when U => return "ultra";
24.       when M => return "mixed";
25.     end case;
26.   end function;
27. end package body;
```

VHDL 代码 6-7

```
1.  type string is array (positive range <>) of character
2.
3.  constant message1 : string(1 to 19) := "hold time violation";
4.  signal letter : character;
5.  signal message2 : string(1 to 10);
6.  ...
7.  message2 <= "Not" & letter;
```

VHDL 代码 6-8

```
1.  --File: sp_ram_v1.vhd
2.  library ieee;
3.  use ieee.std_logic_1164.all;
4.  use ieee.numeric_std.all;
5.  use work.mem_pkg.all;
6.
7.  entity sp_ram_v1 is
8.    generic (
9.      AW             : positive    := 11;
10.     DW             : positive    := 18;
11.     WRITE_MODE     : ram_wmode_t := no_change;
12.     RAM_STYLE_VAL  : ram_style_t := M;
13.     DEPTH          : positive    := 2 ** AW
14.   );
15.   port (
16.     clk : in std_logic;
```

```vhdl
17.        we   : in std_logic;
18.        addr : in std_logic_vector(AW - 1 downto 0);
19.        din  : in std_logic_vector(DW - 1 downto 0);
20.        dout : out std_logic_vector(DW - 1 downto 0)
21.   );
22. end entity;
23.
24. architecture rtl of sp_ram_v1 is
25.   constant ram_style_val_str : string := ram_style_str(RAM_STYLE_VAL);
26.   signal mem     : mem_t(0 to DEPTH - 1)(DW - 1 downto 0);
27.   attribute RAM_STYLE : string;
28.   attribute RAM_STYLE of mem : signal is ram_style_val_str;
29.   signal we_d1   : we'subtype;
30.   signal addr_d1 : unsigned(addr'range);
31.   signal din_d1  : din'subtype;
32. begin
33.   process(clk)
34.   begin
35.     if rising_edge(clk) then
36.       we_d1   <= we;
37.       addr_d1 <= unsigned(addr);
38.       din_d1  <= din;
39.     end if;
40.   end process;
41.
42.   gen0:
43.   case WRITE_MODE generate
44.     when read_first =>
45.       process(clk)
46.       begin
47.         if rising_edge(clk) then
48.           if we_d1 then
49.             mem(to_integer(addr_d1)) <= din_d1;
50.           end if;
51.           dout <= mem(to_integer(addr_d1));
52.         end if;
53.       end process;
54.     when write_first =>
55.       process(clk)
56.       begin
57.         if rising_edge(clk) then
58.           if we_d1 then
59.             mem(to_integer(addr_d1)) <= din_d1;
60.             dout                     <= din_d1;
61.           else
62.             dout <= mem(to_integer(addr_d1));
63.           end if;
64.         end if;
65.       end process;
66.     when no_change =>
```

```
67.        process(clk)
68.        begin
69.          if rising_edge(clk) then
70.            if we_d1 then
71.              mem(to_integer(addr_d1)) <= din_d1;
72.            else
73.              dout <= mem(to_integer(addr_d1));
74.            end if;
75.          end if;
76.        end process;
77.    end generate;
78. end architecture;
```

🔍 VHDL-2008 新特性

VHDL-2008 新增了 case-generate 语句，这使得我们在描述基于条件选取代码块的电路模型时更为便利。不同于 case 语句，case-generate 语句可以没有 when others 分支。

💡 **设计规则 7**：将 component、constant、type、function 和 procedure 定义在 package 中，可提高代码的可复用性，同时可使代码更为简洁。

应用案例 3：桶形缓存器（输入/输出数据位宽相等）

桶形缓存器在各种数字信号处理应用中被广泛使用，如 FIR 滤波器、多通道滤波器等。如图 6-21 所示，数据依次被写入 0~n-1 号地址上，经过 n 个时钟周期后，再从相应地址输出该数据。可见，数据被延迟了 n 个时钟周期。这种延迟可通过单端口 RAM 实现，要求 RAM 工作在读优先模式下。尽管这种延迟也可通过移位寄存器实现，但当数据位宽较大或延迟深度较大时，采用 BRAM 更高效。

桶形缓存器的设计框图如图 6-22 所示，由两部分构成：计数器和单端口 RAM。其中，RAM 的写使能恒为高电平。计数器用于产生 RAM 的地址，也决定了延迟深度。

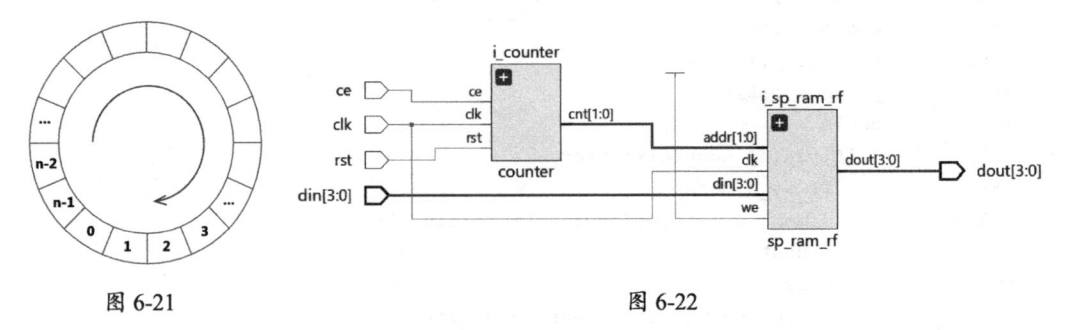

图 6-21   图 6-22

结合表 6-17 的时序图，计数器的计数范围为 0~n-1，但是 D(0) 被延迟了 n+2 个时钟周期，这是因为单端口 RAM 本身从输入到输出就有 2 个时钟周期，所以延迟时钟周期个数与计数器输出最大值相差 3，即 n+2-(n-1)=3。

表 6-17

地　　址	0	1	2	…	n-1	0	1	2	3
输 入 数 据	D(0)	D(1)	D(2)	…	D(n-1)	F(0)	F(1)	F(2)	F(3)
输 出 数 据	0	0	0	0	0	0	0	D(0)	D(1)

若要将数据延迟 6 个时钟周期，则计数器的计数范围为 0~3，相应的输入/输出信号之间的时序关系如图 6-23 所示。

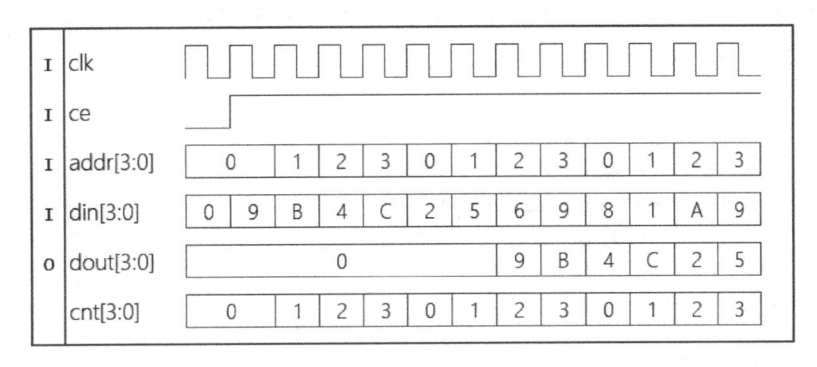

图 6-23

BRAM 支持字节选通功能，因此，写入数据时可以按字节方式写入指定地址。读出数据时依然按照读优先、写优先和保持模式有所不同。同时，设定参数地址位宽 AW、字节个数 NB（决定了数据位宽，数据位宽等于 NB×8）和 RAM_STYLE_VAL（控制 RAM 的实现方式）。VHDL 代码 6-9 为读优先模式下带字节选通功能的单端口 RAM，VHDL 代码 6-10 为写优先模式下带字节选通功能的单端口 RAM，VHDL 代码 6-11 为保持模式下带字节选通功能的单端口 RAM。

📄 VHDL 代码 6-9

```
1.  --File: bytewrite_ram_rf.vhd
2.  library ieee;
3.  use ieee.std_logic_1164.all;
4.  use ieee.numeric_std.all;
5.  use work.mem_pkg.all;
6.
7.  entity bytewrite_ram_rf is
8.    generic (
9.      AW             : positive := 10;
10.     NB             : positive := 4;
11.     RAM_STYLE_VAL  : string   := "block"
12.   );
13.   port (
14.     clk  : in  std_logic;
15.     we   : in  std_logic_vector(NB - 1 downto 0);
16.     addr : in  std_logic_vector(AW - 1 downto 0);
17.     din  : in  std_logic_vector(NB * 8 - 1 downto 0);
18.     dout : out std_logic_vector(NB * 8 - 1 downto 0)
19.   );
```

```vhdl
20. end entity;
21.
22. architecture rtl of bytewrite_ram_rf is
23. signal we_d1   : we'subtype;
24. signal addr_d1 : unsigned(addr'range);
25. signal din_d1  : din'subtype;
26. signal mem: mem_t(0 to 2 ** AW - 1)(NB * 8 - 1 downto 0) := (others => (others => '0'));
27. attribute RAM_STYLE : string;
28. attribute RAM_STYLE of mem : signal is RAM_STYLE_VAL;
29. begin
30.   process(clk)
31.   begin
32.     if rising_edge(clk) then
33.       we_d1   <= we;
34.       addr_d1 <= unsigned(addr);
35.       din_d1  <= din;
36.     end if;
37.   end process;
38.
39.   process(clk)
40.   begin
41.     if rising_edge(clk) then
42.       for i in 0 to NB - 1 loop
43.         if we_d1(i) then
44.           mem(to_integer(addr_d1))(i * 8 + 7 downto i * 8)
45.             <= din_d1(i * 8 + 7 downto i * 8);
46.         end if;
47.       end loop;
48.       dout <= mem(to_integer(addr_d1));
49.     end if;
50.   end process;
51. end architecture;
```

📄 VHDL 代码 6-10

```vhdl
1.  --File: bytewrite_ram_wf.vhd
2.    process(clk)
3.    begin
4.      if rising_edge(clk) then
5.        for i in 0 to NB - 1 loop
6.          if we_d1(i) then
7.            mem(to_integer(addr_d1))(i * 8 + 7 downto i * 8)
8.              <= din_d1(i * 8 + 7 downto i * 8);
9.            dout(i * 8 + 7 downto i * 8)
10.             <= din_d1(i * 8 + 7 downto i * 8);
11.         else
12.           dout(i * 8 + 7 downto i * 8)
13.             <= mem(to_integer(addr_d1))(i * 8 + 7 downto i * 8);
14.         end if;
15.       end loop;
```

```
16.       end if;
17.   end process;
```

> **VHDL 代码 6-11**

```
1.  --File: bytewrite_ram_nc.vhd
2.  process(clk)
3.  begin
4.    if rising_edge(clk) then
5.      for i in 0 to NB - 1 loop
6.        if we_d1(i) then
7.          mem(to_integer(addr_d1))(i * 8 + 7 downto i * 8)
8.            <= din_d1(i * 8 + 7 downto i * 8);
9.        end if;
10.     end loop;
11.   end if;
12. end process;
13.
14. process(clk)
15. begin
16.   if rising_edge(clk) then
17.     if (not (or we_d1)) then
18.       dout <= mem(to_integer(addr_d1));
19.     end if;
20.   end if;
21. end process;
```

当字节个数 NB 为 2、输入/输出数据位宽为 16、地址位宽为 2 时，这三段代码的输入/输出信号之间的时序关系如图 6-24 所示。读者可结合代码和时序图深入理解这三种写操作模式的含义。

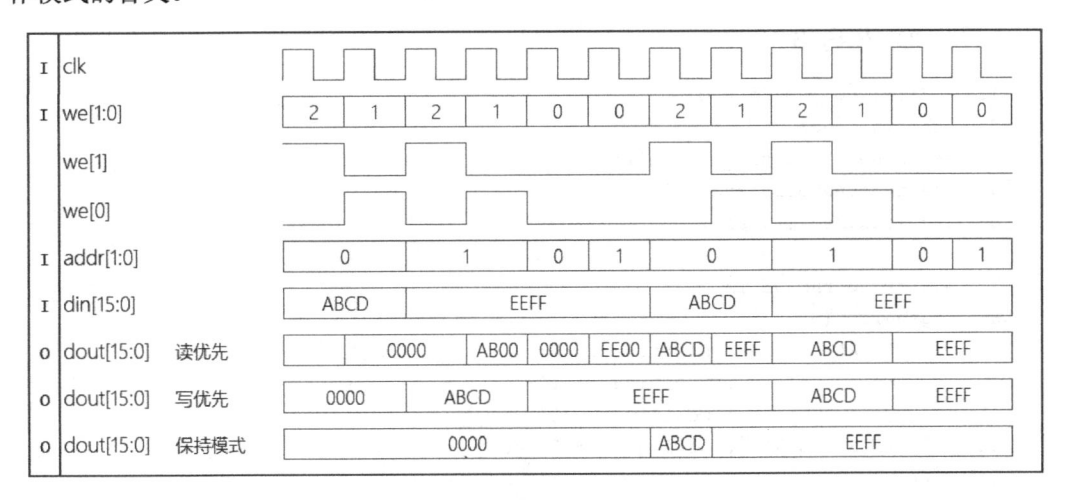

图 6-24

RAM 实际上是一个由深度和宽度构成的二维空间，如果多个 RAM 构成 RAM 组，这就是一个三维空间，如图 6-25 所示。因此，需要在 package 中定义新的数据类型 matrix，如 VHDL 代码 6-12 第 9 行所示。这样，输入（din）/输出（dout）数据均为 mem_t 类型，存储空间为 matrix 类型。Vivado 可以对其进行正确处理，且 RAM_STYLE 依然有效，如

VHDL 代码 6-13 所示，其仿真结果如图 6-26 所示（这里将 RAM 的初始值设置为全 "1"，因此仿真结果中 dout 最开始输出{f, f}）。

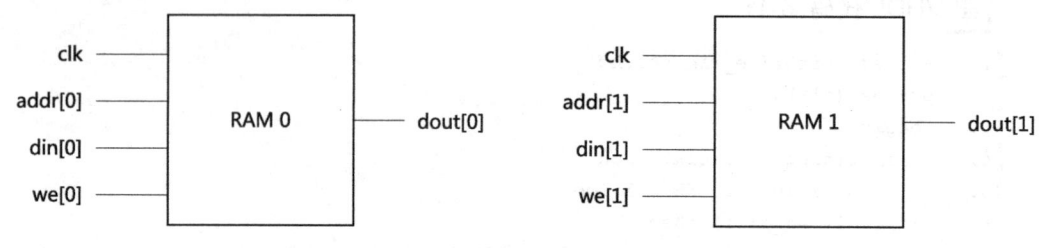

图 6-25

VHDL 代码 6-12

```
1.  --File: mem_pkg.vhd
2.  library ieee;
3.  use ieee.std_logic_1164.all;
4.
5.  package mem_pkg is
6.    type mem_t is array(natural range<>) of std_logic_vector;
7.    type ram_wmode_t is (read_first, write_first, no_change);
8.    type ram_style_t is (B, D, R, U, M);
9.    type matrix is array(natural range<>) of mem_t;
10.   /* B : BRAM, D : Distributed RAM (LUTRAM)
11.      R : Registers, U : UltraRAM (UltraScale/UltraScale+/Versal)
12.      M : Mixed, a mix of BRAM, URAM and LUTRAM
13.   */
14.   function ram_style_str (ram_type : ram_style_t) return string;
15. end package;
```

VHDL 代码 6-13

```
1.  --File: sp_ram_3d.vhd
2.  library ieee;
3.  use ieee.std_logic_1164.all;
4.  use ieee.numeric_std.all;
5.  use work.mem_pkg.all;
6.
7.  entity sp_ram_3d is
8.    generic (
9.      NUM_RAMS      : positive := 2;
10.     AW            : positive := 10;
11.     DW            : positive := 32;
12.     RAM_STYLE_VAL : string   := "block"
13.   );
14.   port (
15.     clk  : in std_logic;
16.     we   : in std_logic_vector(NUM_RAMS - 1 downto 0);
17.     addr : in mem_t(0 to NUM_RAMS - 1)(AW - 1 downto 0);
```

```
18.         din   : in mem_t(0 to NUM_RAMS - 1)(DW - 1 downto 0);
19.         dout  : out mem_t(0 to NUM_RAMS - 1)(DW - 1 downto 0)
20.     );
21. end entity;
22.
23. architecture rtl of sp_ram_3d is
24.   constant DEPTH : positive := 2 ** AW;
25.   signal mem : matrix(0 to NUM_RAMS - 1)(0 to DEPTH - 1)(DW - 1 downto 0);
26.   attribute RAM_STYLE : string;
27.   attribute RAM_STYLE of mem : signal is RAM_STYLE_VAL;
28.   signal we_d1    : we'subtype;
29.   signal addr_d1  : addr'subtype;
30.   signal din_d1   : din'subtype;
31. begin
32.   process(clk)
33.   begin
34.     if rising_edge(clk) then
35.       we_d1   <= we;
36.       addr_d1 <= addr;
37.       din_d1  <= din;
38.     end if;
39.   end process;
40.
41.   process(clk)
42.   begin
43.     if rising_edge(clk) then
44.       for i in 0 to NUM_RAMS - 1 loop
45.         if we_d1(i) then
46.           mem(i)(to_integer(unsigned(addr_d1(i)))) <= din_d1(i);
47.         end if;
48.         dout(i) <= mem(i)(to_integer(unsigned(addr_d1(i))));
49.       end loop;
50.     end if;
51.   end process;
52. end architecture;
```

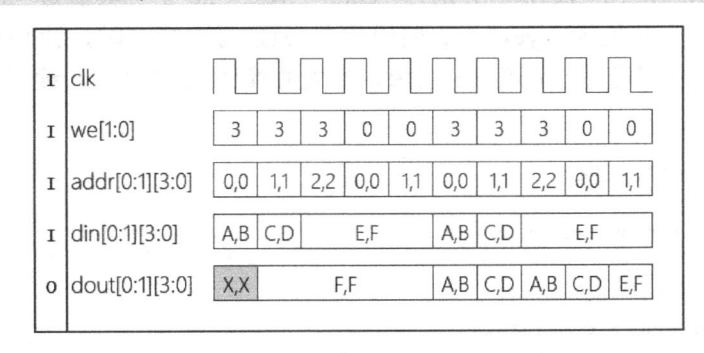

I	clk									
I	we[1:0]	3	3	3	0	0	3	3	3	0	0
I	addr[0:1][3:0]	0,0	1,1	2,2	0,0	1,1	0,0	1,1	2,2	0,0	1,1
I	din[0:1][3:0]	A,B	C,D	E,F			A,B	C,D	E,F		
O	dout[0:1][3:0]	X,X		F,F			A,B	C,D	A,B	C,D	E,F

图 6-26

VHDL 代码 6-13 对应的 Elaborated Design 视图如图 6-27 所示。可以看到所谓的 3D RAM 本质上是由多个独立的 2D RAM 构成的存储空间。

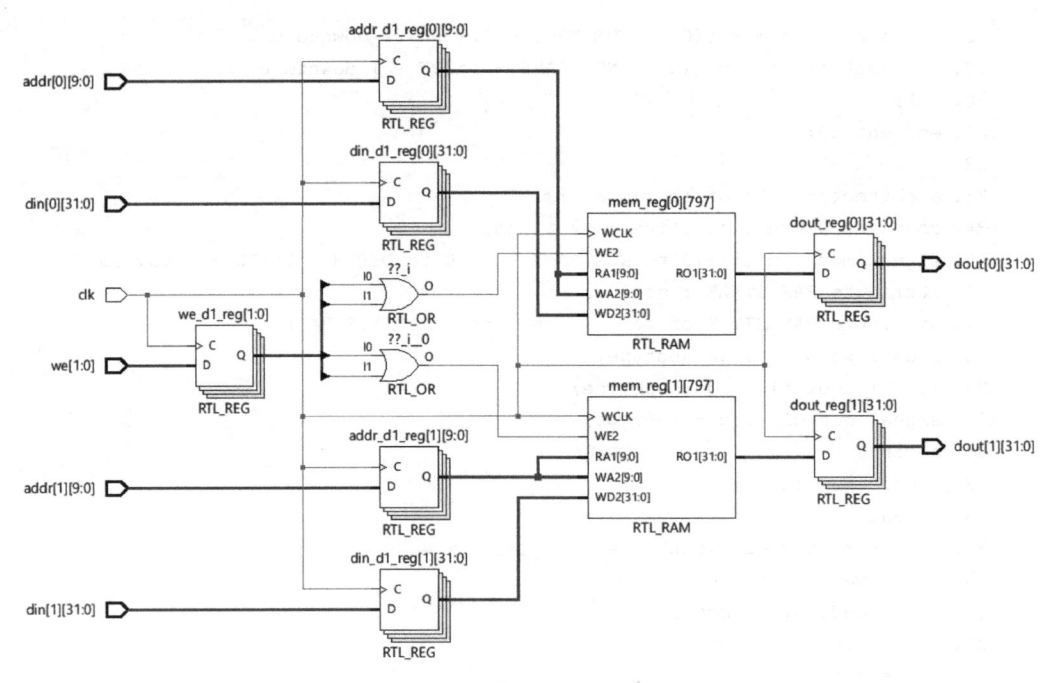

图 6-27

6.3 简单双端口 RAM 代码风格

简单双端口 RAM，顾名思义，有两个端口，只是一个端口仅执行写操作，另一个端口仅执行读操作。这两个端口可以工作在同一时钟下，也可以有独立的时钟。VHDL 代码 6-14 显示的是读/写时钟为同一时钟时的情形。这里将地址位宽 AW、数据位宽 DW、RAM_STYLE 和 RW_ADDR_COLLISION 的值作为参数，以增强代码的可复用性。其中，综合属性 RW_ADDR_COLLISION 只有在 RAM_STYLE 为 "block"、RAM 为简单双端口 RAM 且读地址由寄存器输出时才有效。其目的是管理对同一地址同时进行读/写操作时读端口的工作模式，有三个可选值，分别为 "auto" "yes" "no"。当其值为 "auto" 时，输出是不可预测的。当其值为 "yes"（表明用户关心读/写冲突）时，读端口将被设置为 WRITE_FIRST。当其值为 "no"（表明用户不关心读/写冲突）时，读端口将被设置为 NO_CHANGE。

📄 VHDL 代码 6-14

```
1.  --File: sdp_one_clk.vhd
2.  library ieee;
3.  use ieee.std_logic_1164.all;
4.  use ieee.numeric_std.all;
5.  use work.mem_pkg.all;
6.
7.  entity sdp_one_clk is
8.    generic (
9.      AW            : positive := 4;
10.     DW            : positive := 2;
```

```vhdl
11.       RAM_STYLE_VAL     : string   := "distributed";
12.       RW_ADDR_COL_VAL : string   := "yes"
13.   );
14.   port (
15.     clk   : in std_logic;
16.     wen   : in std_logic;
17.     waddr : in std_logic_vector(AW - 1 downto 0);
18.     din   : in std_logic_vector(DW - 1 downto 0);
19.     ren   : in std_logic;
20.     raddr : in std_logic_vector(AW - 1 downto 0);
21.     dout  : out std_logic_vector(DW - 1 downto 0)
22.   );
23. end entity;
24.
25. architecture rtl of sdp_one_clk is
26. constant DEPTH : positive := 2 ** AW;
27. signal wen_d1   : wen'subtype;
28. signal ren_d1   : ren'subtype;
29. signal din_d1   : din'subtype;
30. signal waddr_d1: waddr'subtype;
31. signal raddr_d1: raddr'subtype;
32. signal mem     : mem_t(0 to DEPTH - 1)(DW - 1 downto 0);
33. attribute RAM_STYLE : string;
34. attribute RW_ADDR_COLLISION : string;
35. attribute RAM_STYLE of mem : signal is RAM_STYLE_VAL;
36. attribute RW_ADDR_COLLISION of mem : signal is RW_ADDR_COL_VAL;
37. begin
38.   process(clk)
39.   begin
40.     if rising_edge(clk) then
41.       wen_d1    <= wen;
42.       waddr_d1 <= waddr;
43.       ren_d1    <= ren;
44.       raddr_d1 <= raddr;
45.       din_d1    <= din;
46.     end if;
47.   end process;
48.
49.   process(clk)
50.   begin
51.     if rising_edge(clk) then
52.       if wen_d1 then
53.         mem(to_integer(unsigned(waddr_d1))) <= din_d1;
54.       end if;
55.     end if;
56.   end process;
57.
58.   process(clk)
59.   begin
60.     if rising_edge(clk) then
61.       if ren_d1 then
```

```
62.            dout <= mem(to_integer(unsigned(raddr_d1)));
63.         end if;
64.      end if;
65.   end process;
66. end architecture;
```

VHDL 代码 6-14 的输入/输出信号之间的时序关系如图 6-28 所示。可以看到，当对同一地址同时进行读/写操作时，读出的是该地址上的原始数据，这意味着该端口其实工作在读优先模式下。如果将 RW_ADDR_COLLISION 设置为 "yes"，那么 Vivado 就会将该端口设置为写优先模式，这样综合后的仿真结果与行为级仿真结果就会有所不同。

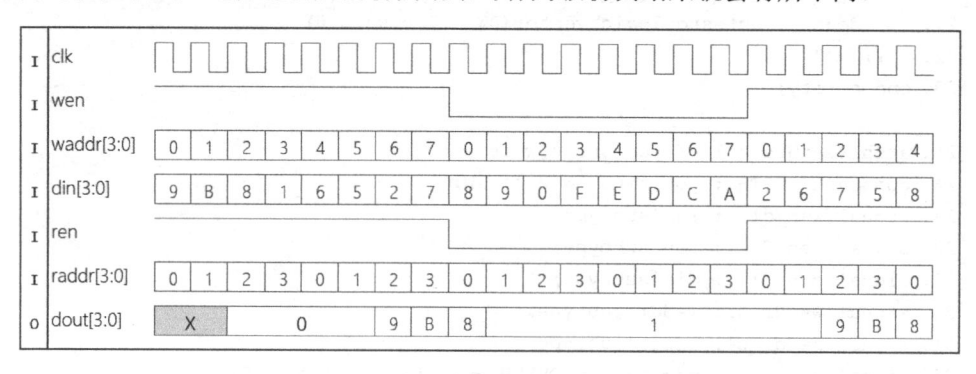

图 6-28

VHDL 代码 6-15 显示了读/写时钟独立的简单双端口 RAM 的描述方式。A 端口为写端口，B 端口为读端口。此时，写使能、写地址和写数据与写时钟同步，读使能、读地址、读数据与读时钟同步。在这种情况下，要避免同时对同一地址进行读/写操作。相应的输入/输出信号之间的时序关系如图 6-29 所示。

VHDL 代码 6-15

```
1.  --File: sdp_two_clk.vhd
2.  process(clka)
3.  begin
4.    if rising_edge(clka) then
5.      wea_d1   <= wea;
6.      addra_d1 <= addra;
7.      dina_d1  <= dina;
8.    end if;
9.  end process;
10.
11. process(clkb)
12. begin
13.   if rising_edge(clkb) then
14.     reb_d1   <= reb;
15.     addrb_d1 <= addrb;
16.   end if;
17. end process;
18.
19. process(clka)
20. begin
21.   if rising_edge(clka) then
```

```
22.        if wea_d1 then
23.          mem(to_integer(unsigned(addra_d1))) <= dina_d1;
24.        end if;
25.      end if;
26.  end process;
27.
28.  process(clkb)
29.  begin
30.      if rising_edge(clkb) then
31.        if reb_d1 then
32.          doutb <= mem(to_integer(unsigned(addrb_d1)));
33.        end if;
34.      end if;
35.  end process;
```

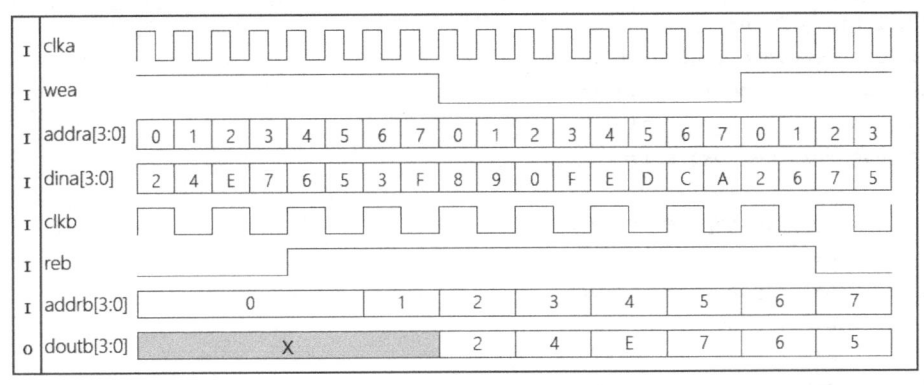

图 6-29

> 💡 **设计规则 8**：使用简单双端口 RAM 时，若读/写时钟独立，那么应避免同时对同一地址进行读/写操作。

BRAM 支持非对称读/写。在简单双端口模式下，非对称读/写分为两种情形：①读出数据位宽是写入数据位宽的整数倍；②写入数据位宽是读出数据位宽的整数倍。无论是哪种情形，本质上存储的数据位数都是一致的，如图 6-30 所示。左边的 RAM 深度为 8、宽度为 4、总存储空间为 32 位。右边的 RAM 深度为 2、宽度为 16、总存储空间仍为 32 位。对比左右两边的 RAM 可发现：就宽度而言（数据位宽），左边的 RAM 较窄，右边的较宽；就深度而言，左边的较深，右边的较浅。如果左边对应写操作，右边对应读操作，就是情形①；如果右边对应写操作，左边对应读操作，就是情形②。

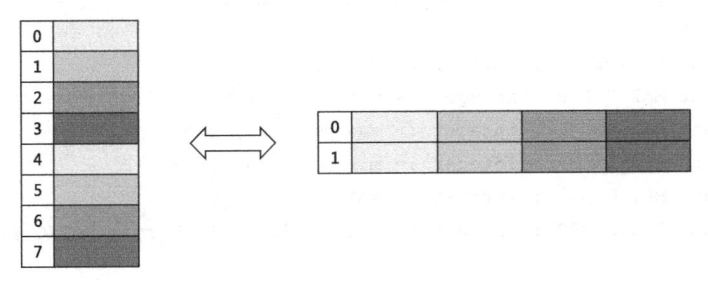

图 6-30

情形①如 VHDL 代码 6-16 所示，相应的输入/输出信号之间的时序关系如图 6-31 所示。

VHDL 代码 6-16

```vhdl
1.  /*File: asym_sdp_ram_read_wider.vhd
2.  AWA : address width of port A
3.  DWA : data width of port A
4.  AWB : address width of port B
5.  DWB : data width of port B
6.  DEPTHA : 2 ** AWA
7.  DEPTHB : 2 ** AWB
8.  DEPTHA * DWA = DEPTHB * DWB
9.  */
10. library ieee;
11. use ieee.std_logic_1164.all;
12. use ieee.numeric_std.all;
13. use ieee.math_real.all;
14. use work.mem_pkg.all;
15.
16. entity asym_sdp_ram_read_wider is
17.   generic (
18.     AWA : positive := 10;
19.     DWA : positive := 4;
20.     AWB : positive := 8;
21.     DWB : positive := 16
22.   );
23.   port (
24.     clka  : in std_logic;
25.     wea   : in std_logic;
26.     addra : in std_logic_vector(AWA - 1 downto 0);
27.     dina  : in std_logic_vector(DWA - 1 downto 0);
28.     clkb  : in std_logic;
29.     reb   : in std_logic;
30.     addrb : in std_logic_vector(AWB - 1 downto 0);
31.     doutb : out std_logic_vector(DWB - 1 downto 0)
32.   );
33. end entity;
34.
35. architecture rtl of asym_sdp_ram_read_wider is
36.   constant DEPTHA    : integer := 2 ** AWA;
37.   constant DEPTHB    : integer := 2 ** AWB;
38.   constant MAX_DEPTH : integer := maximum(DEPTHA, DEPTHB);
39.   constant MAX_DW    : integer := maximum(DWA, DWB);
40.   constant MIN_DW    : integer := minimum(DWA, DWB);
41.   constant RATIO     : integer := MAX_DW / MIN_DW;
42.   constant LOG2RATIO : integer := integer(ceil(log2(real(RATIO))));
```

```vhdl
43.  signal mem : mem_t(0 to MAX_DEPTH - 1)(MIN_DW - 1 downto 0);
44.  signal wea_d1   : wea'subtype;
45.  signal addra_d1 : addra'subtype;
46.  signal dina_d1  : dina'subtype;
47.  signal reb_d1   : reb'subtype;
48.  signal addrb_d1 : addrb'subtype;
49. begin
50.   process(clka)
51.   begin
52.     if rising_edge(clka) then
53.       wea_d1   <= wea;
54.       addra_d1 <= addra;
55.       dina_d1  <= dina;
56.     end if;
57.   end process;
58.
59.   process(clkb)
60.   begin
61.     if rising_edge(clkb) then
62.       reb_d1   <= reb;
63.       addrb_d1 <= addrb;
64.     end if;
65.   end process;
66.
67.   process(clka)
68.   begin
69.     if rising_edge(clka) then
70.       if wea_d1 then
71.         mem(to_integer(unsigned(addra_d1))) <= dina_d1;
72.       end if;
73.     end if;
74.   end process;
75.
76.   process(clkb)
77.     variable addrb_lsb : unsigned(LOG2RATIO - 1 downto 0);
78.   begin
79.     if rising_edge(clkb) then
80.       if reb_d1 then
81.         for i in 0 to RATIO - 1 loop
82.           addrb_lsb := to_unsigned(i, LOG2RATIO);
83.           doutb((i + 1) * MIN_DW - 1 downto i * MIN_DW)
84.             <= mem(to_integer(unsigned(addrb_d1) & addrb_lsb));
85.         end loop;
86.       end if;
87.     end if;
88.   end process;
89. end architecture;
```

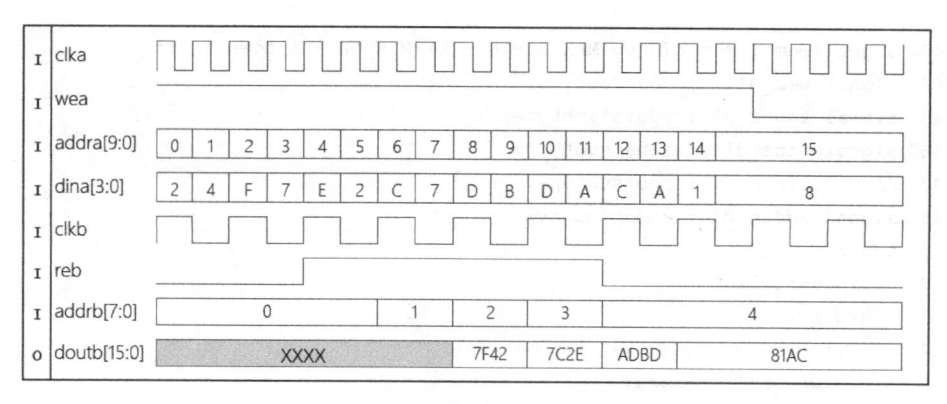

I	clka	
I	wea	
I	addra[9:0]	0 1 2 3 4 5 6 7 8 9 10 11 12 13 14 15
I	dina[3:0]	2 4 F 7 E 2 C 7 D B D A C A 1 8
I	clkb	
I	reb	
I	addrb[7:0]	0 1 2 3 4
O	doutb[15:0]	XXXX 7F42 7C2E ADBD 81AC

图 6-31

🔍 VHDL-2008 新特性

VHDL-2008 引入了 maximum 和 minimum 函数。这两个函数均由 2 个参数和 1 个返回值构成。前者返回 2 个参数中的较大者，后者返回较小者。这两个参数的数据类型可以是 std_logic_vector、unsigned、signed、sfixed、ufixed 等。如果使用 VHDL-87 版或 VHDL-93 版，就需要用户定义函数实现选大和选小功能，如 VHDL 代码 6-17 所示。

VHDL 代码 6-16 第 42 行还使用了 math_real package 中的函数 log2（以 2 为底的对数）和 ceil（向上取整）。实际上，以 2 为底的对数也可以通过用户定义的方式实现，如 VHDL 代码 6-17 第 20 行至第 30 行所示。

📄 **VHDL 代码 6-17**

```
1.  --File: mem_pkg.vhd
2.  function max(L, R : integer) return integer is
3.  begin
4.    if L > R then
5.      return L;
6.    else
7.      return R;
8.    end if;
9.  end function;
10.
11. function min(L, R : integer) return integer is
12. begin
13.   if L < R then
14.     return L;
15.   else
16.     return R;
17.   end if;
18. end function;
19.
20. function log2(val : integer) return natural is
21.   variable res : natural;
22. begin
23.   for i in 0 to 31 loop
24.     if (val <= (2 ** i)) then
```

```
25.          res := i;
26.          exit;
27.       end if;
28.    end loop;
29.    return res;
30. end function;
```

情形②如 VHDL 代码 6-18 所示。这里只给出了关键部分的代码片段，其中，代码第 2 行至第 14 行为写操作。代码第 16 行至第 23 行为读操作。相应的输入/输出信号之间的时序关系如图 6-32 所示。

📄 **VHDL 代码 6-18**

```
1.  --File: asym_sdp_ram_write_wider.vhd
2.  process(clka)
3.    variable addra_lsb : unsigned(LOG2RATIO - 1 downto 0);
4.  begin
5.    if rising_edge(clka) then
6.      if wea_d1 then
7.        for i in 0 to RATIO - 1 loop
8.          addra_lsb := to_unsigned(i, LOG2RATIO);
9.          mem(to_integer(unsigned(addra_d1) & addra_lsb))
10.            <= dina_d1((i + 1) * MIN_DW - 1 downto i * MIN_DW);
11.        end loop;
12.      end if;
13.    end if;
14. end process;
15.
16. process(clkb)
17. begin
18.   if rising_edge(clkb) then
19.     if reb_d1 then
20.       doutb <= mem(to_integer(unsigned(addra_d1)));
21.     end if;
22.   end if;
23. end process;
```

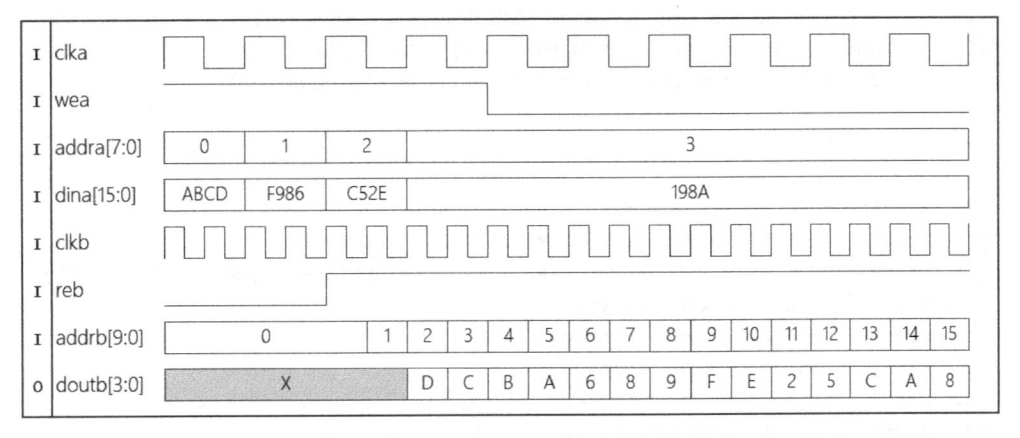

I	clka	
I	wea	
I	addra[7:0]	0 | 1 | 2 | 3
I	dina[15:0]	ABCD | F986 | C52E | 198A
I	clkb	
I	reb	
I	addrb[9:0]	0 | 1 | 2 | 3 | 4 | 5 | 6 | 7 | 8 | 9 | 10 | 11 | 12 | 13 | 14 | 15
O	doutb[3:0]	X | D | C | B | A | 6 | 8 | 9 | F | E | 2 | 5 | C | A | 8

图 6-32

💡 **设计规则 9**：对 variable 赋值时需要使用操作符 ":=" 而不是 "<="，如 VHDL 代码 6-18 第 8 行所示。

多个同规格的简单双端口 RAM 可构成一个三维简单双端口 RAM，如 VHDL 代码 6-19 所示。代码第 19 行定义了读端口的复位信号，该复位信号用于对输出寄存器进行复位，对应代码第 66 行至第 75 行。这里需要注意复位信号与读使能信号（代码第 20 行）的作用对象是不同的，因此没有优先级之分，这种写法不会消耗查找表。但如果复位信号的优先级高于读使能信号，即只要复位信号有效输出，就是全 0。否则检查读使能信号，若其有效，则允许从 RAM 中读取数据，这样将会消耗查找表，这其实并不是良好的代码风格。我们始终强调要保证代码风格与硬件架构相匹配。对于输入/输出信号之间的时序关系，读者可结合一维简单双端口 RAM 的时序图进行理解。

📄 **VHDL 代码 6-19**

```vhdl
1.  --File: sdp_ram_3d.vhd
2.  library ieee;
3.  use ieee.std_logic_1164.all;
4.  use ieee.numeric_std.all;
5.  use work.mem_pkg.all;
6.
7.  entity sdp_ram_3d is
8.    generic (
9.      NUM_RAMS      : positive := 2;
10.     AW            : positive := 10;
11.     DW            : positive := 32
12.   );
13.   port (
14.     clka  : in std_logic;
15.     wea   : in std_logic_vector(NUM_RAMS - 1 downto 0);
16.     addra : in mem_t(0 to NUM_RAMS - 1)(AW - 1 downto 0);
17.     dina  : in mem_t(0 to NUM_RAMS - 1)(DW - 1 downto 0);
18.     clkb  : in std_logic;
19.     rstb  : in std_logic_vector(NUM_RAMS - 1 downto 0);
20.     reb   : in std_logic_vector(NUM_RAMS - 1 downto 0);
21.     addrb : in mem_t(0 to NUM_RAMS - 1)(AW - 1 downto 0);
22.     doutb : out mem_t(0 to NUM_RAMS - 1)(DW - 1 downto 0)
23.   );
24. end entity;
25.
26. architecture rtl of sdp_ram_3d is
27.   constant DEPTH : positive := 2 ** AW;
28.   signal mem : matrix(0 to NUM_RAMS - 1)(0 to DEPTH - 1)(DW - 1 downto 0);
29.   signal wea_d1    : wea'subtype;
30.   signal addra_d1  : addra'subtype;
31.   signal dina_d1   : dina'subtype;
32.   signal reb_d1    : reb'subtype;
33.   signal rstb_d1   : rstb'subtype;
34.   signal addrb_d1  : addrb'subtype;
35.   signal doutb_i   : doutb'subtype;
```

```vhdl
36. begin
37.    process(clka)
38.    begin
39.      if rising_edge(clka) then
40.        wea_d1   <= wea;
41.        addra_d1 <= addra;
42.        dina_d1  <= dina;
43.      end if;
44.    end process;
45.
46.    process(clkb)
47.    begin
48.      if rising_edge(clkb) then
49.        rstb_d1  <= rstb;
50.        reb_d1   <= reb;
51.        addrb_d1 <= addrb;
52.      end if;
53.    end process;
54.
55.    process(clka)
56.    begin
57.      if rising_edge(clka) then
58.        for i in 0 to NUM_RAMS - 1 loop
59.          if wea_d1(i) then
60.            mem(i)(to_integer(unsigned(addra_d1(i)))) <= dina_d1(i);
61.          end if;
62.        end loop;
63.      end if;
64.    end process;
65.
66.    process(clkb)
67.    begin
68.      if rising_edge(clkb) then
69.        for k in 0 to NUM_RAMS - 1 loop
70.          if reb_d1(k) then
71.            doutb_i(k) <= mem(k)(to_integer(unsigned(addrb_d1(k))));
72.          end if;
73.        end loop;
74.      end if;
75.    end process;
76.
77.    process(clkb)
78.    begin
79.      if rising_edge(clkb) then
80.        for k in 0 to NUM_RAMS - 1 loop
81.          doutb(k) <= (others => '0') when rstb_d1(k) else
82.                      doutb_i(k);
83.        end loop;
84.      end if;
85.    end process;
86. end architecture;
```

6.4 真双端口 RAM 代码风格

真双端口 RAM 有两个端口，不同于简单双端口 RAM，真双端口 RAM 的每个端口既可执行写操作又可执行读操作，但这两个端口需要共享存储空间。VHDL 代码 6-20 所示为两个端口均为读优先模式的真双端口。不难看出，每个端口的描述方式跟一个单端口 RAM 是一致的，区别在于这两个端口要共享存储空间，如代码第 58 行和第 68 行所示。相应的输入/输出信号之间的时序关系如图 6-33 所示。

📄 **VHDL 代码 6-20**

```vhdl
1.  --File: tdp_ram_rf.vhd.vhd
2.  library ieee;
3.  use ieee.std_logic_1164.all;
4.  use ieee.numeric_std.all;
5.  use work.mem_pkg.all;
6.
7.  entity tdp_ram_rf is
8.    generic (
9.      AW              : positive := 11;
10.     DW              : positive := 18
11.   );
12.   port (
13.     clka  : in std_logic;
14.     addra : in std_logic_vector(AW - 1 downto 0);
15.     wea   : in std_logic;
16.     dina  : in std_logic_vector(DW - 1 downto 0);
17.     douta : out std_logic_vector(DW - 1 downto 0);
18.     clkb  : in std_logic;
19.     addrb : in std_logic_vector(AW - 1 downto 0);
20.     web   : in std_logic;
21.     dinb  : in std_logic_vector(DW - 1 downto 0);
22.     doutb : out std_logic_vector(DW - 1 downto 0)
23.   );
24. end entity;
25.
26. architecture rtl of tdp_ram_rf is
27.   constant DEPTH  : positive := 2 ** AW;
28.   signal wea_d1    : wea'subtype;
29.   signal addra_d1  : addra'subtype;
30.   signal dina_d1   : dina'subtype;
31.   signal web_d1    : web'subtype;
32.   signal addrb_d1  : addrb'subtype;
33.   signal dinb_d1   : dinb'subtype;
34.   signal mem       : mem_t(0 to DEPTH - 1)(DW - 1 downto 0);
35. begin
36.   process(clka)
37.   begin
38.     if rising_edge(clka) then
```

```vhdl
39.        wea_d1    <= wea;
40.        addra_d1  <= addra;
41.        dina_d1   <= dina;
42.     end if;
43.  end process;
44.
45.  process(clkb)
46.  begin
47.     if rising_edge(clkb) then
48.        web_d1    <= web;
49.        addrb_d1  <= addrb;
50.        dinb_d1   <= dinb;
51.     end if;
52.  end process;
53.
54.  process(clka)
55.  begin
56.     if rising_edge(clka) then
57.        if wea_d1 then
58.           mem(to_integer(unsigned(addra_d1))) <= dina_d1;
59.        end if;
60.        douta <= mem(to_integer(unsigned(addra_d1)));
61.     end if;
62.  end process;
63.
64.  process(clkb)
65.  begin
66.     if rising_edge(clkb) then
67.        if web_d1 then
68.           mem(to_integer(unsigned(addrb_d1))) <= dinb_d1;
69.        end if;
70.        doutb <= mem(to_integer(unsigned(addrb_d1)));
71.     end if;
72.  end process;
73. end architecture;
```

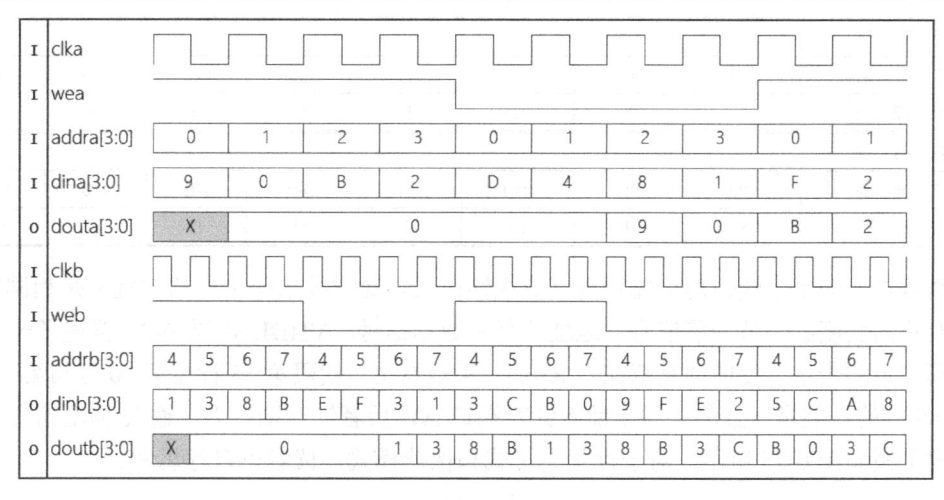

图 6-33

应用案例 4：将一个 18Kb BRAM 配置为两个独立的单端口 RAM

理论上，一个 18Kb BRAM 只可配置为一个不同规格的单端口 RAM。但是，我们注意到也可以将它配置为不同规格的真双端口 RAM。对于真双端口 RAM，其每个端口的行为与单端口 RAM 的行为是一致的，不同之处在于两个端口需要共享存储空间。如果要使两个端口成为独立的单端口 RAM，就需要在物理上将其地址空间隔离开来。于是可采用如图 6-34 所示的方式。A 端口和 B 端口地址的低 n 位是相同的，但它们的最高位是相反的，前者恒接高电平，后者恒接低电平。这样就使得两者的地址空间完全隔离。

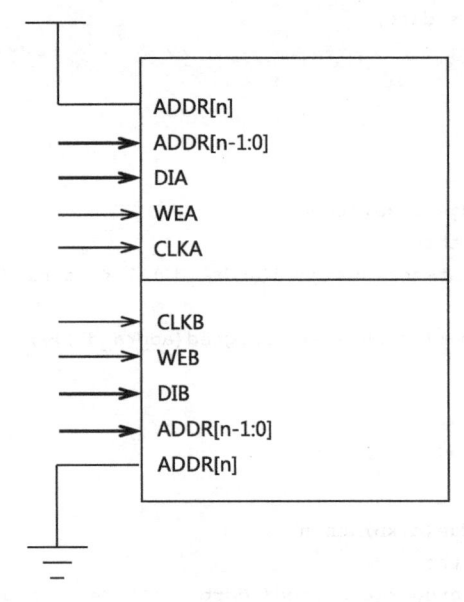

图 6-34

以 4 位地址空间为例，可构成 0~15 共 16 个地址空间，将低 3 位提取出来，最高位分别接 0 和 1，可形成表 6-18，可以看到，最高位的不同导致地址空间完全隔离开来。

表 6-18

最 高 位	0						
二 进 制	000	001	010	011	100	101	110	111
十 进 制	0	1	2	3	4	5	6	7
最 高 位	1						
二 进 制	000	001	010	011	100	101	110	111
十 进 制	8	9	10	11	12	13	14	15

真双端口的每个端口均可配置为字节选通模式，此时要求每个端口的规格是相同的，即具有相同宽度的字节使能信号、地址信号和数据信号。VHDL 代码 6-21 描述了一个写优先模式下的带字节使能的真双端口 RAM。不难看出，代码第 55 行至第 70 行与第 72 行至第 88 行的功能是一致的，只是前者针对端口 A，后者针对端口 B。输入/输出信号之间的时序关系如图 6-35 所示。对于读优先模式和保持模式，读者可自行思考。

VHDL 代码 6-21

```vhdl
1.  --File: bytewrite_tdp_ram_wf.vhd
2.  library ieee;
3.  use ieee.std_logic_1164.all;
4.  use ieee.numeric_std.all;
5.  use work.mem_pkg.all;
6.
7.  entity bytewrite_tdp_ram_wf is
8.    generic (
9.      AW : positive := 10;
10.     NB : positive := 4;
11.     DW : positive := NB * 8
12.   );
13.   port (
14.     clka  : in  std_logic;
15.     wea   : in  std_logic_vector(NB - 1 downto 0);
16.     addra : in  std_logic_vector(AW - 1 downto 0);
17.     dina  : in  std_logic_vector(DW - 1 downto 0);
18.     douta : out std_logic_vector(DW - 1 downto 0);
19.     clkb  : in  std_logic;
20.     web   : in  std_logic_vector(NB - 1 downto 0);
21.     addrb : in  std_logic_vector(AW - 1 downto 0);
22.     dinb  : in  std_logic_vector(DW - 1 downto 0);
23.     doutb : out std_logic_vector(DW - 1 downto 0)
24.   );
25. end entity;
26.
27. architecture rtl of bytewrite_tdp_ram_wf is
28.   constant DEPTH : positive := 2 ** AW;
29.   signal wea_d1    : wea'subtype;
30.   signal addra_d1  : addra'subtype;
31.   signal dina_d1   : dina'subtype;
32.   signal web_d1    : web'subtype;
33.   signal addrb_d1  : addrb'subtype;
34.   signal dinb_d1   : dinb'subtype;
35.   signal mem       : mem_t(0 to DEPTH - 1)(DW - 1 downto 0) := (others => (others => '0'));
36. begin
37.   process(clka)
38.   begin
39.     if rising_edge(clka) then
40.       wea_d1   <= wea;
41.       addra_d1 <= addra;
42.       dina_d1  <= dina;
43.     end if;
```

```vhdl
44.    end process;
45.
46.    process(clkb)
47.    begin
48.      if rising_edge(clkb) then
49.        web_d1   <= web;
50.        addrb_d1 <= addrb;
51.        dinb_d1  <= dinb;
52.      end if;
53.    end process;
54.
55.    process(clka)
56.    begin
57.      if rising_edge(clka) then
58.        for i in 0 to NB - 1 loop
59.          if wea_d1(i) then
60.            mem(to_integer(unsigned(addra_d1)))(i * 8 + 7 downto i * 8)
61.              <= dina_d1(i * 8 + 7 downto i * 8);
62.            douta(i * 8 + 7 downto i * 8)
63.              <= dina_d1(i * 8 + 7 downto i * 8);
64.          else
65.            douta(i * 8 + 7 downto i * 8)
66.              <= mem(to_integer(unsigned(addra_d1)))(i * 8 + 7 downto i * 8);
67.          end if;
68.        end loop;
69.      end if;
70.    end process;
71.
72.    process(clkb)
73.    begin
74.      if rising_edge(clkb) then
75.        for i in 0 to NB - 1 loop
76.          if web_d1(i) then
77.            mem(to_integer(unsigned(addrb_d1)))(i * 8 + 7 downto i * 8)
78.              <= dinb_d1(i * 8 + 7 downto i * 8);
79.            doutb(i * 8 + 7 downto i * 8)
80.              <= dinb_d1(i * 8 + 7 downto i * 8);
81.          else
82.            doutb(i * 8 + 7 downto i * 8)
83.              <= mem(to_integer(unsigned(addrb_d1)))(i * 8 + 7 downto i * 8);
84.          end if;
85.        end loop;
86.      end if;
87.    end process;
88. end architecture;
```

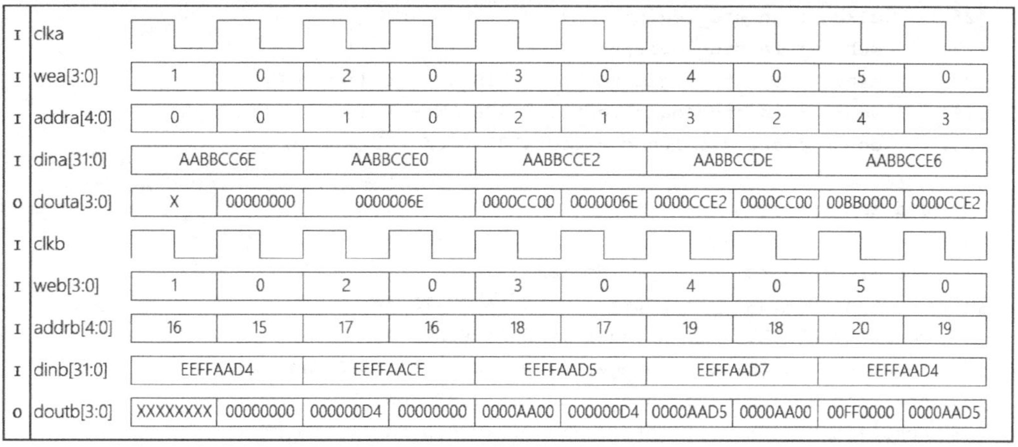

clka										
I wea[3:0]	1	0	2	0	3	0	4	0	5	0
I addra[4:0]	0	0	1	0	2	1	3	2	4	3
I dina[31:0]	AABBCC6E		AABBCCE0		AABBCCE2		AABBCCDE		AABBCCE6	
o douta[3:0]	X	00000000	0000006E	0000CC00	0000006E	0000CCE2	0000CC00	00BB0000	0000CCE2	
I clkb										
I web[3:0]	1	0	2	0	3	0	4	0	5	0
I addrb[4:0]	16	15	17	16	18	17	19	18	20	19
I dinb[31:0]	EEFFAAD4		EEFFAACE		EEFFAAD5		EEFFAAD7		EEFFAAD4	
o doutb[3:0]	XXXXXXXX	00000000	000000D4	00000000	0000AA00	000000D4	0000AAD5	0000AA00	00FF0000	0000AAD5

图 6-35

真双端口 RAM 也支持非对称读/写，即读/写位宽不同，但要求 RAM 类型为 BRAM。VHDL 代码 6-22 描述了一个非对称真双端口 RAM，读优先，可以看到端口 A 和端口 B 的位宽是不同的。输入/输出信号之间的时序关系如图 6-36 所示。端口 A 的 0 号地址对应端口 B 的 0～3 号地址，因此当端口 A 向 0 号地址写入 AAC4 时，端口 B 由 0～3 号地址依次读出 4、C、A、A。代码第 94 行至第 100 行仅用于仿真时观测地址的变化规律和相应的写入数据。

📄 VHDL 代码 6-22

```
1.  /*File: asym_tdp_ram_rf.sv
2.  //PORTA_DW: data width of port A
3.  //PORTA_AW: address width of port A
4.  //PORTB_DW: data width of port B
5.  //PORTB_AW: address width of port B
6.  //(2 ** PORTA_AW) * PORTA_DW = (2 ** PORTB_AW) * PORTB_DW
7.  */
8.  library ieee;
9.  use ieee.std_logic_1164.all;
10. use ieee.numeric_std.all;
11. use work.mem_pkg.all;
12.
13. entity asym_tdp_ram_rf is
14.   generic (
15.     PORTA_DW : positive := 16;
16.     PORTA_AW : positive := 8;
17.     PORTB_DW : positive := 4;
18.     PORTB_AW : positive := 10
19.   );
20.   port (
21.     clka  : in std_logic;
22.     wea   : in std_logic;
23.     dina  : in std_logic_vector(PORTA_DW - 1 downto 0);
24.     addra : in std_logic_vector(PORTA_AW - 1 downto 0);
25.     douta : out std_logic_vector(PORTA_DW - 1 downto 0);
```

```vhdl
26.        clkb  : in std_logic;
27.        web   : in std_logic;
28.        dinb  : in std_logic_vector(PORTB_DW - 1 downto 0);
29.        addrb : in std_logic_vector(PORTB_AW - 1 downto 0);
30.        doutb : out std_logic_vector(PORTB_DW - 1 downto 0)
31.    );
32. end entity;
33.
34. architecture rtl of asym_tdp_ram_rf is
35.    constant DEPTHA    : integer := 2 ** PORTA_AW;
36.    constant DEPTHB    : integer := 2 ** PORTB_AW;
37.    constant MAX_DEPTH : integer := max(DEPTHA, DEPTHB);
38.    constant MAX_DW    : integer := max(PORTA_DW, PORTB_DW);
39.    constant MIN_DW    : integer := min(PORTA_DW, PORTB_DW);
40.    constant RATIO     : integer := MAX_DW / MIN_DW;
41.    constant LOG2RATIO : integer := log2(RATIO);
42.    signal wea_d1      : wea'subtype;
43.    signal addra_d1    : addra'subtype;
44.    signal dina_d1     : dina'subtype;
45.    signal web_d1      : web'subtype;
46.    signal addrb_d1    : addrb'subtype;
47.    signal dinb_d1     : dinb'subtype;
48.    signal mem : mem_t(0 to MAX_DEPTH - 1)(MIN_DW - 1 downto 0) := (others => (others => '0'));
49.    signal xaddr       : integer_vector(RATIO - 1 downto 0) := (others => 0);
50.    signal xdin : mem_t(0 to RATIO - 1)(MIN_DW - 1 downto 0);
51. begin
52.    process(clka)
53.    begin
54.      if rising_edge(clka) then
55.        wea_d1   <= wea;
56.        addra_d1 <= addra;
57.        dina_d1  <= dina;
58.      end if;
59.    end process;
60.
61.    process(clkb)
62.    begin
63.      if rising_edge(clkb) then
64.        web_d1   <= web;
65.        addrb_d1 <= addrb;
66.        dinb_d1  <= dinb;
67.      end if;
68.    end process;
69.
70.    process(clkb)
71.    begin
72.      if rising_edge(clkb) then
73.        if web_d1 then
74.          mem(to_integer(unsigned(addrb_d1))) <= dinb_d1;
75.        end if;
```

```
76.        doutb <= mem(to_integer(unsigned(addrb_d1)));
77.      end if;
78.    end process;
79.
80.    process(clka)
81.    begin
82.      if rising_edge(clka) then
83.        for i in 0 to RATIO - 1 loop
84.          douta((i + 1) * MIN_DW - 1 downto i * MIN_DW)
85.            <= mem(to_integer(unsigned(addra_d1) & to_unsigned(i, LOG2RATIO)));
86.          if wea_d1 then
87.            mem(to_integer(unsigned(addra_d1) & to_unsigned(i, LOG2RATIO)))
88.              <= dina_d1((i + 1) * MIN_DW - 1 downto i * MIN_DW);
89.          end if;
90.        end loop;
91.      end if;
92.    end process;
93.    --Only for debug
94.    process(all)
95.    begin
96.      for k in 0 to RATIO - 1 loop
97.        xaddr(k) <= to_integer(unsigned(addra_d1) & to_unsigned(k, LOG2RATIO));
98.        xdin(k) <= dina_d1((k + 1) * MIN_DW - 1 downto k * MIN_DW);
99.      end loop;
100.   end process;
101.end architecture;
```

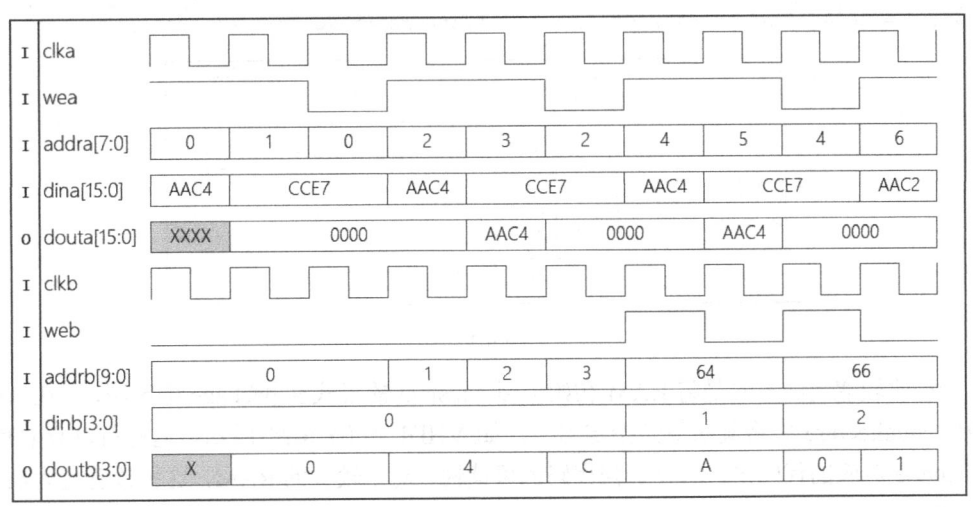

图 6-36

VHDL-2008 新特性

VHDL-2008 引入了 integer_vector 数据类型，本质上是 integer 构成的一维数组，如 VHDL 代码 6-22 第 49 行所示。

类似地，也可以用 VHDL 描述写优先和保持模式的非对称真双端口 RAM。

6.5 RAM 的初始化与 ROM 代码风格

默认情形下，无论是分布式 RAM、BRAM 还是 UltraRAM，其初始值均为 0，这意味着 RAM 的每个地址上存储的数据均为 0，也意味着 FPGA 一旦上电工作，在执行写操作之前，从 RAM 中读取到的数据均为 0。另一方面，我们也可以对 RAM 进行可综合的初始化，使其初始值符合预期要求。RAM 的初始化有两种方法，第一种方法是将初始值写在 RTL 代码里，第二种方法是将初始值写在外部文件中。

第一种方法如 VHDL 代码 6-23 所示。代码第 2 行表示 RAM 所有地址上的数据均为 0，第 3 行则表示所有地址上的数据均为 F，代码第 4 行和第 5 行表示 RAM 的 0 号地址上的数据为 A（十六进制），1 号地址上的数据为 C，3 号地址上的数据为 1，其余地址上的数据均为 0，其仿真结果如图 6-37 所示。可以看到，该方法适用于仅对 RAM 个别地址进行初始化（代码第 5 行）或 RAM 所有地址的初始值相同时的初始化（代码第 2 行和第 3 行）。如果 RAM 较大，且要对每个地址指定初始值，使用该方法就会使代码变得冗长啰唆，大大降低可读性。

📄 **VHDL 代码 6-23**

```
1.  --File: ram_init_v1.vhd
2.  signal mem1 : mem_t(0 to 15)(3 downto 0) := (others => (others => '0'));
3.  signal mem2 : mem_t(0 to 15)(3 downto 0) := (others => (others => '1'));
4.  signal mem  : mem_t(0 to 15)(3 downto 0)
5.    := (0 => X"A", 1 => X"C", 3 => X"1",others => (others => '0'));
```

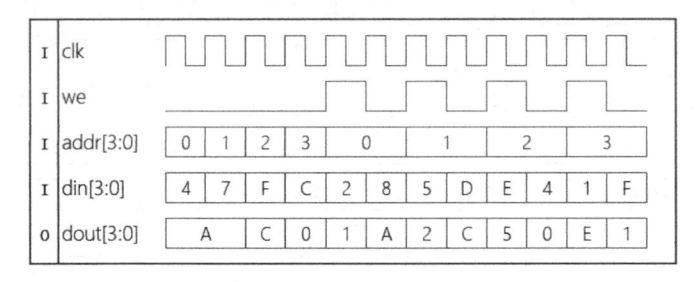

图 6-37

一种比较取巧的方式是将 RAM 初始值以 constant 的形式在 package 中声明，而在 entity 对应的 architecture 中将其赋值给指定信号，如 VHDL 代码 6-24 所示。这样可以有效减少 architecture 对应的代码量，同时增强代码的可读性，对于较大的 RAM，这种改善尤为显著。

📄 **VHDL 代码 6-24**

```
1.  --File: mem_pkg.vhd
2.    subtype mem4_t is mem_t(0 to 3)(3 downto 0);
3.    constant RAM_INIT : mem4_t := (0 => X"A", 1 => X"B", 2 => X"1", 3 => X"5");
4.  --File: ram_init_v1_opt.vhd
5.  signal mem : mem4_t := RAM_INIT;
```

通过外部文件初始化 RAM 需要设计者编写函数，如 VHDL 代码 6-25 所示。图 6-38 显示了 myram.dat 的内容（注意图中第一列行号是编辑器本身显示的行号，并非 myram.dat

的内容)。图 6-39 显示了相应的输入/输出信号之间的时序关系。这里,文件"myram.dat"中的数据以十六进制的形式表示。若以二进制的形式表示,则要将 VHDL 代码 6-25 第 9 行的 hread 替换为 bread。使用 Vivado 自带仿真器 XSIM 仿真时,需要将 myram.dat 文件以设计文件的形式添加到 Vivado 工程中,如图 6-40 所示。

📄 VHDL 代码 6-25

```
1.  --File: ram_init_v2.vhd
2.  impure function init_ram_hex(ram_file_name : in string) return mem16_t is
3.    file ram_file : text open read_mode is ram_file_name;
4.    variable ram_file_line : line;
5.    variable ram           : mem16_t;
6.  begin
7.    for i in mem16_t'range loop
8.      readline(ram_file, ram_file_line);
9.      hread(ram_file_line, ram(i));
10.   end loop;
11.   return ram;
12. end function
13. signal mem : mem16_t := init_ram_hex("myram.dat");
```

0	A
1	B
2	C
3	D
4	E
5	F
6	E
7	D
8	C
9	B
10	A
11	9
12	8
13	7
14	6
15	5

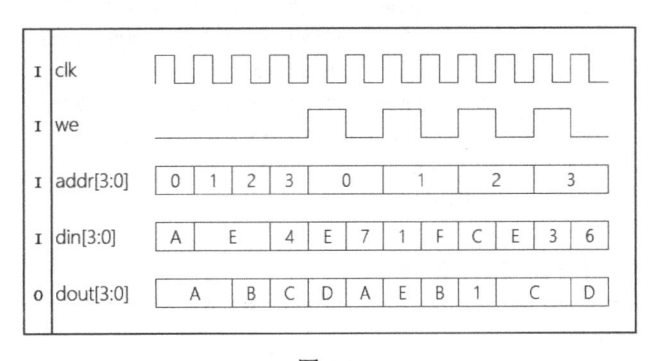

图 6-38

图 6-39

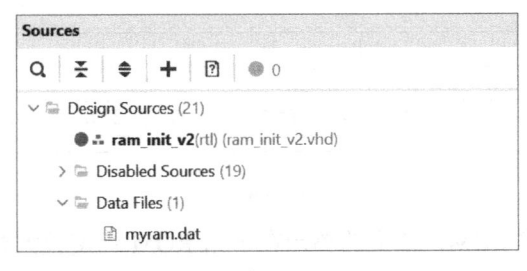

图 6-40

🔍 VHDL-2008 新特性

在 VHDL-2008 中,需要将 text is in ram_file_name 替换为 text open read_mode is ram_file_name,如 VHDL 代码 6-25 第 3 行所示。

在 FPGA 设计中还会经常用到 ROM（Read-Only Memory，只读存储器），如在 FIR 滤波器设计中存储滤波器系数或在快速傅里叶变换中存储旋转因子等。可以采用 case 语句描述，如 VHDL 代码 6-26 所示，输入/输出信号之间的时序关系如图 6-41 所示。

VHDL 代码 6-26

```
1.  --File: rom_case.vhd
2.  process(clk)
3.    begin
4.      if rising_edge(clk) then
5.        if en then
6.          case addr is
7.            when 4D"0"    => dout <= X"200A";
8.            when 4D"1"    => dout <= X"0300";
9.            when 4D"2"    => dout <= X"8101";
10.           when 4D"3"    => dout <= X"4000";
11.           when 4D"4"    => dout <= X"8601";
12.           when 4D"5"    => dout <= X"233A";
13.           when 4D"6"    => dout <= X"0300";
14.           when 4D"7"    => dout <= X"8602";
15.           when 4D"8"    => dout <= X"2222";
16.           when 4D"9"    => dout <= X"4001";
17.           when 4D"10"   => dout <= X"0342";
18.           when 4D"11"   => dout <= X"232B";
19.           when 4D"12"   => dout <= X"0900";
20.           when 4D"13"   => dout <= X"0302";
21.           when 4D"14"   => dout <= X"0102";
22.           when others   => dout <= X"4002";
23.         end case;
24.       end if;
25.     end if;
26.   end process;
```

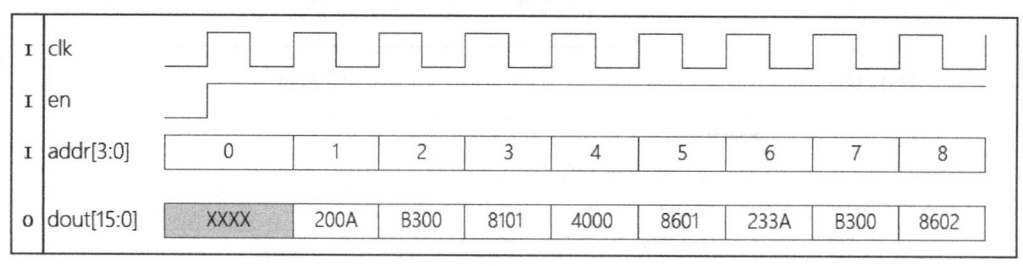

I	clk									
I	en									
I	addr[3:0]	0	1	2	3	4	5	6	7	8
O	dout[15:0]	XXXX	200A	B300	8101	4000	8601	233A	B300	8602

图 6-41

采用 case 语句描述 ROM，最大的弊端在于当 ROM 深度较大时，会导致代码冗长并降低代码的可读性。考虑到 RAM 可借助外部文件初始化，因此也可用此方法描述 ROM。毕竟，对于 RAM 而言，初始化之后，如果只读不写，其功能就是 ROM。同时，Vivado 提供了综合属性 ROM_STYLE，可管理 ROM 的映射方式，其值可以为 distributed、block 和 ultra（仅支持 Versal 器件）。ROM 可以是单端口，也可以是双端口。当 ROM 是双端口时，两个端口共享存储空间，这意味着它们共享其中的存储数据。VHDL 代码 6-27 描述了一个 16×16 的双端口 ROM，ROM_STYLE 的值为 block，因此最终会消耗一个 18Kb BRAM。代码第 3 行调用了函数 init_ram_hex，该函数的具体内容如 VHDL 代码 6-25 所示，只是将返回类型替换为 mem16x16_t，并将整个函数放在 package 中声明。输入/输出信号之间的时序关系可参考图 6-41 进行理解。

VHDL 代码 6-27

```vhdl
1.  --File: rom_dual_port.vhd
2.  architecture rtl of rom_dual_port is
3.  signal rom : mem16x16_t := init_ram_hex("myrom.dat");
4.  attribute ROM_STYLE : string;
5.  attribute ROM_STYLE of rom : signal is ROM_STYLE_VAL;
6.  begin
7.    process(clka)
8.    begin
9.      if rising_edge(clka) then
10.       if ena then
11.         douta <= rom(to_integer(unsigned(addra)));
12.       end if;
13.     end if;
14.   end process;
15.
16.   process(clkb)
17.   begin
18.     if rising_edge(clkb) then
19.       if enb then
20.         doutb <= rom(to_integer(unsigned(addrb)));
21.       end if;
22.     end if;
23.   end process;
24. end architecture;
```

6.6 同步 FIFO 代码风格

FIFO（First Input First Output）简单地说就是先进先出，由存储单元（通常为简单双端口 RAM）和控制单元构成，其特征是第一个被写入队列的数据也是第一个从队列中被读出的数据。FIFO 被广泛应用于 FPGA 设计中，常见的应用场景如下。

- 跨时钟域操作：A 时钟域产生的数据要被 B 时钟域使用。可先将数据用 A 时钟域的时钟写入 FIFO，再通过 B 时钟域的时钟从 FIFO 中读出。
- 当发送数据速率与接收数据速率不匹配时，作为临时存储单元。
- 当输入数据和输出数据的宽度不匹配时，可用于数据宽度调整。

根据 FIFO 读/写时钟的关系，可将其分为同步 FIFO 和异步 FIFO。同步 FIFO 是指读/写时钟为同一时钟的 FIFO。我们从同步 FIFO 说起，如图 6-42 所示，显示了 FIFO 的输入端口和输出端口，端口含义如表 6-19 所示。

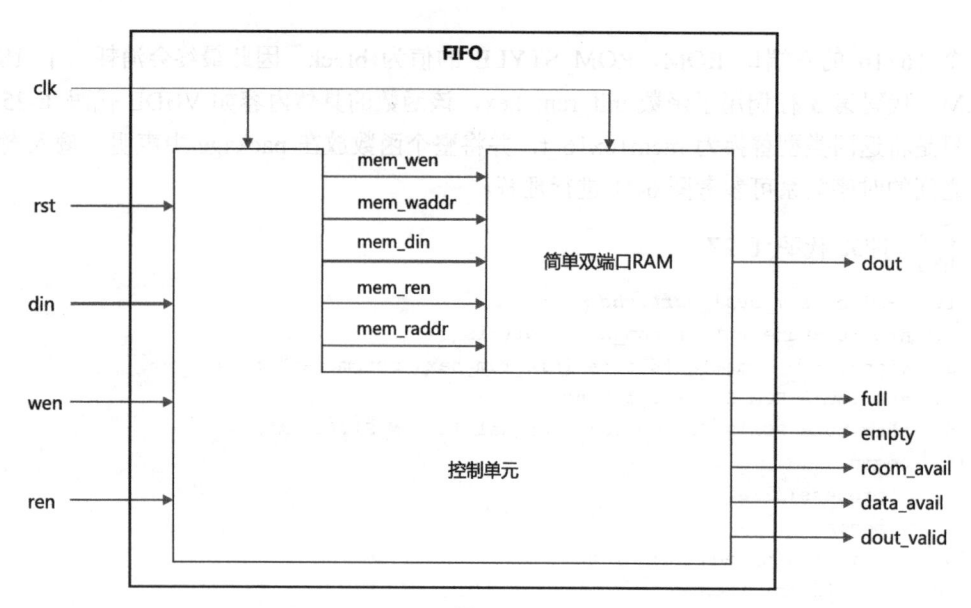

图 6-42

表 6-19

信号名称	方向	含义
clk	输入	时钟信号，控制单元和存储单元共用
rst	输入	控制单元复位信号
wen	输入	写使能信号，高电平有效
din	输入	输入数据
full	输出	满标记信号，高电平有效
ren	输入	读使能信号，高电平有效
dout	输出	输出数据
empty	输出	空标记信号，高电平有效
room_avail	输出	可写入地址空间的个数
data_avail	输出	可供读取的数据个数
dout_valid	输出	输出数据有效标记信号，高电平有效

表面上看，FIFO 只需要读/写使能信号而不需要读/写地址信号，但实际上，控制单元要根据读/写使能信号生成存储单元需要用到的控制信号，包括简单双端口 RAM 的读/写使能信号（对应图 6-42 中的 mem_wen 和 mem_ren）和读/写地址（对应图 6-42 中的 mem_waddr 和 mem_raddr）。相比存储单元，控制单元的设计更为复杂。

为便于说明，我们以深度为 8 的 FIFO 为例，从写操作的角度而言，如图 6-43 所示。在 T1 时间段，写使能 wen 为 1，因此写指针 wr_ptr 执行加 1 操作。在 T2 时间段，wen 为 0，因此 wr_ptr 保持不变。在 T3 时间段，wen 再次为 1，wr_ptr 继续执行加 1 操作，但当 wr_ptr 为 7 时，表明 FIFO 已写满，因此 full 信号抬高，wr_ptr 回到 0 后将保持不变，如 T4 时间段所示。在开始读数之前，full 始终为 1。一旦读使能 ren 为 1，full 将由高电平变为低电平，如 T5 时间段所示。在生成 mem_wen 时，需要注意，当 wen 有效且 full 为高电平时，不能写入数据。考虑到时序性能，给存储单元提供的控制信号最好是寄存器输出，因此 mem_wen、mem_waddr 和 mem_din 分别与 wen、wr_ptr 和 din 相差一个时钟周期。根据图 6-43 可知 FIFO 写时序具备如下特征：①写指针 wr_ptr 在写使能 wen 为高电平且写满标记 full 为低电平时执行加 1 计数，当计数到 FIFO 深度减 1 时回到 0；②存储单元用到的写使能信号 mem_wen 由 wen 和 full 译码生成。

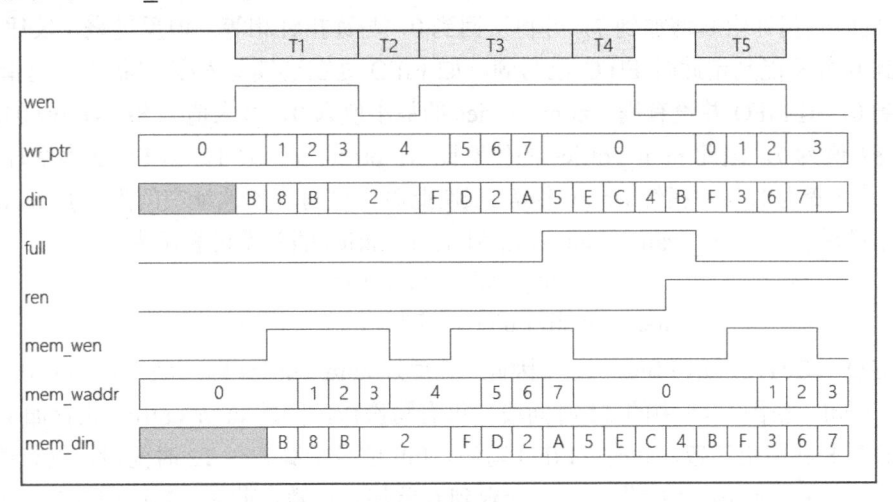

图 6-43

再看读操作，如图 6-44 所示。假定在 T1 时间段之前，FIFO 已写满数据。当读使能 ren 为 1 时，读指针 rd_ptr 开始执行加 1 操作，一旦 rd_ptr 变更为 7，表明此时已读到最后一个数据。之后一个时钟周期 FIFO 将变为空，因此 empty 将抬高，同时 rd_ptr 回到 0，且保持不变。在写使能有效之前，empty 一直保持为高电平。一旦 wen 为 1，empty 将由高电平变为低电平，如 T2 时间段所示。在 T3 时间段，rd_ptr 在 ren 的控制下在原值基础上执行加 1 操作。在生成 mem_ren 时，需要注意，当 ren 为高电平且 empty 为高电平时，mem_ren 将为低电平，不再执行读操作。同样，考虑到时序性能，mem_ren、mem_raddr 分别和 ren、rd_ptr 相差一个时钟周期。为了获得有效输出数据，还需要信号 dout_valid。该信号是 mem_ren 延迟两个时钟周期的结果（读操作的 Latency 为 2）。根据图 6-44 可知 FIFO 读操作具备如下特征：①读指针 rd_ptr 在读使能 ren 为 1 且 empty 为 0 时计数，计数到 FIFO 深度减 1 时回到 0；②存储单元用到的读使能信号 mem_ren 由 ren 和 empty 译码生成。

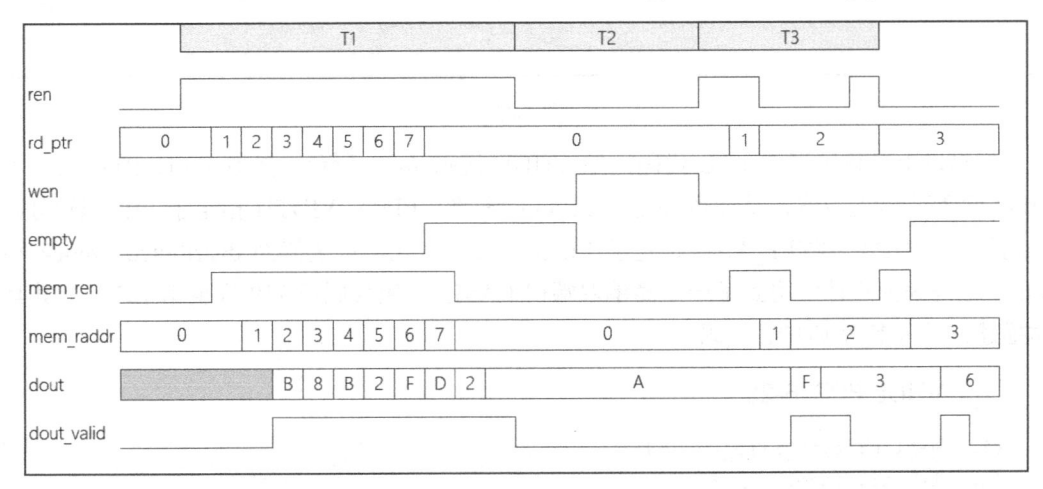

图 6-44

接下来我们看一下空满标记信号如何生成，如图 6-45 所示。当 FIFO 写满时，表明此时 FIFO 中可供读出的数据有 8 个，以 num_entries 表示（data_avail 与 num_entries 的含义相同），同时写指针由 0 增加到 7，并再次回到 0，与读指针相等。但反过来，仅凭写指针再次回到 0 并不能判定此时 FIFO 已写满，如 FIFO 处在连续写入/读出状态，写指针由 0 到 7 再到 0，但 FIFO 并没有满。num_entries 的最小值为 0，最大值为 8；wr_ptr 的最小值为 0，最大值为 7。因此 num_entries 的位宽比 wr_ptr 的位宽大 1。当 FIFO 为空时，读指针 rd_ptr 由 7 变为 0，与写指针 wr_ptr 相等。表明此时可供写入的地址空间为 0，用 room_avail 表示。不难看出，room_avail、data_avail 和 num_entries 的关系可表示为

$$data\_avail = num\_entries$$
$$room\_avail = FIFO\_DEPTH - num\_entries$$

式中，FIFO_DEPTH 表示 FIFO 深度。因此，处理好 num_entries 即可获得正确的 data_avail 和 room_avail。在图 6-45 中的 T1 时间段，仅有写操作，因此 num_entries 执行加 1 操作，一旦计数到 8 就停止计数。此时，FIFO 写满，full 信号抬高。在 T2 时间段，仅有读操作，因此 num_entries 执行减 1 操作，一旦计数到 0 就停止计数。此时，FIFO 读空，empty 信号抬高。在 T3 时间段，读/写操作同时进行，num_entries 保持不变。

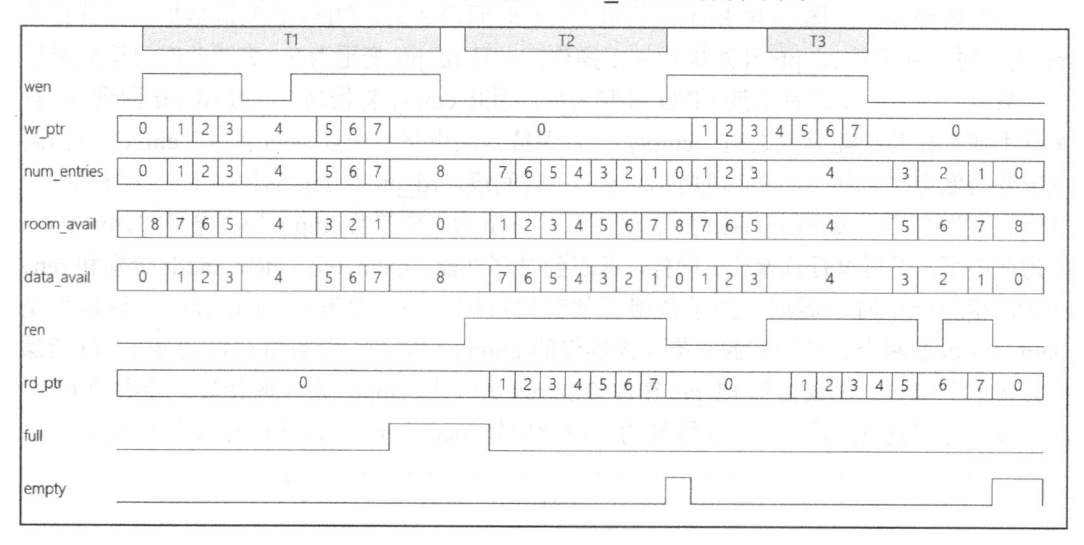

图 6-45

根据以上分析可形成 RTL 代码，如 VHDL 代码 6-28 所示。参数 FIFO_PTR 指定了 FIFO 读/写指针的位宽，从而也确定了 FIFO 的深度；FIFO_WIDTH 指定了 FIFO 读/写数据的位宽；RAM_STYLE_VAL 指定了 FIFO 的实现方式，可选值为 distributed、block 和 ultra。这是因为设计中用到了一个简单双端口 RAM，对应代码第 119 行至第 133 行，其具体描述方式可参考 VHDL 代码 6-14。

📄 **VHDL 代码 6-28**

```
1.   --File: sync_fifo_v1.vhd
2.   library ieee;
```

```vhdl
3.  use ieee.std_logic_1164.all;
4.  use ieee.numeric_std.all;
5.
6.  entity sync_fifo_v1 is
7.    generic (
8.      FIFO_PTR       : positive := 3;    --address width
9.      FIFO_WIDTH     : positive := 4;   --data width
10.     RAM_STYLE_VAL : string    := "block"
11.   );
12.   port (
13.     clk         : in std_logic;
14.     rst         : in std_logic;
15.     wen         : in std_logic;
16.     din         : in std_logic_vector(FIFO_WIDTH - 1 downto 0);
17.     ren         : in std_logic;
18.     full        : out std_logic;
19.     empty       : out std_logic;
20.     room_avail  : out std_logic_vector(FIFO_PTR downto 0);
21.     data_avail  : out std_logic_vector(FIFO_PTR downto 0);
22.     dout_valid  : out std_logic;
23.     dout        : out std_logic_vector(FIFO_WIDTH - 1 downto 0)
24.   );
25. end entity;
26.
27. architecture rtl of sync_fifo_v1 is
28.   constant FIFO_DEPTH   : positive := 2 ** FIFO_PTR;
29.   constant FIFO_DEPTH_1 : unsigned(FIFO_PTR - 1 downto 0)
30.                         := to_unsigned(FIFO_DEPTH - 1, FIFO_PTR);
31.   signal wr_ptr                   : unsigned(FIFO_PTR - 1 downto 0);
32.   signal rd_ptr                   : unsigned(FIFO_PTR - 1 downto 0);
33.   signal num_entries              : unsigned(FIFO_PTR    downto 0) := (others => '0');
34.   signal full_nxt, empty_nxt      : std_logic;
35.   signal room_avail_nxt           : unsigned(FIFO_PTR downto 0);
36.   signal dout_valid_i             : std_logic;
37.   signal dout_valid_i_d1          : std_logic;
38.   signal mem_wen, mem_ren         : std_logic;
39.   signal mem_wen_i, mem_ren_i     : std_logic;
40.   signal mem_waddr, mem_raddr     : std_logic_vector(FIFO_PTR - 1 downto 0);
41.   signal mem_din                  : std_logic_vector(FIFO_WIDTH - 1 downto 0);
42. begin
43.   full_nxt        <= num_entries ?= to_unsigned(FIFO_DEPTH, FIFO_PTR + 1);
44.   empty_nxt       <= num_entries ?= to_unsigned(0, FIFO_PTR + 1);
45.   data_avail      <= std_logic_vector(num_entries);
46.   room_avail_nxt  <= to_unsigned(FIFO_DEPTH, FIFO_PTR + 1) - num_entries;
47.   mem_wen_i       <= wen and (not full_nxt);
48.   mem_ren_i       <= ren and (not empty_nxt);
49.
50.   process(clk)
51.   begin
52.     if rising_edge(clk) then
53.       case? std_logic_vector'(rst, mem_wen_i) is
```

```vhdl
54.         when "1-" => wr_ptr <= (others => '0');
55.         when "01" => wr_ptr <= (others => '0') when wr_ptr ?= FIFO_DEPTH_1 else
56.                      wr_ptr + 1;
57.         when others => wr_ptr <= wr_ptr;
58.       end case?;
59.     end if;
60.   end process;
61.
62.   process(clk)
63.   begin
64.     if rising_edge(clk) then
65.       case? std_logic_vector'(rst, mem_ren_i) is
66.         when "1-" => rd_ptr <= (others => '0');
67.         when "01" => rd_ptr <= (others => '0') when rd_ptr ?= FIFO_DEPTH_1 else
68.                      rd_ptr + 1;
69.         when others => rd_ptr <= rd_ptr;
70.       end case?;
71.     end if;
72.   end process;
73.
74.   process(clk)
75.   begin
76.     if rising_edge(clk) then
77.       case? std_logic_vector'(rst, wen, ren) is
78.         when "1--" => num_entries <= (others => '0');
79.         when "010" => num_entries
80. <= num_entries + (num_entries ?< to_unsigned(FIFO_DEPTH, num_entries'LENGTH));
81.         when "001" => num_entries
82. <= num_entries - (num_entries ?> to_unsigned(0, num_entries'LENGTH));
83.         when others => num_entries <= num_entries;
84.       end case?;
85.     end if;
86.   end process;
87.
88.
89.   process(clk)
90.   begin
91.     if rising_edge(clk) then
92.       if rst then
93.         full             <= '0';
94.         empty            <= '1';
95.         room_avail       <= std_logic_vector(to_unsigned(FIFO_DEPTH, FIFO_PTR + 1));
96.         dout_valid_i_d1 <= '0';
97.         dout_valid       <= '0';
98.       else
99.         full             <= full_nxt;
100.        empty            <= empty_nxt;
101.        room_avail       <= std_logic_vector(room_avail_nxt);
102.        dout_valid_i_d1 <= mem_ren;
103.        dout_valid       <= dout_valid_i_d1;
```

```vhdl
104.       end if;
105.     end if;
106.   end process;
107.
108.   process(clk)
109.   begin
110.     if rising_edge(clk) then
111.       mem_wen   <= mem_wen_i;
112.       mem_din   <= din;
113.       mem_waddr <= std_logic_vector(wr_ptr);
114.       mem_ren   <= mem_ren_i;
115.       mem_raddr <= std_logic_vector(rd_ptr);
116.     end if;
117.   end process;
118.
119.   i_sdp_one_clk : entity work.sdp_one_clk
120.     generic map (
121.       AW            => FIFO_PTR,
122.       DW            => FIFO_WIDTH,
123.       RAM_STYLE_VAL => RAM_STYLE_VAL
124.     )
125.     port map (
126.       clk    => clk,
127.       wen    => mem_wen,
128.       waddr  => mem_waddr,
129.       din    => mem_din,
130.       ren    => mem_ren,
131.       raddr  => mem_raddr,
132.       dout   => dout
133.     );
134. end architecture;
```

🔍 VHDL-2008 新特性

VHDL-2008 引入了 case?语句，用于处理无关态，这实际上使得 case?的条件分支有了优先级，如 VHDL 代码 6-28 第 66 行所示，不难看出，rst 的优先级最高。VHDL-2008 支持 unsigned 类型的数据与一位 std_logic 类型的数据相加，如 VHDL 代码 6-28 第 80 行所示。这里操作符?<的返回值为 std_logic 类型。

再次分析空满标记信号的生成条件，如图 6-46 所示。假定 FIFO 深度为 4，在 T1 时刻，FIFO 为空，写使能由 0 变为 1，读使能保持为 0 不变，因此写指针 wptr 递增，直至再次回到 0（图 6-46 中用二进制数表示读/写指针数值），表明已向 FIFO 写入 4 个数据，FIFO 达到满状态。可以看到此时读/写指针相等，均为 0。在 T2 时刻，写使能由 1 变为 0，读使能由 0 变为 1，读指针 rptr 递增，直至再次回到 0，FIFO 达到空状态，读/写指针再次相等。

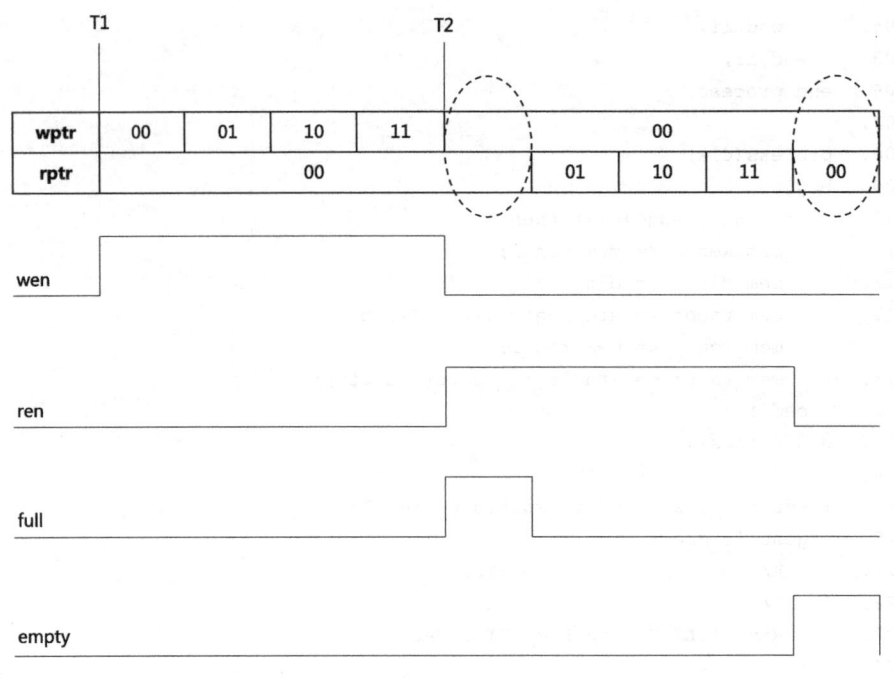

图 6-46

在图 6-47 中，T1 时刻，写使能由 0 变为 1，T2 时刻，写使能由 1 变为 0，同时读使能由 0 变为 1，直至将 FIFO 读空。T3 时刻，写使能又由 0 变为 1，直至将 FIFO 写满。结合图 6-46 和图 6-47，不难看出，当读指针追赶上写指针，两者相等时，FIFO 达到空状态；当写指针追赶上读指针，两者相等时，FIFO 达到满状态。

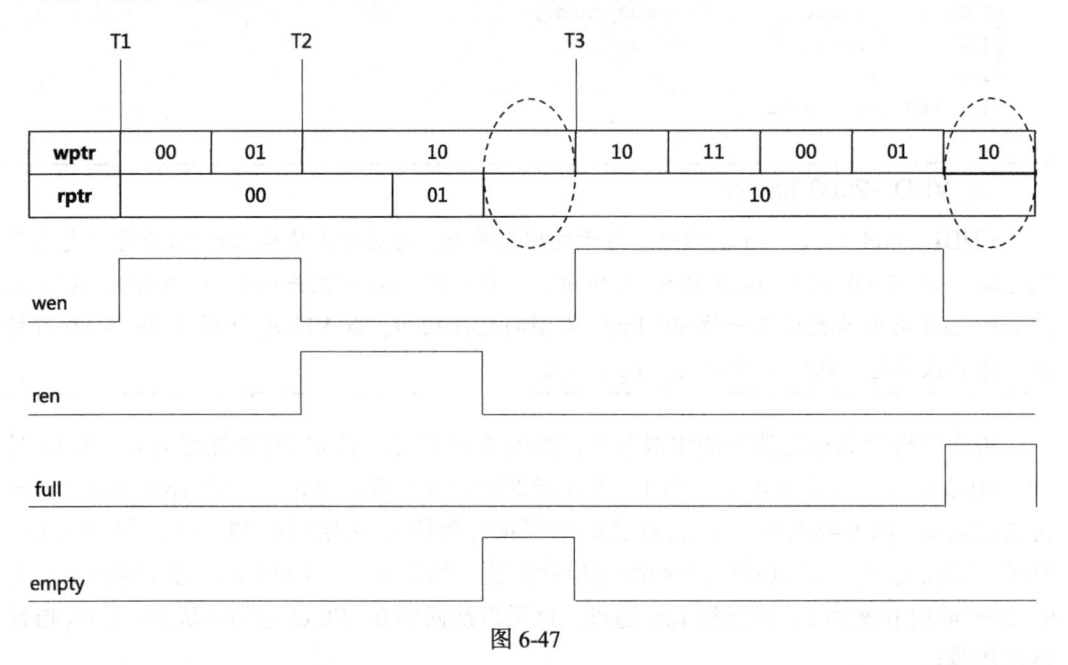

图 6-47

读/写指针相等只能断定 FIFO 达到空状态或满状态，但不能区分是空状态还是满状态。为此，我们将读/写指针增加一位，这样，图 6-46 和图 6-47 就分别变为如图 6-48 和图 6-49

所示的情形。观察可知，当 FIFO 为满状态时，读指针和写指针的最高位相反，其余位相等；而当 FIFO 为空状态时，读指针和写指针的所有位均相等。在此基础上，也很容易生成存储单元所需要的读/写地址：只需要取读/写指针的低两位即可。

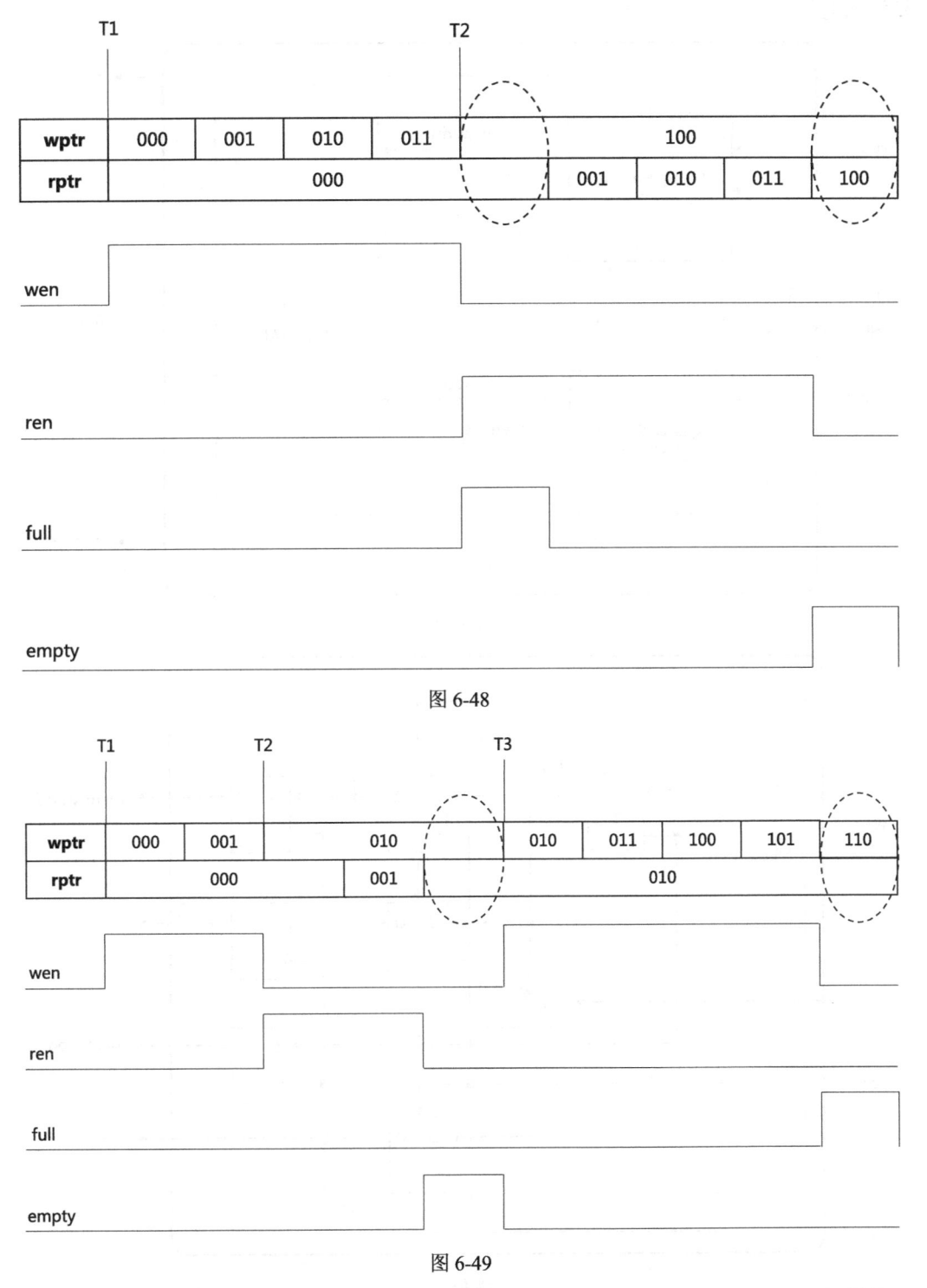

图 6-48

图 6-49

依据此方法，我们将 FIFO 的控制单元分为两部分：写操作相关控制信号逻辑模块和读操作相关控制信号逻辑模块，存储单元仍为一个简单双端口 RAM，如图 6-50 所示。写操作相关控制信号逻辑模块电路如图 6-51 所示，当 full 为高电平时，mem_wen 将为低电平。读操作相关控制信号逻辑模块电路如图 6-52 所示，当 empty 为高电平时，mem_ren 将为低电平。

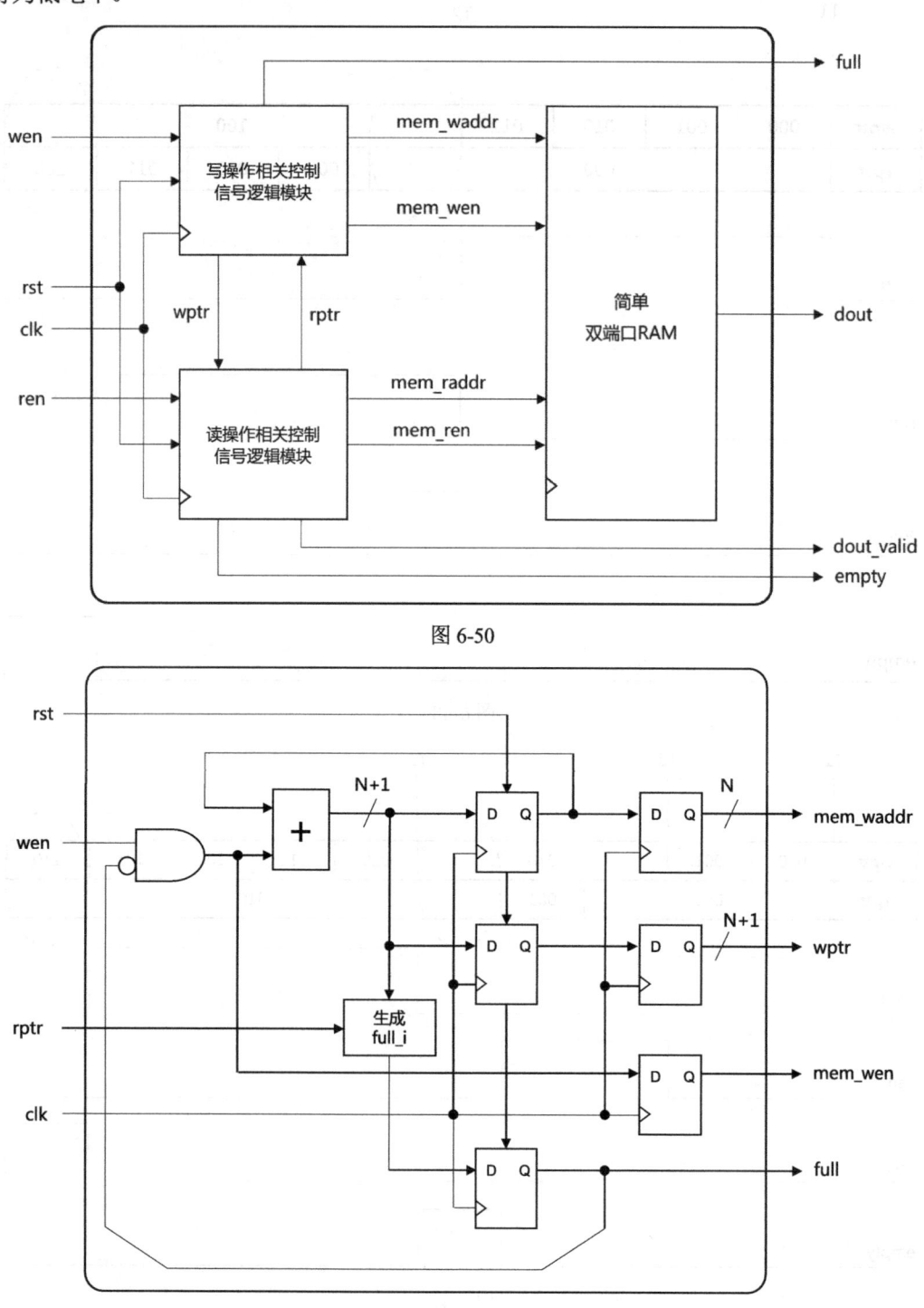

图 6-50

图 6-51

图 6-52

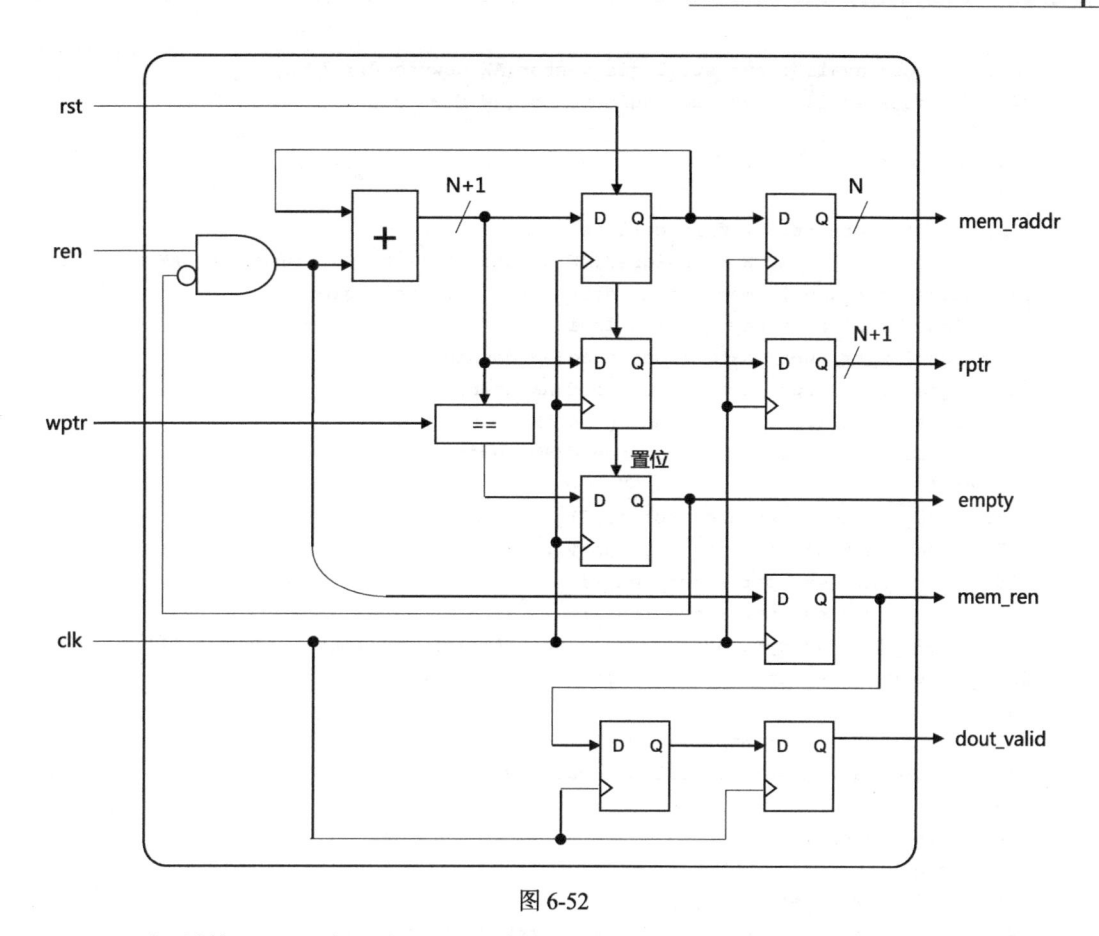

FIFO 控制单元 RTL 代码如 VHDL 代码 6-29 所示，FIFO 顶层如 VHDL 代码 6-30 所示。

📄 VHDL 代码 6-29

```
1.   --File: fifo_ctrl.vhd
2.   library ieee;
3.   use ieee.std_logic_1164.all;
4.   use ieee.numeric_std.all;
5.   
6.   entity fifo_ctrl is
7.     generic ( AW : positive := 4 );
8.     port (
9.       clk        : in std_logic;
10.      rst        : in std_logic;
11.      wen        : in std_logic;
12.      ren        : in std_logic;
13.      mem_wen    : out std_logic;
14.      mem_ren    : out std_logic;
15.      mem_waddr  : out std_logic_vector(AW - 1 downto 0);
16.      mem_raddr  : out std_logic_vector(AW - 1 downto 0);
17.      full       : out std_logic;
18.      empty      : out std_logic;
19.      dout_valid : out std_logic;
```

```vhdl
20.         room_avail : out std_logic_vector(AW downto 0);
21.         data_avail : out std_logic_vector(AW downto 0)
22.     );
23. end entity;
24.
25. architecture rtl of fifo_ctrl is
26. constant FIFO_DEPTH : unsigned(AW downto 0) := to_unsigned(2 ** AW, AW + 1);
27. signal mem_wen_i, mem_ren_i, dout_valid_i : std_logic;
28. signal full_i, empty_i : std_logic;
29. signal wptr, wptr_nxt  : unsigned(AW downto 0);
30. signal rptr, rptr_nxt  : unsigned(AW downto 0);
31. signal data_avail_i    : unsigned(AW downto 0);
32. signal room_avail_i    : unsigned(AW downto 0);
33. begin
34.   mem_wen_i <= wen and (not full);
35.   mem_ren_i <= ren and (not empty);
36.   wptr_nxt  <= wptr + mem_wen_i;
37.   rptr_nxt  <= rptr + mem_ren_i;
38.   full_i    <= wptr_nxt ?= ((not rptr(AW)) & rptr(AW - 1 downto 0));
39.   empty_i   <= rptr_nxt ?= wptr;
40.
41.   data_avail_i <= wptr_nxt - rptr_nxt;
42.   room_avail_i <= FIFO_DEPTH - data_avail_i;
43.
44.   process(clk)
45.   begin
46.     if rising_edge(clk) then
47.       wptr <= (others => '0') when rst else wptr_nxt;
48.       rptr <= (others => '0') when rst else rptr_nxt;
49.     end if;
50.   end process;
51.
52.   process(clk)
53.   begin
54.     if rising_edge(clk) then
55.       mem_wen     <= mem_wen_i;
56.       mem_ren     <= mem_ren_i;
57.       mem_waddr   <= std_logic_vector(wptr(AW - 1 downto 0));
58.       mem_raddr   <= std_logic_vector(rptr(AW - 1 downto 0));
59.       dout_valid_i <= mem_ren;
60.       dout_valid  <= dout_valid_i;
61.       data_avail  <= std_logic_vector(data_avail_i);
62.       room_avail  <= std_logic_vector(room_avail_i);
63.     end if;
64.   end process;
65.
66.   process(clk)
67.   begin
68.     if rising_edge(clk) then
69.       full  <= '0' when rst else full_i;
70.       empty <= '1' when rst else empty_i;
```

```
71.     end if;
72.   end process;
73. end architecture;
```

VHDL 代码 6-30

```
1.  --File: sync_fifo_v2.vhd
2.  library ieee;
3.  use ieee.std_logic_1164.all;
4.  use ieee.numeric_std.all;
5.
6.  entity sync_fifo_v2 is
7.    generic (
8.      AW             : positive := 3;   --address width
9.      DW             : positive := 4;   --data width
10.     RAM_STYLE_VAL : string   := "block"
11.   );
12.   port (
13.     clk         : in  std_logic;
14.     rst         : in  std_logic;
15.     wen         : in  std_logic;
16.     din         : in  std_logic_vector(DW - 1 downto 0);
17.     ren         : in  std_logic;
18.     full        : out std_logic;
19.     empty       : out std_logic;
20.     room_avail  : out std_logic_vector(AW downto 0);
21.     data_avail  : out std_logic_vector(AW downto 0);
22.     dout_valid  : out std_logic;
23.     dout        : out std_logic_vector(DW - 1 downto 0)
24.   );
25. end entity;
26.
27. architecture rtl of sync_fifo_v2 is
28. signal mem_wen, mem_ren          : std_logic;
29. signal mem_waddr, mem_raddr : std_logic_vector(AW - 1 downto 0);
30. signal mem_din                   : std_logic_vector(DW - 1 downto 0);
31. begin
32.   process(clk)
33.   begin
34.     if rising_edge(clk) then
35.       mem_din <= din;
36.     end if;
37.   end process;
38.
39.   i_fifo_ctrl   : entity work.fifo_ctrl
40.   generic map ( AW => AW )
41.   port map (
42.     clk         => clk,
43.     rst         => rst,
44.     wen         => wen,
45.     ren         => ren,
```

```vhdl
46.      mem_wen        => mem_wen,
47.      mem_ren        => mem_ren,
48.      mem_waddr      => mem_waddr,
49.      mem_raddr      => mem_raddr,
50.      full           => full,
51.      empty          => empty,
52.      dout_valid     => dout_valid,
53.      room_avail     => room_avail,
54.      data_avail     => data_avail
55.    );
56.
57.    i_sdp_one_clk : entity work.sdp_one_clk
58.    generic map (
59.      AW             => AW,
60.      DW             => DW,
61.      RAM_STYLE_VAL  => RAM_STYLE_VAL
62.    )
63.    port map (
64.      clk            => clk,
65.      wen            => mem_wen,
66.      waddr          => mem_waddr,
67.      din            => mem_din,
68.      ren            => mem_ren,
69.      raddr          => mem_raddr,
70.      dout           => dout
71.    );
72. end architecture;
```

最后，我们比较一下 VHDL 代码 6-28 和 VHDL 代码 6-30 的时序性能，目标芯片为 xcvu3p-ffvc1517-1-i，FIFO 宽度为 18，Vivado 版本为 2021.2，时钟周期为 2.0ns。在不同地址位宽下，资源利用率与时序指标（WNS 和 WHS）如表 6-20 所示。可以看到，这两个版本有各自的优势。图 6-53 和图 6-54 分别显示了 sync_fifo_v1 和 sync_fifo_v2 布线后关键路径的时序报告（对应地址位宽为 15），可以看到，关键路径均与信号 room_avail 相关。这一点不难理解，对于 sync_fifo_v1，room_avail 与 num_entries 相关，而 num_entries 又和 num_entries_nxt 相关。对于 sync_fifo_2，room_avail 与 data_avail 相关，而 data_avail 又与读/写指针相关。

表 6-20

版　　本	地址位宽	FIFO 深度	FIFO 宽度	LUT	FF	BRAM	WNS	WHS	最大逻辑级数
sync_fifo_v1	12	4096	18	71	144	2	0.178	0.021	7
sync_fifo_v2	12	4096	18	86	143	2	0.167	0.024	6
sync_fifo_v1	13	8192	18	85	152	4	0.175	0.030	7
sync_fifo_v2	13	8192	18	97	151	4	0.115	0.020	7
sync_fifo_v1	14	16384	18	98	160	8	0.106	0.021	6
sync_fifo_v2	14	16384	18	115	159	8	0.169	0.020	6
sync_fifo_v1	15	32768	18	130	172	16	0.119	0.021	7
sync_fifo_v2	15	32768	18	107	174	16	0.012	0.039	8

Name	Slack ^1	Levels	High Fanout	From	To	Total Delay	Logic Delay
Path 1	0.119	7	9	num_entries_reg[9]/C	room_avail_reg[14]/D	1.779	1.086
Path 2	0.135	7	9	num_entries_reg[9]/C	room_avail_reg[15]/D	1.763	1.069
Path 3	0.161	7	9	num_entries_reg[9]/C	room_avail_reg[13]/D	1.737	1.045
Path 4	0.187	7	9	num_entries_reg[9]/C	room_avail_reg[12]/D	1.711	1.015
Path 5	0.195	7	9	num_entries_reg[9]/C	room_avail_reg[10]/D	1.703	1.008
Path 6	0.205	7	9	num_entries_reg[9]/C	room_avail_reg[11]/D	1.693	0.997
Path 7	0.212	0	16	i_sdp_one_..._reg[2]/C	i_sdp_one...DADDR[6]	1.469	0.094

图 6-53

Name	Slack ^1	Levels	High Fanout	From	To	Total Delay	Logic Delay
Path 1	0.012	7	3	i_fifo_ctrl/rptr_reg[0]_replica/C	i_fifo_ctrl/room_avail_reg[14]/D	1.992	1.283
Path 2	0.027	7	3	i_fifo_ctrl/rptr_reg[0]_replica/C	i_fifo_ctrl/room_avail_reg[15]/D	1.977	1.267
Path 3	0.032	8	3	i_fifo_ctrl/rptr_reg[0]_replica/C	i_fifo_ctrl/room_avail_reg[12]/D	1.911	1.215
Path 4	0.040	7	3	i_fifo_ctrl/rptr_reg[0]_replica/C	i_fifo_ctrl/room_avail_reg[10]/D	1.903	1.208
Path 5	0.050	7	3	i_fifo_ctrl/rptr_reg[0]_replica/C	i_fifo_ctrl/room_avail_reg[11]/D	1.893	1.197
Path 6	0.053	7	3	i_fifo_ctrl/rptr_reg[0]_replica/C	i_fifo_ctrl/room_avail_reg[13]/D	1.951	1.243
Path 7	0.066	8	3	i_fifo_ctrl/rptr_reg[0]_replica/C	i_fifo_ctrl/room_avail_reg[9]/D	1.877	1.183

图 6-54

6.7 异步 FIFO 代码风格

异步 FIFO 是指读/写时钟是异步关系的 FIFO。例如，读/写时钟分别来自不同的 MMCM。相比同步 FIFO，异步 FIFO 有更为广阔的应用场合，常常用于多位数据的跨时钟域操作。设计的关键仍然是如何判定 FIFO 是满还是空。与同步 FIFO 的设计思路一致，当读/写指针最高位相反、其余位相等时，FIFO 达到满状态；当读/写指针完全相等时，FIFO 达到空状态，如图 6-48 和图 6-49 所示。但问题是对于异步 FIFO，读/写指针隶属于不同的时钟域（前者属于读时钟域，后者属于写时钟域），而设计中需要将读指针传递到写时钟域，同时要将写指针传递到读时钟域，本身就涉及跨时钟域操作。对于由二进制计数器生成的读/写指针，其本身是一个多位矢量，为了保证安全地跨时钟域传送，需要将二进制码转换为格雷码。这样，图 6-48 和图 6-49 就分别变为如图 6-55 和图 6-56 所示的形式，图中，wptr 第一行为二进制码，第二行为其对应的格雷码。rptr 也是如此。判定 FIFO 是空状态还是满状态就需要依据 wptr 和 rptr 的格雷码特征，从而可以得出如下两个结论。

（1）当 FIFO 为满状态时，读/写指针的高两位相反，其余位相等。

（2）当 FIFO 为空状态时，读/写指针完全相等。

图 6-55

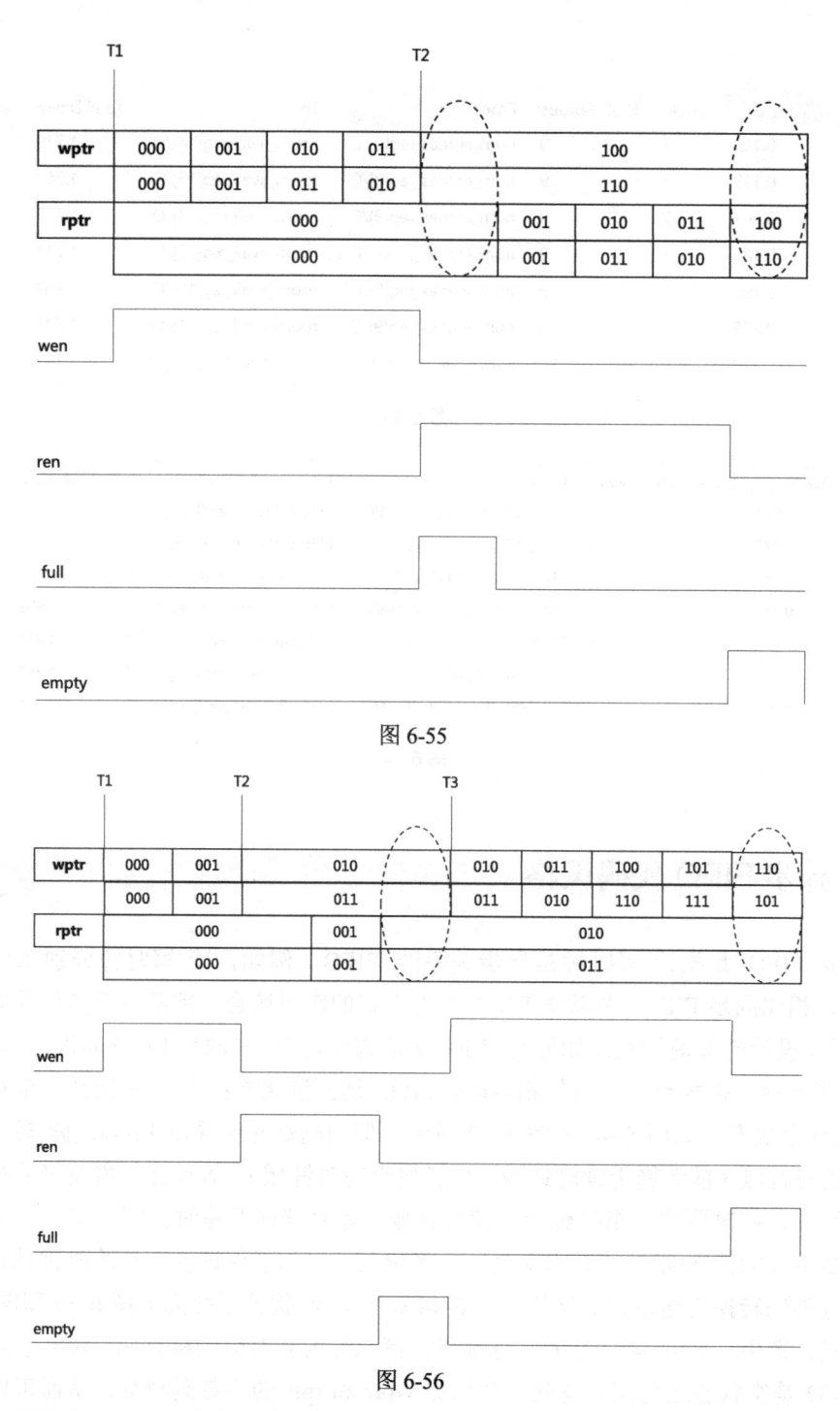

图 6-56

考虑到异步 FIFO 本身有两个不同的时钟域，因此将与写操作相关的控制逻辑（如生成存储单元写地址、写使能信号、生成 FIFO 满标记信号）作为一个独立的模块，命名为 wptr_full；将与读操作相关的控制逻辑（如生成存储单元读地址、读使能信号、输出数据有效标记信号、生成 FIFO 空标记信号）作为一个独立的模块，命名为 rptr_empty，如图 6-57 所示，图中，虚线框内为经典的双触发器跨时钟域电路。

第 6 章 优化存储器 | 303

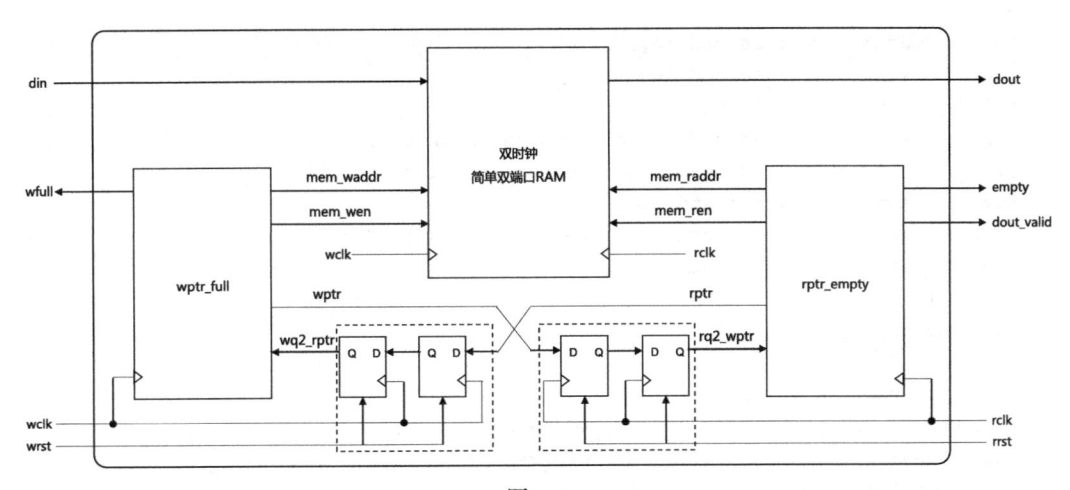

图 6-57

模块 wptr_full 与同步 FIFO 中的对应电路相比（见图 6-51），增加了二进制码到格雷码的转换电路，如图 6-58 所示，相应的 RTL 代码如 VHDL 代码 6-31 所示。

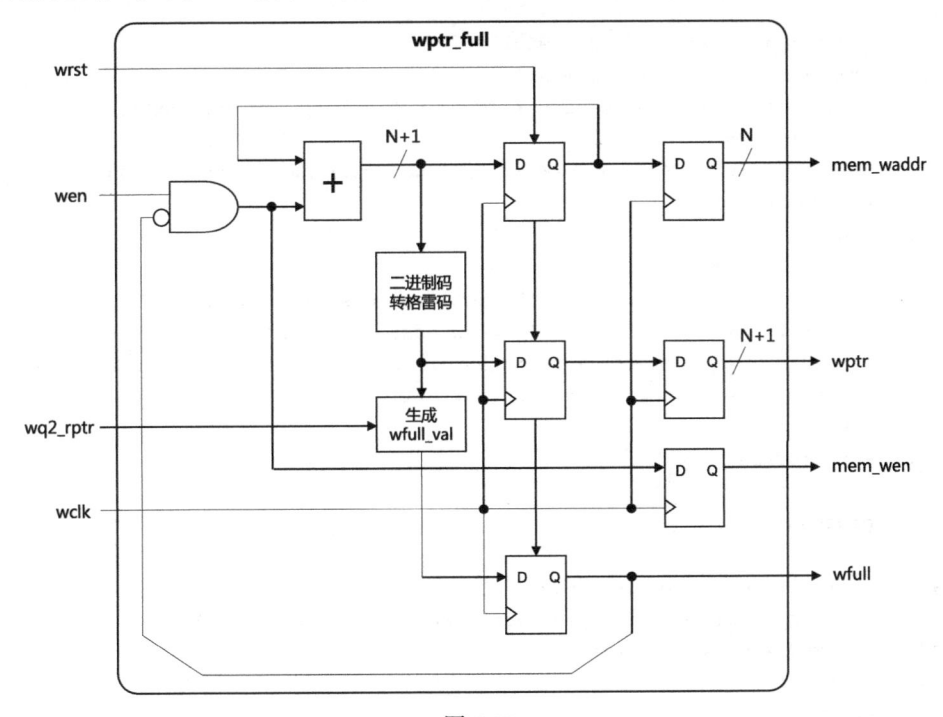

图 6-58

📄 VHDL 代码 6-31

```
1.   --File: wptr_full.vhd
2.   library ieee;
3.   use ieee.std_logic_1164.all;
4.   use ieee.numeric_std.all;
5.
6.   entity wptr_full is
7.     generic ( AW : positive := 4 );
8.     port (
```

```
9.      wclk        : in std_logic;
10.     wrst        : in std_logic;
11.     wen         : in std_logic;
12.     wq2_rptr    : in std_logic_vector(AW downto 0);
13.     mem_waddr   : out std_logic_vector(AW - 1 downto 0);
14.     mem_wen     : out std_logic;
15.     wptr        : out std_logic_vector(AW downto 0);
16.     wfull       : out std_logic
17.     );
18. end entity;
19.
20. architecture rtl of wptr_full is
21.   signal wbin, wbin_nxt : unsigned(AW downto 0);
22.   signal wgray_nxt      : std_logic_vector(AW downto 0);
23.   signal mem_wen_i      : std_logic;
24.   signal wfull_val      : std_logic;
25.   signal wptr_i         : std_logic_vector(AW downto 0);
26. begin
27.
28.   mem_wen_i <= wen and (not wfull);
29.   wbin_nxt  <= wbin + mem_wen_i;
30.   wgray_nxt <= std_logic_vector((wbin_nxt srl 1) xor wbin_nxt);
31.   wfull_val <= wgray_nxt ?=
32.     ((not wq2_rptr(AW downto AW - 1)) & wq2_rptr(AW - 2 downto 0));
33.   process(wclk)
34.   begin
35.     if rising_edge(wclk) then
36.       wbin    <= (others => '0') when wrst else wbin_nxt;
37.       wptr_i  <= (others => '0') when wrst else wgray_nxt;
38.       wptr    <= (others => '0') when wrst else wptr_i;
39.       wfull   <= '0' when wrst else wfull_val;
40.     end if;
41.   end process;
42.
43.   process(wclk)
44.   begin
45.     if rising_edge(wclk) then
46.       mem_waddr <= std_logic_vector(wbin(AW - 1 downto 0));
47.       mem_wen   <= mem_wen_i;
48.     end if;
49.   end process;
50. end architecture;
```

🔍 VHDL-2008 新特性

VHDL-2008 支持 unsigned 类型与 std_logic 类型相加或相减, 如 VHDL 代码 6-31 第 29 行所示。这里, wbin 为 unsigned 类型, mem_wen_i 为 std_logic 类型。对于带使能信号的计数器, 如果步进为 1, 那么本质上是将当前计数值与使能信号相加, 获得下一个计数值。

> 💡 **设计规则 10**：尽可能将相关信号放在一个 process 中描述，这会增强代码的可读性，如 VHDL 代码 6-31 第 36 行至第 39 行所示。同时，使用 when-else 语句描述同步触发器相比 if 语句更为简洁。

模块 rptr_empty 与同步 FIFO 中的对应电路相比（见图 6-52），增加了二进制码到格雷码的转换电路，如图 6-59 所示，相应的 RTL 代码如 VHDL 代码 6-32 所示。

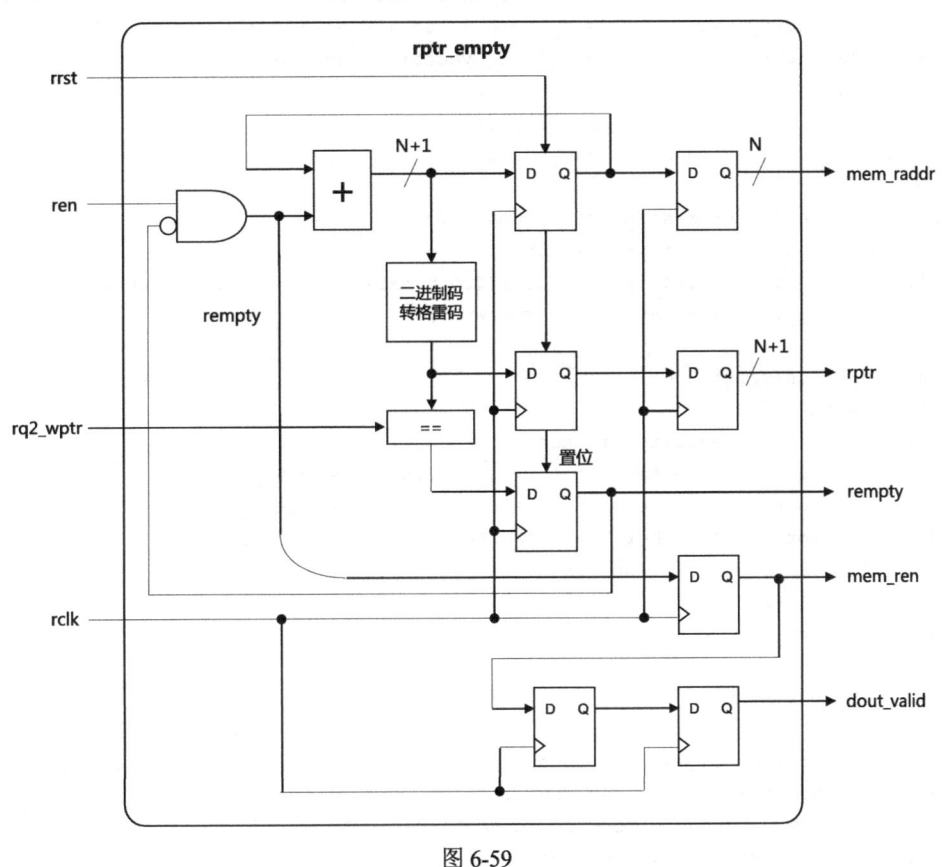

图 6-59

📄 **VHDL 代码 6-32**

```
1.  --File: rptr_empty.vhd
2.  library ieee;
3.  use ieee.std_logic_1164.all;
4.  use ieee.numeric_std.all;
5.
6.  entity rptr_empty is
7.    generic ( AW : positive := 4 );
8.    port (
9.      rclk      : in std_logic;
10.     rrst      : in std_logic;
11.     ren       : in std_logic;
12.     rq2_wptr  : in std_logic_vector(AW downto 0);
13.     rptr      : out std_logic_vector(AW downto 0);
14.     mem_raddr : out std_logic_vector(AW - 1 downto 0);
```

```vhdl
15.        rempty       : out std_logic;
16.        mem_ren      : out std_logic;
17.        dout_valid   : out std_logic
18.   );
19. end entity;
20.
21. architecture rtl of rptr_empty is
22.   signal rbin, rbin_nxt : unsigned(AW downto 0);
23.   signal rgray_nxt      : std_logic_vector(AW downto 0);
24.   signal rptr_i         : std_logic_vector(AW downto 0);
25.   signal mem_ren_i      : std_logic;
26.   signal rempty_val     : std_logic;
27.   signal dout_valid_i   : std_logic;
28. begin
29.   mem_ren_i  <= ren and (not rempty);
30.   rbin_nxt   <= rbin + mem_ren_i;
31.   rgray_nxt  <= std_logic_vector((rbin_nxt srl 1) xor rbin_nxt);
32.   rempty_val <= rgray_nxt ?= rq2_wptr;
33.
34.   process(rclk)
35.   begin
36.     if rising_edge(rclk) then
37.       rbin   <= (others => '0') when rrst else rbin_nxt;
38.       rptr_i <= (others => '0') when rrst else rgray_nxt;
39.       rptr   <= (others => '0') when rrst else rptr_i;
40.       rempty <= '1' when rrst else rempty_val;
41.     end if;
42.   end process;
43.
44.   process(rclk)
45.   begin
46.     if rising_edge(rclk) then
47.       mem_ren      <= mem_ren_i;
48.       mem_raddr    <= std_logic_vector(rbin(AW - 1 downto 0));
49.       dout_valid_i <= mem_ren;
50.       dout_valid   <= dout_valid_i;
51.     end if;
52.   end process;
53. end architecture;
```

双触发器跨时钟域电路 RTL 代码如 VHDL 代码 6-33 所示。这里对两组触发器均施加了 Vivado 综合属性 ASYNC_REG，以告知综合工具这两个触发器均位于同步链上，以接收异步数据，这样工具在综合阶段及布局布线阶段就不会将其优化掉。

VHDL 代码 6-33

```vhdl
1. --File: cdc_sync.vhd
2. library ieee;
3. use ieee.std_logic_1164.all;
4.
5. entity cdc_sync is
6.   generic ( AW : positive := 4 );
```

```vhdl
7.    port (
8.      clk : in std_logic;
9.      rst : in std_logic;
10.     din : in std_logic_vector(AW downto 0);
11.     dout : out std_logic_vector(AW downto 0)
12.   );
13. end entity;
14.
15. architecture rtl of cdc_sync is
16. signal din_d1, din_d2 : std_logic_vector(AW downto 0);
17. attribute ASYNC_REG : boolean;
18. attribute ASYNC_REG of din_d1 : signal is TRUE;
19. attribute ASYNC_REG of din_d2 : signal is TRUE;
20. begin
21.   dout <= din_d2;
22.   process(clk)
23.   begin
24.     if rising_edge(clk) then
25.       din_d1 <= (others => '0') when rst else din;
26.       din_d2 <= (others => '0') when rst else din_d1;
27.     end if;
28.   end process;
29. end architecture;
```

异步 FIFO 顶层 RTL 代码如 VHDL 代码 6-34 所示。代码第 81 行实例化的两个时钟的简单双端口 RAM 对应的 RTL 代码如 VHDL 代码 6-15 所示。

📄 VHDL 代码 6-34

```vhdl
1.  --File: async_fifo.vhd
2.  library ieee;
3.  use ieee.std_logic_1164.all;
4.
5.  entity async_fifo is
6.    generic (
7.      DW             : positive := 18;
8.      AW             : positive := 14;
9.      RAM_STYLE_VAL : string := "block"
10.   );
11.   port (
12.     wclk, wrst, wen : in std_logic;
13.     rclk, rrst, ren : in std_logic;
14.     din             : in std_logic_vector(DW - 1 downto 0);
15.     wfull           : out std_logic;
16.     rempty          : out std_logic;
17.     dout            : out std_logic_vector(DW - 1 downto 0);
18.     dout_valid      : out std_logic
19.   );
20. end entity;
21.
22. architecture rtl of async_fifo is
23. signal wptr, rptr            : std_logic_vector(AW downto 0);
```

```vhdl
24. signal rq2_wptr, wq2_rptr    : std_logic_vector(AW downto 0);
25. signal mem_waddr, mem_raddr  : std_logic_vector(AW - 1 downto 0);
26. signal mem_din               : std_logic_vector(DW - 1 downto 0);
27. signal mem_wen, mem_ren      : std_logic;
28. begin
29.   process(wclk)
30.   begin
31.     if rising_edge(wclk) then
32.       mem_din <= din;
33.     end if;
34.   end process;
35.
36.   r2w_cdc_sync : entity work.cdc_sync
37.   generic map ( AW => AW )
38.   port map (
39.     clk  => wclk,
40.     rst  => wrst,
41.     din  => rptr,
42.     dout => wq2_rptr
43.   );
44.
45.   w2r_cdc_sync : entity work.cdc_sync
46.   generic map ( AW => AW )
47.   port map (
48.     clk  => rclk,
49.     rst  => rrst,
50.     din  => wptr,
51.     dout => rq2_wptr
52.   );
53.
54.   i_wptr_full : entity work.wptr_full
55.   generic map ( AW => AW )
56.   port map (
57.     wclk      => wclk,
58.     wrst      => wrst,
59.     wen       => wen,
60.     wq2_rptr  => wq2_rptr,
61.     mem_waddr => mem_waddr,
62.     mem_wen   => mem_wen,
63.     wptr      => wptr,
64.     wfull     => wfull
65.   );
66.
67.   i_rptr_empty : entity work.rptr_empty
68.   generic map ( AW => AW )
69.   port map (
70.     rclk      => rclk,
71.     rrst      => rrst,
72.     ren       => ren,
73.     rq2_wptr  => rq2_wptr,
74.     rptr      => rptr,
```

```
75.      mem_raddr    => mem_raddr,
76.      rempty       => rempty,
77.      mem_ren      => mem_ren,
78.      dout_valid   => dout_valid
79.    );
80.
81.    i_sdp_two_clk : entity work.sdp_two_clk
82.    generic map ( AW => AW, DW => DW, RAM_STYLE_VAL => RAM_STYLE_VAL )
83.    port map (
84.      clka  => wclk,
85.      wea   => mem_wen,
86.      addra => mem_waddr,
87.      dina  => mem_din,
88.      clkb  => rclk,
89.      reb   => mem_ren,
90.      addrb => mem_raddr,
91.      doutb => dout
92.    );
93. end architecture;
```

对于异步 FIFO，时序约束也很重要，因为涉及如何对跨时钟域路径进行约束，如 Tcl 代码 6-4 所示。代码第 2 行和第 3 行分别设置了写时钟频率和读时钟频率，单位均为 MHz，代码第 8 行获取读/写时钟周期的最小值，将此值设置为跨时钟域路径所允许的最大延迟，如代码第 9 行至第 12 行所示。

Tcl 代码 6-4

```
1.  #File: async_fifo_timing.tcl
2.  set wclk_freq 500
3.  set rclk_freq 200
4.  set wclk_period [expr {1.0 / double($wclk_freq) * 1000}]
5.  set rclk_period [expr {1.0 / double($rclk_freq) * 1000}]
6.  create_clock -name wclk -period $wclk_period [get_ports wclk]
7.  create_clock -name rclk -period $rclk_period [get_ports rclk]
8.  set delay [expr {min($wclk_period, $rclk_period)}]
9.  set_max_delay -from [get_pins i_wptr_full/wptr_reg[*]/C] \
10.            -to [get_pins w2r_cdc_sync/din_d1_reg[*]/D] $delay
11. set_max_delay -from [get_pins i_rptr_empty/rptr_reg[*]/C] \
12.            -to [get_pins r2w_cdc_sync/din_d1_reg[*]/D] $delay
```

基于上述约束，选取目标芯片为 xcvu3p-ffvc1517-1-i，FIFO 地址位宽为 14，数据位宽为 18，Vivado 版本为 2021.2，布线后的时序报告如图 6-60 所示，可以看到，时序能够收敛。关键路径如图 6-61 所示，可以看到，关键路径均为跨时钟域路径。

Setup		Hold		Pulse Width	
Worst Negative Slack (WNS):	0.308 ns	Worst Hold Slack (WHS):	0.021 ns	Worst Pulse Width Slack (WPWS):	0.450 ns
Total Negative Slack (TNS):	0.000 ns	Total Hold Slack (THS):	0.000 ns	Total Pulse Width Negative Slack (TPWS):	0.000 ns
Number of Failing Endpoints:	0	Number of Failing Endpoints:	0	Number of Failing Endpoints:	0
Total Number of Endpoints:	597	Total Number of Endpoints:	597	Total Number of Endpoints:	267

All user specified timing constraints are met.

图 6-60

Name	Slack ^1	Levels	High Fanout	From	To	Total Delay	Logic Delay
Path 41	0.308	0	1	i_rptr_empty/rptr_reg[14]/C	r2w_cdc_sync/din_d1_reg[14]/D	1.266	0.094
Path 42	0.326	0	1	i_rptr_empty/rptr_reg[2]/C	r2w_cdc_sync/din_d1_reg[2]/D	1.242	0.099
Path 43	0.332	0	1	i_rptr_empty/rptr_reg[1]/C	r2w_cdc_sync/din_d1_reg[1]/D	1.227	0.096
Path 44	0.339	0	1	i_rptr_empty/rptr_reg[10]/C	r2w_cdc_sync/din_d1_reg[10]/D	1.238	0.095
Path 45	0.391	0	1	i_rptr_empty/rptr_reg[13]/C	r2w_cdc_sync/din_d1_reg[13]/D	1.145	0.099
Path 46	0.397	0	1	i_rptr_empty/rptr_reg[0]/C	r2w_cdc_sync/din_d1_reg[0]/D	1.148	0.095
Path 47	0.414	0	1	i_rptr_empty/rptr_reg[12]/C	r2w_cdc_sync/din_d1_reg[12]/D	1.154	0.096
Path 48	0.415	0	1	i_rptr_empty/rptr_reg[3]/C	r2w_cdc_sync/din_d1_reg[3]/D	1.153	0.098
Path 49	0.419	0	1	i_rptr_empty/rptr_reg[6]/C	r2w_cdc_sync/din_d1_reg[6]/D	1.133	0.095
Path 50	0.434	0	1	i_rptr_empty/rptr_reg[5]/C	r2w_cdc_sync/din_d1_reg[5]/D	1.134	0.099

图 6-61

6.8 平衡 BlockRAM 的功耗与性能

UltraScale/UltraScale+ FPGA 中的 BRAM 具备级联功能，同时，Vivado 提供了综合属性 CASCADE_HEIGHT，控制 BRAM 的级联个数。以 4K×32（1K=1024）的单端口 RAM 为例，如 VHDL 代码 6-35 所示。若 CASCADE_HEIGHT 值为 1，工具将采用 4 个 4K×8 的 RAM 拼接为一个 4K×32 的 RAM，其中，每个 4K×8 的 RAM 占用一个 36Kb BRAM。这样，在执行读/写操作时，每个 BRAM 都处于激活状态，因此在这种情形下，功耗最大，但 F_{max} 最高。若 CASCADE_HEIGHT 值为 4，工具将采用 4 个 1K×32 的 RAM 级联构成一个 4K×32 的 RAM，同样，每个 1K×32 的 RAM 占用一个 36Kb BRAM。这样，在执行读/写操作时，只有一个 BRAM 处于激活状态，因此在这种情形下，功耗最低，但 F_{max} 也最小（级联路径过长，成为性能瓶颈）。若 CASCADE_HEIGHT 值为 2，工具将采用 4 个 2K×16 的 RAM 构成一个 4K×32 的 RAM，其中，每两个 RAM 级联在一起。同样，每个 2K×16 的 RAM 占用一个 36Kb BRAM。这样，在执行读/写操作时，只有两个 BRAM 处于激活状态，因此这种情形是前两种情形的折中，既能满足 F_{max} 的需求，也能将功耗控制在一定范围内。

📄 VHDL 代码 6-35

```
1.  --File: sp_ram_rf_us.vhd
2.  library ieee;
3.  use ieee.std_logic_1164.all;
4.  use ieee.numeric_std.all;
5.  use work.mem_pkg.all;
6.
7.  entity sp_ram_rf_us is
8.    generic (
9.      AW                 : positive := 12;
10.     DW                 : positive := 32;
11.     RAM_STYLE_VAL      : string   := "block";
12.     CASCADE_HEIGHT_VAL : integer  := 2;
```

```
13.     DEPTH              : positive := 2 ** AW
14.   );
15.   port (
16.     clk  : in std_logic;
17.     we   : in std_logic;
18.     addr : in std_logic_vector(AW - 1 downto 0);
19.     din  : in std_logic_vector(DW - 1 downto 0);
20.     dout : out std_logic_vector(DW - 1 downto 0)
21.   );
22. end entity;
23.
24. architecture rtl of sp_ram_rf_us is
25. signal mem: mem_t(0 to DEPTH - 1)(DW - 1 downto 0) := (others => (others => '0'));
26. signal we_d1    : std_logic;
27. signal addr_d1 : unsigned(AW - 1 downto 0);
28. signal din_d1   : std_logic_vector(DW - 1 downto 0);
29. attribute RAM_STYLE : string;
30. attribute RAM_STYLE of mem : signal is RAM_STYLE_VAL;
31. attribute CASCADE_HEIGHT: integer;
32. attribute CASCADE_HEIGHT of mem : signal is CASCADE_HEIGHT_VAL;
33. begin
34.   process(clk)
35.   begin
36.     if rising_edge(clk) then
37.       we_d1   <= we;
38.       addr_d1 <= unsigned(addr);
39.       din_d1  <= din;
40.     end if;
41.   end process;
42.
43.   process(clk)
44.   begin
45.     if rising_edge(clk) then
46.       if we_d1 then
47.         mem(to_integer(addr_d1)) <= din_d1;
48.       end if;
49.       dout <= mem(to_integer(addr_d1));
50.     end if;
51.   end process;
52. end architecture;
```

表 6-21 和图 6-62 对上述三种情形进行了对比。在图 6-62 中，括号中的数字(1)、(4)、(2)表示 CASCADE_HEIGHT 的值。

表 6-21

CASCADE_HEIGHT	每个 BRAM 的规格	36Kb BRAM 消耗量	功　耗	时 序 性 能
1	4K×8	4	1.106	最好
4	1K×32	4	0.909	最差
2	2K×16	4	0.971	居中

Name		Part	Total Power	BRAM	LUT	FF	WNS	TNS	WHS	THS
∨ synth_1	(1)	xcvu3p-ffvc1517-1-e		4.0	0	0				
impl_1		xcvu3p-ffvc1517-1-e	1.106	4.0	0	0	NA	NA	NA	NA
∨ synth_2	(4)	xcvu3p-ffvc1517-1-e		4.0	6	0				
impl_2		xcvu3p-ffvc1517-1-e	0.909	4.0	6	0	-0.021	-0.021	0.058	0.000
∨ synth_3	(2)	xcvu3p-ffvc1517-1-e		4.0	2	0				
impl_3		xcvu3p-ffvc1517-1-e	0.971	4.0	2	0	0.288	0.000	0.177	0.000

图 6-62

Vivado 还提供了综合属性 RAM_DECOMP，该综合属性只有一个值 "power"。其目的是对于较大的 RAM，当映射为 BRAM 时，使其功耗最低。对于 4K×32 的单端口 RAM，RAM_DECOMP 可使工具将其采用 4 个 1K×32 的 RAM 实现。RAM_DECOMP 可与 CASCADE_HEIGHT 联合使用，以获得性能与功耗的折中。需要注意的是 CASCADE_HEIGHT 和 RAM_DECOMP 仅适用于 UltraScale/UltraScale+ FPGA 芯片。

6.9 异构 RAM

通常情况下，对于 RTL 代码描述的 RAM，Vivado 会将其映射为分布式 RAM、BRAM 或 UltraRAM 中的一种。同时，Vivado 提供了综合属性 RAM_SYLTE，用于管理 RAM 的实现方式。RAM_STYLE 有一个值为 "mixed"，可将 RAM 映射为三种 RAM 资源的组合，以节省资源。

例如，对于 2K×20 的单端口 RAM，如果 RAM_STYLE 值为 "block"，会消耗 1.5 个 36Kb BRAM。若选择 "mixed"，则 2K×20 的单端口 RAM 被分解为 2K×18 和 2K×2 两部分，前者消耗 1 个 36Kb BRAM，后者消耗 64 个 LUTRAM，从而节省了 BRAM 资源。类似地，对于 4K×80 的单端口 RAM，若选择 "block"，则消耗 9 个 36Kb BRAM；若选择 "ultra"，则消耗 2 个 UltraRAM；若选择 "mixed"，则消耗 1 个 36Kb BRAM 和 1 个 UltraRAM，从而节省了 UltraRAM 资源。进行对比，如表 6-22 所示。

表 6-22

RAM 规格	RAM_STYLE 可选值		
	block	mixed	ultra
2K x 20	1.5 BRAM	1 BRAM + 64 LUTRAM + 6 LUT	1 URAM
4K x 80	9 BRAM	1 BRAM + 1 URAM	2 URAM

要使 Vivado 将 RAM 映射为异构 RAM，必须使用 RAM_STYLE，并将其设置为 "mixed"。默认情形下，只会映射为单一资源。

6.10 以 IP 方式使用 RAM 和 FIFO

Vivado 提供了 IP，可使用户方便地定制各种类型的 RAM。对于 7 系列和 UltraScale/UltraScale+ FPGA，有两个 IP 可供选择：Distributed Memory Generator 和 Block Memory Generator。前者对应分布式 RAM，后者对应 BRAM，如图 6-63 所示。UltraRAM 不支持

IP 方式调用。对于 Versal ACAP，仅有一个 IP 可供选择，即 Embedded Memory Generator，如图 6-64 所示。但该 IP 可支持 LUTRAM、BRAM、UltraRAM 及这三者的组合。

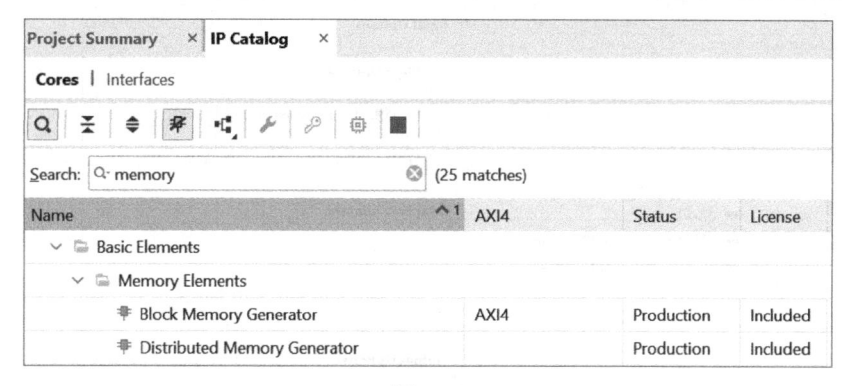

图 6-63

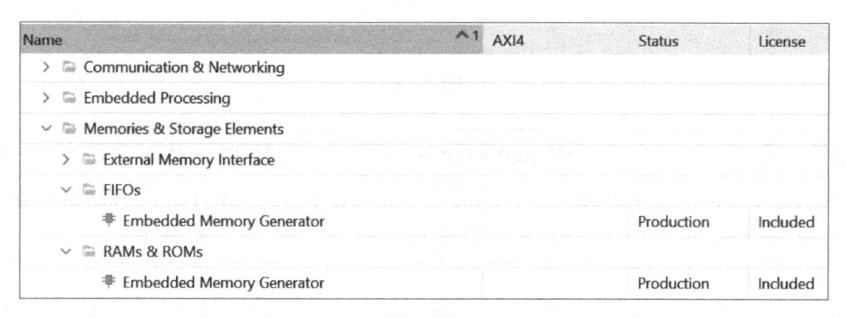

图 6-64

Distributed Memory Generator IP 用于定制基于 LUTRAM 的单端口 ROM、单端口 RAM、简单双端口 RAM 和真双端口 RAM，这些 RAM/ROM 均只有一个时钟。在 IP 定制首页，可对 RAM 的深度、位宽和类型进行设置，如图 6-65 所示。

图 6-65

在端口配置页面，可以选择是否对输入/输出寄存，如图 6-66 所示。这里需要注意的是，当将 Output Options 选择为 Non Registered 时，输出只有一个端口 spo；当选择 Registered 时，输出只有一个端口 qspo；当选择 both 时，输出有两个端口，即 spo 和 qspo。

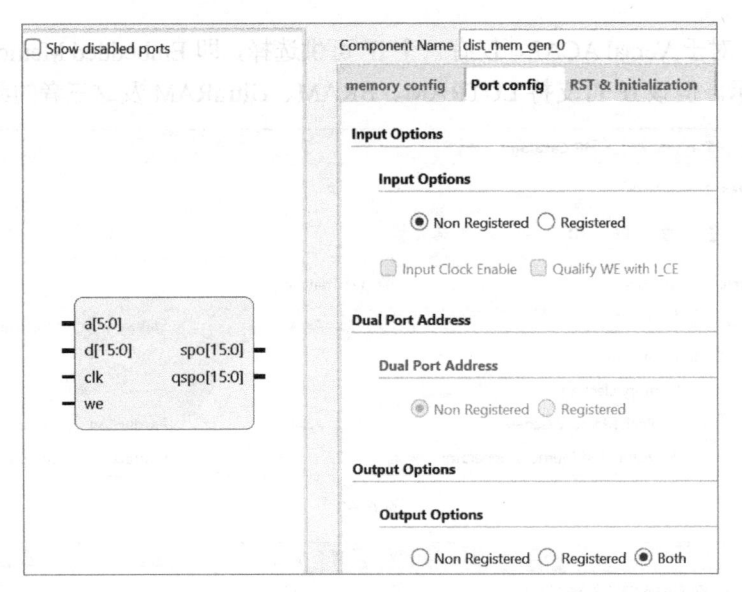

图 6-66

> **设计规则 11**：使用 Distributed Memory Generator IP 时，从时序收敛的角度考虑，Output Options 应选择 Registered，即对 LUTRAM 的输出寄存后再使用，这意味着管脚 qspo 可见，同时会消耗 SLICE 中的触发器。在高速设计中，这一点尤为重要。

在复位和初始值设置页面，如图 6-67 所示。可以借助 .coe 文件设置 RAM 的初始值。在 IP 生成过程中，.coe 文件会被转换为 .mif（memory initialization file）文件。该文件在综合和仿真时会被用到。

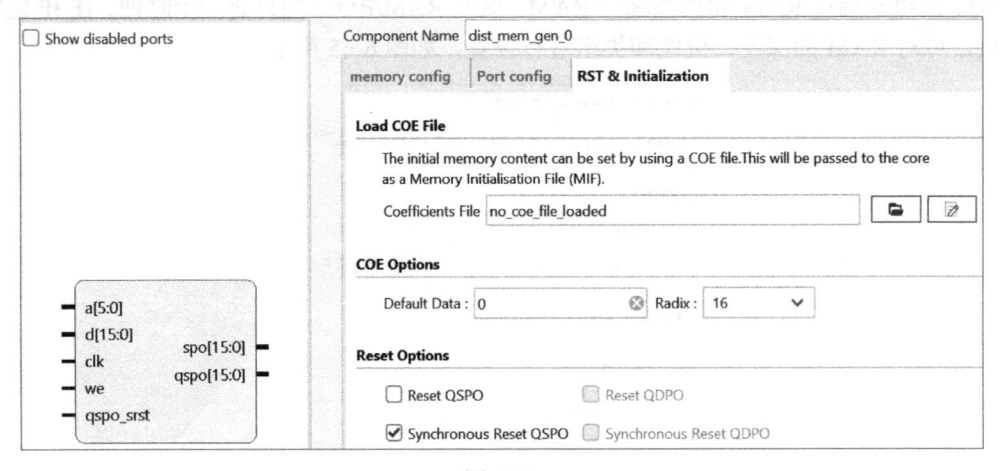

图 6-67

> **设计规则 12**：.coe 文件有固定的格式，由两个参数构成。参数 memory_initialization_radix 定义了数据的进制，有三个可选值：2、10 和 16。参数 memory_initialization_vector 定义了具体与进制相匹配的数值，也就是 RAM 的初始值，从 0 号地址到最后一个地址，相邻两个数据之间可用空格、换行符或逗号隔开，最后一个数据需要用分号结束。

memory_initialization_radix = 16;

memory_initialization_vector =

12, 34, 56, 78, AB, CD, EF, 12, 34, 56, 78, 90, AA, A5, 5A, BA;

以 64×16 的单端口 RAM 为例，采用 Distributed Memory Generator 实现，输入/输出信号之间的时序关系如图 6-68 所示。可以看到，spo 和 qspo 相差一个时钟周期。

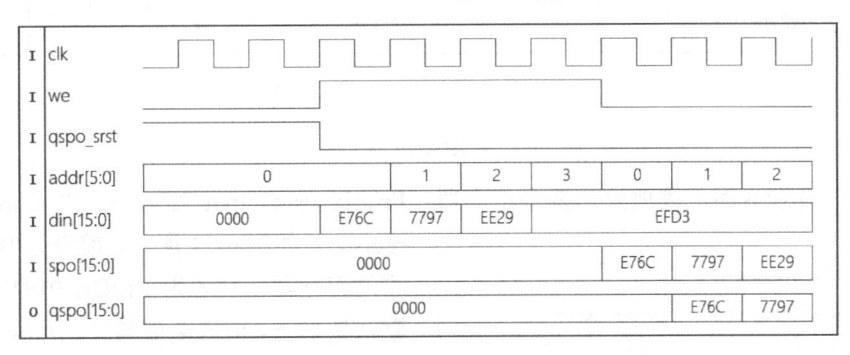

图 6-68

Block Memory Generator IP 用于定制基于 BRAM 的单端口 RAM、简单双端口 RAM、真双端口 RAM、单端口 ROM 和双端口 ROM。在 IP 基础页面，如图 6-69 所示，可选择存储器类型（Memory Type）。需要注意的是，当选择简单双端口 RAM、真双端口 RAM 或双端口 ROM 时，紧随其后的 Common Clock，即图 6-69 中的标记(1)是可选的。若两个端口的时钟相同，则一定要将此选项勾选。该 IP 还支持写操作字节选通功能，如图 6-69 中的标记(2)。在实现算法（Algorithm Options）方面，该 IP 提供了三个可选项：最小面积（Minimum Area）、低功耗（Low Power）和固定原语（Fixed Primitives）。

图 6-69

以 3K×16 的单端口 RAM 为例，三种算法采用的实现方式如图 6-70 所示。可以看到，前两种算法均消耗 3 个 18Kb BRAM，但第二种算法在同一时刻只有一个 18Kb BRAM 处于激活状态，因此其功耗最低。第三种算法使用了原语 4K×4，最终消耗 4 个 18Kb BRAM。

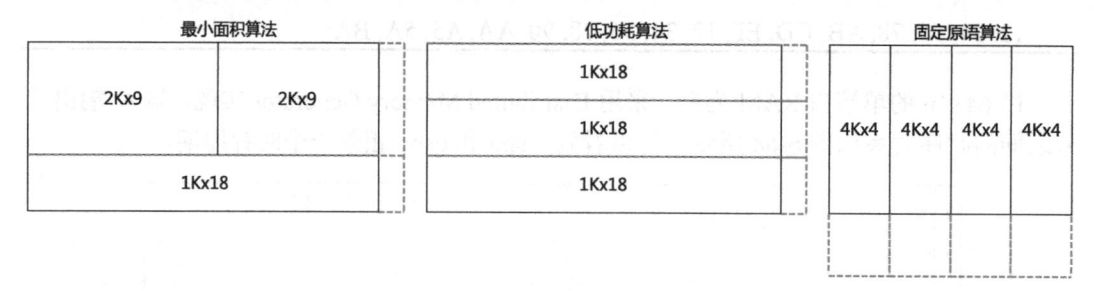

最小面积算法	低功耗算法	固定原语算法
2Kx9 | 2Kx9	1Kx18	4Kx4 | 4Kx4 | 4Kx4 | 4Kx4
1Kx18	1Kx18	
	1Kx18	

图 6-70

在端口设置界面，需要注意两个寄存器：Primitives Output Register 和 Core Output register，如图 6-71 所示，前者为 BRAM 自带的寄存器，后者为 SLICE 中的寄存器。默认情形下，前者会被勾选，因其对时钟到输出延迟有很大的改善作用，这在表 6-16 中已经阐述过。对于高速设计（时钟频率大于 300MHz），建议将两者都勾选。当仅勾选前者时，从输入到输出需要 2 个时钟周期（Latency=2）；当将两者都勾选时，从输入到输出需要 3 个时钟周期（Latency=3）。它们对 Latency 是有影响的，因此，在设计规划初期就要结合时钟频率和系统 Latency 需求决定如何选择这两个寄存器。

图 6-71

💡 **设计规则 13**：使用 Block Memory Generator IP 时，应确保 Primitives Output Register 被勾选，即优先使用 BRAM 自带的寄存器，这在高速设计中尤为重要。

与 Distributed Memory Generator IP 一样，也可借助 .coe 文件对 BRAM 进行初始化。同时在最后的 Summary 页面可以看到设计所消耗的 BRAM 的个数及输入到输出的时钟周期个数。

如果目标芯片为 Versal ACAP，使用 RAM 时可以借助 Embedded Memory Generator IP。这个 IP 集成了 Distributed Memory Generator 和 Block Memory Generator 两个 IP，并在此基础上增加了 UltraRAM（简写为 URAM）和异构 RAM，这在基础界面的 Memory Primitive 中可以看到，如图 6-72 所示。Memory Primitive 可支持 AUTO、LUTRAM、BRAM、URAM 和 MIXED，如图 6-72 中的标记(1)所示。同时，用户要选择时钟模式，如图 6-72 中的标记(2)所示，可支持 Common Clock（共享时钟）和 Independent Clock（独立时钟），还可控制 BRAM 或 URAM 的级联高度，如图 6-72 中的标记(3)所示，通过选项 Cascade Height 设置，这实质上对应综合属性 CASCADE_HEIGHT。对于输入到输出所需要的时钟周期个数，可通过选项 Read Latency Port B 设置，如图 6-72 中的标记(4)所示，其默认值为 2。这里是一个简单双端口 RAM，A 端口只执行写操作，B 端口只执行读操作，因此只有 B 端口有选项 Read Latency，若为单端口 RAM 或真双端口 RAM，则 A 端口也会出现选项 Read Latency。

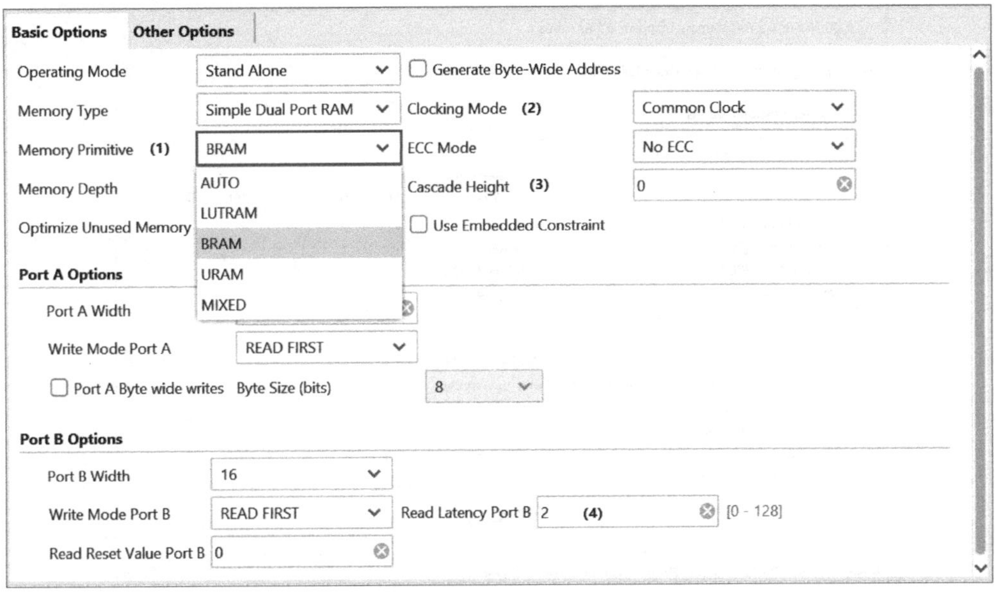

图 6-72

> 💡 **设计规则 14**：使用 Embedded Memory Generator IP 时，要合理设置输入到输出的时钟周期个数（Read Latency）。就高速设计而言，此值至少为 2。

Embedded Memory Generator IP 也可支持 RAM/ROM 的初始化，不同的是，它需要.mem 文件。相比.coe 文件，.mem 文件的格式更为简单，但其要求数据必须为十六进制，如图 6-73 所示，这是一个深度为 4、宽度为 16 的 RAM 的初始值，每行一个数据，对应一个地址号，地址为 0~3。这里，第一列是文本编辑器本身显示的行号，并非.mem 文件的内容。

0	AABB
1	CCDD
2	EEFF
3	9900

图 6-73

使用 FIFO 时，若芯片为 7 系列 FPGA 或 UltraScale/UltraScale+ FPGA，则可以借助 FIFO Generator IP，其基础界面如图 6-74 所示。从时钟的角度而言，可支持同步 FIFO，读/写时钟相同，对应 Commong Clock；也可支持异步 FIFO，读/写时钟不同，对应 Independent Clocks（RD_CLK, WR_CLK）。从资源的角度而言，可支持内嵌 FIFO（FIFO 控制逻辑由硬核实现，不会消耗 SLICE 中的资源，存储资源为 BRAM）、BRAM（FIFO 控制逻辑由 SLICE 中的资源实现）、分布式 RAM 或移位寄存器。每种资源支持的特性在基础界面中都可看到。

端口配置界面，如图 6-75 所示。对于 UltraScale/UltraScale+ FPGA，可支持非对称读/写，如图 6-75 中的标记(1)所示。当使用内嵌 FIFO 或 BRAM 实现 FIFO 时，选项 Output Registers 可供勾选，如图 6-75 中的标记(2)所示。该选项有 3 个可选值，其中，Embedded Registers 为 BRAM 自带的寄存器，Fabric Registers 是 SLICE 中的寄存器。

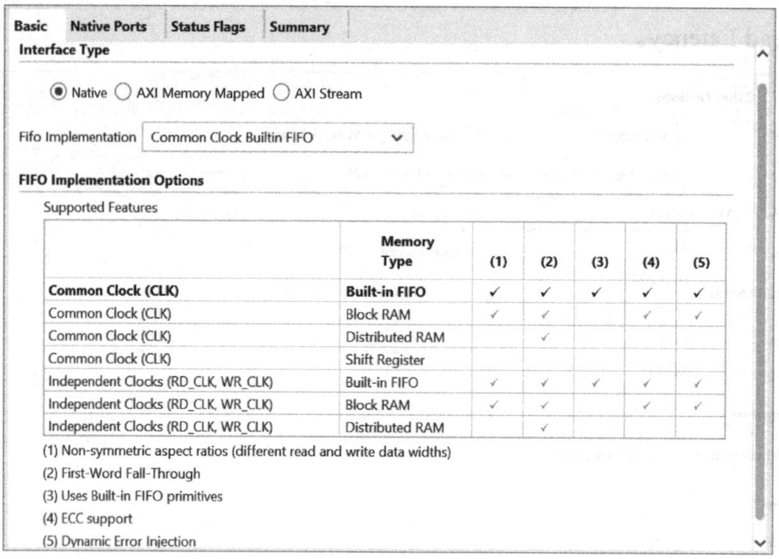

图 6-74

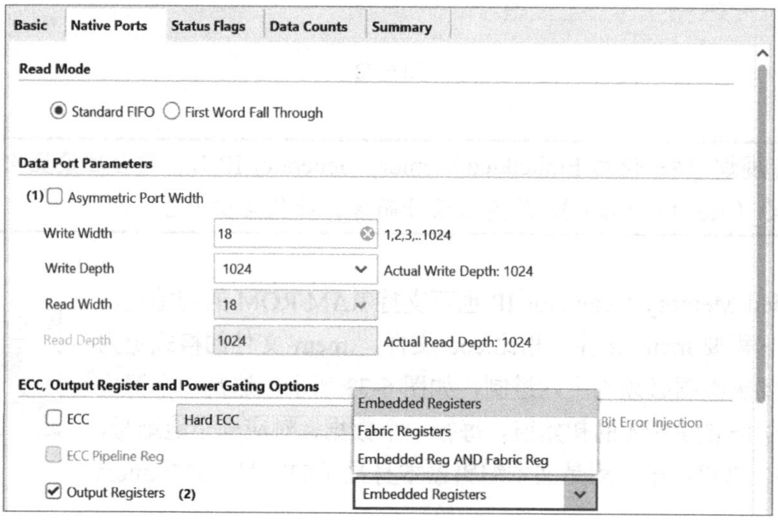

图 6-75

> 💡 **设计规则 15**：使用 FIFO Generator IP 时，若选择内嵌 FIFO 或 BRAM 实现 FIFO，要勾选 Output Registers，并将其值设置为 Embedded Registers（内嵌 FIFO 只有这一个可选值）或 Embedded Reg AND Fabric Reg，这在高速设计时尤为重要，对时序收敛有很大影响。

当使用 Versal ACAP 时，可借助 Embedded FIFO Generator IP 使用 FIFO。其基础界面如图 6-76 所示。FIFO 的存储类型可以是 BRAM、LUTRAM 或 URAM，如图 6-76 中的标记(1)所示。当存储类型为 BRAM 或 URAM 时，通过选项 Cascade Height 可设定级联高度，如图 6-76 中的标记(2)所示。通过选项 FIFO Read Latency 可设定输入到输出所需要的时钟周期个数，如图 6-76 中的标记(3)所示。当存储类型为 BRAM 时，可支持非对称读/写，如图 6-76 中的标记(4)所示。

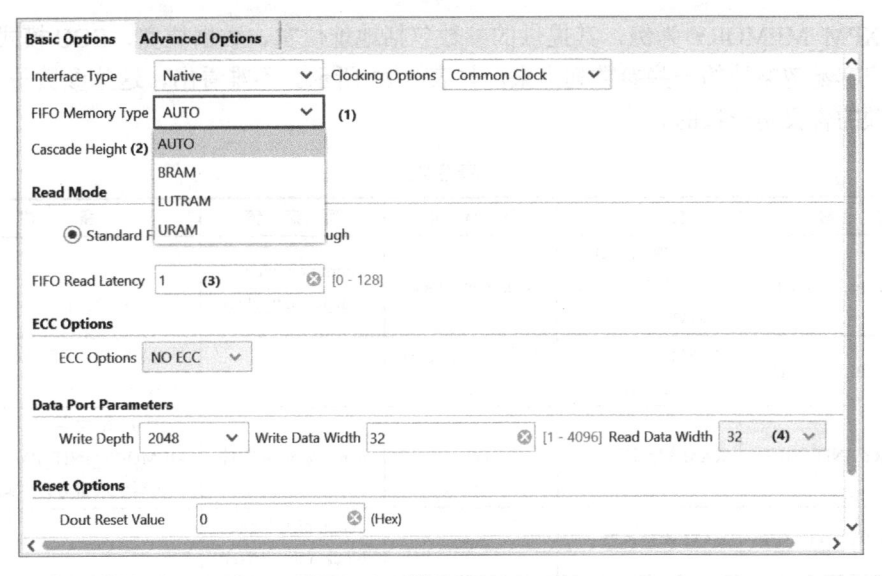

[图 6-76]

> **设计规则 16**：使用 Embedded FIFO Generator 时，要合理设定 FIFO Read Latency 的值。就高速设计而言，此值至少为 2。

6.11 以 XPM 方式使用 RAM 或 FIFO

XPM（Xilinx Parameterized Macro）是 Xilinx 提供的一种参数化宏，在 Vivado 下可直接使用。安装 Vivado 后，在 Language Templates 下可以看到 XPM，有两个版本：Verilog 版和 VHDL 版，如图 6-77 所示。就存储器而言，有 XPM_FIFO 和 XPM_MEMORY，两者的具体内容如图 6-78 所示。

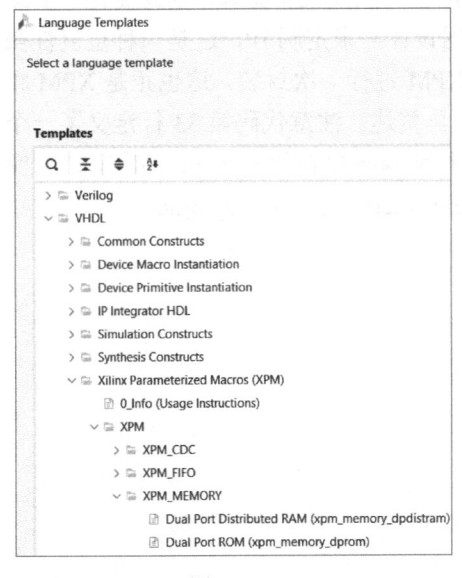

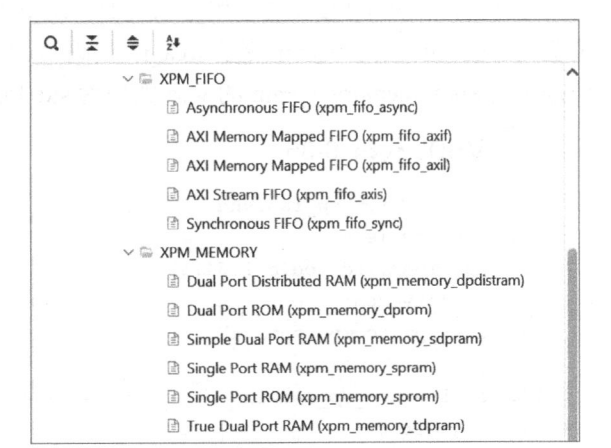

图 6-77　　　　　　　　　　　　　　　图 6-78

以 XPM_MEMORY 为例，其提供的参数包括地址位宽、数据位宽、ECC 模式等。这里将用户容易忽略掉的一些参数列出来，如表 6-23 所示。不难看出，这些参数与 IP 中的对应选项的含义是一致的。

表 6-23

参　数	含　义	默认值	可选值	备　注
CLOCKING_MODE	端口 A/B 的时钟是同一时钟还是不同时钟	common_clock	common_clock, independent_clock	
CASCADE_HEIGHT	BRAM 或 URAM 的允许级联高度	0（由 Vivado 自行决定级联高度）	0~64	针对 UltraScale/UltraScale+ FPGA 和 Versal ACAP
MEMORY_INIT_FILE	RAM 初始值	none	none 或 .mem 文件	当为 .mem 文件时，要求参数 MEMORY_INIT_PARAM 值为 ""（两个双引号，中间无空格）
MEMORY_PRIMITIVE	RAM 实现类型	auto	auto, distributed block, ultra, mixed	
MEMORY_SIZE	RAM 大小，单位为 bit	2048	2~150994944	该值实际上是 RAM 深度与宽度的乘积
READ_LATENCY_A/B	端口 A/B 的输入到输出的时钟周期个数	2	0~100	高速设计时，对于 BRAM 和 URAM，建议此值最小为 2，分布式 RAM 建议此值最小为 1
WRITE_MODE_A/B	端口 A/B 的写操作模式	no_change	no_change, read_first, write_first	

为便于说明，我们以单端口 RAM 为例，如 VHDL 代码 6-36 所示。代码第 36 行至 75 行为 XPM 内容。同时，我们将感兴趣的 XPM 参数提取出来，如代码第 9 行至第 16 行所示。这实质上是对 XPM 进行了一次封装，封装后的模块从外部看只有个别参数和部分管脚。这样做的好处是：当设计中需要用到多个不同规格的单端口 RAM 时，只需要对此封装后的模块进行实例化即可。由于此模块已经过滤了一些参数和管脚，因此简化了后续的实例化。如果以 IP 方式使用 RAM，就要反复根据设计需求定制 IP，这是一种重复性操作，而 XPM 避免了这一操作，但其前提是需要对 XPM 进行一次封装。这也正是 XPM 相比 IP 的优势。XPM_FIFO 的使用方法类似，这里不再赘述。注意代码第 33 行定义了一个位宽为 1 的 std_logic_vector 变量 wea_i，代码第 35 行将 wea 赋值给 wea_i 的 0 位，这么操作是因为 xpm_memory_spram 的 wea 类型为 std_logic_vector 而不是 std_logic。

📄 **VHDL 代码 6-36**

```
1.  --File: xpm_sp_ram.vhd
2.  library ieee;
3.  use ieee.std_logic_1164.all;
4.  library xpm;
5.  use xpm.vcomponents.all;
6.
7.  entity xpm_sp_ram is
8.    generic (
9.      AW              : positive := 4;
10.     DW              : positive := 4;
```

```vhdl
11.      CASCADE_HEIGHT     : integer := 8;
12.      ECC_MODE           : string  := "no_ecc";
13.      MEMORY_INIT_FILE   : string  := "sp_ram_8x4.mem";
14.      MEMORY_PRIMITIVE   : string  := "distributed";
15.      READ_LATENCY_A     : integer := 3;
16.      WRITE_MODE_A       : string  := "read_first"
17.   );
18.   port (
19.      clka   : in std_logic;
20.      rsta   : in std_logic;
21.      addra  : in std_logic_vector(AW - 1 downto 0);
22.      dina   : in std_logic_vector(DW - 1 downto 0);
23.      ena    : in std_logic;
24.      wea    : in std_logic;
25.      regcea : in std_logic;
26.      douta  : out std_logic_vector(DW - 1 downto 0)
27.   );
28. end entity;
29.
30. architecture rtl of xpm_sp_ram is
31. constant MEMORY_DEPTH : integer := 2 ** AW;
32. constant MEMORY_SIZE  : integer := MEMORY_DEPTH * DW;
33. signal wea_i : std_logic_vector(0 downto 0);
34. begin
35.   wea_i(0) <= wea;
36.   i_xpm_memory_spram : xpm_memory_spram
37.   generic map (
38.      ADDR_WIDTH_A        => AW,
39.      AUTO_SLEEP_TIME     => 0,
40.      BYTE_WRITE_WIDTH_A  => DW,
41.      CASCADE_HEIGHT      => CASCADE_HEIGHT,
42.      ECC_MODE            => ECC_MODE,
43.      MEMORY_INIT_FILE    => MEMORY_INIT_FILE,
44.      MEMORY_INIT_PARAM   => open,
45.      MEMORY_OPTIMIZATION => "true",
46.      MEMORY_PRIMITIVE    => MEMORY_PRIMITIVE,
47.      MEMORY_SIZE         => MEMORY_SIZE,
48.      MESSAGE_CONTROL     => 0,
49.      READ_DATA_WIDTH_A   => AW,
50.      READ_LATENCY_A      => READ_LATENCY_A,
51.      READ_RESET_VALUE_A  => "0",
52.      RST_MODE_A          => "SYNC",
53.      SIM_ASSERT_CHK      => 0,
54.      USE_MEM_INIT        => 0,
55.      USE_MEM_INIT_MMI    => 0,
56.      WAKEUP_TIME         => "disable_sleep",
57.      WRITE_DATA_WIDTH_A  => DW,
58.      WRITE_MODE_A        => WRITE_MODE_A,
59.      WRITE_PROTECT       => 1
60.   )
61.   port map (
```

```
62.        dbiterra        => open,
63.        douta           => douta,
64.        sbiterra        => open,
65.        addra           => addra,
66.        clka            => clka,
67.        dina            => dina,
68.        ena             => ena,
69.        injectdbiterra  => '0',
70.        injectsbiterra  => '0',
71.        regcea          => regcea,
72.        rsta            => rsta,
73.        sleep           => '1',
74.        wea             => wea_i
75.     );
76. end architecture;
```

6.12 管理时序路径上的 BRAM 和 UltraRAM

在高速设计中（时钟频率大于 300MHz），对于以 BRAM 或 UltraRAM 为终点单元的时序路径（本质上是以其内部寄存器为终点单元），要确保该路径的起点单元为寄存器，尤其是终点单元对应的管脚为控制管脚，如读/写地址、读/写使能等。若 BRAM 的输入数据由其他 Block（这里的 Block 指 BRAM、UltraRAM、DSP48/DSP58 等）提供，则要确保 Block 和 Block 之间由寄存器中转；同样地，BRAM 的输出数据若提供给其他 Block，也要先通过寄存器中转，如图 6-79 所示。这是因为不同类型的 Block 位于 FPGA 中的不同列，且往往不是相邻的，通过寄存器中转可有效降低线延迟，但会增加 Latency。对于同类型的 Block，即使工具将其放置在同列，也必然位于不同行，若直接相连，布线延迟依然可能会很大，所以，还是需要借助寄存器中转。

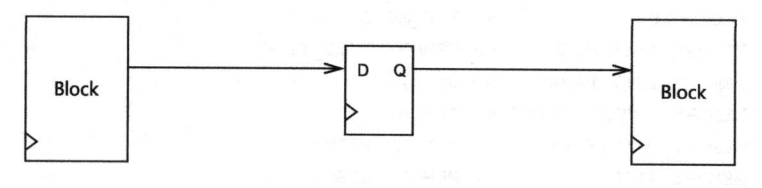

Block: BRAM, UltraRAM, DSP48/DSP58

图 6-79

> **设计规则 17**：高速设计中（时钟频率大于 300MHz），对于 Block 到 Block 的路径，要将其优化为 Block 到寄存器再到 Block，借助寄存器中转可以降低线延迟。这里的 Block 包括 BRAM、UltraRAM、DSP48 或 DSP58。

对于设计中的 RAM，当其以 BRAM 或 UltraRAM 实现时，要确保使用了内部自带的寄存器。BRAM 的端口 A 和端口 B 各自有一个属性，即 DOA_REG 和 DOB_REG，只要选择 BRAM，在其属性窗口中就能看到，如图 6-80 所示。当其为 1 时，表明该 BRAM 使用了自带寄存器,否则就没有使用自带寄存器。UltraRAM 也有类似的属性,名为 OREG_A 和 OREG_B,当其为 TRUE 时,表明 UltraRAM 使用了自带寄存器,否则就没有使用自带寄存器,如图 6-81 所示。

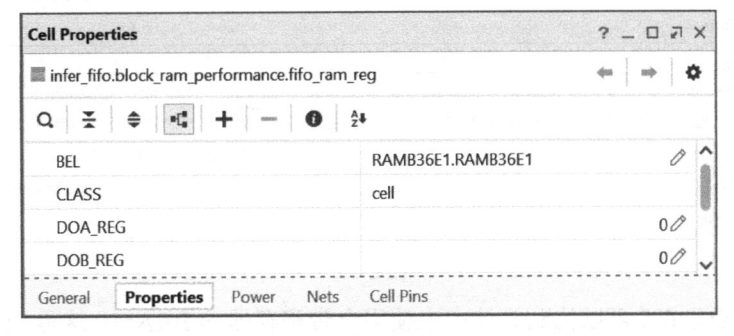

图 6-80

图 6-81

BRAM 和 UltraRAM 自带寄存器对时序的影响很大,如图 6-82 所示。由于没有使用自带寄存器,BRAM 时钟到输出延迟达到了 1.800ns,而使用自带寄存器之后,其延迟降低到 0.622ns。同样地,UltraRAM 未使用自带寄存器时,时钟到输出延迟达到了 2.211ns,而使用自带寄存器之后,延迟降低到 0.844ns,如图 6-83 所示。

图 6-82

Data Path			
Delay Type	Incr (ns)	Path (ns)	Location
URAM288 (Prop_URAM_288K_INST...M288_CLK_CAS_OUT_DOUT_B[71])	(r) 2.211	6.748	Site: URAM288_X0Y48
net (fo=1, estimated)		0.001	6.749	
URAM288			Site: URAM288_X0Y49
Arrival Time		6.749	

Data Path			
Delay Type	Incr (ns)	Path (ns)	Location
URAM288 (Prop_URAM_288K_INST...M288_CLK_CAS_OUT_DOUT_A[0])	(r) 0.844	5.381	Site: URAM288_X0Y48
net (fo=1, estimated)		0.001	5.382	
URAM288			Site: URAM288_X0Y49
Arrival Time		5.382	

图 6-83

借助 Tcl 代码 6-5 可以找到设计中未使用自带寄存器的 BRAM，这里，BRAM 仅用于实现各种类型的 RAM，包括单端口 RAM、简单双端口 RAM 和真双端口 RAM，不包括 FIFO。参数 family 的可选值为 7 和 us，前者表示 7 系列 FPGA，后者表示 UltraScale/UltraScale+ FPGA。fn 用于指定生成文件名。由于 RAM 的类型并不确定，这就意味着端口 A 或端口 B 可能存在未被使用（对应输出管脚未连接）的情形，这时就无须关注 DOA_REG 或 DOB_REG，因此代码第 25 行至第 28 行用于判断端口 A 和端口 B 的数据输出管脚是否被使用。最终会生成一个 .csv 文件，其内容如图 6-84 所示。若 DOA_REG 或 DOB_REG 值为 X，则表明对应端口的数据输出管脚未被使用；若其值为 0，则表明该端口未使用自带寄存器。

Tcl 代码 6-5

```
1.   #File: check_bram_oreg_status.tcl
2.   #fn: file name
3.   #family: 7 or us => 7 stands for 7 series FPGA
4.   #              => us stands for UltraScale/UltraScale+
5.   proc check_bram_oreg_status {family fn} {
6.       switch -exact -- $family {
7.           7 {
8.               set prim_type BMEM.bram.*
9.               set dout_name [list DOADO DOBDO]
10.          }
11.          default {
12.              set prim_type BLOCKRAM.BRAM.*
13.              set dout_name [list DOUTADOUT DOUTBDOUT]
14.          }
15.      }
16.      set fid [open ${fn}.csv w]
17.      puts $fid "BRAM_NAME, DOA_REG, CLKA_FREQ(MHZ), DOB_REG, CLKB_FREQ(MHZ)"
18.
19.      set bram [get_cells -hierarchical -filter "PRIMITIVE_TYPE =~ $prim_type"]
20.      set bram_without_oreg {}
21.      foreach i_bram $bram {
22.          foreach i_dout_name $dout_name {
23.              set dout_pin [get_pins $i_bram/${i_dout_name}*]
```

```
24.            set is_connected 0
25.            foreach i_dout_pin $dout_pin {
26.                set connected_status [get_property IS_CONNECTED $i_dout_pin]
27.                set is_connected [expr {$is_connected || $connected_status}]
28.            }
29.            if {[string first A $i_dout_name] != -1} {
30.                if {$is_connected == 1} {
31.                    puts "$i_bram: $i_dout_name is connected"
32.                    set doa_reg_status [get_property DOA_REG $i_bram]
33.                    set clka [get_clocks -of [get_pins $i_bram/CLKARDCLK]]
34.                    set clka_period [get_property PERIOD $clka]
35.                    set clka_freq [expr {1 / $clka_period * 1000.0}]
36.                } else {
37.                    puts "$i_bram: $i_dout_name is unconnected"
38.                    set doa_reg_status X
39.                    set clka_freq U
40.                }
41.            } else {
42.                if {$is_connected == 1} {
43.                    puts "$i_bram: $i_dout_name is connected"
44.                    set dob_reg_status [get_property DOB_REG $i_bram]
45.                    set clkb [get_clocks -of [get_pins $i_bram/CLKBWRCLK]]
46.                    set clkb_period [get_property PERIOD $clkb]
47.                    set clkb_freq [expr {1 / $clkb_period * 1000.0}]
48.                } else {
49.                    puts "$i_bram: $i_dout_name is unconnected"
50.                    set dob_reg_status X
51.                    set clka_freq U
52.                }
53.            }
54.        }
55.        if {$doa_reg_status == 0 || $dob_reg_status == 0} {
56.            puts $fid "$i_bram, $doa_reg_status, $clka_freq, $dob_reg_status, $clkb_freq"
57.            lappend bram_without_oreg $i_bram
58.        }
59.    }
60.    set num_bram [llength $bram_without_oreg]
61.    if {$num_bram > 0} {
62.        show_objects -name BRAM_REVIEW $bram_without_oreg
63.        puts "$num_bram BRAMs should be reviewed!"
64.    } else {
65.        puts "${fn}.csv is empty and can be deleted."
66.    }
67.    close $fid
68. }
69.
70. set family 7
71. set fn bram_review_0912
72. check_bram_oreg_status $family $fn
```

	A	B	C	D	E
1	BRAM_NAME	DOA_REG	CLKA_FREQ(MHZ)	DOB_REG	CLKB_FREQ(MHZ)
2	u_f3_role_top/u_role_mem_if/u1_ddr	0	150.060024	0	150.060024
3	u_f3_role_top/u_role_mem_if/u1_ddr	0	150.060024	0	150.060024
4	u_f3_role_top/u_role_mem_if/u1_ddr	0	150.060024	0	150.060024
5	u_f3_role_top/u_role_mem_if/u1_ddr	0	150.060024	0	150.060024
6	u_f3_role_top/u_role_mem_if/u1_ddr	0	150.060024	0	150.060024
7	u_f3_role_top/u_role_mem_if/u1_ddr	0	150.060024	0	150.060024
8	u_f3_role_top/u_role_mem_if/u1_ddr	0	150.060024	0	150.060024

图 6-84

对于 UltraRAM，可借助 Tcl 代码 6-6 找到设计中未使用自带寄存器的 UltraRAM。该代码与 Tcl 代码 6-5 类似，这里不再赘述。

Tcl 代码 6-6

```
1.  #File: check_uram_oreg_status.tcl
2.  #fn: file name
3.  proc check_uram_oreg_status {fn} {
4.      set prim_type BLOCKRAM.URAM.*
5.      set dout_name [list DOUT_A DOUT_B]
6.      set fid [open ${fn}.csv w]
7.      puts $fid "URAM_NAME, OREG_A, OREG_B, CLK_FREQ(MHZ)"
8.
9.      set uram [get_cells -hierarchical -filter "PRIMITIVE_TYPE =~ $prim_type"]
10.     set uram_without_oreg {}
11.     foreach i_uram $uram {
12.         foreach i_dout_name $dout_name {
13.             set dout_pin [get_pins $i_uram/${i_dout_name}*]
14.             set is_connected 0
15.             foreach i_dout_pin $dout_pin {
16.                 set connected_status [get_property IS_CONNECTED $i_dout_pin]
17.                 set is_connected [expr {$is_connected || $connected_status}]
18.             }
19.             if {[string first A $i_dout_name] != -1} {
20.                 if {$is_connected == 1} {
21.                     puts "$i_uram: $i_dout_name is connected"
22.                     set doa_reg_status [get_property OREG_A $i_uram]
23.                 } else {
24.                     puts "$i_uram: $i_dout_name is unconnected"
25.                     set doa_reg_status X
26.                 }
27.             } else {
28.                 if {$is_connected == 1} {
29.                     puts "$i_uram: $i_dout_name is connected"
30.                     set dob_reg_status [get_property OREG_B $i_uram]
31.                 } else {
32.                     puts "$i_uram: $i_dout_name is unconnected"
33.                     set dob_reg_status X
34.                 }
35.             }
36.         }
37.         if {$doa_reg_status == FALSE || $dob_reg_status == FALSE} {
```

```
38.            set clk [get_clocks -of [get_pins $i_uram/CLK]]
39.            set clk_period [get_property PERIOD $clk]
40.            set clk_freq [expr {1 / $clk_period * 1000.0}]
41.            puts $fid "$i_uram, $doa_reg_status, $dob_reg_status, $clk_freq"
42.            lappend uram_without_oreg $i_uram
43.        }
44.    }
45.    set num_uram [llength $uram_without_oreg]
46.    if {$num_uram > 0} {
47.        show_objects -name BRAM_REVIEW $uram_without_oreg
48.        puts "$num_uram BRAMs should be reviewed!"
49.    } else {
50.        puts "${fn}.csv is empty and can be deleted."
51.    }
52.    close $fid
53. }
54.
55. set fn uram_review_0912
56. check_uram_oreg_status $fn
```

对于 FIFO，若其以 BRAM 实现，可借助 Tcl 代码 6-7 找到设计中未使用自带寄存器的 FIFO 单元。与 Tcl 代码 6-5 相比，这里的代码更简单，这是因为 FIFO 只有一个输出数据管脚。

Tcl 代码 6-7

```
1.  #File: check_fifo_oreg_status.tcl
2.  set family us
3.  if {$family == 7} {
4.      set prim_type BMEM.fifo.*
5.  } else {
6.      set prim_type BLOCKRAM.FIFO.*
7.  }
8.  set myfifo [get_cells -hier -filter "PRIMITIVE_TYPE =~ $prim_type"]
9.  set myfifo_without_oreg {}
10. if {[llength $myfifo] > 0} {
11.     foreach i_myfifo $myfifo {
12.         set oreg_status [get_property DO_REG $i_myfifo]
13.         if {$oreg_status == 0} {
14.             lappend myfifo_without_oreg $i_myfifo
15.         }
16.     }
17. }
18. if {[llength $myfifo_without_oreg] > 0} {
19.     show_objects $myfifo_without_oreg -name FIFO_REVIEW
20. } else {
21.     puts "All FIFOs use embedded registers"
22. }
```

LUTRAM、BRAM 和 UltraRAM 都可用于存储数据，那么如何判定采用哪种资源是最佳选择呢？数据规模是优先要考虑的重要因素。当数据规模小于 8Kb 时，优先选用 LUTRAM。时钟频率也是一个重要因素，以 UltraScale+ FPGA 为例，对于 Virtex 速度等级为-1 的芯片，BRAM 所能运行的最高时钟频率为 645MHz，而 UltraRAM 所能运行的最高时钟频率为 575MHz。

6.13 思考空间

1. 对于深度为 512、宽度为 72 的单端口 RAM，若采用 LUTRAM 实现，会消耗多少个 LUTRAM？若采用 BRAM 实现，会消耗多少个 BRAM？若采用 UltraRAM 实现，会消耗多少个 UltraRAM？这三种方案相比较，哪种实现方式最优？

2. 试利用原语 RAM64x1S 实现一个深度为 64、宽度为 4 的单端口 RAM，并进行功能仿真，观察读/写地址相同时的输出数据，判断其是读优先、写优先还是保持模式。

3. 对于一个 4K×72 的简单双端口 RAM，读/写时钟彼此独立，能否采用 UltraRAM 实现？若能，请给出实现代码，若不能，请解释原因。

4. 某设计中的 FIFO 写数据位宽为 18，读数据位宽为 9，需要调用 FIFO Generator IP，其目标芯片能否为 7 系列 FPGA？并给出原因。

5. 请分别用 Block Memory Generator IP、XPM_MEMORY 和 RTL 代码三种方式实现一个 4K×16 的简单双端口 RAM，通过比较，给出这三种实现方式的优缺点。

6. 请分别用 URAM288 原语、XPM_MEMORY 和 RTL 代码三种方式实现一个 8K×72 的单端口 RAM，通过比较，给出这三种实现方式的优缺点。

7. 对于 5K×17 的单端口 RAM，若采用 BRAM 实现，试给出最小面积算法和低功耗算法的实现方式。

8. 对于一个深度为 32K、宽度为 72 的简单双端口 RAM，若采用 UltraRAM 实现，时钟频率为 400MHz，则 Latency 设置为多少较为合适？

9. 对于 VHDL 代码 6-28 和 VHDL 代码 6-30 所描述的同步 FIFO，试在此基础上增加输出信号 almost_full 和 almost_empty，分别用于判定 FIFO 将近满（FIFO 当前可写入的地址空间的个数是 FIFO 深度的 1/4）和 FIFO 将近空（FIFO 当前可供读取的数据个数是 FIFO 深度的 1/4）。

10. 对于 VHDL 代码 6-34，试在此基础上增加输出信号 almost_full，同时要求 FIFO 将近满的判定标准是参数化的。

11. 对于 VHDL 代码 6-34，试在此基础上增加输出信号 almost_empty，同时要求 FIFO 将近空的判定标准是参数化的。

12. 试解释 VHDL 代码 6-33 中综合属性 ASYNC_REG 的含义，并说明其作用对象与应用场合。

13. 试给出 VHDL 代码 6-35 中的参数 RAM_STYLE_VAL 的可选值，并比较在其他参数保持不变的情况下，不同 RAM_STYLE_VAL 值对应的资源利用率与时序性能。

14. 发现某时序路径上的 BRAM 未使用自带寄存器，如何快速检验使用自带寄存器对时序性能改善的力度？

第 7 章

优化乘加运算单元

7.1 乘加器资源

7.1.1 7 系列 FPGA 中的乘加器资源

7 系列 FPGA 中的乘加器是 DSP48E1，其主要端口如图 7-1 所示，简化结构如图 7-2 所示，其核心部分包括预加器（既可执行加法运算也可执行减法运算）、乘法器和算术逻辑单元（ALU）。主要输入/输出端口及模式配置端口的位宽与流水深度如表 7-1 所示。除此之外，还有专用级联端口，包括 ACIN/ACOUT、BCIN/BCOUT、PCIN/PCOUT、CARRYCASCIN/CARRYCASCOUT 和 MULTSIGNIN/MULTSIGNOUT（级联输入/级联输出）端口。同列且位于同一个 SLR（Super Logic Region）之内的 DSP48E1 可通过专用级联端口和专用布线资源实现级联。

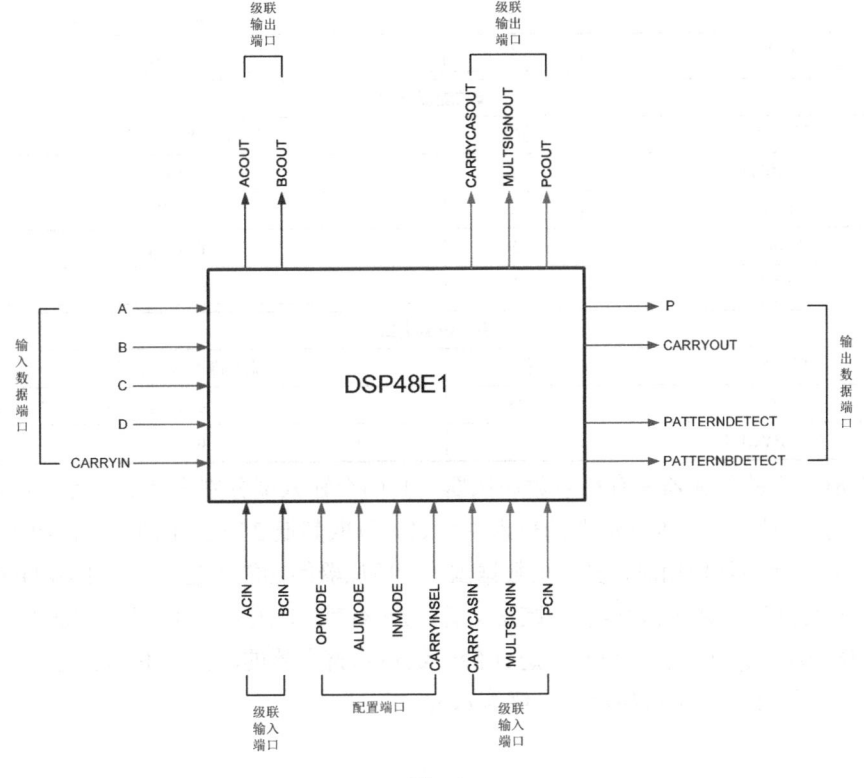

图 7-1

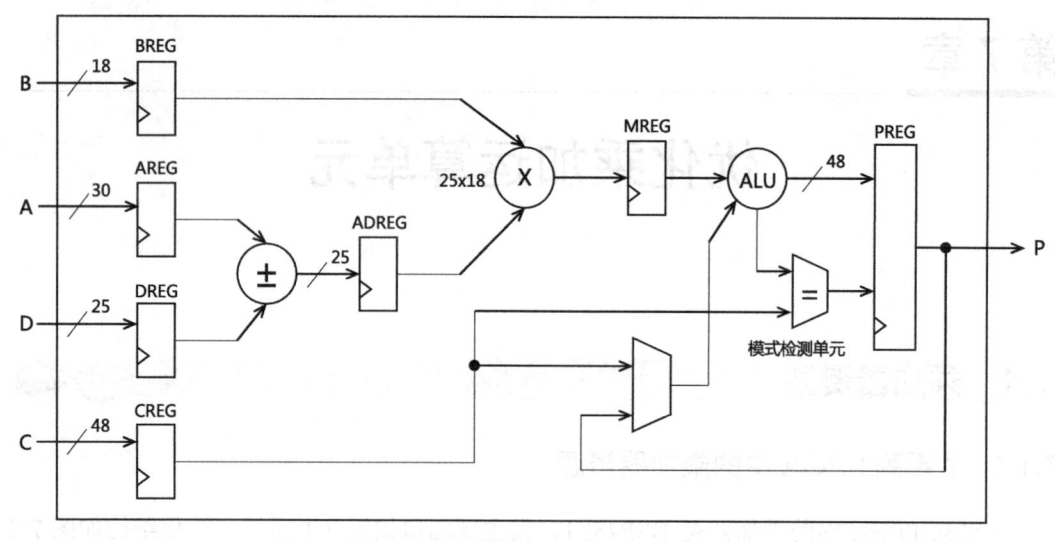

图 7-2

表 7-1

输入数据端口		
端口名称	位宽	流水深度（默认值）
A	30	0, 1, 2 (1)
B	18	0, 1, 2 (1)
C	48	0, 1 (1)
D	25	0, 1 (1)
CARRYIN	1	0, 1 (1)
模式配置端口		
端口名称	位宽	流水深度（默认值）
OPMODE	7	0, 1 (1)
ALUMODE	4	0, 1 (1)
INMODE	5	0, 1 (1)
CARRYINSEL	3	0, 1 (1)
输出数据端口		
端口名称	位宽	流水深度（默认值）
P	48	0, 1 (1)
CARRYOUT	4	0, 1 (1)

DSP48E1 中的预加器为有符号数加法器，其两个输入端分别为 A 端口和 D 端口。尽管 A 端口为 30 位，但作为预加器的输入端口时，只取其低 25 位。由于预加器的输出数据仍为 25 位，因此实际使用时，要防止数据溢出。预加器的功能由控制端口 INMODE 管理，可执行加法运算（$D+A_1$ 或 $D+A_2$）、减法运算（$D-A_1$ 或 $D-A_2$），同时可使预加器直接输出 A_1、A_2、D、0、$-A_1$ 或 $-A_2$（这里 A_1 是 A 经一级流水寄存器的输出，A_2 是 A 经二级流水寄存器的输出，简化结构中仅显示了一级流水）。

DSP48E1 中的乘法器是 25×18 的有符号数乘法器，对于无符号数，要先转换为有符号数后再使用。当使用乘法器时，从时序角度而言，尽可能使从输入到输出为三级流水，即与输入端口直接相连的输入寄存器 AREG 和 BREG；紧随乘法器输出端口的寄存器 MREG；ALU 输出端口寄存器 PREG。以 Virtex-7 为例，表 7-2 给出了使用 MREG 和未使用 MREG 时，DSP48E1 实现乘法运算所能达到的最高频率。由此可见，MREG 对设计性能的影响是很大的。

表 7-2

最高频率（MHz）	速度等级		
特征	-3	-2	-1
三级流水	741.84	650.2	547.95
二级流水（未使用 MREG）	412.2	360.75	303.77

💡 **设计规则 1**：当使用 DSP48E1 中的乘法器时，要确保 MREG 被使用。在高速设计中，要确保从输入到输出为三级流水。如果设计对从输入到输出的时钟周期个数（Latency）的要求为 2，那么在设计本身的 F_{max} 满足要求的前提下，应优先去除输入寄存器（AREG/BREG）或输出寄存器（PREG）。

DSP48E1 中的 ALU 为 48 位，可执行加/减法运算、累加运算和按位逻辑运算。同时，ALU 可工作在单指令多数据模式（Single Instruction Multiple Data，SIMD）下。在此模式下，ALU 可被配置为 4 个 12 位加法器/减法器/累加器、2 个 24 位加法器/减法器/累加器或 1 个 48 位加法器/减法器/累加器。数据源由 A:B（表示 A 端口和 B 端口通过位拼接构成 48 位数据，只有在这种情况下，A 端口的 30 位数据才全部有效）、C 端口、P 端口（构成累加器）或 PCIN（级联端口）提供。输出数据由 P 端口输出，具体如表 7-3 所示。ALU 还可执行按位逻辑运算，包括按位与、按位或、按位异或、按位同或、按位与非、按位或非等。

表 7-3

SIMD 模式	输出数据	进位输出端
4 个 12 位加法器	P[11:0]	CARRYOUT[0]
	P[23:12]	CARRYOUT[1]
	P[35:24]	CARRYOUT[2]
	P[47:36]	CARRYOUT[3]
2 个 24 位加法器	P[23:0]	CARRYOUT[1]
	P[47:24]	CARRYOUT[3]
1 个 48 位加法器	P[47:0]	CARRYOUT[3]

ALU 之所以具备如此丰富的功能，是因为其数据源来自 3 个数据选择器 X、Y 和 Z。这三个数据选择器受 OPMODE 的控制（OPMODE 位宽为 7）。其中，OPMODE 低两位控制 X，高 3 位控制 Z，中间两位控制 Y。同时，ALU 接收 ALUMODE 给的控制字（ALUMODE 位宽为 4），实现功能的切换。无论是 OPMODE 还是 ALUMODE，都可以实时更新，ALU 功能也可以实时动态切换。

模式检测单元执行的操作可表示为

$$PATTERNDETECT = \&((P{=}{=}PATTERN)\|MASK)$$

即按位比较 P 是否与 PATTERN 相等，其计算结果为 48 位数据，将此数据与 MASK 按位或计算，计算结果仍为 48 位，再将此 48 位数据执行与缩减运算，得到 1 位布尔类型数据，输出给 PATTERNDETECT。这里，PATTERN 可以来自 C 端口，或由内部 PATTERN 属性设置；同样地，MASK 可来自 C 端口，或由内部 MASK 属性设置。MASK 起屏蔽作用，即若不对 P 中的某一位或某几位进行模式检测，则将相应的 MASK 位置 "1"。模式检测电路还可执行操作

$$PATTERNBDETECT = \&((P{=}{=}{\sim}PATTERN)\|MASK)$$

式中，~PATTERN 表示对 PATTERN 按位取反。

DSP48E1 内部所有寄存器都有复位端口和时钟使能端口。复位高电平有效且仅支持同步复位。从 RTL 代码风格的角度而言，描述 DSP48E1 所支持的各种运算时，若需要对其内部寄存器复位，则要写成同步高电平有效，以保证工具能将其完全映射到 DSP48E1 内。

7.1.2 UltraScale/UltraScale+ FPGA 中的乘加器资源

UltraScale/UltraScale+ FPGA 中的乘加器为 DSP48E2，其简化结构如图 7-3 所示。与 DSP48E1 相比，预加器的位宽由 25 位变为 27 位，其两个输入数据一个来自 D 端口，另一个既可由 A 端口提供，也可由 B 端口提供。乘法器由 25×18 变为 27×18，结合预加器，很容易实现 $(A \pm D)^2$ 或 $(B \pm D)^2$。同时，新增了数据选择器 W 和异或运算单元 XOR。

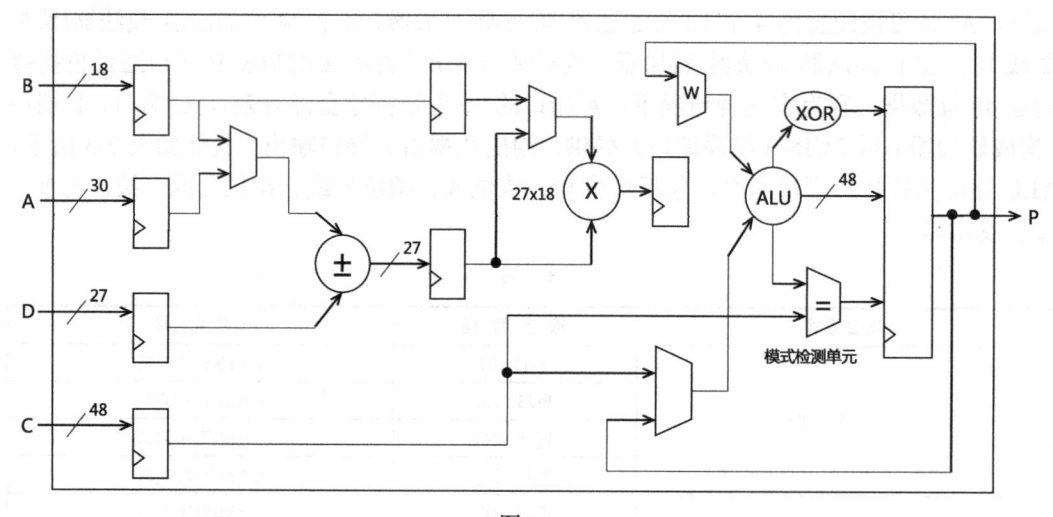

图 7-3

7.1.3 Versal ACAP 中的乘加器资源

Versal ACAP 中的乘加器为 DSP58，相比 DSP48E2，DSP58 的结构更复杂，但其功能也更强大。DSP58 有多种工作模式。当工作在标量模式下时，其简化结构如图 7-4 所示。这里，A 端口为 34 位，D 端口为 27 位，预加器输出为 27 位，乘法器为 27×24（两个操作数的位宽分别为 27 和 24），同时增加了控制端口 NEGATE，可将乘法器的输出取反。ALU 为 58 位。

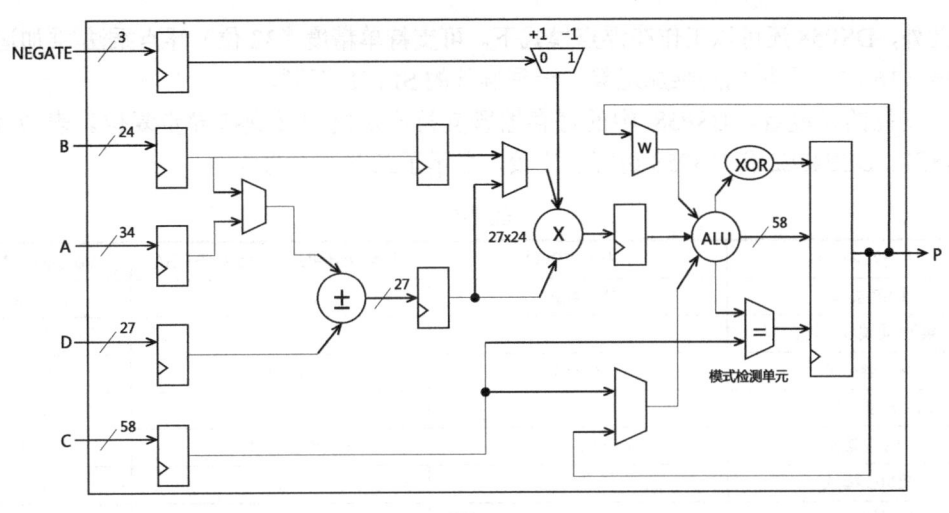

图 7-4

当其工作在向量模式下时，其简化结构如图 7-5 所示。可实现向量 $\boldsymbol{a}=\begin{bmatrix}a_0 & a_1 & a_2\end{bmatrix}$ 和向量 $\boldsymbol{b}=\begin{bmatrix}b_0 & b_1 & b_2\end{bmatrix}$ 的内积，即

$$\boldsymbol{a}\cdot\boldsymbol{b}=\pm a_0b_0\pm a_1b_1\pm a_2b_2$$

式中，向量 $\boldsymbol{a}$ 中每个元素的位宽为 9，向量 $\boldsymbol{b}$ 中每个元素的位宽为 8。

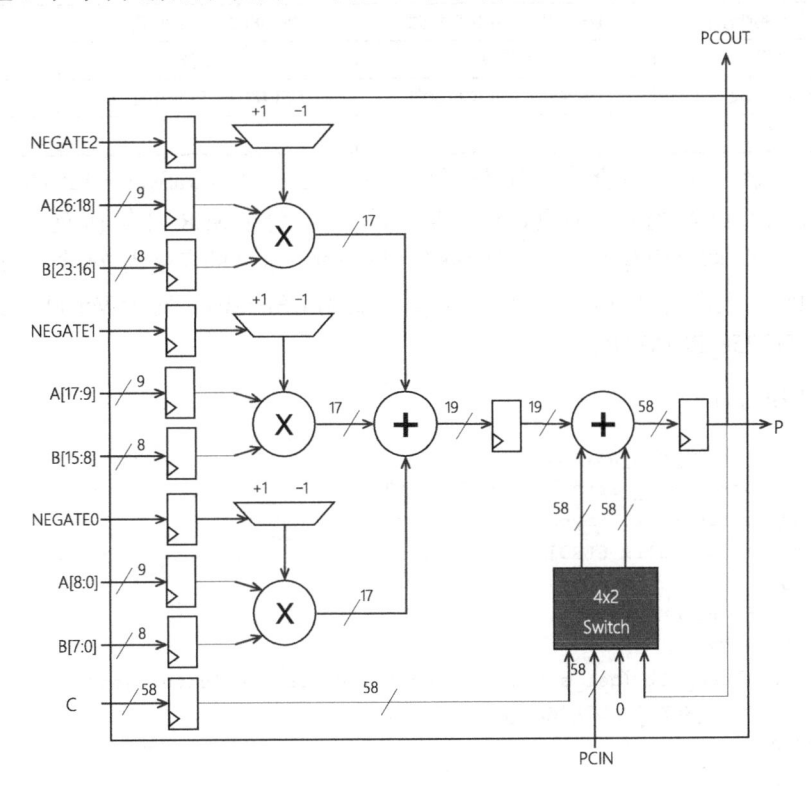

图 7-5

DSP58 可工作在复数乘法模式下，此时，两个 DSP58 即可实现 18×18（两个复数的实部和虚部均为 18 位）的复数运算。

此外，DSP58 还可以工作在浮点模式下。可支持单精度（32 位）浮点乘法/乘加运算、半精度（16 位）浮点乘法/乘加运算，无须额外的 SLICE 资源。

从复位的角度看，DSP58 中的寄存器既支持同步复位又支持异步复位。表 7-4 对 DSP48E1、DSP48E2 和 DSP58 进行了比较，显示了三者之间的差异。

表 7-4

比较内容	7 系列 FPGA	UltraScale/UltraScale+ FPGA	Versal ACAP
DSP 类型	DSP48E1	DSP48E2	DSP58
乘加/乘减/乘累加	√	√	√
乘法器位宽	25×18	27×18	27×24
$(A\pm D)^2$ 或 $(B\pm D)^2$	×	√	√
18×18 复数乘法	3 个 DSP48E1	3 个 DSP48E2	2 个 DSP58
SIMD 模式	√	√	√
模式检测电路	√	√	√
大位宽数据选择器	√	√	√
大位宽异或运算	×	96	116
专用级联布线资源	√	√	√
流水寄存器	√	√	√
累加器位宽	48	48	58
32/16 位浮点乘加运算	DSP48E1 + SLICE 资源	DSP48E2 + SLICE 资源	√
向量内积	软支持	软支持	√
乘法器输出取反	DSP48E1 + SLICE 资源	DSP48E2 + SLICE 资源	√
异步复位	×	×	√

采用 Tcl 代码 7-1 可以查看乘加器的分布状况。这里以 UltraScale+ FPGA 芯片为例，从代码第 38 行至第 44 行可以看到，对于 VU5P 而言，每个 SLR 内有 19 列 DSP48E2，每列的高度为 120，即每列有 120 个 DSP48E2，每个时钟区域的高度为 24 个 DSP48E2。对于 7 系列 FPGA，要将代码第 6 行的 site_type 设置为 DSP48E1。对于 Versal ACAP，则要将其设置为 DSP58_PRIMARY。

Tcl 代码 7-1

```
1.  #File: check_dsp_us.tcl
2.  set mypart [get_parts xcvu5p-flva2104-2-i]
3.  link_design -part $mypart
4.  set slr [get_slrs SLR0]
5.  => SLR0
6.  set site_type DSP48E2
7.  => DSP48E2
8.  set dsp_one_slr [get_sites -filter "SITE_TYPE == $site_type" -of $slr]
9.  => DSP48E2_X0Y118 DSP48E2_X0Y119 ...
10. set x {}
11. set y {}
12. foreach dsp_one_slr_i $dsp_one_slr {
13.     set loc [lindex [split $dsp_one_slr_i _] 1]
14.     set xy  [split $loc XY]
15.     lappend x [lindex $xy 1]
```

```
16.         lappend y [lindex $xy 2]
17.     }
18.     set col_index_slr [lsort -integer -unique $x]
19.     => 0 1 2 3 4 5 6 7 8 9 10 11 12 13 14 15 16 17 18
20.     set row_index_slr [lsort -integer -unique $y]
21.     => 0 1 2 3 4 5 6 7 8 9 10 11 12 13 14 15 16 17 18 19 ...
22.     set cr [get_clock_regions X0Y0]
23.     => X0Y0
24.     set dsps_one_cr [get_sites -filter "SITE_TYPE == $site_type" -of $cr]
25.     => DSP48E2_X0Y22 DSP48E2_X0Y23 DSP48E2_X1Y22 DSP48E2_X1Y23 ...
26.     set x {}
27.     set y {}
28.     foreach dsps_one_cr_i $dsps_one_cr {
29.         set loc [lindex [split $dsps_one_cr_i _] 1]
30.         set xy [split $loc XY]
31.         lappend x [lindex $xy 1]
32.         lappend y [lindex $xy 2]
33.     }
34.     set col_index [lsort -integer -unique $x]
35.     => 0 1 2 3 4
36.     set row_index [lsort -integer -unique $y]
37.     => 0 1 2 3 4 5 6 7 8 9 10 11 12 13 14 15 16 17 18 19 20 21 22 23
38.     puts "Number of $site_type columns in $slr : [llength $col_index_slr]"
39.     => Number of DSP48E2 columns in SLR0 : 19
40.     puts "Height of SLR by $site_type : [llength $row_index_slr]"
41.     => Height of SLR by DSP48E2 : 120
42.     puts "Number of $site_type columns in CR $cr : [llength $col_index]"
43.     => Number of DSP48E2 columns in CR X0Y0 : 5
44.     puts "Height of CR by $site_type : [llength $row_index]"
45.     => Height of CR by DSP48E2 : 24
```

7.2 以乘法为核心运算的代码风格

乘法运算可分为有符号数乘法和无符号数乘法，且无符号数乘法可转化为有符号数乘法，因此这里以有符号数乘法运算为例进行阐述。同时，考虑到 DSP48E1、DSP48E2 和 DSP58 结构的相似性，若无特殊声明，所示代码均适用于这三种乘加器，只是需要注意，这三者的输入数据和输出数据的位宽有所不同。这里以 DSP48E2 为例进行说明。本节介绍的运算如表 7-5 所示。

表 7-5

功能	备注	代码	适用对象 DSP48E1	适用对象 DSP48E2	适用对象 DSP58
$A \times B$		VHDL 代码 7-1	√	√	√
$A \times B + C$	可改写为 $A \times B - C$，$C - A \times B$	VHDL 代码 7-2	√	√	√
$A \times B + P$	可改写为 $A \times B - P$	VHDL 代码 7-3	√	√	√
$(A+D) \times B$	可改写为 $(D-A) \times B$	VHDL 代码 7-4	√	√	√

续表

功 能	备 注	代 码	适 用 对 象		
			DSP48E1	DSP48E2	DSP58
$(A\pm D)\times B$	A 恒接地,可实现 $(\pm D)\times B$	VHDL 代码 7-5	√	√	√
$(A+D)^2$		VHDL 代码 7-6		√	√
$(A\pm D)^2$		VHDL 代码 7-7		√	√
$(A+D)^2+P$	可改写为 $(A-D)^2+P$	VHDL 代码 7-8		√	√
$-(A\times B)$		VHDL 代码 7-9			√
$C\pm(A\times B)$	动态控制乘积取反	VHDL 代码 7-10			√
$A\times B$ (27×19)	可改写为 28×18	VHDL 代码 7-11	25×19 或 26×18	27×19 或 28×18	27×25 或 28×24

> 💡 **设计规则 2**：一个 M 位有符号定点数和一个 N 位有符号定点数相乘，其乘积的位宽为 $M+N$。

仅就乘法运算而言，考虑到 DSP48E2 的架构及系统性能，通常需要三级流水，如图 7-6 所示。乘法器输入端有一级流水寄存器，输出端有二级流水寄存器，其实，最后一级流水寄存器位于 ALU 输出端。对 DSP48E2 而言，计算 $A\times B$，本质上计算的是 $A\times B+0$。采用 VHDL 语言描述，如 VHDL 代码 7-1 所示。这里要特别注意：输入数据和乘积结果要转换为 signed 类型，如代码第 30 行至第 32 行所示，否则，Vivado 不会将其映射为 DSP48E2。此代码定义了三个参数：AW 为输入数据 ain 的位宽，BW 为输入数据 bin 的位宽，MW 等于 AW 和 BW 之和，是乘积位宽。默认值为单个 DSP48E2 实现乘法运算时可支持的最大位宽。代码使用了时钟使能信号来控制输入/输出数据的传输速率，由于 DSP48E2 内部寄存器支持时钟使能，因此这段代码可完全映射为 DSP48E2 而不会消耗额外的逻辑资源。输入/输出信号之间的时序关系如图 7-7 所示。可以看出，当时钟使能信号 ce 恒为高电平时，从输入到输出需要 3 个时钟周期。

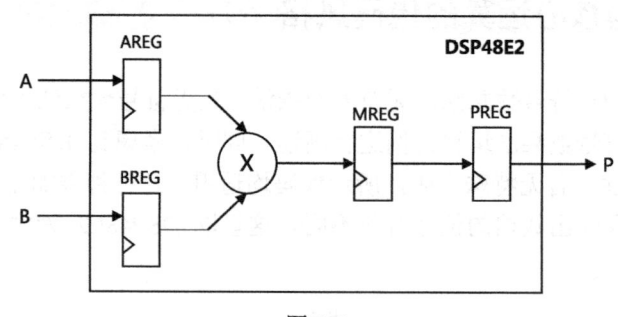

图 7-6

📄 **VHDL 代码 7-1**

```
1.  --File: basic_mult.vhd
2.  library ieee;
3.  use ieee.std_logic_1164.all;
4.  use ieee.numeric_std.all;
5.
```

```vhdl
6.  entity basic_mult is
7.    generic (
8.      AW : integer := 27;
9.      BW : integer := 18;
10.     MW : integer := AW + BW
11.   );
12.   port (
13.     clk  : in std_logic;
14.     ce   : in std_logic;
15.     ain  : in std_logic_vector(AW - 1 downto 0);
16.     bin  : in std_logic_vector(BW - 1 downto 0);
17.     prod : out std_logic_vector(MW - 1 downto 0)
18.   );
19. end entity;
20.
21. architecture rtl of basic_mult is
22. signal ain_d1 : signed(ain'RANGE);
23. signal bin_d1 : signed(bin'RANGE);
24. signal mreg   : signed(prod'RANGE);
25. begin
26.   process(clk)
27.   begin
28.     if rising_edge(clk) then
29.       if ce then
30.         ain_d1 <= signed(ain);
31.         bin_d1 <= signed(bin);
32.         mreg   <= ain_d1 * bin_d1;
33.         prod   <= std_logic_vector(mreg);
34.       end if;
35.     end if;
36.   end process;
37. end architecture;
```

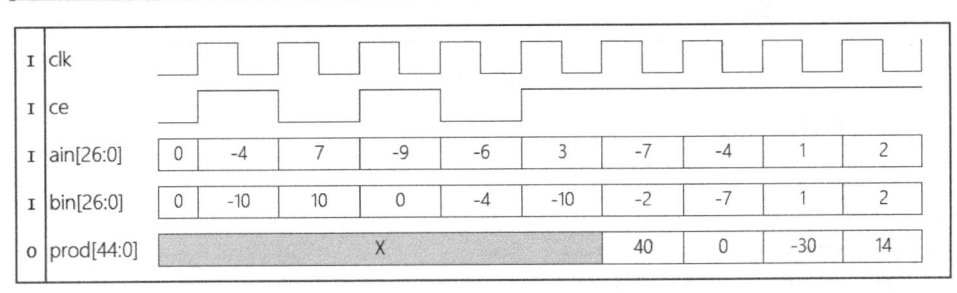

I	clk
I	ce
I	ain[26:0]	0	-4	7	-9	-6	3	-7	-4	1	2
I	bin[26:0]	0	-10	10	0	-4	-10	-2	-7	1	2
O	prod[44:0]	X					40	0	-30	14

图 7-7

🔍 VHDL-2008 新特性

对于 if 或 when-else 语句中的条件表达式，VHDL-2008 允许其返回值的数据类型为 boolean 或 std_logic，如 VHDL 代码 7-1 第 29 行所示。在 VHDL-87 版或 VHDL-93 版中，该行应写为 if ce='1' then。

> 💡 **设计规则 3**：使用 VHDL 描述算术运算（如加法、减法和乘法运算）时，需要将数据类型声明为 unsigned 或 signed，但 entity 的输入/输出端口仍然使用 std_logic_vector 类型，这是因为 Xilinx IP 的输入/输出大多为 std_logic_vector 类型，而 VHDL 又是强类型语言，只有同一类型才可相互赋值。声明为 std_logic_vector 类型便于与其他模块相连接。

由于 DSP48E2 内部不仅有乘法器，还有加法器，因此可以很方便地实现乘加运算，如图 7-8 所示。由于从 A/B 端口到加法器输入端需要二级流水寄存器，因此 C 端口也需要二级流水寄存器以实现数据对齐，但 DSP48E2 C 端口内部仅有一级流水寄存器，因此需要在其外部再添加一级流水寄存器，如 VHDL 代码 7-2 所示。这里使用了同步复位和时钟使能，DSP48E2 内部寄存器是支持的，因此这段代码最终会映射为 DSP48E2，同时会消耗 48 个寄存器。代码第 49 行表示的 mreg+cin_d2，也可写为 mreg-cin_d2 或 cin_d2-mreg，即分别实现 $A×B-C$ 和 $C-A×B$，均可映射为 DSP48E2。注意这里都要使用 resize 函数对 mreg 和 cin_d2 进行符号位扩展，这是因为代码第 8 行至第 11 行中指定的参数值是映射为 DSP48E2 时所允许的最大值，实际使用时可能会出现 CW 与 MW（AW+BW）不相等的情形，resize 可使加号左右两侧及二者之和的位宽相等，从而满足 VHDL 的语法要求（同类型、同位宽可赋值）。输入/输出信号之间的时序关系如图 7-9 所示。

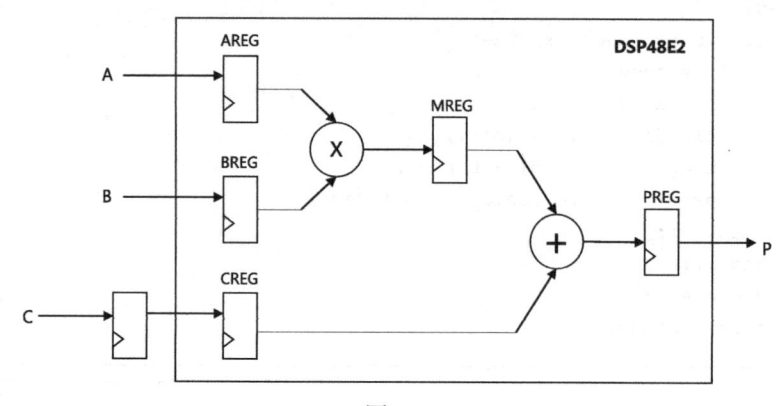

图 7-8

📄 **VHDL 代码 7-2**

```
1.  --File: mult_add_v2.vhd
2.  library ieee;
3.  use ieee.std_logic_1164.all;
4.  use ieee.numeric_std.all;
5.
6.  entity mult_add_v2 is
7.    generic (
8.      AW : integer := 27;
9.      BW : integer := 18;
10.     CW : integer := 48;
11.     PW : integer := 48
12.   );
13.   port (
14.     clk : in std_logic;
```

```vhdl
15.        rst   : in std_logic;
16.        ce    : in std_logic;
17.        ain   : in std_logic_vector(AW - 1 downto 0);
18.        bin   : in std_logic_vector(BW - 1 downto 0);
19.        cin   : in std_logic_vector(CW - 1 downto 0);
20.        pout  : out std_logic_vector(PW - 1 downto 0)
21.    );
22. end entity;
23.
24. architecture rtl of mult_add_v2 is
25.    constant MW : integer := AW + BW;
26.    signal ain_d1           : signed(ain'RANGE);
27.    signal bin_d1           : signed(bin'RANGE);
28.    signal cin_d1, cin_d2   : signed(cin'RANGE);
29.    signal mreg             : signed(MW - 1 downto 0);
30.    signal pout_i           : signed(PW - 1 downto 0);
31. begin
32.    pout <= std_logic_vector(pout_i);
33.    process(clk)
34.    begin
35.        if rising_edge(clk) then
36.            if rst then
37.                ain_d1 <= (others => '0');
38.                bin_d1 <= (others => '0');
39.                cin_d1 <= (others => '0');
40.                cin_d2 <= (others => '0');
41.                mreg   <= (others => '0');
42.                pout_i <= (others => '0');
43.            elsif ce then
44.                ain_d1 <= signed(ain);
45.                bin_d1 <= signed(bin);
46.                cin_d1 <= signed(cin);
47.                cin_d2 <= signed(cin_d1);
48.                mreg   <= ain_d1 * bin_d1;
49.                pout_i <= resize(mreg, PW) + resize(cin_d2, PW);
50.            end if;
51.        end if;
52.    end process;
53. end architecture;
```

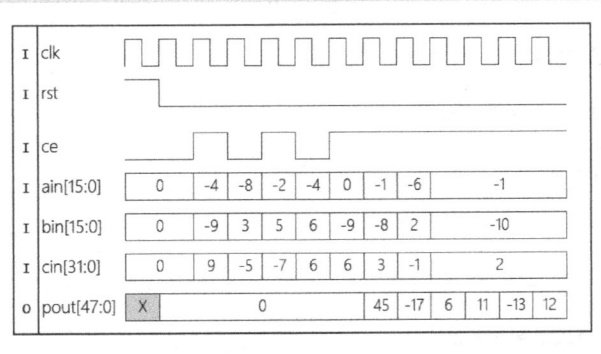

I	clk									
I	rst									
I	ce									
I	ain[15:0]	0	-4	-8	-2	-4	0	-1	-6	-1
I	bin[15:0]	0	-9	3	5	6	-9	-8	2	-10
I	cin[31:0]	0	9	-5	-7	6	6	3	-1	2
O	pout[47:0]	X		0			45	-17	6	11	-13	12

图 7-9

> **设计规则 4**：DSP48E1 和 DSP48E2 内部寄存器仅支持同步复位，因此，使用 RTL 代码描述乘法或乘加、乘累加运算时，若需要复位，则应采用同步复位，以保证工具能将相关寄存器映射到 DSP48E1 或 DSP48E2 内部。

由于 ALU 还支持累加功能，因此 DSP48E2 也可以方便地实现乘累加功能。如果数据均为 18 位有符号数，那么其乘积就是 36 位有符号数，而 DSP48E2 累加器输出为 48 位，这样可以实现 2^{12} 也就是 4096 次累加。乘累加器的电路图如图 7-10 所示。图中，sload 为同步加载信号，当其值为 1 时，数据选择器输出为 0，意味着 DSP48E2 执行 $A \times B + 0$，这其实就是待累加的第一个数据；当其值为 0 时，数据选择器输出为 P，意味着 DSP48E2 执行 $A \times B + P$。采用 VHDL 代码描述，如 VHDL 代码 7-3 所示。最终会消耗一个 DSP48E2 和一个触发器（sload 端口的寄存器）。而实际上，在物理优化阶段（phys_opt_design），Vivado 会根据时序需求，把该寄存器移到 DSP48E2 内部，映射为 OPMODEREG（模式控制字 OPMODE 端口的寄存器）。图 7-10 中的数据选择器也会映射为 DSP48E2 内部的数据选择器。如果要实现 $A \times B - P$ 或 $P - A \times B$，即乘累减，只需要把代码第 40 行改写为 mreg-old_res 或 old_res-mreg 即可。

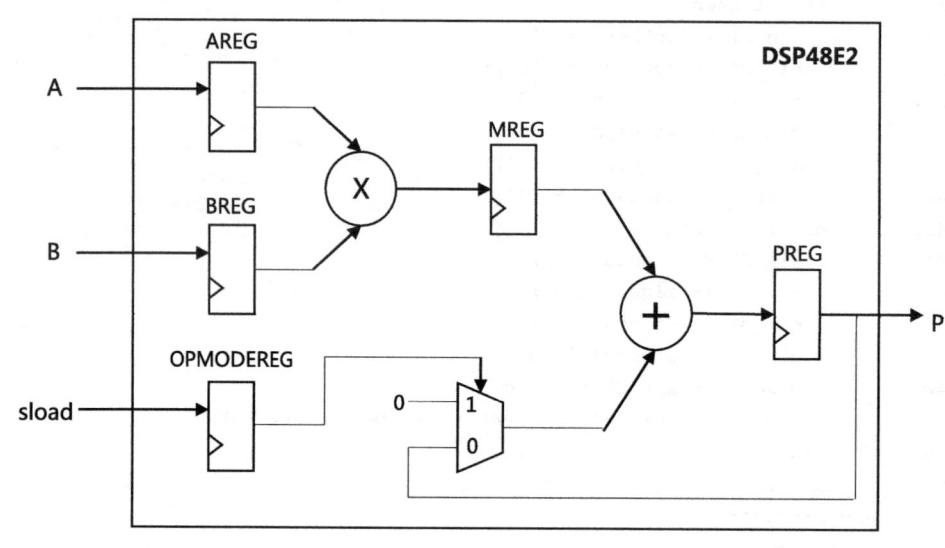

图 7-10

VHDL 代码 7-3

```
1.  --File: mult_acc_v2.vhd
2.  library ieee;
3.  use ieee.std_logic_1164.all;
4.  use ieee.numeric_std.all;
5.
6.  entity mult_acc_v2 is
7.    generic (
8.      AW : positive := 27;
```

```vhdl
9.      BW : positive := 18;
10.     PW : positive := 48
11.   );
12.   port (
13.     clk   : in std_logic;
14.     sload : in std_logic; --sync load
15.     ain   : in std_logic_vector(AW - 1 downto 0);
16.     bin   : in std_logic_vector(BW - 1 downto 0);
17.     pout  : out std_logic_vector(PW - 1 downto 0)
18.   );
19. end entity;
20.
21. architecture rtl of mult_acc_v2 is
22.   constant MW : positive := AW + BW;
23.   signal sload_d1   : sload'subtype;
24.   signal ain_d1     : signed(ain'RANGE);
25.   signal bin_d1     : signed(bin'RANGE);
26.   signal mreg       : signed(MW - 1 downto 0);
27.   signal adder_out  : signed(PW - 1 downto 0);
28.   signal old_res    : signed(PW - 1 downto 0);
29. begin
30.   old_res <= (others => '0') when sload_d1 else adder_out;
31.   pout    <= std_logic_vector(adder_out);
32.
33.   process(clk)
34.   begin
35.     if rising_edge(clk) then
36.       sload_d1  <= sload;
37.       ain_d1    <= signed(ain);
38.       bin_d1    <= signed(bin);
39.       mreg      <= ain_d1 * bin_d1;
40.       adder_out <= old_res + resize(mreg, PW);
41.     end if;
42.   end process;
43. end architecture;
```

VHDL 代码 7-3 的输入/输出信号之间的时序关系如图 7-11 所示。图中，sload 周期为 4，意味着每帧数据长度为 4，可理解为 DSP48E2 将执行两个长度为 4 的向量的内积。这里需要注意 sload 和起始数据的时序关系，即两者相差一个时钟周期。图中，pout 输出数据-26（pout 对应灰色部分）为第一帧数据的内积，119 为第二帧数据的内积。可用带时钟使能的寄存器捕获内积，时钟使能信号为 sload_d1（sload 延迟一个时钟周期的结果，如代码第 36 行所示）。

DSP48E2 内部还有预加器，位于乘法器之前，这样可实现先加再乘的操作，这在系数对称的 FIR 滤波器设计中非常有用。电路结构如图 7-12 所示，计算 $(A+D) \times B$。采用 VHDL 描述，如 VHDL 代码 7-4 所示。这里使用了同步复位和时钟使能信号，与 DSP48E2 架构匹配。从输入到输出需要 4 个时钟周期，即 Latency 为 4。

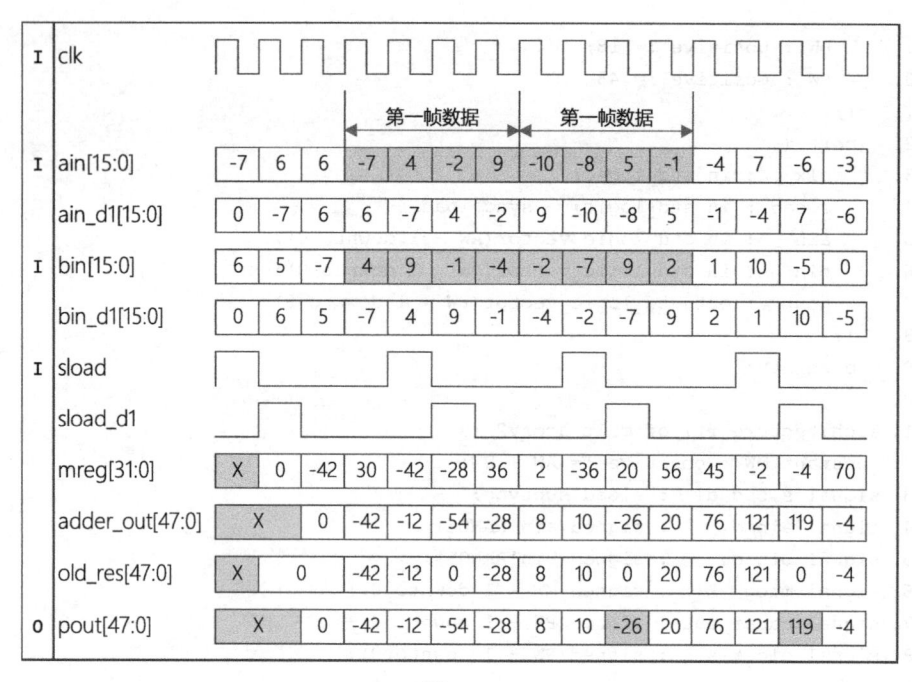

图 7-11

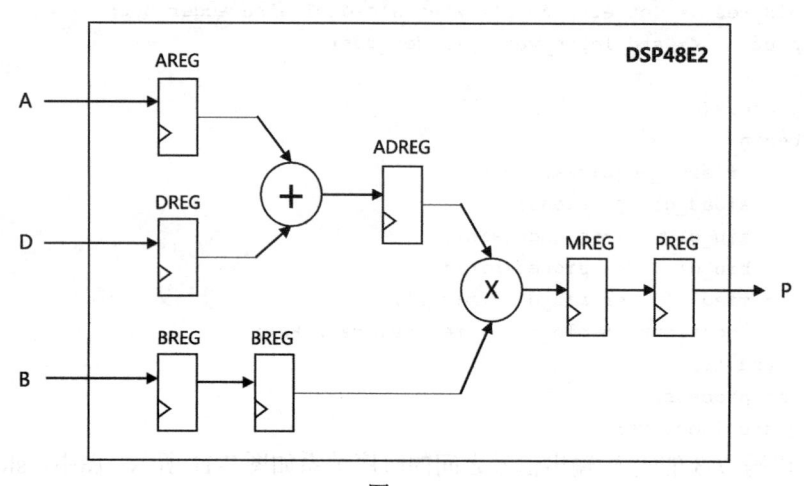

图 7-12

VHDL 代码 7-4

```
1.   --File: preadd_mult.vhd
2.   library ieee;
3.   use ieee.std_logic_1164.all;
4.   use ieee.numeric_std.all;
5.
6.   entity preadd_mult is
7.     generic (
8.       AW : positive := 16;
9.       BW : positive := 18;
10.      MW : positive := AW + 1 + BW
11.    );
12.    port (
```

```vhdl
13.      clk   : in std_logic;
14.      rst   : in std_logic;
15.      ce    : in std_logic;
16.      ain   : in std_logic_vector(AW - 1 downto 0);
17.      din   : in std_logic_vector(AW - 1 downto 0);
18.      bin   : in std_logic_vector(BW - 1 downto 0);
19.      pout  : out std_logic_vector(MW - 1 downto 0)
20.   );
21. end entity;
22.
23. architecture rtl of preadd_mult is
24.   signal ain_d1  : signed(ain'RANGE);
25.   signal din_d1  : signed(din'RANGE);
26.   signal bin_d1  : signed(bin'RANGE);
27.   signal bin_d2  : signed(bin'RANGE);
28.   signal add_reg : signed(AW downto 0);
29.   signal mreg    : signed(MW - 1 downto 0);
30. begin
31.   process(clk)
32.   begin
33.     if rising_edge(clk) then
34.       if rst then
35.         ain_d1  <= (others => '0');
36.         din_d1  <= (others => '0');
37.         bin_d1  <= (others => '0');
38.         bin_d2  <= (others => '0');
39.         add_reg <= (others => '0');
40.         mreg    <= (others => '0');
41.         pout    <= (others => '0');
42.       elsif ce then
43.         ain_d1  <= signed(ain);
44.         din_d1  <= signed(din);
45.         bin_d1  <= signed(bin);
46.         bin_d2  <= bin_d1;
47.         add_reg <= resize(ain_d1, AW + 1) + resize(din_d1, AW + 1);
48.         mreg    <= add_reg * bin_d2;
49.         pout    <= std_logic_vector(mreg);
50.       end if;
51.     end if;
52.   end process;
53. end architecture;
```

🔍 VHDL-2008 新特性

VHDL-2008 支持在 generic 中引用已定义参数，如 VHDL 代码 7-4 第 10 行所示，参数 MW 依赖 AW 和 BW。这种写法要求在 MW 之前声明 AW 和 BW。

DSP48E2 内部的预加器既可以实现加法运算，也可以实现减法运算，且可以通过外部信号动态控制预加器功能，从而实现 $(A \pm D) \times B$，其电路结构如图 7-13 所示。从输入到输出需要 4 个时钟周期，即 Latency 为 4。采用 VHDL 描述，如 VHDL 代码 7-5 所示。

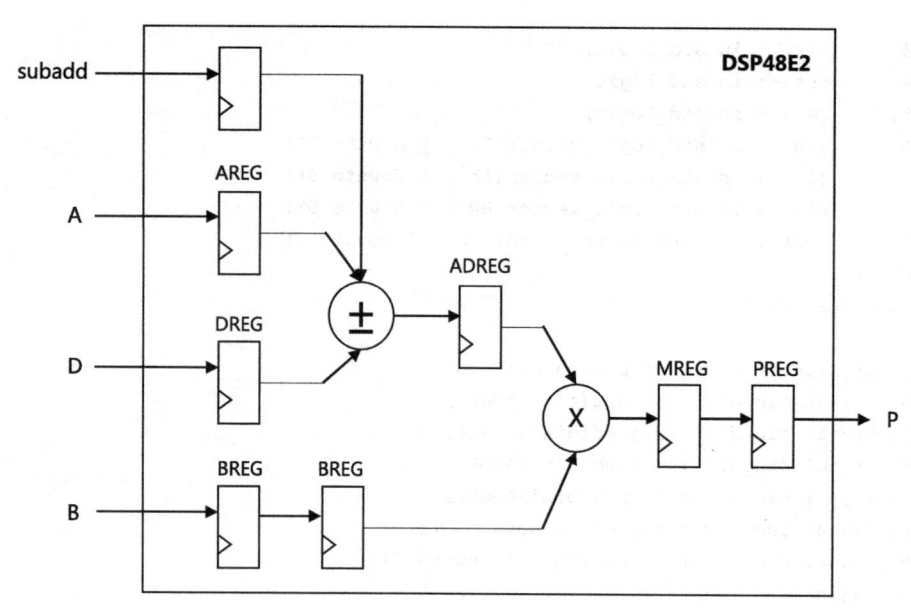

图 7-13

> **VHDL 代码 7-5**

```vhdl
1.  --File: dynamic_preadd_mult_v2.vhd
2.  --P = (A±D)XB
3.  --subadd 1 : A-D; 0 : A+D
4.  library ieee;
5.  use ieee.std_logic_1164.all;
6.  use ieee.numeric_std.all;
7.
8.  entity dynamic_preadd_mult_v2 is
9.    generic (
10.     AW : positive := 16;
11.     BW : positive := 16;
12.     MW : positive := AW + 1 + BW
13.   );
14.   port (
15.     clk    : in std_logic;
16.     subadd : in std_logic;
17.     ain    : in std_logic_vector(AW - 1 downto 0);
18.     din    : in std_logic_vector(AW - 1 downto 0);
19.     bin    : in std_logic_vector(BW - 1 downto 0);
20.     pout   : out std_logic_vector(MW - 1 downto 0)
21.   );
22. end entity;
23.
24. architecture rtl of dynamic_preadd_mult_v2 is
25. signal subadd_d1 : subadd'subtype;
26. signal ain_d1    : signed(ain'RANGE);
27. signal din_d1    : signed(din'RANGE);
28. signal bin_d1    : signed(bin'RANGE);
29. signal bin_d2    : signed(bin'RANGE);
30. signal add_reg   : signed(AW downto 0);
```

```
31. signal MREG         : signed(MW - 1 downto 0);
32. signal sum          : signed(add_reg'RANGE);
33. signal sub          : signed(add_reg'RANGE);
34. begin
35.     sum <= resize(ain_d1, AW + 1) + resize(din_d1, AW + 1);
36.     sub <= resize(ain_d1, AW + 1) - resize(din_d1, AW + 1);
37.     process(clk)
38.     begin
39.       if rising_edge(clk) then
40.         subadd_d1 <= subadd;
41.         ain_d1    <= signed(ain);
42.         din_d1    <= signed(din);
43.         bin_d1    <= signed(bin);
44.         bin_d2    <= bin_d1;
45.         add_reg   <= sub when subadd_d1 else sum;
46.         mreg      <= add_reg * bin_d2;
47.         pout      <= std_logic_vector(mreg);
48.       end if;
49.     end process;
50. end architecture;
```

对于 DSP48E2，由于 A 端口和 B 端口均可作为预加器的输入端口，因此可以方便地实现 $(D \pm A)^2$、$(D \pm B)^2$、$(D \pm A) \times A$、$(D \pm A) \times B$、$(D \pm B) \times B$ 和 $(D \pm B) \times A$。这里以两个数之差的平方（适用于 DSP48E2 和 DSP58）为例进行说明，相应的电路架构如图 7-14 所示。图中，加法器执行的是 A-D 的功能。由于 DSP48E2 内部乘法器是 27×18，因此预加器输出数据最大位宽为 18。采用 VHDL 描述，如 VHDL 代码 7-6 所示，从输入到输出的 Latency 为 4。

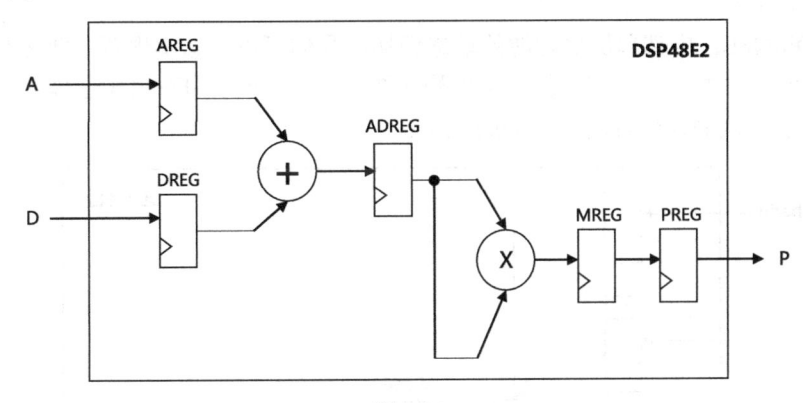

图 7-14

VHDL 代码 7-6

```
1. --File: square.sv
2. --input data width: W
3. --pre-adder output width: W+1
4. --Mult output width: 2(W+1)
5. --For DSP48E2: W+1 <= 18 => W < 17
6. --For DSP58: W+1 <= 24 => W < 23
7. library ieee;
```

```vhdl
8.  use ieee.std_logic_1164.all;
9.  use ieee.numeric_std.all;
10.
11. entity square is
12.   generic ( W : positive := 17 );
13.   port (
14.     clk  : in std_logic;
15.     ain  : in std_logic_vector(W - 1 downto 0);
16.     bin  : in std_logic_vector(W - 1 downto 0);
17.     pout : out std_logic_vector(2 * W + 1 downto 0)
18.   );
19. end entity;
20.
21. architecture rtl of square is
22. signal ain_d1, bin_d1 : signed(W - 1 downto 0);
23. signal diff           : signed(W downto 0);
24. signal mreg           : signed(2 * W + 1 downto 0);
25. begin
26.   process(clk)
27.   begin
28.     if rising_edge(clk) then
29.       ain_d1 <= signed(ain);
30.       bin_d1 <= signed(bin);
31.       diff   <= resize(ain_d1, W + 1) - resize(bin_d1, W + 1);
32.       mreg   <= diff * diff;
33.       pout   <= std_logic_vector(mreg);
34.     end if;
35.   end process;
36. end architecture;
```

基于 DSP48E2 内部预加器功能的动态切换，我们可以动态实现两个数之和的平方与两个数之差的平方的计算，相应的电路架构如图 7-15 所示。采用 VHDL 描述，如 VHDL 代码 7-7 所示，从输入到输出的 Latency 为 4。

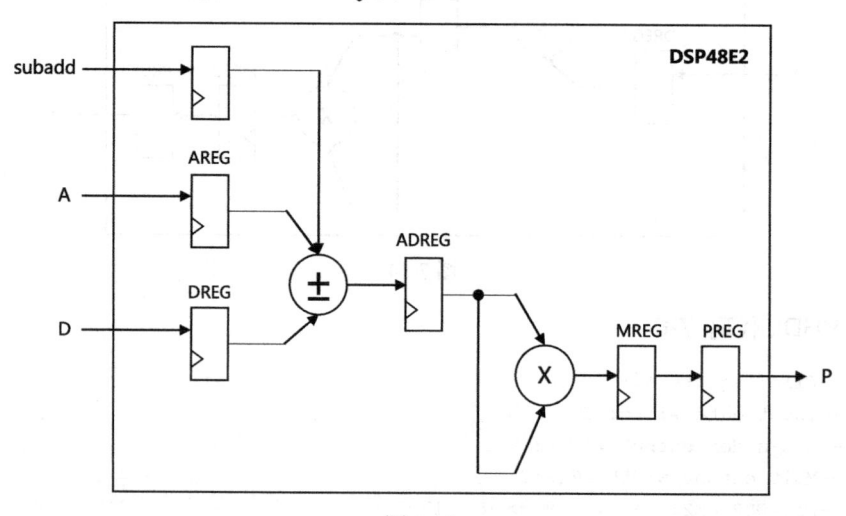

图 7-15

VHDL 代码 7-7

```vhdl
1.  --File: dynamic_square.sv
2.  --input data width: W
3.  --pre-adder output width: W+1
4.  --Mult output width: 2(W+1)
5.  --For DSP48E2: W+1 <= 18 => W < 17
6.  --For DSP58: W+1 <= 24 => W < 23
7.  library ieee;
8.  use ieee.std_logic_1164.all;
9.  use ieee.numeric_std.all;
10.
11. entity dynamic_square is
12.   generic ( W : positive := 17 );
13.   port (
14.     clk    : in std_logic;
15.     subadd : in std_logic;
16.     ain    : in std_logic_vector(W - 1 downto 0);
17.     bin    : in std_logic_vector(W - 1 downto 0);
18.     pout   : out std_logic_vector(2 * W + 1 downto 0)
19.   );
20. end entity;
21.
22. architecture rtl of dynamic_square is
23.   signal subadd_d1      : std_logic;
24.   signal ain_d1, bin_d1 : signed(W - 1 downto 0);
25.   signal diff           : signed(W downto 0);
26.   signal mreg           : signed(2 * W + 1 downto 0);
27.   signal sum, sub       : signed(W downto 0);
28. begin
29.   sum <= resize(ain_d1, W + 1) + resize(bin_d1, W + 1);
30.   sub <= resize(ain_d1, W + 1) - resize(bin_d1, W + 1);
31.   process(clk)
32.   begin
33.     if rising_edge(clk) then
34.       subadd_d1 <= subadd;
35.       ain_d1 <= signed(ain);
36.       bin_d1 <= signed(bin);
37.       diff   <= sub when subadd_d1 else sum;
38.       mreg   <= diff * diff;
39.       pout   <= std_logic_vector(mreg);
40.     end if;
41.   end process;
42. end architecture;
```

> 💡 **设计规则 5**：通常数据路径上的流水寄存器并不需要复位，因为"老的数据"总会被"新的数据"冲走。移除这类不必要的复位信号可以降低复位信号的扇出。

DSP48E2 可在计算两个数之和的平方或两个数之差的平方的基础上，利用其内部乘法器之后的加法器继续计算平方的累加和（适用于 DSP48E2 和 DSP58），相应的电路架构如

图 7-16 所示。注意图中乘法器之后的加法器，其中，一个端口来自乘法器输出，另一个端口受 sload 控制，但 sload 从输入到数据选择器控制端口经过了 1 个时钟周期，而数据 A 和 D 从输入到乘法器的输入端口经历了 3 个时钟周期，因此在此结构之外，应对 sload 做 2 个时钟周期的延迟补偿。采用 VHDL 描述如图 7-16 所示的结构，如 VHDL 代码 7-8 所示。从输入到输出需要 4 个时钟周期，即 Latency 为 4。

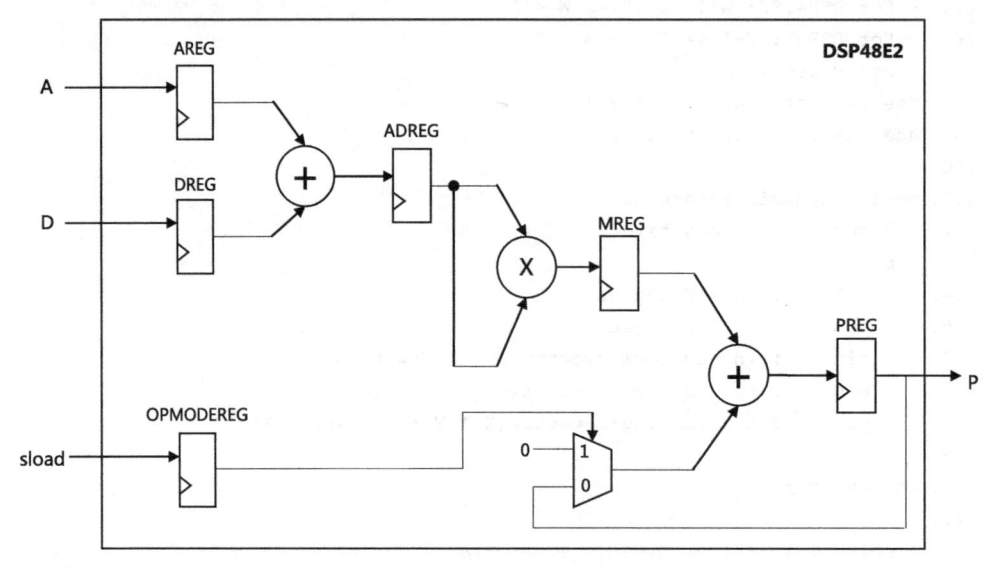

图 7-16

VHDL 代码 7-8

```vhdl
1.  --File: square_mac.vhd
2.  library ieee;
3.  use ieee.std_logic_1164.all;
4.  use ieee.numeric_std.all;
5.
6.  entity square_mac is
7.    generic (
8.      W  : positive := 17; --input data width
9.      PW : positive := 48  --accumulator output width
10.   );
11.   port (
12.     clk  : in std_logic;
13.     sload : in std_logic;
14.     ain  : in std_logic_vector(W - 1 downto 0);
15.     bin  : in std_logic_vector(W - 1 downto 0);
16.     pout : out std_logic_vector(PW - 1 downto 0)
17.   );
18. end entity;
19.
20. architecture rtl of square_mac is
21.   signal sload_d1          : std_logic;
22.   signal ain_d1, bin_d1    : signed(W - 1 downto 0);
23.   signal diff              : signed(W     downto 0);
24.   signal mreg              : signed(2 * W + 1 downto 0);
```

```
25. signal adder_out, old_res : signed(PW - 1 downto 0);
26. begin
27.    old_res   <= (others => '0') when sload_d1 else adder_out;
28.    pout      <= std_logic_vector(adder_out);
29.    process(clk)
30.    begin
31.       if rising_edge(clk) then
32.          ain_d1    <= signed(ain);
33.          bin_d1    <= signed(bin);
34.          diff      <= resize(ain_d1, W + 1) - resize(bin_d1, W + 1);
35.          mreg      <= diff * diff;
36.          sload_d1  <= sload;
37.          adder_out <= old_res + mreg;
38.       end if;
39.    end process;
40. end architecture;
```

对于 DSP48E2，其电路结构如图 7-13 所示，其功能为 $(A \pm D) \times B$，若将 A 端口恒接地，那么其功能就变为 $(\pm D) \times B$，这意味着乘法器输出可以是 $D \times B$，也可以是 $-D \times B$，从而实现对乘积取反的操作，但这种操作借用了预加器。DSP58 无须预加器就可以直接实现对乘积的取反，其电路结构如图 7-17 所示。一个 M 位有符号整数和一个 N 位有符号整数相乘，乘积的最大值为

$$-2^{M-1} \times \left(-2^{N-1}\right) = 2^{M+N-2}$$

乘积的最小值为

$$-2^{M-1} \times \left(2^{N-1} - 1\right) = -2^{M+N-2} + 2^{M-1}$$

或

$$-2^{N-1} \times \left(2^{M-1} - 1\right) = -2^{M+N-2} + 2^{N-1}$$

由此可知，两者乘积取反的最小值为 -2^{M+N-2}，采用二进制补码的形式表示时，需要 $M+N-1$ 位。采用 VHDL 描述如图 7-17 所示的电路，如 VHDL 代码 7-9 所示，注意代码第 30 行的写法，这是 Vivado 要求的方式，即将取反操作和乘积写在一个表达式里，否则工具无法正确推断出 DSP58。从输入到输出需要 3 个时钟周期，即 Latency 为 3。

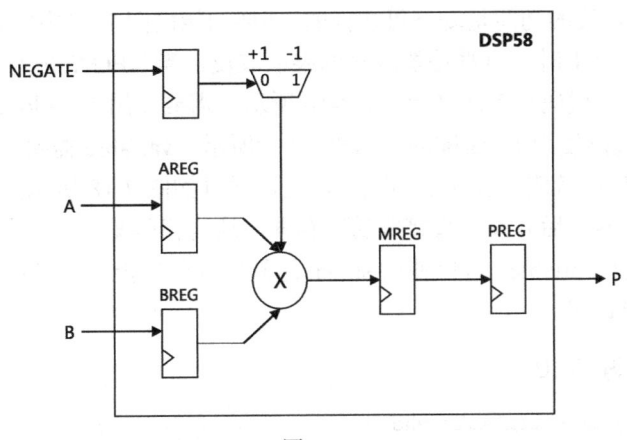

图 7-17

VHDL 代码 7-9

```
1.  --File: neg_mult.vhd
2.  library ieee;
3.  use ieee.std_logic_1164.all;
4.  use ieee.numeric_std.all;
5.
6.  entity neg_mult is
7.    generic (
8.      AW : positive := 27;
9.      BW : positive := 24;
10.     MW : positive := AW + BW
11.   );
12.   port (
13.     clk  : in std_logic;
14.     ain  : in std_logic_vector(AW - 1 downto 0);
15.     bin  : in std_logic_vector(BW - 1 downto 0);
16.     prod : out std_logic_vector(MW - 1 downto 0)
17.   );
18. end entity;
19.
20. architecture rtl of neg_mult is
21. signal ain_d1 : signed(AW - 1 downto 0);
22. signal bin_d1 : signed(BW - 1 downto 0);
23. signal mreg   : signed(MW - 1 downto 0);
24. begin
25.   process(clk)
26.   begin
27.     if rising_edge(clk) then
28.       ain_d1 <= signed(ain);
29.       bin_d1 <= signed(bin);
30.       mreg   <= -(ain_d1 * bin_d1);
31.       prod   <= std_logic_vector(mreg);
32.     end if;
33.   end process;
34. end architecture;
```

DSP58 对乘积取反的功能是动态可切换的,切换功能由图 7-17 中的输入端口 NEGATE 控制。当 NEGATE 为 1 时,对乘法器输出的乘积取反,当 NEGATE 为 0 时,输出两个数的乘积。相应的 RTL 代码如 VHDL 代码 7-10 所示(需要使用 Vivado 2022.2 才可以映射为一个 DSP58)。代码使用了 variable,如第 29 行所示。variable 需要在 process 中声明,需要通过 ":=" 操作符赋值,且应立即赋值,这可通过如图 7-18 所示的电路进一步理解。在图 7-18 中,乘法器之后并没有直接跟随寄存器,这与代码第 35 行所示的功能相匹配。这里并不是必须使用 variable,可以使用 signal 达到同样的功能,只需要把乘法器描述部分放置在 process 之外即可。

VHDL 代码 7-10

```
1.  --File: dynamic_neg_mult.vhd
2.  library ieee;
3.  use ieee.std_logic_1164.all;
```

```vhdl
4.  use ieee.numeric_std.all;
5.
6.  entity dynamic_neg_mult is
7.    generic (
8.      AW : positive := 27;
9.      BW : positive := 24;
10.     MW : positive := AW + BW
11.   );
12.   port (
13.     clk  : in std_logic;
14.     neg  : in std_logic;
15.     ain  : in std_logic_vector(AW - 1 downto 0);
16.     bin  : in std_logic_vector(BW - 1 downto 0);
17.     pout : out std_logic_vector(MW - 1 downto 0)
18.   );
19. end entity;
20.
21. architecture rtl of dynamic_neg_mult is
22.   signal neg_d1 : neg'subtype;
23.   signal ain_d1 : signed(ain'RANGE);
24.   signal bin_d1 : signed(bin'RANGE);
25.   signal mreg   : signed(MW - 1 downto 0);
26. begin
27.   pout <= std_logic_vector(mreg);
28.   process(clk)
29.     variable mreg_i : signed(MW - 1 downto 0);
30.   begin
31.     if rising_edge(clk) then
32.       neg_d1 <= neg;
33.       ain_d1 <= signed(ain);
34.       bin_d1 <= signed(bin);
35.       mreg_i := ain_d1 * bin_d1;
36.       mreg   <= -mreg_i when neg_d1 else mreg_i;
37.     end if;
38.   end process;
39. end architecture;
```

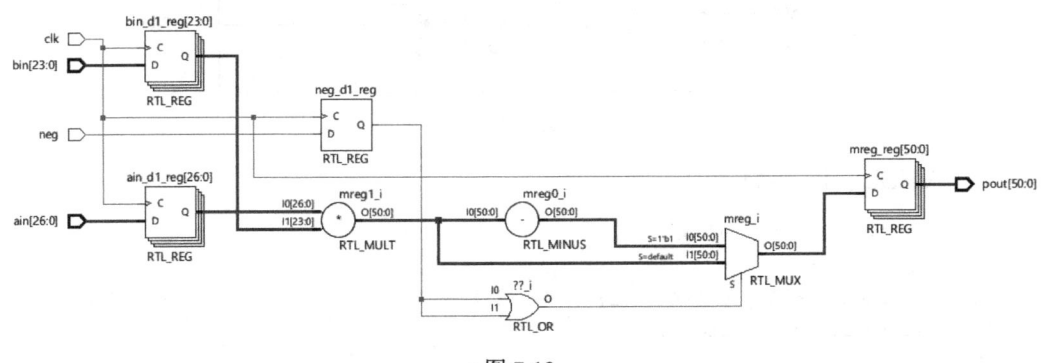

图 7-18

若乘法运算的两个操作数的位宽分别为 27 和 19，那么可进行如图 7-19 所示的分解，从而使该乘法运算可以使用 1 个 DSP48E2 加 1 个与门实现。具体电路结构如图 7-20 所示。

采用 VHDL 描述，如 VHDL 代码 7-11 所示。这里要特别注意，代码第 29 行和第 39 行使用了 signed 将 std_ulogic_vector 转换为 signed。VHDL 代码 7-11 只是为了描述如图 7-20 所示的电路。在实际工程中，可直接使用 VHDL 代码 7-1，将 AW 和 BW 分别设置为 27 和 19 即可。Vivado 会综合为如图 7-20 所示的电路。从输入到输出需要 3 个时钟周期，即 Latency 为 3。

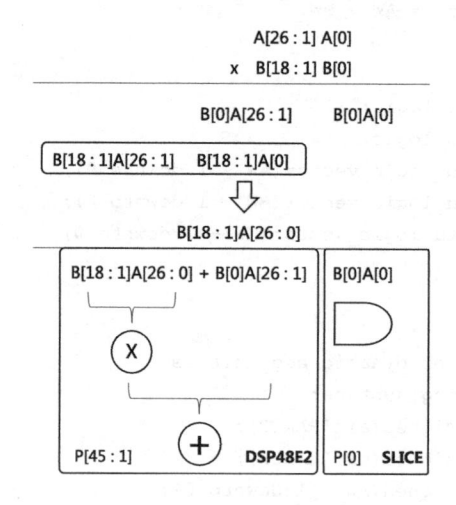

图 7-19

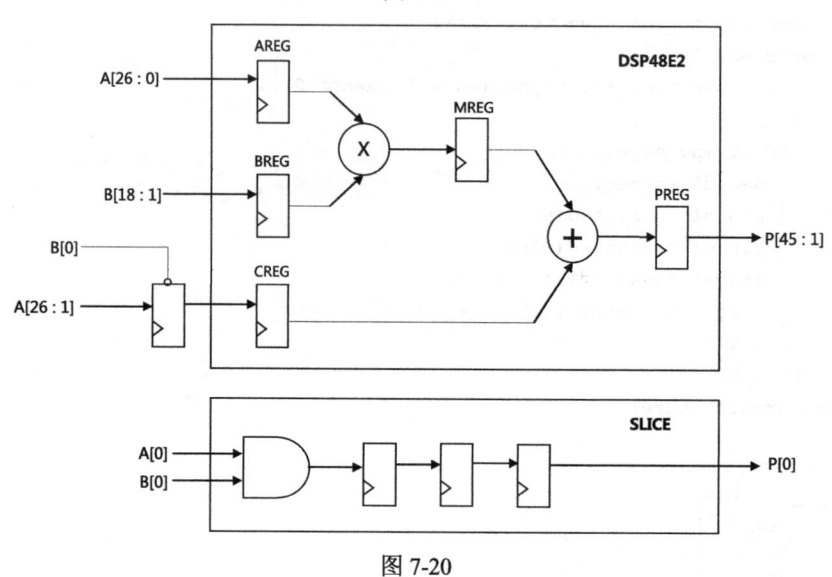

图 7-20

VHDL 代码 7-11

```
1.  --File: mult_27x19.vhd
2.  library ieee;
3.  use ieee.std_logic_1164.all;
4.  use ieee.numeric_std.all;
5.
6.  entity mult_27x19 is
7.    port (
8.      clk  : in std_logic;
```

```vhdl
9.        ain  : in std_logic_vector(26 downto 0);
10.       bin  : in std_logic_vector(18 downto 0);
11.       pout : out std_logic_vector(45 downto 0)
12.  );
13. end entity;
14.
15. architecture rtl of mult_27x19 is
16. signal ain_d1 : signed(ain'RANGE);
17. signal bin_d1 : signed(bin'RANGE);
18. signal mreg   : signed(44 downto 0);
19. signal cin_d1 : signed(25 downto 0);
20. signal cin_d2 : signed(25 downto 0);
21. signal pout_h : signed(44 downto 0);
22. signal p0, p0_d1, p0_d2 : std_logic;
23. begin
24.   pout <= std_logic_vector(pout_h & p0_d2);
25.   process(clk)
26.   begin
27.     if rising_edge(clk) then
28.       cin_d1 <= (others => '0') when (not bin(0)) else
29.                 signed(ain(26 downto 1));
30.     end if;
31.   end process;
32.
33.   process(clk)
34.   begin
35.     if rising_edge(clk) then
36.       ain_d1 <= signed(ain);
37.       bin_d1 <= signed(bin);
38.       cin_d2 <= cin_d1;
39.       mreg   <= ain_d1 * signed(bin_d1(18 downto 1));
40.       pout_h <= mreg + cin_d2;
41.     end if;
42.   end process;
43.
44.   process(clk)
45.   begin
46.     if rising_edge(clk) then
47.       p0    <= ain(0) and bin(0);
48.       p0_d1 <= p0;
49.       p0_d2 <= p0_d1;
50.     end if;
51.   end process;
52. end architecture;
```

💡 **设计规则 6**：VHDL 中定义的 signed 或 unsigned 数据，其每一位的数据类型为 std_ulogic。因此，若取其中的几位构成一个新的数据，则新数据的数据类型为 std_ulogic_vector。若将该数据赋值给一个 signed 类型的数据，则需要先通过函数 signed 进行数据类型转换，再赋值。

应用案例 1：脉动 FIR（Finite Impulse Response，有限冲激响应）滤波器

为便于说明，这里以 4 阶 FIR 滤波器为例，输出信号 $y(n)$ 与输入信号 $x(n)$ 之间的关系可表示为

$$y(n)=\sum_{k=0}^{N-1}x(n-k)h(k)$$

式中，N 为 4，表征了滤波器阶数。$h(k)(k=0,1,2,3)$ 为滤波器系数。根据此式不难得出如表 7-6 所示的数据关系。可以看到，每次滤波运算（本质上就是卷积运算）需要执行 4 次乘法运算和 3 次加法运算。

表 7-6

$h(0)$	$h(1)$	$h(2)$	$h(3)$		
$x(0)$	0	0	0	=	$y(0)$
$x(1)$	$x(0)$	0	0	=	$y(1)$
$x(2)$	$x(1)$	$x(0)$	0	=	$y(2)$
$x(3)$	$x(2)$	$x(1)$	$x(0)$	=	$y(3)$
$x(4)$	$x(3)$	$x(2)$	$x(1)$	=	$y(4)$
$x(5)$	$x(4)$	$x(3)$	$x(2)$	=	$y(5)$

第一步：根据输入数据和滤波器系数的位宽确定输出数据位宽。

假定输入数据位宽为 XIN_W，系数位宽为 COE_W，那么输出数据的最大位宽可表示为

$$\text{YOUT\_W}=\text{XIN\_W}+\text{COE\_W}+\text{ceil}\left(\log_2^N\right)$$

输出数据的位宽也决定了卷积运算过程中累加器的位宽。在此案例中，若输入数据和滤波器系数的位宽均为 16，那么可得

$$\text{YOUT\_W}=16+16+\text{ceil}\left(\log_2^4\right)=34$$

采用该式计算的位宽必然不会出现溢出，但可能会过大，从而造成资源浪费，因为并未考虑滤波器系数的具体数值，所以，更准确的计算方式是根据滤波器系数先计算出输出信号的理论最大值 YOUT_MAX，在此基础上，确定 YOUT_MAX 用二进制补码表示时所需要的位宽，即输出信号位宽。

$$\text{YOUT\_MAX}=2^{\text{XIN\_W}-1}\cdot\sum_{i=0}^{N-1}|h(n)|$$

在此案例中，如果滤波器系数为 7、14、−138 和 129，输入数据为 16 位有符号整数，那么可得

$$\text{YOUT\_MAX}=2^{15}\times(7+14+138+129)=9437184$$

YOUT_MAX 用二进制补码可表示为 0_1001_0000_0000_0000_0000_0000，共 25 位。对比这两种方式计算所得的输出数据的位宽可知，后者更精确。尽管 DSP48E2 输出数据的位宽可达 48，但当要将滤波器输出结果给后续单元进行处理时，显然 25 位数据比 34 位数据更省资源。

第二步：确定硬件电路结构。

4 阶 FIR 滤波器直接型结构如图 7-21 所示。可以看到，路径 A0→P0→P1 与 A0→A1→P1 相差一个延迟单元。在此基础上，对直接型结构稍加改动，即在等效路径上插入相同个数的延迟单元，结果如图 7-22 所示。不难看出，路径 A0→P0→P1 与 A0→A1→P1 仍然相差一个延迟单元，因此这两个结构是等效的。

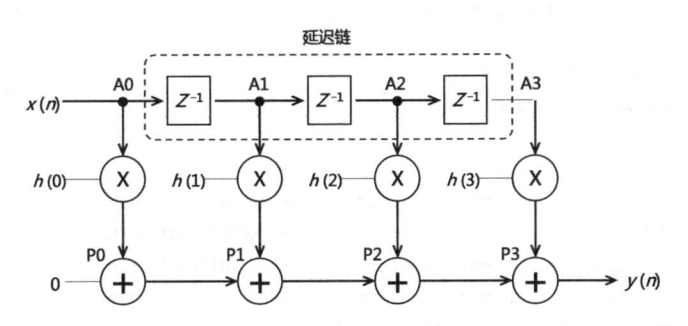

图 7-21

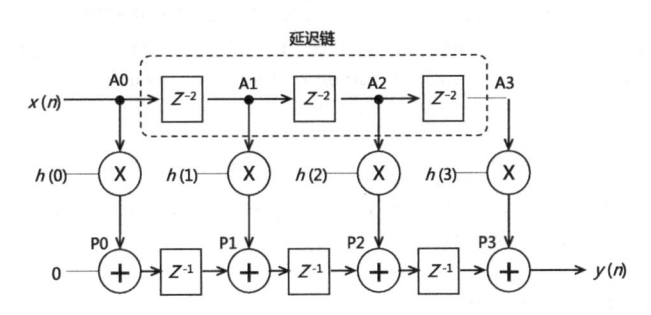

图 7-22

考虑到 DSP48E2 的架构，在乘法器和加法器的输出端添加寄存器，从而形成如图 7-23 所示的硬件电路架构，这就是脉动结构的 FIR 滤波器。在此结构中，相邻 DSP48E2 通过专用级联端口 ACOUT/ACIN 构成延迟链，通过 PCOUT/PCIN 传递累加结果，因此，这 4 个 DSP48E2 必然位于同一列。

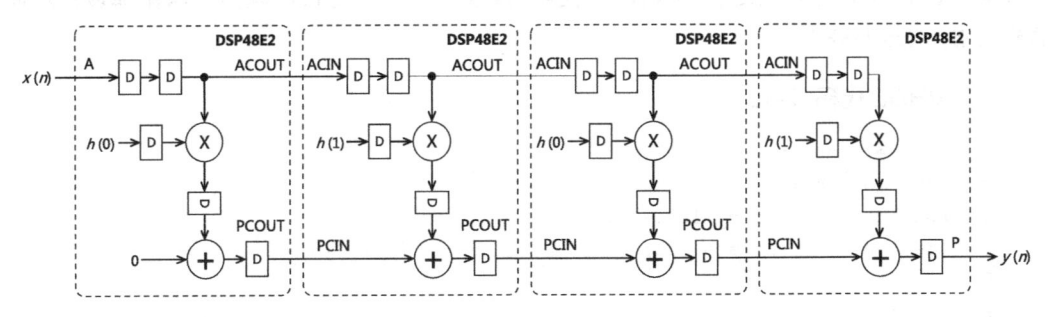

图 7-23

第三步：采用 VHDL 描述电路模型。

在如图 7-23 所示的硬件电路结构的基础上，采用 VHDL 对其进行描述。首先将系统参数定义在 package 中，如 VHDL 代码 7-12 所示，包括滤波器阶数（TAP）、输入数据位宽（XIN_W）、滤波器系数位宽（COE_W）、累加器位宽（ACC_W）和输出数据位宽

（YOUT_W）。同时，以数组形式定义滤波器系数，如代码第 14 行所示。这里使用了数据类型 integer_vector，其每个组成元素的类型为 integer，因此，需要将其转换成 std_logic_vector 类型，这可以通过用户定义函数 intv2logic 实现。

VHDL 代码 7-12

```vhdl
1.  --File: systolic_fir_pkg.vhd
2.  library ieee;
3.  use ieee.std_logic_1164.all;
4.  use ieee.numeric_std.all;
5.
6.  package systolic_fir_pkg is
7.    constant TAP      : positive := 4;  --length of FIR
8.    constant XIN_W    : positive := 16; -- width of x(n)
9.    constant COE_W    : positive := 16; -- width of h(n)
10.   constant ACC_W    : positive := 48; -- width of accumulator
11.   constant YOUT_W   : positive := 25; -- width of y(n)
12.   type logic_vec is array(0 to TAP - 1) of std_logic_vector;
13.   subtype coe_t is logic_vec(open)(COE_W - 1 downto 0);
14.   constant COE_DVEC : integer_vector := (7, 14, -138, 129); -- coefficients
15.   function intv2logic (intv : integer_vector) return coe_t;
16. end package;
17.
18. package body systolic_fir_pkg is
19.   function intv2logic (intv : integer_vector) return coe_t is
20.     variable res : coe_t;
21.   begin
22.     for i in intv'RANGE loop
23.       res(i) := std_logic_vector(to_signed(intv(i), COE_W));
24.     end loop;
25.     return res;
26.   end function;
27. end package body;
```

图 7-23 所示的硬件电路结构的核心单元是乘加器，每个乘加器有两个输入端口和两个输出端口。采用 VHDL 描述，如 VHDL 代码 7-13 所示。注意代码风格，以保证该模块能准确映射为 DSP48E2。

VHDL 代码 7-13

```vhdl
1.  --File: mac.vhd
2.  library ieee;
3.  use ieee.std_logic_1164.all;
4.  use ieee.numeric_std.all;
5.
6.  entity mac is
7.    generic (
8.      AW : positive := 27;
9.      BW : positive := 18;
10.     PW : positive := 48;
11.     MW : positive := AW + BW
12.   );
```

```
13.   port (
14.     clk   : in std_logic;
15.     ain   : in std_logic_vector(AW - 1 downto 0);
16.     bin   : in std_logic_vector(BW - 1 downto 0);
17.     cin   : in std_logic_vector(PW - 1 downto 0);
18.     acout : out std_logic_vector(AW - 1 downto 0);
19.     pout  : out std_logic_vector(PW - 1 downto 0)
20.   );
21. end entity;
22.
23. architecture rtl of mac is
24.   signal ain_d1 : signed(ain'RANGE) := (others => '0');
25.   signal ain_d2 : signed(ain'RANGE) := (others => '0');
26.   signal bin_d1 : signed(bin'RANGE) := (others => '0');
27.   signal mreg   : signed(MW - 1 downto 0) := (others => '0');
28.   signal pout_i : signed(pout'RANGE);
29. begin
30.   acout <= std_logic_vector(ain_d2);
31.   pout  <= std_logic_vector(pout_i);
32.   process(clk)
33.   begin
34.     if rising_edge(clk) then
35.       ain_d1 <= signed(ain);
36.       ain_d2 <= ain_d1;
37.       bin_d1 <= signed(bin);
38.       mreg   <= ain_d2 * bin_d1;
39.       pout_i <= resize(mreg, PW) + signed(cin);
40.     end if;
41.   end process;
42. end architecture;
```

有了核心单元，即可通过 for generate 语句构造顶层单元，如 VHDL 代码 7-14 所示。注意代码第 5 行，其目的是使该模块可以访问 package 中的参数。

📄 **VHDL 代码 7-14**

```
1.  --File: systolic_fir.vhd
2.  library ieee;
3.  use ieee.std_logic_1164.all;
4.  use ieee.numeric_std.all;
5.  use work.systolic_fir_pkg.all;
6.
7.  entity systolic_fir is
8.    port (
9.      clk : in std_logic;
10.     xin : in std_logic_vector(XIN_W - 1 downto 0);
11.     yout : out std_logic_vector(YOUT_W - 1 downto 0)
12.   );
13. end entity;
14.
15. architecture rtl of systolic_fir is
16.   constant COE : coe_t := intv2logic(COE_DVEC);
```

```
17.     signal acout : logic_vec(open)(XIN_W - 1 downto 0);
18.     signal pout  : logic_vec(open)(ACC_W - 1 downto 0);
19.   begin
20.     yout <= pout(TAP - 1)(YOUT_W - 1 downto 0);
21.     i0_mac : entity work.mac
22.       generic map ( AW => XIN_W, BW => COE_W, PW => ACC_W)
23.       port map (
24.         clk  => clk,
25.         ain  => xin,
26.         bin  => COE(0),
27.         cin  => (others => '0'),
28.         acout => acout(0),
29.         pout  => pout(0)
30.       );
31.
32.     gen0:
33.     for i in 1 to TAP - 1 generate
34.       i1_mac : entity work.mac
35.       generic map ( AW => XIN_W, BW => COE_W, PW => ACC_W)
36.       port map (
37.         clk  => clk,
38.         ain  => acout(i - 1),
39.         bin  => COE(i),
40.         cin  => pout(i - 1),
41.         acout => acout(i),
42.         pout  => pout(i)
43.       );
44.     end generate;
45.   end architecture;
```

第四步：功能仿真。

对 FIR 滤波器而言，其冲激响应就是滤波器系数，根据此特征可以将输入数据设置为冲激函数，这样可以很快验证滤波器的功能是否正确。读者可根据此思路编写测试方案。但仅有此测试是不够的，因为无法验证是否存在数据溢出问题，所以还要考虑极端情况，即将输入数据分别设置为最大值和最小值。在设计初期已经确定了位宽，以保证设计不会发生溢出，但这里的验证侧重点在于代码描述是否准确。

> 💡 **设计规则 7**：使用 subtype 定义数据类型时，需要注意其与 type 定义的类型之间的关系，即 subtype 定义的类型是 type 定义的类型的子类型，两者有继承关系，如 VHDL 代码 7-12 第 12 行和第 13 行所示，同时，由于第 12 行定义的数组深度已经是固定值，因此第 13 行使用 subtype 时深度用 open 来代替，表明继承了第 12 行 logic_vec 的深度。

应用案例 2：系数对称的脉动 FIR 滤波器

若滤波器系数对称，则可利用此对称性节省乘加器资源。这里以 8 阶偶对称 FIR 滤波器为例，其直接型结构如图 7-24 所示，滤波器系数的对称性如图 7-25 所示。据此可知进入预加器的两个输入数据相差的延迟单元个数。

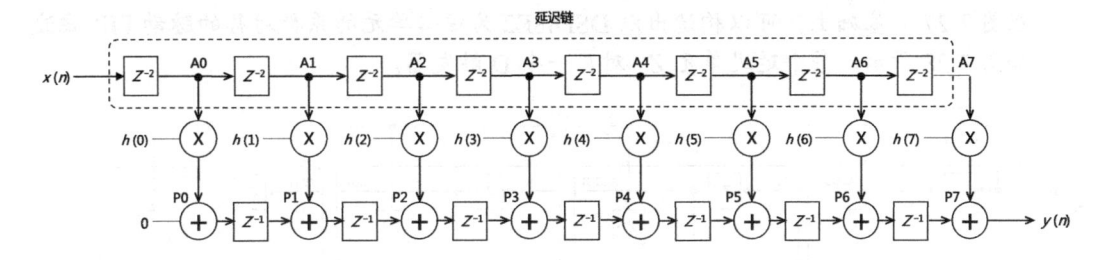

图 7-24

图 7-25

仅考虑延迟链，可得如图 7-26 所示的延迟结构。可以看到，最左侧的预加器的两个输入数据分别为 $x(n-8)$ 和 $x(n-1)$，两者相差 7 个延迟单元；最右侧预加器的两个输入数据分别为 $x(n-8)$ 和 $x(n-7)$，两者相差 1 个延迟单元。这与如图 7-25 所示的结论是一致的。由于进入 4 个预加器的数据都包含 $x(n-8)$，因此将其对应的延迟单元提取出来并稍加改动，可得如图 7-27 所示的延迟链路。

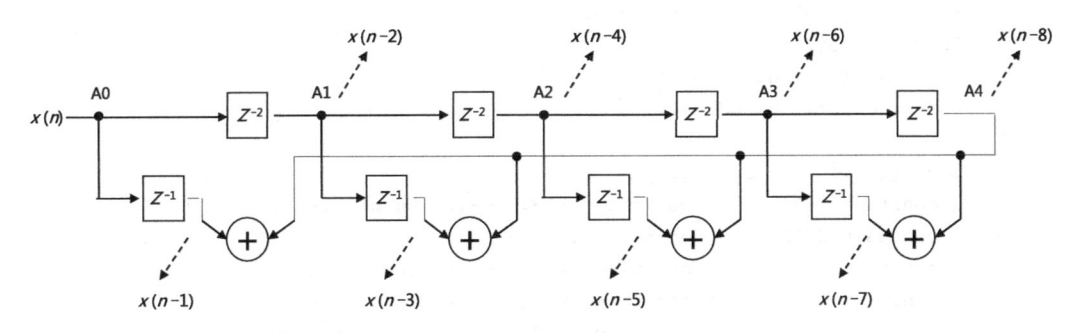

图 7-26

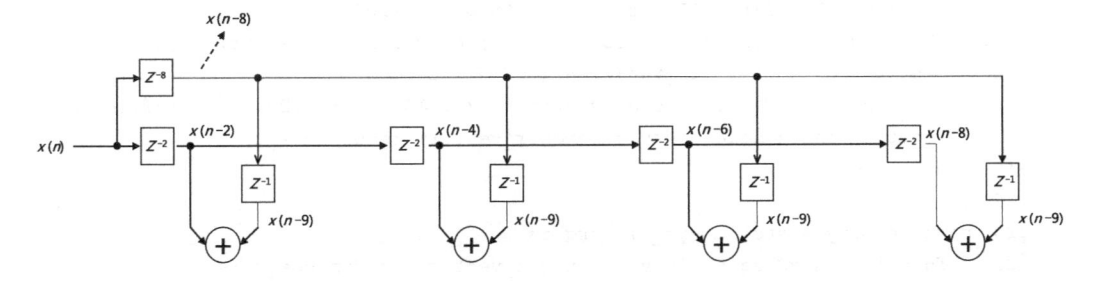

图 7-27

在图 7-27 的基础上，可以构造出以 DSP48E2 为核心单元的系数对称的脉动 FIR 滤波器，如图 7-28 所示。每个延迟单元 Z^{-1} 对应一个 D 触发器。

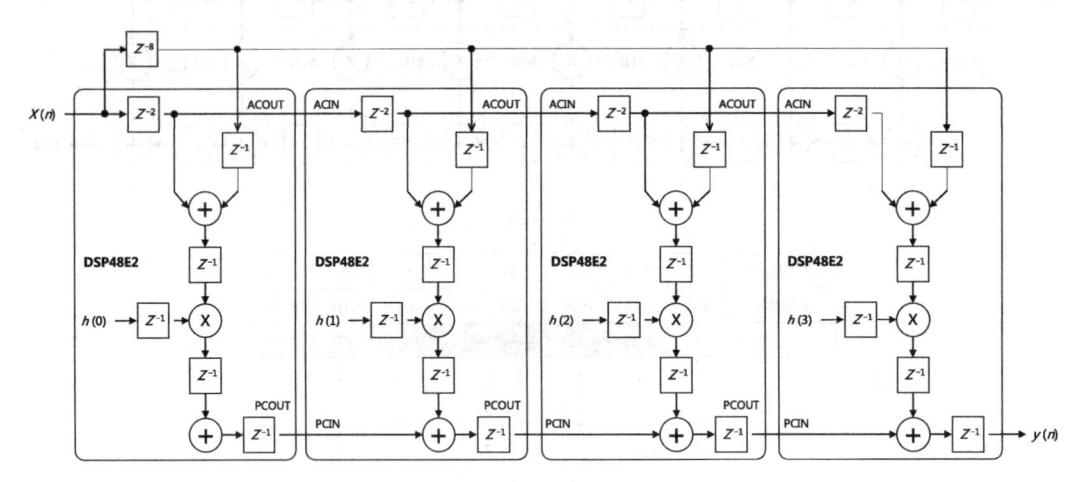

图 7-28

VHDL 代码 7-15 中创建了 package，其中定义了一些参数，包括滤波器系数，如代码第 12 行所示。8 阶滤波器偶对称，因此这里只给出前 4 个系数。VHDL 代码 7-16 为包含预加器的乘加单元，最终会映射为 DSP48E2。VHDL 代码 7-17 为顶层设计单元，通过 for generate 描述如图 7-28 所示的脉动结构，其中，代码第 22 行的移位寄存器的具体描述可参考第 5 章 VHDL 代码 5-5。

VHDL 代码 7-15

```vhdl
1.  --File: systolic_sym_fir_pkg.vhd
2.  library ieee;
3.  use ieee.std_logic_1164.all;
4.  use ieee.numeric_std.all;
5.
6.  package systolic_sym_fir_pkg is
7.    constant TAP         : positive := 8;  --length of FIR
8.    constant HTAP        : positive := TAP / 2;
9.    constant XIN_W       : positive := 16; -- width of x(n)
10.   constant COE_W       : positive := 16; -- width of h(n)
11.   constant ACC_W       : positive := 48; -- width of accumulator
12.   constant YOUT_W      : positive := 26; -- width of y(n)
13.   constant SRL_STYLE_VAL : string := "reg_srl_reg";
14.   type logic_vec is array(0 to HTAP - 1) of std_logic_vector;
15.   subtype coe_t is logic_vec(open)(COE_W - 1 downto 0);
16.   constant COE_DVEC : integer_vector := (7, 14, -138, 129); -- coefficients
17.   function intv2logic (intv : integer_vector) return coe_t;
18. end package;
19.
20. package body systolic_sym_fir_pkg is
21.   function intv2logic (intv : integer_vector) return coe_t is
```

```
22.      variable res : coe_t;
23.   begin
24.      for i in intv'RANGE loop
25.         res(i) := std_logic_vector(to_signed(intv(i), COE_W));
26.      end loop;
27.      return res;
28.   end function;
29. end package body;
```

VHDL 代码 7-16

```
1.  --File: preadder_mac.vhd
2.  library ieee;
3.  use ieee.std_logic_1164.all;
4.  use ieee.numeric_std.all;
5.
6.  entity preadder_mac is
7.    generic (
8.      AW : positive := 26;
9.      BW : positive := 18;
10.     PW : positive := 48;
11.     MW : positive := AW + BW + 1
12.   );
13.   port (
14.     clk   : in std_logic;
15.     ain   : in std_logic_vector(AW - 1 downto 0);
16.     din   : in std_logic_vector(AW - 1 downto 0);
17.     bin   : in std_logic_vector(BW - 1 downto 0);
18.     cin   : in std_logic_vector(PW - 1 downto 0);
19.     acout : out std_logic_vector(AW - 1 downto 0);
20.     pout  : out std_logic_vector(PW - 1 downto 0)
21.   );
22. end entity;
23.
24. architecture rtl of preadder_mac is
25. signal ain_d1 : signed(ain'RANGE) := (others => '0');
26. signal ain_d2 : signed(ain'RANGE) := (others => '0');
27. signal din_d1 : signed(din'RANGE) := (others => '0');
28. signal addreg : signed(AW downto 0) := (others => '0');
29. signal bin_d1 : signed(bin'RANGE) := (others => '0');
30. signal mreg   : signed(MW - 1 downto 0) := (others => '0');
31. signal preg   : signed(PW - 1 downto 0) := (others => '0');
32. begin
33.    acout <= std_logic_vector(ain_d2);
34.    pout  <= std_logic_vector(preg);
35.    process(clk)
36.    begin
37.      if rising_edge(clk) then
38.        ain_d1 <= signed(ain);
39.        ain_d2 <= ain_d1;
```

```vhdl
40.         din_d1 <= signed(din);
41.         bin_d1 <= signed(bin);
42.         addreg <= resize(ain_d2, AW + 1) + resize(din_d1, AW + 1);
43.         mreg   <= addreg * bin_d1;
44.         preg   <= resize(mreg, PW) + signed(cin);
45.       end if;
46.    end process;
47. end architecture;
```

📄 **VHDL 代码 7-17**

```vhdl
1.  --File: systolic_sym_fir.vhd
2.  library ieee;
3.  use ieee.std_logic_1164.all;
4.  use ieee.numeric_std.all;
5.  use work.systolic_sym_fir_pkg.all;
6.
7.  entity systolic_sym_fir is
8.    port (
9.      clk  : in std_logic;
10.     xin  : in std_logic_vector(XIN_W - 1 downto 0);
11.     yout : out std_logic_vector(YOUT_W - 1 downto 0)
12.    );
13. end entity;
14.
15. architecture rtl of systolic_sym_fir is
16. constant COE     : coe_t := intv2logic(COE_DVEC);
17. signal xin_udx : xin'subtype;
18. signal acout   : logic_vec(open)(XIN_W - 1 downto 0);
19. signal pout    : logic_vec(open)(ACC_W - 1 downto 0);
20. begin
21.    yout <= pout(HTAP - 1)(YOUT_W - 1 downto 0);
22.    i_xin_delay_line : entity work.static_multi_bit_sreg_v3
23.      generic map (
24.        DEPTH         => TAP,
25.        WIDTH         => XIN_W,
26.        SRL_STYLE_VAL => SRL_STYLE_VAL
27.      )
28.      port map (
29.        clk => clk, ce => '1', si => xin, so => xin_udx
30.      );
31.
32.    i0_preadder_mac : entity work.preadder_mac
33.      generic map ( AW => XIN_W, BW => COE_W, PW => ACC_W )
34.      port map (
35.        clk => clk, ain => xin, din => xin_udx, bin => COE(0),
36.        cin => (others => '0'), acout => acout(0), pout => pout(0)
37.      );
38.
39.    gen0:
```

```
40.    for i in 1 to HTAP - 1 generate
41.        i_preadder_mac : entity work.preadder_mac
42.        generic map ( AW => XIN_W, BW => COE_W, PW => ACC_W )
43.        port map (
44.            clk => clk, ain => acout(i - 1), din => xin_udx, bin => coe(i),
45.            cin => pout(i - 1), acout => acout(i), pout => pout(i)
46.        );
47.    end generate;
48. end architecture;
```

输入/输出及内部关键信号之间的时序关系如图 7-29 所示，图中，从上至下每对 {addreg, mreg} 依次对应图 7-28 从左至右每个预加器和乘法器的输出。

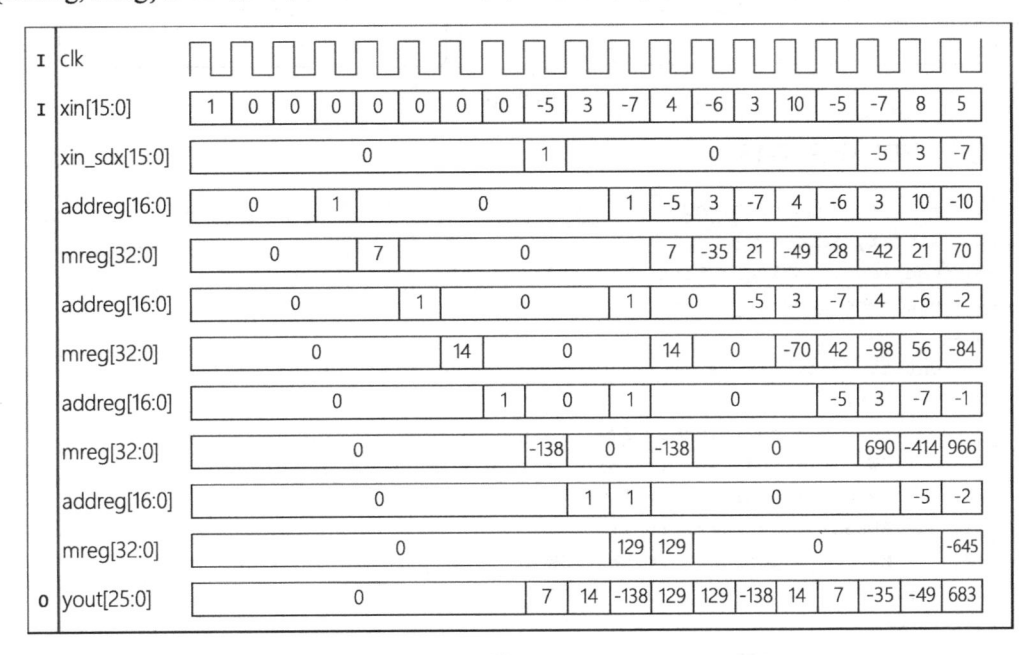

I	clk																	
I	xin[15:0]	1	0	0	0	0	0	0	-5	3	-7	4	-6	3	10	-5	-7	8	5
	xin_sdx[15:0]	0					1				0					-5	3	-7
	addreg[16:0]	0		1			0			1	-5	3	-7	4	-6	3	10	-10
	mreg[32:0]	0		7			0			7	-35	21	-49	28	-42	21	70
	addreg[16:0]	0			1			0			1	0	-5	3	-7	4	-6	-2
	mreg[32:0]	0			14			0			14	0	-70	42	-98	56	-84
	addreg[16:0]		0			1	0	1			0			-5	3	-7	-1
	mreg[32:0]		0			-138	0	-138			0			690	-414	966
	addreg[16:0]		0				1	1			0				-5	-2
	mreg[32:0]		0				129	129			0				-645
o	yout[25:0]		0			7	14	-138	129	129	-138	14	7	-35	-49	683

图 7-29

7.3 复数乘法运算代码风格

复数 a 的实部和虚部分别为 a_r 和 a_i，复数 b 的实部和虚部分别为 b_r 和 b_i，以 p 表示这两个复数相乘的结果，则 p 的实部和虚部可表示为

$$\begin{cases} p_r = a_r \times b_r - a_i \times b_i \\ p_i = a_r \times b_i + a_i \times b_r \end{cases}$$

采用此算法可知：一次复数乘法需要 4 次实数乘法运算和 2 次实数加法运算。相应的硬件电路架构如图 7-30 所示。显然，a_r、a_i 和 b_r、b_i 的位宽受限于乘法器的位宽。若采用 DSP48E1 实现，则 a_r 和 a_i 的位宽最大为 25，若采用 DSP48E2 实现，则其位宽最大为 27。另一个复数的最大位宽为 18。在图 7-30 中，上侧的 DSP48E2 执行的功能是 $a \times b + c$（$c=0$），下侧的 DSP48E2 执行的功能是 $a \times b$+PCIN。观察每个模块，可以看到乘法器输入端或为一级流水寄存器，或为二级流水寄存器，同时，乘法器之后的加法器或执行加法运算，或执行减法运算。因此可将流水寄存器级数及切换加法器功能的控制信号作为参数。这样，这

4个模块只用一套代码即可完成，如 VHDL 代码 7-18 所示。用于描述 4 个模块之间连接关系的顶层设计单元如 VHDL 代码 7-19 所示。从输入到输出需要 4 个时钟周期（Latency=4）。

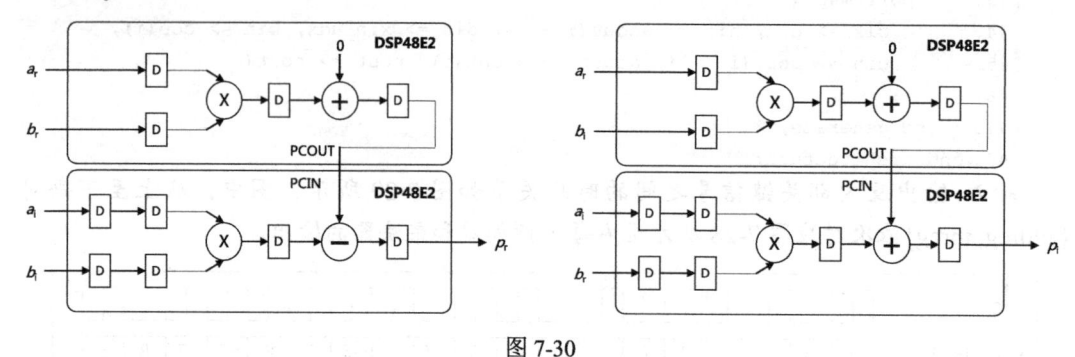

图 7-30

VHDL 代码 7-18

```vhdl
1.  --File: cmac_unit.vhd
2.  library ieee;
3.  use ieee.std_logic_1164.all;
4.  use ieee.numeric_std.all;
5.
6.  entity cmac_unit is
7.    generic (
8.      AW     : positive := 27;
9.      BW     : positive := 18;
10.     AREG   : positive := 1;
11.     ADDSUB : integer  := 1;
12.     MW     : positive := AW + BW
13.   );
14.   port (
15.     clk : in std_logic;
16.     ain : in std_logic_vector(AW - 1 downto 0);
17.     bin : in std_logic_vector(BW - 1 downto 0);
18.     cin : in std_logic_vector(MW - 1 downto 0);
19.     pout : out std_logic_vector(MW downto 0)
20.   );
21. end entity;
22.
23. architecture rtl of cmac_unit is
24. signal amult : signed(ain'RANGE);
25. signal bmult : signed(bin'RANGE);
26. signal ain_d1: signed(ain'RANGE);
27. signal bin_d1: signed(bin'RANGE);
28. signal prod  : signed(MW - 1 downto 0);
29. signal pout_i: signed(pout'RANGE);
30. begin
31.   pout <= std_logic_vector(pout_i);
32.   gen0:
```

```
33.   if AREG = 1 generate
34.     process(clk)
35.     begin
36.       if rising_edge(clk) then
37.         amult <= signed(ain);
38.         bmult <= signed(bin);
39.       end if;
40.     end process;
41.   else generate
42.     process(clk)
43.     begin
44.       if rising_edge(clk) then
45.         ain_d1 <= signed(ain);
46.         bin_d1 <= signed(bin);
47.         amult   <= ain_d1;
48.         bmult   <= bin_d1;
49.       end if;
50.     end process;
51.   end generate;
52.
53.   process(clk)
54.   begin
55.     if rising_edge(clk) then
56.       prod <= amult * bmult;
57.     end if;
58.   end process;
59.
60.   gen1:
61.   if ADDSUB = 1 generate
62.     process(clk)
63.     begin
64.       if rising_edge(clk) then
65.         pout_i <= resize(signed(cin), MW + 1) - resize(prod, MW + 1);
66.       end if;
67.     end process;
68.   else generate
69.     process(clk)
70.     begin
71.       if rising_edge(clk) then
72.         pout_i <= resize(signed(cin), MW + 1) + resize(prod, MW + 1);
73.       end if;
74.     end process;
75.   end generate;
76. end architecture;
```

VHDL 代码 7-19

```
1. --File: cplx_mul_v1.vhd
2. library ieee;
3. use ieee.std_logic_1164.all;
```

```vhdl
4.  use ieee.numeric_std.all;
5.
6.  entity cplx_mul_v1 is
7.    generic (
8.      AW : positive := 27;
9.      BW : positive := 18;
10.     MW : positive := AW + BW
11.   );
12.   port (
13.     clk : in std_logic;
14.     ar  : in std_logic_vector(AW - 1 downto 0);
15.     ai  : in std_logic_vector(AW - 1 downto 0);
16.     br  : in std_logic_vector(BW - 1 downto 0);
17.     bi  : in std_logic_vector(BW - 1 downto 0);
18.     pr  : out std_logic_vector(MW downto 0);
19.     pi  : out std_logic_vector(MW downto 0)
20.   );
21. end entity;
22.
23. architecture rtl of cplx_mul_v1 is
24. signal ar_br : std_logic_vector(MW downto 0);
25. signal ar_bi : std_logic_vector(MW downto 0);
26. begin
27.   i_ar_br_cmac_unit : entity work.cmac_unit
28.   generic map ( AW => AW, BW => BW, AREG => 1, ADDSUB => 0 )
29.   port map (
30.     clk => clk, ain => ar, bin => br, cin => (others => '0'), pout => ar_br
31.   );
32.
33.   i_ai_bi_cmac_unit : entity work.cmac_unit
34.   generic map ( AW => AW, BW => BW, AREG => 2, ADDSUB => 1 )
35.   port map (
36.     clk => clk, ain => ai, bin => bi, cin => ar_br(MW - 1 downto 0), pout => pr
37.   );
38.
39.   i_ar_bi_cmac_unit : entity work.cmac_unit
40.   generic map ( AW => AW, BW => BW, AREG => 1, ADDSUB => 0 )
41.   port map (
42.     clk => clk, ain => ar, bin => bi, cin => (others => '0'), pout => ar_bi
43.   );
44.
45.   i_ai_br_cmac_unit : entity work.cmac_unit
46.   generic map ( AW => AW, BW => BW, AREG => 2, ADDSUB => 0 )
47.   port map (
48.     clk => clk, ain => ai, bin => br, cin => ar_bi(MW - 1 downto 0), pout => pi
49.   );
50. end architecture;
```

对于如图 7-30 所示的电路结构，也可将其作为一个整体进行描述，如 VHDL 代码 7-20 所示。最终，Vivado 也会将其映射为 4 个 DSP48E2，而不会消耗其他 SLICE 内的资源。

VHDL 代码 7-20

```vhdl
1.  --File: cplx_mul_v3.vhd
2.  library ieee;
3.  use ieee.std_logic_1164.all;
4.  use ieee.numeric_std.all;
5.
6.  entity cplx_mul_v3 is
7.    generic (
8.      AW : positive := 27;
9.      BW : positive := 18;
10.     MW : positive := AW + BW
11.   );
12.   port (
13.     clk : in std_logic;
14.     ar  : in std_logic_vector(AW - 1 downto 0);
15.     ai  : in std_logic_vector(AW - 1 downto 0);
16.     br  : in std_logic_vector(BW - 1 downto 0);
17.     bi  : in std_logic_vector(BW - 1 downto 0);
18.     pr  : out std_logic_vector(MW downto 0);
19.     pi  : out std_logic_vector(MW downto 0)
20.   );
21. end entity;
22.
23. architecture rtl of cplx_mul_v3 is
24.   signal ar_d1, ai_d1, ai_d2        : signed(AW - 1 downto 0);
25.   signal br_d1, br_d2, bi_d1, bi_d2 : signed(BW - 1 downto 0);
26.   signal ar_br, ai_bi, ar_bi, ai_br : signed(MW - 1 downto 0);
27.   signal ar_br_d1, ar_bi_d1         : signed(MW - 1 downto 0);
28.   signal pr_i                       : signed(MW     downto 0);
29.   signal pi_i                       : signed(MW     downto 0);
30. begin
31.   pr <= std_logic_vector(pr_i);
32.   pi <= std_logic_vector(pi_i);
33.   process(clk)
34.   begin
35.     if rising_edge(clk) then
36.       ar_d1    <= signed(ar);
37.       br_d1    <= signed(br);
38.       ar_br    <= ar_d1 * br_d1;
39.       ar_br_d1 <= ar_br;
40.     end if;
41.   end process;
42.
43.   process(clk)
44.   begin
45.     if rising_edge(clk) then
46.       ai_d1 <= signed(ai);
47.       ai_d2 <= ai_d1;
```

```
48.        bi_d1 <= signed(bi);
49.        bi_d2 <= bi_d1;
50.        ai_bi <= ai_d2 * bi_d2;
51.        pr_i  <= resize(ar_br_d1, MW + 1) - resize(ai_bi, MW + 1);
52.      end if;
53.   end process;
54.
55.   process(clk)
56.   begin
57.      if rising_edge(clk) then
58.        ar_bi    <= ar_d1 * bi_d1;
59.        ar_bi_d1 <= ar_bi;
60.      end if;
61.   end process;
62.
63.   process(clk)
64.   begin
65.      if rising_edge(clk) then
66.        br_d2 <= br_d1;
67.        ai_br <= ai_d2 * br_d2;
68.        pi_i  <= resize(ar_bi_d1, MW + 1) + resize(ai_br, MW + 1);
69.      end if;
70.   end process;
71. end architecture;
```

复数乘法的另一种算法可表示为

$$\text{multcommon} = (a_r - a_i) \times b_i$$

$$\begin{cases} p_r = \text{multcommon} + (b_r - b_i) \times a_r \\ p_i = \text{multcommon} + (b_r + b_i) \times a_i \end{cases}$$

可以看到，基于此算法，完成一次复数乘法运算需要 3 次实数乘法运算和 5 次加法运算，与前一种算法相比，减少了乘法运算的次数，增加了加法运算的次数。相应的硬件电路如图 7-31 所示。此时会消耗 3 个 DSP48E2 和一些额外的触发器。对于位于中间的 DSP48E2，其输出结果要给其上方和下方的 DSP48E2，因此，只可能和其中一个 DSP48E2 通过 PCOUT/PCIN 专用级联通道连接。这样，中间的 DSP48E2 就有两个输出端口：POUT 和 PCOUT。3 个 DSP48E2 均使用了预加器。预加器的两个输入数据或进行一级流水，或进行二级流水。这样，乘法器的输入数据或为二级流水寄存器，或为三级流水寄存器。预加器或执行减法运算，或执行加法运算。同时，还应注意，预加器的两个输入数据或为 a_r、a_i，或为 b_r、b_i，而预加器的位宽为 27，因此乘法器的另一输入数据的位宽不能超过 18，由此可知，此电路结构要求两个复数实部和虚部的位宽均不能超过 18。根据这些特征并考虑到设计复用性，将预加器输入端的数据流水寄存器的级数、乘法器输入端的流水寄存器的级数和切换预加器功能的控制信号作为参数，形成 VHDL 代码 7-21。顶层设计单元如 VHDL 代码 7-22 所示。从输入到输出需要 5 个时钟周期（Latency=5），与前面的算法相比，该算法增加了 1 个时钟周期。

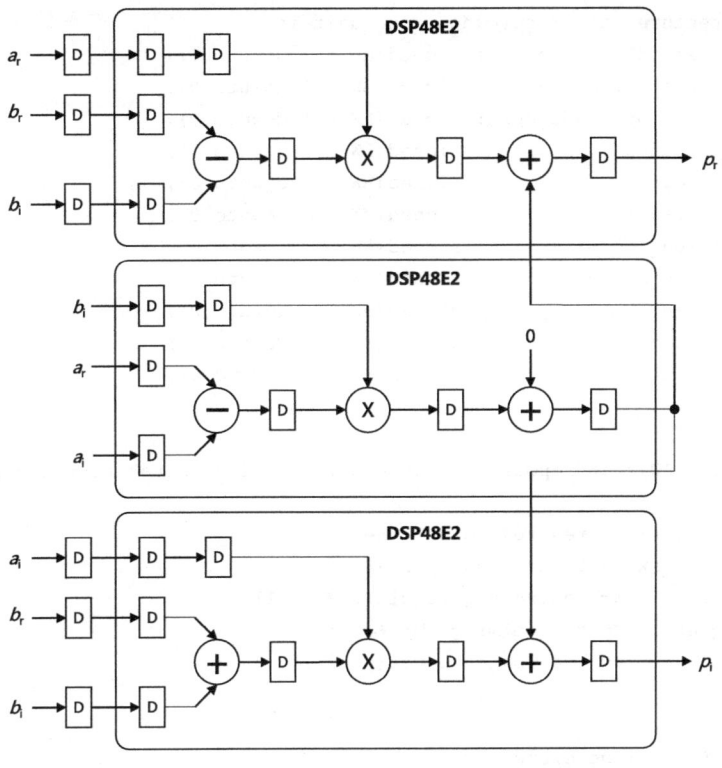

图 7-31

VHDL 代码 7-21

```
1.  --File: cpreadder_mac_unit.vhd
2.  library ieee;
3.  use ieee.std_logic_1164.all;
4.  use ieee.numeric_std.all;
5.
6.  entity cpreadder_mac_unit is
7.    generic (
8.      AW     : positive := 18;
9.      BW     : positive := 18;
10.     AREG   : integer  := 2;
11.     BREG   : integer  := 3;
12.     ADDSUB : bit      := '0';
13.     MW     : positive := AW + BW + 1
14.   );
15.   port (
16.     clk : in std_logic;
17.     ain : in std_logic_vector(AW - 1 downto 0);
18.     din : in std_logic_vector(AW - 1 downto 0);
19.     bin : in std_logic_vector(BW - 1 downto 0);
20.     cin : in std_logic_vector(MW - 1 downto 0);
21.     pout: out std_logic_vector(MW downto 0)
22.   );
23. end entity;
24.
```

```vhdl
25. architecture rtl of cpreadder_mac_unit is
26. signal ain_d1, din_d1      : signed(AW - 1 downto 0);
27. signal bin_d1, bin_d2      : signed(BW - 1 downto 0);
28. signal add_op1, add_op2    : signed(AW - 1 downto 0);
29. signal addreg              : signed(AW     downto 0);
30. signal bmult               : signed(BW - 1 downto 0);
31. signal mreg                : signed(MW - 1 downto 0);
32. signal add_op1_ex          : signed(AW     downto 0);
33. signal add_op2_ex          : signed(AW     downto 0);
34. signal cin_ex              : signed(MW     downto 0);
35. signal mreg_ex             : signed(MW     downto 0);
36. signal pout_i              : signed(MW     downto 0);
37. begin
38.
39.    assert AW > 26 report "AW exceeds limit of 26 bits" severity ERROR;
40.
41.    add_op1_ex <= resize(add_op1, AW + 1);
42.    add_op2_ex <= resize(add_op2, AW + 1);
43.    cin_ex     <= resize(signed(cin), MW + 1);
44.    mreg_ex    <= resize(mreg, MW + 1);
45.    pout       <= std_logic_vector(pout_i);
46.
47.    gen0:
48.    if AREG = 1 generate
49.      process(clk)
50.      begin
51.        if rising_edge(clk) then
52.          add_op1 <= signed(ain);
53.          add_op2 <= signed(din);
54.        end if;
55.      end process;
56.    else generate
57.      process(clk)
58.      begin
59.        if rising_edge(clk) then
60.          ain_d1 <= signed(ain);
61.          add_op1<= ain_d1;
62.          din_d1 <= signed(din);
63.          add_op2<= din_d1;
64.        end if;
65.      end process;
66.    end generate;
67.
68.    gen1:
69.    if BREG = 2 generate
70.      process(clk)
71.      begin
72.        if rising_edge(clk) then
73.          bin_d1 <= signed(bin);
74.          bmult  <= bin_d1;
75.        end if;
```

```
76.      end process;
77.    else generate
78.      process(clk)
79.      begin
80.        if rising_edge(clk) then
81.          bin_d1 <= signed(bin);
82.          bin_d2 <= bin_d1;
83.          bmult  <= bin_d2;
84.        end if;
85.      end process;
86.    end generate;
87.
88.    process(clk)
89.    begin
90.      if rising_edge(clk) then
91.        addreg <= add_op1_ex - add_op2_ex when ADDSUB else
92.                  add_op1_ex + add_op2_ex;
93.        mreg   <= addreg * bmult;
94.        pout_i <= mreg_ex + cin_ex;
95.      end if;
96.    end process;
97. end architecture;
```

> **VHDL 代码 7-22**

```
1.  --File: cplx_mul_v2.vhd
2.  library ieee;
3.  use ieee.std_logic_1164.all;
4.  use ieee.numeric_std.all;
5.
6.  entity cplx_mul_v2 is
7.    generic (
8.      AW : positive := 18;
9.      BW : positive := 18;
10.     MW : positive := AW + BW + 1
11.   );
12.   port (
13.     clk : in std_logic;
14.     ar  : in std_logic_vector(AW - 1 downto 0);
15.     ai  : in std_logic_vector(AW - 1 downto 0);
16.     br  : in std_logic_vector(BW - 1 downto 0);
17.     bi  : in std_logic_vector(BW - 1 downto 0);
18.     pr  : out std_logic_vector(MW downto 0);
19.     pi  : out std_logic_vector(MW downto 0)
20.   );
21. end entity;
22.
23. architecture rtl of cplx_mul_v2 is
24.   signal multcommon_ex : std_logic_vector(MW     downto 0);
25.   signal multcommon    : std_logic_vector(MW - 1 downto 0);
26. begin
27.   multcommon <= multcommon_ex(MW - 1 downto 0);
```

```
28.    middle_cpreadder_mac_unit : entity work.cpreadder_mac_unit
29.      generic map (AW => AW, BW => BW, AREG => 1, BREG => 2, ADDSUB => '1')
30.      port map (
31.        clk => clk, ain => ar, din => ai, bin => bi,
32.        cin => (others => '0'), pout => multcommon_ex
33.      );
34.
35.    top_cpreadder_mac_unit : entity work.cpreadder_mac_unit
36.      generic map (AW => BW, BW => AW, AREG => 2, BREG => 3, ADDSUB => '1')
37.      port map (
38.        clk => clk, ain => br, din => bi, bin => ar,
39.        cin => multcommon, pout => pr
40.      );
41.
42.    bottom_cpreadder_mac_unit : entity work.cpreadder_mac_unit
43.      generic map (AW => BW, BW => AW, AREG => 2, BREG => 3, ADDSUB => '0')
44.      port map (
45.        clk => clk, ain => br, din => bi, bin => ai,
46.        cin => multcommon, pout => pi
47.      );
48. end architecture;
```

注意 VHDL 代码 7-22 第 36 行和第 43 行，generic map 中的形参 AW 和 BW 的实际值分别为 BW 和 AW。这是因为此时 ain 和 din 端口进入的数据分别为 b_r 和 b_i，而 bin 端口进入的数据是 a_r 或 a_i。

VHDL 代码 7-21 和 VHDL 代码 7-22 的描述方式是先将算法分解，并确保分解后的功能单元能够用独立的 DSP48E2 实现。这里我们将算法作为整体进行描述，由 Vivado 完成功能单元的分割和到 DSP48E2 的映射，相应的代码风格如 VHDL 代码 7-23 所示。该模块如果实现 18×18 的复数乘法（两个复数的实部和虚部均为 18 位有符号整数），将消耗 3 个 DSP48E2；若采用 DSP58 实现，则只消耗 2 个 DSP58（对于 DSP58，应采用此代码风格）。从输入到输出，在全流水模式下需要 4 个时钟周期(Latency=4)。如果需要降低 Latency，可以将输入数据的二级流水寄存器降为一级流水寄存器。

📄 VHDL 代码 7-23

```
1.  library ieee;
2.  use ieee.std_logic_1164.all;
3.  use ieee.numeric_std.all;
4.
5.  entity dspcplx_fully_pipeline is
6.    generic (
7.      AW : positive := 18;
8.      BW : positive := 18;
9.      MW : positive := AW + BW
10.   );
11.   port (
12.     clk : in std_logic;
13.     ar  : in std_logic_vector(AW - 1 downto 0);
14.     ai  : in std_logic_vector(AW - 1 downto 0);
15.     br  : in std_logic_vector(BW - 1 downto 0);
```

```
16.        bi  : in std_logic_vector(BW - 1 downto 0);
17.        pr  : out std_logic_vector(MW downto 0);
18.        pi  : out std_logic_vector(MW downto 0)
19.    );
20. end entity;
21.
22. architecture rtl of dspcplx_fully_pipeline is
23. signal ar_d1, ar_d2, ai_d1, ai_d2 : signed(AW - 1 downto 0);
24. signal br_d1, bi_d1, bi_d2        : signed(BW - 1 downto 0);
25. signal addcommon                  : signed(AW     downto 0);
26. signal addr, addi                 : signed(BW     downto 0);
27. signal multcommon, multr, multi   : signed(MW     downto 0);
28. signal pr_int, pi_int             : signed(MW     downto 0);
29. begin
30.    pr <= std_logic_vector(pr_int);
31.    pi <= std_logic_vector(pi_int);
32.    process(clk)
33.    begin
34.       if rising_edge(clk) then
35.          --Inputs are registered AREG=BREG=2
36.          ar_d1 <= signed(ar);
37.          ar_d2 <= ar_d1;
38.          ai_d1 <= signed(ai);
39.          ai_d2 <= ai_d1;
40.          bi_d1 <= signed(bi);
41.          bi_d2 <= bi_d1;
42.          br_d1 <= signed(br);
43.          --Pre-adders are registered ADREG=1
44.          addcommon <= resize(ar_d1, AW + 1) - resize(ai_d1, AW + 1);
45.          addr      <= resize(br_d1, BW + 1) - resize(bi_d1, BW + 1);
46.          addi      <= resize(br_d1, BW + 1) + resize(bi_d1, BW + 1);
47.       end if;
48.    end process;
49.
50.    --Multiplier output is registered MREG=1
51.    process(clk)
52.    begin
53.       if rising_edge(clk) then
54.          multcommon <= bi_d2 * addcommon;
55.          multr      <= ar_d2 * addr;
56.          multi      <= ai_d2 * addi;
57.       end if;
58.    end process;
59.
60.    --Complex output is registered PREG=1
61.    process(clk)
62.    begin
63.       if rising_edge(clk) then
64.          pr_int <= multcommon + multr;
65.          pi_int <= multcommon + multi;
66.       end if;
```

```
67.    end process;
68. end architecture;
```

在复数乘法的基础上很容易实现复数乘累加，其电路结构如图 7-32 所示。复数乘累加需要先计算复数乘法，再分别对乘积的实部和虚部进行累加，因此累加只发生在图 7-32 中的上部和下部的两个 DSP48E2 中。可以看到，从输入到输出需要 6 个时钟周期（Latency=6）。采用 VHDL 描述，如 VHDL 代码 7-24 所示（该代码风格仅适用于 DSP48E1 和 DSP48E2）。

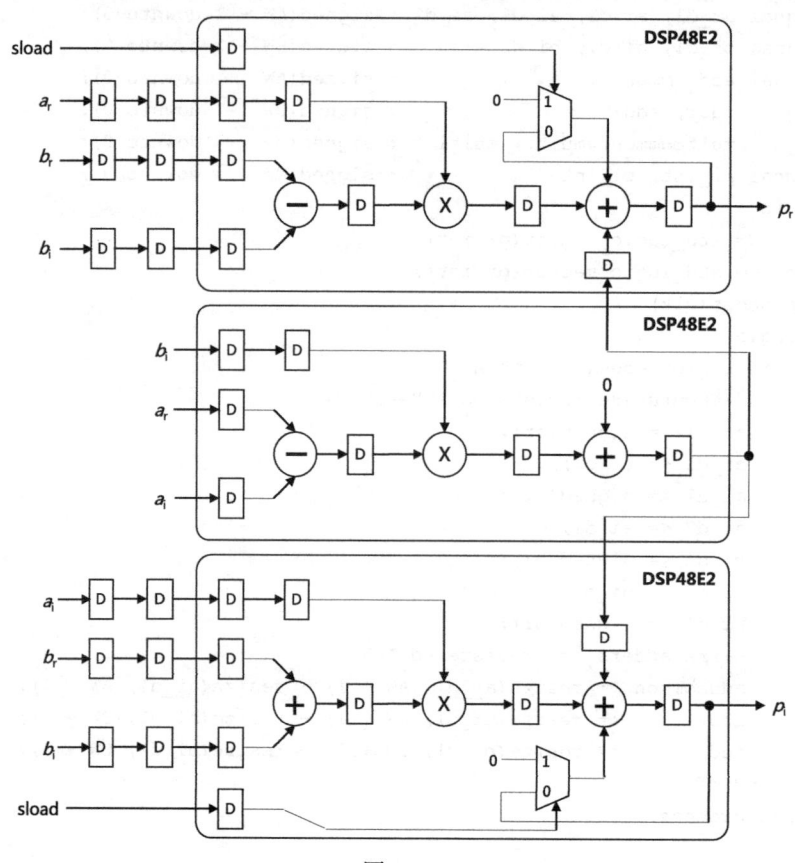

图 7-32

VHDL 代码 7-24

```
1.  --File: cplx_acc_dsp48.vhd
2.  library ieee;
3.  use ieee.std_logic_1164.all;
4.  use ieee.numeric_std.all;
5.
6.  entity cplx_acc_dsp48 is
7.    generic (
8.      AW : positive := 18;
9.      BW : positive := 18;
10.     PW : positive := 40;
11.     MW : positive := AW + BW
12.   );
13.   port (
14.     clk   : in std_logic;
```

```
15.      sload : in std_logic;
16.      ar    : in std_logic_vector(AW - 1 downto 0);
17.      ai    : in std_logic_vector(AW - 1 downto 0);
18.      br    : in std_logic_vector(BW - 1 downto 0);
19.      bi    : in std_logic_vector(BW - 1 downto 0);
20.      pr    : out std_logic_vector(PW - 1 downto 0);
21.      pi    : out std_logic_vector(PW - 1 downto 0)
22.    );
23. end entity;
24.
25. architecture rtl of cplx_acc_dsp48 is
26. signal sload_d1                       : std_logic;
27. signal ar_d1, ar_d2, ar_d3, ar_d4 : signed(AW - 1 downto 0);
28. signal ai_d1, ai_d2, ai_d3, ai_d4 : signed(AW - 1 downto 0);
29. signal br_d1, br_d2, br_d3        : signed(BW - 1 downto 0);
30. signal bi_d1, bi_d2, bi_d3        : signed(BW - 1 downto 0);
31. signal addcommon                      : signed(AW     downto 0);
32. signal addr, addi                     : signed(BW     downto 0);
33. signal mult0, multr, multi            : signed(MW     downto 0);
34. signal common, commonr1, commonr2 : signed(MW     downto 0);
35. signal pr_int, pi_int                 : signed(PW - 1 downto 0);
36. signal old_res_re, old_res_im         : signed(PW - 1 downto 0);
37. begin
38.    pr <= std_logic_vector(pr_int);
39.    pi <= std_logic_vector(pi_int);
40.    process(clk)
41.    begin
42.      if rising_edge(clk) then
43.         ar_d1    <= signed(ar);
44.         ar_d2    <= ar_d1;
45.         ai_d1    <= signed(ai);
46.         ai_d2    <= ai_d1;
47.         br_d1    <= signed(br);
48.         br_d2    <= br_d1;
49.         br_d3    <= br_d2;
50.         bi_d1    <= signed(bi);
51.         bi_d2    <= bi_d1;
52.         bi_d3    <= bi_d2;
53.         sload_d1 <= sload;
54.      end if;
55.    end process;
56.
57.    process(clk)
58.    begin
59.      if rising_edge(clk) then
60.         addcommon <= resize(ar_d1, AW + 1) - resize(ai_d1, AW + 1);
61.         mult0     <= addcommon * bi_d2;
62.         common    <= mult0;
63.      end if;
64.    end process;
65.
```

```vhdl
66.    old_res_re <= (others => '0') when sload_d1 else pr_int;
67.
68.    process(clk)
69.    begin
70.      if rising_edge(clk) then
71.        ar_d3    <= ar_d2;
72.        ar_d4    <= ar_d3;
73.        addr     <= resize(br_d3, BW + 1) - resize(bi_d3, BW + 1);
74.        multr    <= addr * ar_d4;
75.        commonr1 <= common;
76.        pr_int   <= resize(multr, PW) + resize(commonr1, PW) + resize(old_res_re, PW);
77.      end if;
78.    end process;
79.
80.    old_res_im <= (others => '0') when sload_d1 else pi_int;
81.
82.    process(clk)
83.    begin
84.      if rising_edge(clk) then
85.        ai_d3    <= ai_d2;
86.        ai_d4    <= ai_d3;
87.        addi     <= resize(br_d3, BW + 1) + resize(bi_d3, BW + 1);
88.        multi    <= addi * ai_d4;
89.        commonr2 <= common;
90.        pi_int   <= resize(multi, PW) + resize(commonr2, PW) + resize(old_res_im, PW);
91.      end if;
92.    end process;
93. end architecture;
```

对于 DSP58，实现复数乘累加应采用如 VHDL 代码 7-25 所示的描述方式，最终，Vivado 将采用 2 个 DSP58 实现。从输入到输出需要 2 个时钟周期（Latency=2）。

VHDL 代码 7-25

```vhdl
1.  --File: cplx_acc_dsp58.vhd
2.  library ieee;
3.  use ieee.std_logic_1164.all;
4.  use ieee.numeric_std.all;
5.
6.  entity cplx_acc_dsp58 is
7.    generic (
8.      AW : positive := 18;
9.      BW : positive := 18;
10.     PW : positive := 40;
11.     MW : positive := AW + BW
12.   );
13.   port (
14.     clk   : in std_logic;
15.     sload : in std_logic;
16.     ar    : in std_logic_vector(AW - 1 downto 0);
```

```vhdl
17.        ai       : in std_logic_vector(AW - 1 downto 0);
18.        br       : in std_logic_vector(BW - 1 downto 0);
19.        bi       : in std_logic_vector(BW - 1 downto 0);
20.        pr       : out std_logic_vector(PW - 1 downto 0);
21.        pi       : out std_logic_vector(PW - 1 downto 0)
22.    );
23. end entity;
24.
25. architecture rtl of cplx_acc_dsp58 is
26. signal sload_d1              : std_logic;
27. signal ar_d1, ar_d2          : signed(AW - 1 downto 0);
28. signal ai_d1, ai_d2          : signed(AW - 1 downto 0);
29. signal br_d1                 : signed(BW - 1 downto 0);
30. signal bi_d1, bi_d2          : signed(BW - 1 downto 0);
31. signal addcommon             : signed(AW     downto 0);
32. signal addr, addi            : signed(BW     downto 0);
33. signal multcommon            : signed(MW     downto 0);
34. signal multcommon_d          : signed(MW     downto 0);
35. signal multr, multi          : signed(MW     downto 0);
36. signal multr_d, multi_d      : signed(MW     downto 0);
37. signal pr_int, pr_old        : signed(PW - 1 downto 0);
38. signal pi_int, pi_old        : signed(PW - 1 downto 0);
39. begin
40.    pr <= std_logic_vector(pr_int);
41.    pi <= std_logic_vector(pi_int);
42.    process(clk)
43.    begin
44.       if rising_edge(clk) then
45.          ar_d1     <= signed(ar);
46.          ar_d2     <= ar_d1;
47.          ai_d1     <= signed(ai);
48.          ai_d2     <= ai_d1;
49.          br_d1     <= signed(br);
50.          bi_d1     <= signed(bi);
51.          bi_d2     <= bi_d1;
52.          sload_d1 <= sload;
53.       end if;
54.    end process;
55.
56.    process(clk)
57.    begin
58.       if rising_edge(clk) then
59.          addcommon <= resize(ar_d1, AW + 1) - resize(ai_d1, AW + 1);
60.          addr      <= resize(br_d1, BW + 1) - resize(bi_d1, BW + 1);
61.          addi      <= resize(br_d1, BW + 1) + resize(bi_d1, BW + 1);
62.       end if;
63.    end process;
64.
65.    process(clk)
66.    begin
67.       if rising_edge(clk) then
```

```
68.         multcommon <= bi_d2 * addcommon;
69.         multr      <= ar_d2 * addr;
70.         multi      <= ai_d2 * addi;
71.      end if;
72.   end process;
73.
74.   pr_old <= (others => '0') when sload_d1 else pr_int;
75.   pi_old <= (others => '0') when sload_d1 else pi_int;
76.
77.   process(clk)
78.   begin
79.      if rising_edge(clk) then
80.         pr_int <= resize(multcommon, PW) + resize(multr, PW) + pr_old;
81.         pi_int <= resize(multcommon, PW) + resize(multi, PW) + pi_old;
82.      end if;
83.   end process;
84. end architecture;
```

7.4 向量内积代码风格

DSP58 支持向量模式，可计算两个长度为 3 的向量的内积，硬件电路结构如图 7-5 所示。用 VHDL 描述，如 VHDL 代码 7-26 所示。注意被注释掉的第 44 行和第 45 行，尽管 DSP58 也支持该功能，但 Vivado 2022.1 版本无法将其映射为 DSP58，而会采用 LUT 和 FF 实现。

VHDL 代码 7-26

```
1.  --File: dot_prod.vhd
2.  library ieee;
3.  use ieee.std_logic_1164.all;
4.  use ieee.numeric_std.all;
5.
6.  entity dot_prod is
7.    generic (
8.      AW : positive := 9;
9.      BW : positive := 8;
10.     MW : positive := AW + BW
11.   );
12.   port (
13.     clk          : in std_logic;
14.     a0, a1, a2   : in std_logic_vector(AW - 1 downto 0);
15.     b0, b1, b2   : in std_logic_vector(BW - 1 downto 0);
16.     pout         : out std_logic_vector(MW + 1 downto 0)
17.   );
18. end entity;
19.
20. architecture rtl of dot_prod is
21.   signal a0_d1, a1_d1, a2_d1 : signed(AW - 1 downto 0);
```

```
22.  signal b0_d1, b1_d1, b2_d1    : signed(BW - 1 downto 0);
23.  signal mult0, mult1, mult2    : signed(MW - 1 downto 0);
24.  signal dotpr, dotpr_d1        : signed(MW + 1 downto 0);
25. begin
26.    --Input registers
27.    process(clk)
28.    begin
29.      if rising_edge(clk) then
30.        a0_d1 <= signed(a0);
31.        a1_d1 <= signed(a1);
32.        a2_d1 <= signed(a2);
33.        b0_d1 <= signed(b0);
34.        b1_d1 <= signed(b1);
35.        b2_d1 <= signed(b2);
36.      end if;
37.    end process;
38.
39.    --Multiplier
40.    mult0 <= a0_d1 * b0_d1;
41.    mult1 <= a1_d1 * b1_d1;
42.    mult2 <= a2_d1 * b2_d1;
43.    dotpr <= resize(mult0, MW+2) + resize(mult1, MW+2) + resize(mult2, MW+2);
44.    --dotpr <= resize(-mult0, MW+2) + resize(-mult1, MW+2) + resize(-mult2, MW+2);
45.    --dotpr <= resize(mult0, MW+2) + resize(-mult1, MW+2) + resize(mult2, MW+2);
46.
47.    --Registering dot product output MREG=PREG=1
48.    process(clk)
49.    begin
50.      if rising_edge(clk) then
51.        dotpr_d1 <= dotpr;
52.        pout     <= std_logic_vector(dotpr_d1);
53.      end if;
54.    end process;
55. end architecture;
```

向量内积的一个典型应用是计算矩阵乘法。例如，两个 3×3 的矩阵 U 和矩阵 V 相乘，可表示为

$$W = U \times V = \begin{bmatrix} u_0 \\ u_1 \\ u_2 \end{bmatrix} \times \begin{bmatrix} v_0 & v_1 & v_2 \end{bmatrix}$$

这里，将 U 按行向量的方式表示，将 V 按列向量的方式表示。采用 3 个 DSP58 并行计算 3 个向量内积，每个 DSP58 端口的数据流如表 7-7 所示。由此数据流可知该结构具有以下几个特征：①3 个 DSP58 的端口 a0 的输入数据完全相同，端口 a1 和 a2 亦是如此；②0 号 DSP58 的端口 b0、b1 和 b2 对应 V 的第 0 列，1 号 DSP58 对应第 1 列，2 号 DSP58 对应第 2 列；③3 个 DSP58 的端口 b0、b1 和 b2 的输入数据持续的时钟周期个数是端口 a0、a1 和 a2 的 3 倍。

表 7-7

端口	DSP58 #0			DSP58 #1			DSP58 #2		
a0	u_{00}	u_{10}	u_{20}	u_{00}	u_{10}	u_{20}	u_{00}	u_{10}	u_{20}
a1	u_{01}	u_{11}	u_{21}	u_{01}	u_{11}	u_{21}	u_{01}	u_{11}	u_{21}
a2	u_{02}	u_{12}	u_{22}	u_{02}	u_{12}	u_{22}	u_{02}	u_{12}	u_{22}
b0	v_{00}	v_{00}	v_{00}	v_{01}	v_{01}	v_{01}	v_{02}	v_{02}	v_{02}
b1	v_{10}	v_{10}	v_{10}	v_{11}	v_{11}	v_{11}	v_{12}	v_{12}	v_{12}
b2	v_{20}	v_{20}	v_{20}	v_{21}	v_{21}	v_{21}	v_{22}	v_{22}	v_{22}
p	w_{00}	w_{10}	w_{20}	w_{01}	w_{11}	w_{21}	w_{02}	w_{12}	w_{22}

7.5 以加法为核心运算的电路结构

DSP48E1、DSP48E2 和 DSP58 内部都包含一个算术逻辑单元，可执行加法、累加运算（这里以 DSP48E2 为例，若无特殊说明，相应应用也适用于 DSP48E1 和 DSP58），对于大位宽的加法运算是非常适用的。例如，一个 48 位的有符号数加法器用 DSP48E2 实现，具体电路结构如图 7-33 所示。此时，两个操作数由 A、B、C 三个端口进入。其中，A、B 两个端口拼接为 48 位，C 端口为 48 位。

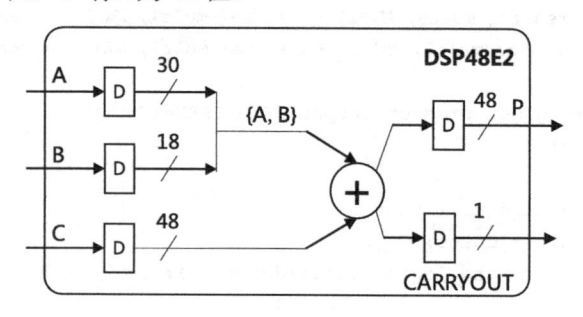

图 7-33

默认情况下，对于加法运算，Vivado 都会采用查找表和进位链的方式实现，而不会将其映射为 DSP48E2。此时，可用综合属性 USE_DSP，并将其值设置为 "yes" 来指导 Vivado 采用 DSP48E2 实现加法运算，如 VHDL 代码 7-27 所示。该描述方式对于 DIW（输入数据位宽）小于 48 的情形是适用的，当 DIW 大于或等于 48 时，尽管会采用 DSP48E2 实现，但会消耗额外的查找表和触发器，此时可采用 DSP48E2 原语，合理设置 OPMODE、ALUMODE 和 INMODE 实现加法功能（后续应用均无相应代码风格支持，需要用原语方式实现）。图 7-33 所示的电路可以实现加法运算，也可以实现减法运算，还可以实现动态切换加/减功能。

VHDL 代码 7-27

```
1.  --File: adder.vhd
2.  library ieee;
3.  use ieee.std_logic_1164.all;
```

```vhdl
4.  use ieee.numeric_std.all;
5.
6.  entity adder is
7.    generic (
8.      DIW         : positive := 32;
9.      USE_DSP_VAL : string := "yes";
10.     DOW         : positive := DIW + 1
11.   );
12.   port (
13.     clk : in std_logic;
14.     ain : in std_logic_vector(DIW - 1 downto 0);
15.     bin : in std_logic_vector(DIW - 1 downto 0);
16.     sum : out std_logic_vector(DOW - 1 downto 0)
17.   );
18. end entity;
19.
20. architecture rtl of adder is
21.   signal ain_d1   : signed(ain'RANGE);
22.   signal bin_d1   : signed(bin'RANGE);
23.   signal sum_int  : signed(sum'RANGE);
24.   attribute USE_DSP : string;
25.   attribute USE_DSP of sum_int : signal is USE_DSP_VAL;
26. begin
27.   sum <= std_logic_vector(sum_int);
28.   process(clk)
29.   begin
30.     if rising_edge(clk) then
31.       ain_d1  <= signed(ain);
32.       bin_d1  <= signed(bin);
33.       sum_int <= resize(ain_d1, DOW) + resize(bin_d1, DOW);
34.     end if;
35.   end process;
36. end architecture;
```

在查找表利用率较高或加法器要工作在较高时钟频率下的情况中，采用 DSP48E2 实现大位宽加法运算是一个可选的优化方法。例如，一个 32 位加法器，用 VHDL 代码 7-27 的方式实现，将 USE_DSP 设置为 "no"，那么会消耗 32 个查找表、97 个触发器、5 个进位链（CARRY8），运行在 500MHz 下，功耗为 0.832W；如果将 USE_DSP 设置为 "yes"，那么仅会消耗 1 个 DSP48E2，运行在 500MHz 下，功耗为 0.829W。

> 设计规则 8：对于大位宽的加法器，如果需要运行在 300MHz 及 300MHz 以上的频率下，可以尝试采用 DSP48 实现。在加法运算较为密集而乘法运算较少甚至没有乘法运算的情况下，也可以采用 DSP48 实现加法运算，以节省查找表资源。

DSP48E2 还可以实现 4 个 46 位整数相加。此时要用到两个 DSP48E2，如图 7-34 所示。4 个操作数分别为 op1、op2、op3 和 op4。考虑到 4 个数相加位宽会增加 2 位（$\log_2 4$），同时，在这种情形下，DSP48E2 的进位输出不可用，这就是这里的数据的最大位宽是 46 位的原因。

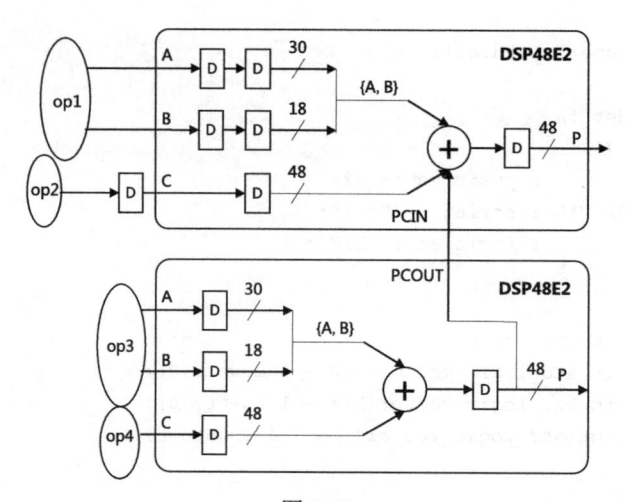

图 7-34

超过 48 位的大位宽加法运算也可以采用 DSP48E2 实现，如图 7-35 所示，两个 96 位数据相加（符合位扩展之后为 96 位），借助 CARRYCASOUT 和 CARRYCASIN 端口，将其分解为低 48 位和高 48 位。

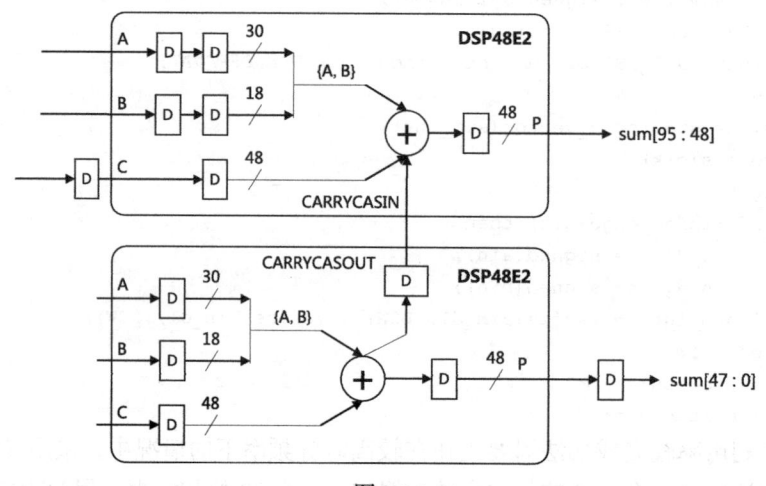

图 7-35

DSP48E2 还可以实现累加运算。一种情形是累加器仅有一个输入端口和一个输出端口，其电路结构如图 7-36（a）所示，电路执行的功能是 $C+P$。若输入数据为 32 位，则可执行 2^{16} 次累加（16=48-32）。另一种情形是累加器有两个输入端口和一个输出端口，其电路结构如图 7-36（b）所示。此时，电路执行的功能是 $\{A,B\}+C+P$，其中，$\{A,B\}$ 构成一个操作数，C 是另一个操作数。

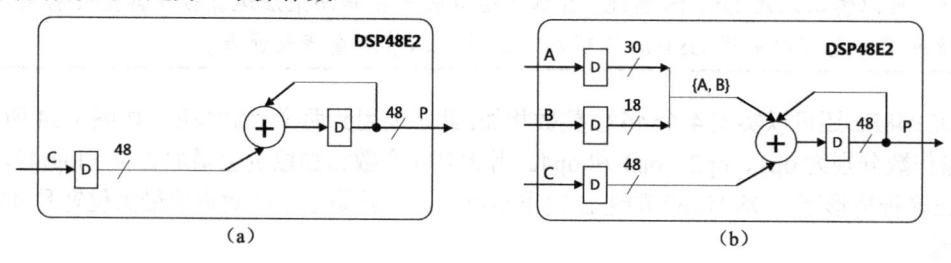

图 7-36

对于更大位宽的累加器，如 96 位累加器，则需要用两个 DSP48E2 实现，如图 7-37 所示。与图 7-35 中的电路结构类似，仍然将 96 位数据分解为低 48 位和高 48 位。

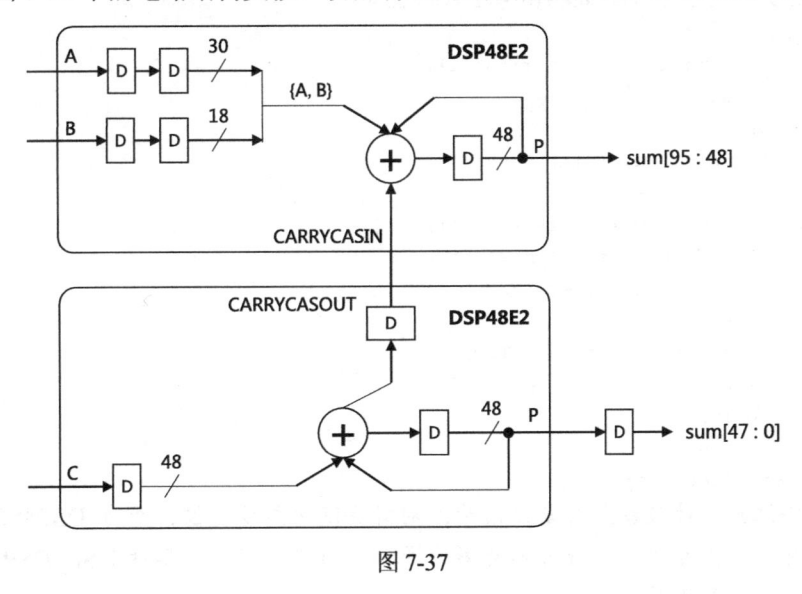

图 7-37

应用案例 3：用 DSP48E2 实现大位宽计数器

计数器在 FPGA 设计中被广泛使用，如 RAM 的读/写地址，本质上都是计数器。VHDL 代码 7-28 描述了一个位宽（W）、计数最大值（CNT_MAX）和计数步进（STEP）可参数化的计数器。设置 USE_DSP 为"yes"，可采用 DSP48E2 实现。

VHDL 代码 7-28

```
1.  --File: counter.vhd
2.  library ieee;
3.  use ieee.std_logic_1164.all;
4.  use ieee.numeric_std.all;
5.
6.  entity counter is
7.    generic (
8.      W            : positive := 48;
9.      STEP         : integer  := 2;
10.     CNT_MAX      : integer  := 16;
11.     USE_DSP_VAL  : string   := "yes"
12.   );
13.   port (
14.     clk : in std_logic;
15.     rst : in std_logic;
16.     cnt : out std_logic_vector(W - 1 downto 0)
17.   );
18. end entity;
19.
20. architecture rtl of counter is
21.   attribute USE_DSP : string;
```

```
22.   attribute USE_DSP of rtl : architecture is USE_DSP_VAL;
23.   constant CNT_MAXI : unsigned(W - 1 downto 0) := to_unsigned(CNT_MAX, W);
24.   signal cnt_i : unsigned(W - 1 downto 0);
25. begin
26.   cnt <= std_logic_vector(cnt_i);
27.   process(clk)
28.   begin
29.     if rising_edge(clk) then
30.       if rst then
31.         cnt_i <= (others => '0');
32.       elsif cnt_i = CNT_MAXI then
33.         cnt_i <= (others => '0');
34.       else
35.         cnt_i <= cnt_i + STEP;
36.       end if;
37.     end if;
38.   end process;
39. end architecture;
```

这里需要注意，计数器若有复位信号，则需要使用同步复位，因为 DSP48E2 的所有复位端口仅支持同步复位。步进并不会影响资源利用率。同时，属性 USE_DSP 的作用对象是 architecture，而不是 signal。

DSP48E2 中的 ALU 还支持 SIMD 模式，从而可实现并行加法器，具体电路结构如图 7-38 所示。图中，左侧表明 ALU 可配置为 4 个 12 位的加法器，右侧表明 ALU 可配置为 2 个 24 位的加法器。

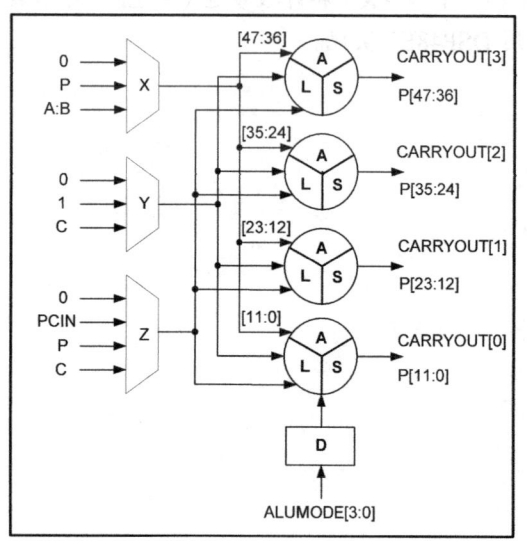

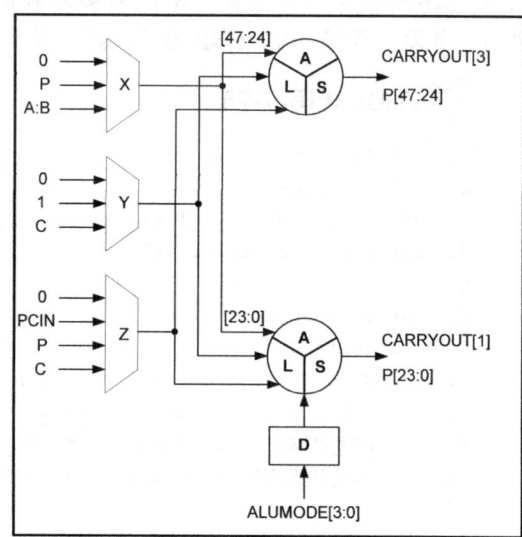

图 7-38

SIMD 模式有相应的代码风格支持，如 VHDL 代码 7-29 所示。代码第 31 行将 USE_DSP（use_dsp 可全部大写，也可全部小写）设置为 "simd"，在代码第 18 行中，参数 N 定义了加法器的个数，可以是 1、2 或 4。代码第 19 行定义了加法器输入数据位宽。从输入到输出需要 2 个时钟周期（Latency=2）。

VHDL 代码 7-29

```vhdl
1.  --File: parallel_adder.vhd
2.  library ieee;
3.  use ieee.std_logic_1164.all;
4.  use ieee.numeric_std.all;
5.
6.  package dsp_pkg is
7.    type logic_array is array(natural range<>) of std_logic_vector;
8.    type signed_array is array(natural range<>) of signed;
9.  end package;
10.
11. library ieee;
12. use ieee.std_logic_1164.all;
13. use ieee.numeric_std.all;
14. use work.dsp_pkg.all;
15.
16. entity parallel_adder is
17.   generic (
18.     N : positive := 4;
19.     W : positive := 10
20.   );
21.   port (
22.     clk : in std_logic;
23.     a   : in logic_array(0 to N - 1)(W - 1 downto 0);
24.     b   : in logic_array(0 to N - 1)(W - 1 downto 0);
25.     sum : out logic_array(0 to N - 1)(W - 1 downto 0)
26.   );
27. end entity;
28.
29. architecture rtl of parallel_adder is
30.   attribute USE_DSP : string;
31.   attribute USE_DSP of rtl : architecture is "simd";
32.   signal a_d1 : signed_array(0 to N - 1)(W - 1 downto 0);
33.   signal b_d1 : signed_array(0 to N - 1)(W - 1 downto 0);
34. begin
35.   process(clk)
36.   begin
37.     if rising_edge(clk) then
38.       for i in a'RANGE loop
39.         a_d1(i) <= signed(a(i));
40.         b_d1(i) <= signed(b(i));
41.         sum(i)  <= std_logic_vector(a_d1(i) + b_d1(i));
42.       end loop;
43.     end if;
44.   end process;
45. end architecture;
```

应用案例 4：16 个 8 位有符号数相加

这 16 个数据是并行的。16 个数据相加会使位宽增加 4 位（$\log_2^{16}$），因此可将 parallel_adder 中的参数 W 设置为 12，相应的系统框图如图 7-39 所示。这里，并不是每个 parallel_adder 都需要用 DSP48E2 实现，而是根据设计需求来决定。例如，可将第一级（最左侧）两个 parallel_adder 用 DSP48E2 实现，而其余用查找表实现，以获取 DSP48E2 和查找表资源利用率的折中。

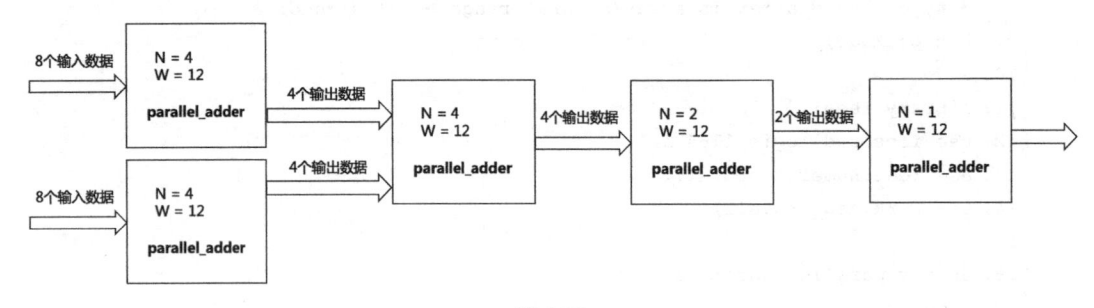

图 7-39

7.6 管理时序路径上的乘加器

表 7-2 表明当 DSP48E1、DSP48E2 或 DSP58 执行乘法运算时，紧随乘法器之后的寄存器对设计性能有很大的影响，当未使用乘法器而只使用 ALU 时，紧随 ALU 之后的寄存器对设计性能有很大影响，尤其是当系统时钟频率大于 300MHz 时。对于已有的网表文件（.dcp），从中找到这些有"缺陷"的 DSP 单元尤为关键。这可通过 Tcl 脚本实现，如 Tcl 代码 7-2 所示。

Tcl 代码 7-2

```
1.  #File: get_dsp.tcl
2.  set dsp_mult [get_cells -hier -filter "REF_NAME == DSP48E1 && USE_MULT != NONE"]
3.  set dsp_no_mreg [filter $dsp_mult "MREG == 0"]
4.  set dsp_no_preg [get_cells -hier -filter "REF_NAME == DSP48E1 && PREG == 0"]
5.  if {[llength $dsp_no_mreg] > 0} { show_objects $dsp_no_mreg -name dsp_no_mreg}
6.  if {[llength $dsp_no_preg] > 0} { show_objects $dsp_no_preg -name dsp_no_preg}
```

其中，代码第 2 行用于获取使用了乘法器的 DSP48E1（对于 DSP48E2 和 DSP58，只需要将 REF_NAME 的值换为 DSP48E2 和 DSP58 即可），代码第 3 行则是从中过滤出没有使用 MREG 的 DSP48E1。代码第 4 行用于获取未使用 PREG 的 DSP48E1。代码第 5 行和第 6 行则将这些结果显示出来。

在高速设计中，以 DSP48E1、DSP48E2 或 DSP58 为起点的时序路径，如果输出数据要给 BRAM 或 UltraRAM，那么最好确保路径为 DSP→FF→BRAM/URAM。

7.7 思考空间

1. 试描述 DSP48E1、DSP48E2 和 DSP58 在结构上的关键差异。
2. 试用 DSP Macro IP 实现乘累加运算。
3. 试用浮点 IP 实现单精度浮点乘加运算，并比较 DSP48E2 和 DSP58 在实现浮点运算时的主要差异。
4. 将 VHDL 代码 7-2 描述的乘加运算中的同步复位改为异步复位，并用 DSP58 实现，比较采用 DSP58 实现与采用 DSP48E2 实现时的主要差异。
5. 一个 8 阶 FIR 滤波器，若输入数据为 12 位有符号整数，滤波器系数为 16 位有符号整数，试判断输出数据的最大位宽。若采用 DSP48E2 以脉动形式实现，试判断需要消耗多少个 DSP48E2。若系数为偶对称，试判断需要多少个 DSP48E2。
6. 一个 8 阶滤波器，滤波器系数为 {−16,−24,−32,−48,48,32,24,16}，输入数据为 16 位有符号整数，在此基础上，输出数据的最大位宽为多少比较合适？
7. 现用 DSP58 实现两个 6×6 的矩阵乘法，两个矩阵元素均为 8 位有符号整数，需要执行多少次乘法运算和多少次加法运算？给出可行的方案并列出每种方案消耗的 DSP58 的个数。
8. 试用 DSP48E2 中的预加器实现一个 2 选 1 的数据选择器。
9. 试用 DSP48E2 实现一个 8 选 1 的数据选择器。
10. 试用 DSP58 实现一个 3×3 的矩阵乘法。
11. 对于如图 7-34 和图 7-35 所示的电路结构，试给出 OPMODE、ALUMODE 和 INMODE 的具体数值。
12. 对于 VHDL 代码 7-28 所描述的计数器，增加一个输入信号 incdec，当其值为 1 时，进行增计数，当其值为 0 时，进行减计数，并观察能否用 DSP48E2 实现。
13. 试用 DSP48E2 实现两个 48 位数据按位与运算，给出 OPMODE、ALUMODE 和 INMODE 的具体数值。
14. 用 VHDL 描述一个累加器，输入数据为 16 位，输出数据为 24 位，该累加器可累加多少次而不会导致数据溢出。若用 DSP48E2 实现，则一个 DSP48E2 可实现多少个这样的累加器？
15. 试用 VHDL 分别描述一个 8×8 的有符号整数乘法运算和无符号整数乘法运算，在 Vivado 下，观察其综合结果，并推断在默认情形下，是否乘法运算一定会采用乘加器资源实现。

第 8 章

优化状态机

8.1 基本概念

组合逻辑的输出仅取决于当前输入值,而时序逻辑的输出不仅与当前输入值有关,还与之前的存储信息有关。这些存储信息位于存储单元,即触发器内。换言之,触发器是构成时序逻辑电路不可或缺的元素。存储在触发器内的信息可以被视为系统的一个状态,如果系统会在有限个状态之间转移,那么我们就可以用有限状态机(Finite State Machine,FSM,以下简称状态机)来设计这个系统。

状态机的本质是对具有逻辑顺序或时序规律的事件的一种描述方式。这意味着凡是具有逻辑顺序或时序规律的系统都适合采用状态机的方式实现,这也意味着这个系统将会在不同的状态之间转移,这就涉及状态机的三个基本要素:状态、输出和输入。状态也叫状态变量。在逻辑设计中,使用状态划分逻辑顺序或时序规律。输出是指在某种状态时特定发生的事件。输入是指状态机进入每种状态的条件,有的状态机没有输入条件,其中,状态转移较为简单,有的状态机有输入条件,当某个输入条件存在时,才能转移到相应的状态。

这三个要素共同构成了状态机的两大功能:第一个功能是根据外部变化实现状态转移;第二个功能是根据特定状态和输入数值来驱动输出。针对第一个功能,我们把状态机的状态分为现态(Current State,CS)和次态(Next State,NS)。现态是状态机在某个特定时刻的状态,次态是状态机将要转移到达的状态。次态跟现态的关系类似于触发器输入跟输出之间的关系。在状态转移条件满足前,状态机会一直驻留在某种状态下。有时,一种状态的转移路径可以有多条。在这种情况下,具有最高优先级的转移路径决定了状态如何转移。对于简单的状态机,其状态个数通常为 2 个或 3 个;但对于复杂的状态机,其状态个数可能为 10 个以上,有时还会包含嵌套的子状态机。默认或复位状态通常称为空闲(Idle)状态。

状态机可分为摩尔型(Moore)状态机和米勒型(Mealy)状态机。摩尔型状态机的输出只取决于状态,独立于输入。换言之,输出表达式中仅包含状态变量,这样,只有当状态变化时输出才会变化。米勒型状态机的输出不只与状态有关,还与输入有关,如图 8-1 所示。从图 8-1 中也可以看出,状态机是组合逻辑和时序逻辑的混合电路。

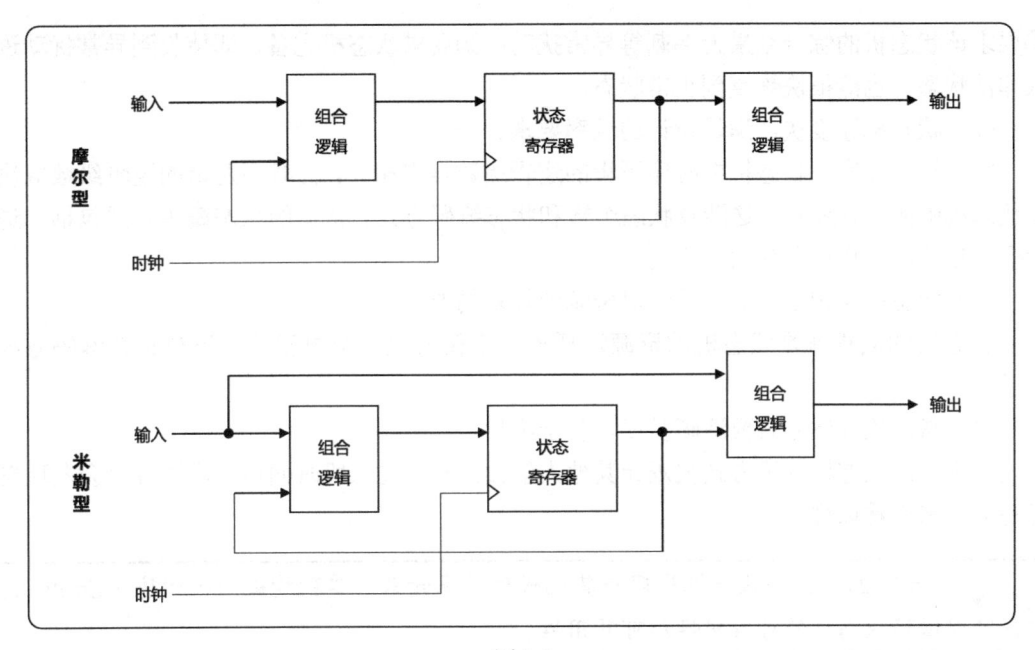

图 8-1

由于摩尔型状态机的输出来自组合逻辑，因此可能会有毛刺，从而产生不利影响。而米勒型状态机的输出还与输入有关，这很可能会恶化时序，从而限制系统的工作频率。更好的方法是将这两种状态机混合使用，也就是说输出既与状态有关，又与输入有关，并且采用寄存器输出，如图 8-2 所示。

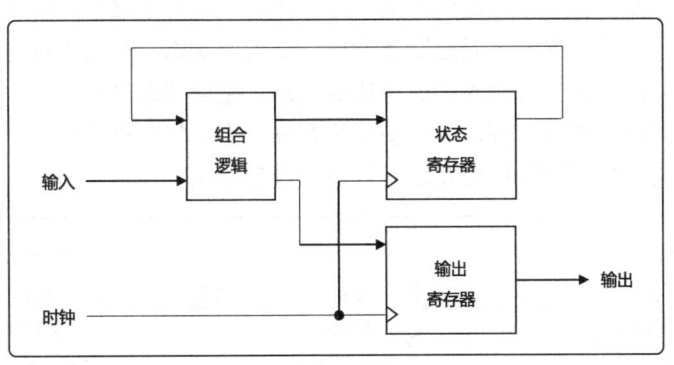

图 8-2

状态机的设计思路有两种：一种是从状态变量入手。如果一个系统具有逻辑顺序或时序规律，那么我们就可以自然而然地规划出状态，从这些状态入手，分析每种状态的输入、状态转移条件和输出，从而完成电路功能。另一种是从输出关系入手。这些输出相当于状态的输出，不同的输出对应不同的状态，据此可回溯规划每种状态、状态转移条件和状态输入。

从 RTL 代码风格的角度而言，好的状态机的描述方式要遵循以下标准。
（1）状态机要足够安全，而且具有高稳定性。

所谓状态机安全，是指状态机不会进入死循环，特别是不会进入非法状态，而且一旦因某些扰动进入非法状态，也能很快恢复，进入合法的状态循环中。这里面有两层含义：

①要求该状态机的综合结果无毛刺等异常扰动；②要求状态机完备，即使收到异常扰动进入非法状态，也能很快恢复到正常状态。

（2）状态机速度快，满足设计的频率要求。

状态机的速度与状态机中时序逻辑的逻辑级数紧密相关，而时序逻辑的逻辑级数取决于状态机中的组合逻辑。这涉及状态个数和状态编码方式。常见的状态编码方式包括二进制码、独热码、格雷码和约翰逊码。

（3）状态机面积小，满足设计对资源消耗量的要求。

状态机的面积就是状态机的资源消耗量，不仅与状态个数相关，还与状态编码方式相关。

（4）状态机描述方式要清晰易懂，易于维护。

不规范的状态机描述方式很难让其他人解读，甚至过一段时间后，设计者也会发现自己写的状态机难以维护。

> 💡 **设计规则 1**：将状态机代码与其他代码隔离开来，单独构成一个模块（Entity），这样可以增强状态机的可维护性和可复用性。

8.2 状态机代码风格

如 8.1 节所述，状态机是组合逻辑和时序逻辑的混合体。绝大多数状态机包含三个功能模块：状态发生器、次态译码逻辑和输出译码逻辑，这三者之间的关系如图 8-3 所示。其中，状态发生器由时序逻辑单元触发器构成，输入为次态，输出为现态。次态译码逻辑是纯粹的组合逻辑，根据当前输入和现态决定次态，完成状态转移。输出译码逻辑则根据现态（摩尔型状态机）和当前输入（米勒型状态机）决定状态机的输出，这部分可以是纯粹的组合逻辑输出，也可以添加触发器，形成寄存器输出。

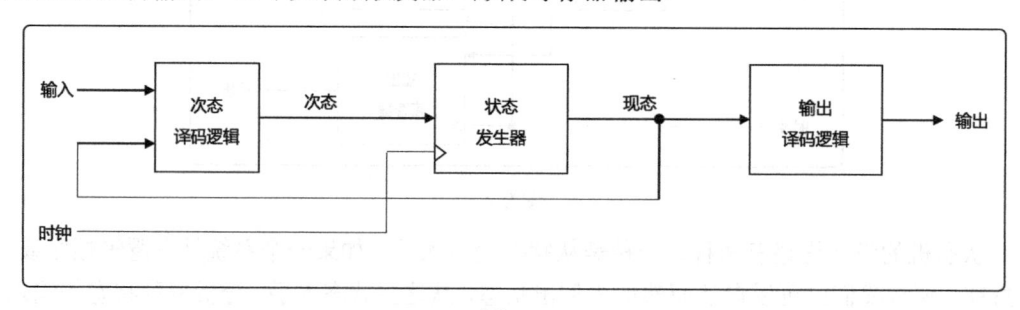

图 8-3

为便于说明，我们以一个序列检测器为例，该序列检测器用于检测输入序列是否包含"101"，一旦包含"101"（允许序列重叠，即序列"10101"会被认为包含两个"101"），就将输出置 1 并保持一个时钟周期，其状态转移图如图 8-4 所示。图中，圆圈内横线上方为状态名称，横线下方为该状态下对应的输出。带箭头的直线或曲线的起始端为现态，终止端为次态，线上的标记为当前输入值。不难判断，这是一个摩尔型状态机。

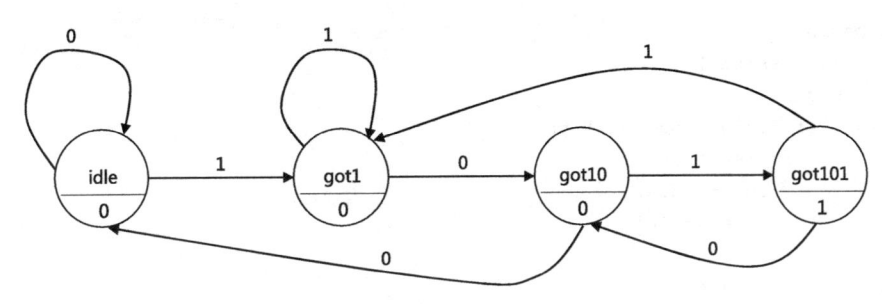

图 8-4

根据状态机的三个功能模块，我们可以采用三种描述方法。第一种描述方法为单进程状态机，如图 8-5 所示。顾名思义，就是将三个功能模块放在一个 process 中描述。相应的 RTL 代码如 VHDL 代码 8-1 所示。代码第 16 行采用用户定义的枚举类型定义状态寄存器。括号中的值为该枚举类型可能出现的所有值（枚举值），在这里就是不同的状态。之所以采用用户定义（User Defined）类型，是因为现态和次态均属于同一类型，这样就可以使用同一枚举类型声明多个枚举变量。代码第 19 行是一个 process 进程，整个状态机只有这一个进程（这就是单进程名字的由来），也只定义了一个状态寄存器 cs。

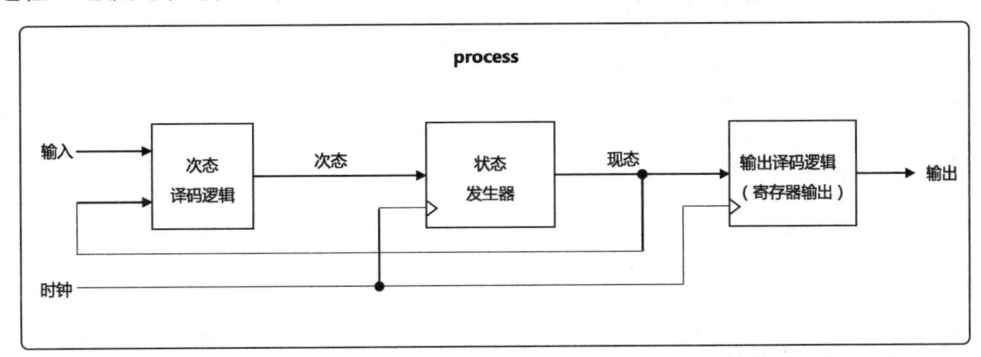

图 8-5

VHDL 代码 8-1

```vhdl
1.  --File: moore_detector_v1.vhd
2.  library ieee;
3.  use ieee.std_logic_1164.all;
4.  use ieee.numeric_std.all;
5.
6.  entity moore_detector_v1 is
7.    port (
8.      clk  : in std_logic;
9.      rst  : in std_logic;
10.     sin  : in std_logic;
11.     done : out std_logic
12.   );
13. end entity;
14.
15. architecture rtl of moore_detector_v1 is
16.   type state_t is (idle, got1, got10, got101);
17.   signal cs : state_t;
```

```vhdl
18. begin
19.   process(clk)
20.   begin
21.     if rising_edge(clk) then
22.       if rst then
23.         cs <= idle;
24.         done <= '0';
25.       else
26.         case cs is
27.           when idle =>
28.             done <= '0';
29.             cs   <= got1 when sin else idle;
30.           when got1 =>
31.             done <= '0';
32.             cs   <= got10 when sin else got1;
33.           when got10 =>
34.             done <= '0';
35.             cs <= got101 when sin else idle;
36.           when got101 =>
37.             done <= '1';
38.             cs <= got1 when sin else got10;
39.           when others =>
40.             done <= '0';
41.             cs <= idle;
42.         end case;
43.       end if;
44.     end if;
45.   end process;
46. end architecture;
```

> **🔍 VHDL-2008 新特性**
>
> VHDL-2008 支持在 process 进程中使用 when-else 语句。就功能而言,when-else 和 if-else 等效,但 when-else 语句可以使代码变得更为简洁。

> **💡 设计规则 2**:在 VHDL 中,对于状态寄存器,应将其声明为用户定义的枚举类型。

第二种描述方法为双进程状态机,如图 8-6 所示。状态发生器作为时序逻辑单独用一个进程 process(clk),而次态译码逻辑和输出译码逻辑为组合逻辑,将两者放在一个进程 process(all)里,如 VHDL 代码 8-2 所示。代码第 15 行定义了枚举类型作为状态寄存器的数据类型。代码第 16 行定义了现态 cs 和次态 ns 两个状态寄存器,这里就体现了用户定义类型的好处。代码第 18 行至第 23 行是固定格式:复位有效时,将现态置为 idle,否则将次态赋值给现态。这其实就是一个同步复位触发器。代码第 25 行至第 46 行是一个 process(all)进程。在这个进程里,通过一个 case 语句完成次态译码逻辑和输出译码逻辑。与单进程描述方式相比,这种描述方式非常清晰、直观,而且可以作为固定模板。

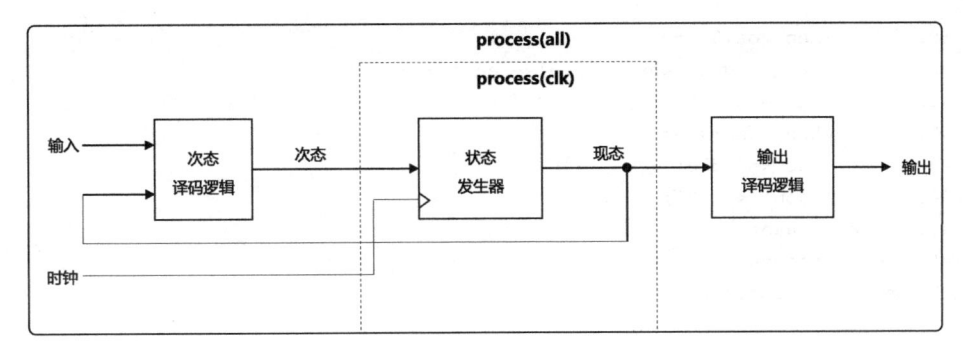

图 8-6

VHDL 代码 8-2

```vhdl
1.  --File: moore_detector_v2.vhd
2.  library ieee;
3.  use ieee.std_logic_1164.all;
4.
5.  entity moore_detector_v2 is
6.    port (
7.      clk : in std_logic;
8.      rst : in std_logic;
9.      sin : in std_logic;
10.     done : out std_logic
11.   );
12. end entity;
13.
14. architecture rtl of moore_detector_v2 is
15.   type state_t is (idle, got1, got10, got101);
16.   signal cs, ns : state_t;
17. begin
18.   process(clk)
19.   begin
20.     if rising_edge(clk) then
21.       cs <= idle when rst else ns;
22.     end if;
23.   end process;
24.
25.   process(all)
26.   begin
27.     ns <= idle;
28.     done <= '0';
29.     case cs is
30.       when idle =>
31.         ns <= got1 when sin else idle;
32.         done <= '0';
33.       when got1 =>
34.         ns <= got1 when sin else got10;
35.         done <= '0';
36.       when got10 =>
37.         ns <= got101 when sin else idle;
38.         done <= '0';
```

```
39.        when got101 =>
40.          ns <= got1 when sin else got10;
41.          done <= '1';
42.        when others =>
43.          ns <= idle;
44.          done <= '0';
45.      end case;
46.   end process;
47. end architecture;
```

> 🔍 **VHDL-2008 新特性**
>
> 如果 process 进程中描述的是纯组合逻辑，那么需要完整地描述敏感变量列表。VHDL-2008 使用关键字 all 避免了这一烦琐的过程。

第三种描述方式为多进程状态机，如图 8-7 所示，继承双进程描述方式中的 process(clk) 进程，然后将其中的 process(all) 进程拆分为两个 process(all) 进程：一个负责次态译码逻辑，另一个负责输出译码逻辑。这样的好处是可以很容易地对输出译码逻辑添加寄存器，使其以寄存器形式输出，从而可以降低相关路径的逻辑级数，改善设计的 F_{max}。相应的 RTL 代码如 VHDL 代码 8-3 所示。代码第 18 行至第 23 行为 process(clk) 进程，与双进程描述方式中的 process(clk) 完全一致。代码第 25 行至第 40 行为 process(all) 进程，可以看到，其实就是提取出双进程描述方式 process(all) 中的状态转移部分。代码第 42 行至第 47 行对应输出译码逻辑，采用寄存器输出。因为该逻辑较为简单，用一个 if-else 语句即可完成。对应到电路上就是一个 2 选 1 的 MUX 加上触发器。有些状态机的输出译码逻辑较为复杂，当时钟频率比较高时，采用寄存器输出非常有必要。

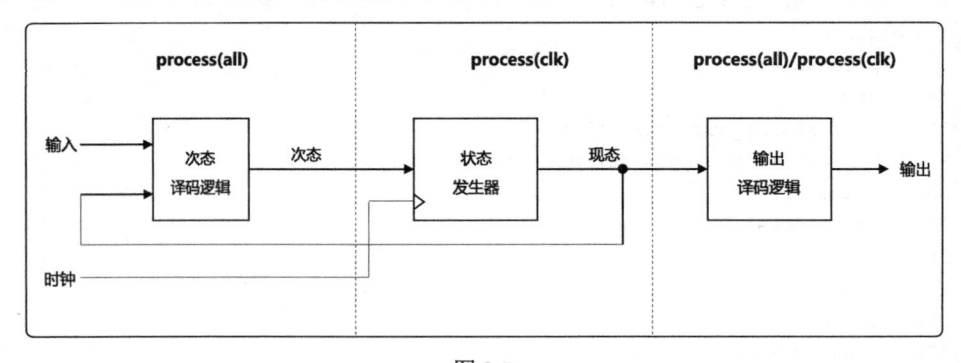

图 8-7

📄 **VHDL 代码 8-3**

```
1.  --File: moore_detector_v3.vhd
2.  library ieee;
3.  use ieee.std_logic_1164.all;
4.
5.  entity moore_detector_v3 is
6.    port (
7.      clk   : in std_logic;
8.      rst   : in std_logic;
```

```vhdl
9.      sin : in std_logic;
10.     done : out std_logic
11.  );
12. end entity;
13.
14. architecture rtl of moore_detector_v3 is
15.   type state_t is (idle, got1, got10, got101);
16.   signal cs, ns : state_t;
17. begin
18.   process(clk)
19.   begin
20.     if rising_edge(clk) then
21.       cs <= idle when rst else ns;
22.     end if;
23.   end process;
24.
25.   process(all)
26.   begin
27.     ns <= idle;
28.     case cs is
29.       when idle =>
30.         ns <= got1 when sin else idle;
31.       when got1 =>
32.         ns <= got1 when sin else got10;
33.       when got10 =>
34.         ns <= got101 when sin else idle;
35.       when got101 =>
36.         ns <= got1 when sin else got10;
37.       when others =>
38.         ns <= idle;
39.     end case;
40.   end process;
41.
42.   process(clk)
43.   begin
44.     if rising_edge(clk) then
45.       done <= '1' when cs = got101 else '0';
46.     end if;
47.   end process;
48. end architecture;
```

> **🔍 VHDL-2008 新特性**
>
> 尽管 VHDL-2008 引入了匹配相等操作符 "?=", 但代码第 45 行不能写为
> done <= cs ?= got101
> 这是因为此操作符不支持枚举类型。

对上述三种描述方式进行仿真, 可得输入/输出信号之间的时序关系, 如图 8-8 所示。可以看到, 单进程和多进程描述方式的输出是在同一时刻由低电平变为高电平的, 双进程则提前一个时钟周期改变, 这是因为双进程输出译码逻辑是纯粹的组合逻辑, 而其余两者

均采用寄存器输出。事实上，做一些微小的代码层面的改动就可以使双进程和多进程输出在同一时钟周期有效，即将 VHDL 代码 8-3 第 45 行的 cs 换为 ns，也就是将基于现态的输出译码逻辑替换为基于次态的输出译码逻辑。

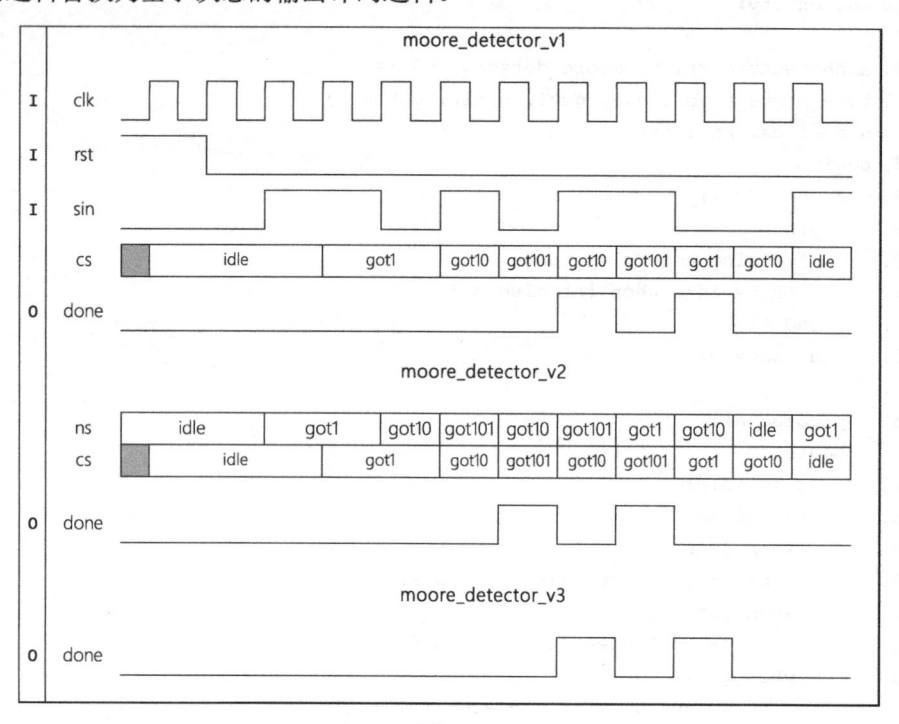

图 8-8

从资源利用率的角度看，这三种描述方式消耗的 LUT 和触发器相当，双进程的触发器会少一些，也是源于其未使用寄存器输出。

对这三种描述方式进行比较，可以发现，单进程描述方式比较混乱，在描述状态个数较多的复杂状态机时更为糟糕。双进程和多进程的描述思路很清晰，每个 process(clk)/process(all)进程有明确功能和设计意图。三者的详细比较如表 8-1 所示。

表 8-1

比较内容	单进程	双进程	多进程
推荐等级	不推荐	推荐	推荐
代码简洁程度（尤其是对于复杂的状态机）	冗长	最简洁	简洁
代码可靠性和可维护性	低	较高	高
代码风格的规范性	低	格式化，规范	格式化，规范
进程个数	1	2	3
是否有组合逻辑输出	无	有	有
是否有利于综合和布局布线	不利于	利于	利于

事实上，"101" 序列检测器也可以采用米勒型状态机实现，状态转移图如图 8-9 所示。图中，圆圈表示状态，直线或曲线上会以 "数字 A/数字 B"（A、B 或为 0，或为 1）的形式标记，其中，数字 A 为当前输入，数字 B 为对应的状态输出。相应的 RTL 代码如 VHDL 代码 8-4 所示。代码第 16 行仍采用用户定义的枚举类型作为状态寄存器的数据类型，第 17 行定义了现态和次态寄存器。采用多进程描述方式，第一个进程对应代码第 19 行至第

24 行,为状态发生器;第 26 行至第 39 行为次态译码逻辑;第 41 行至第 48 行为输出译码逻辑。可以看到,这里的输出与输入有关,如代码第 45 行所示。输入/输出及内部信号的时序关系如图 8-10 所示。

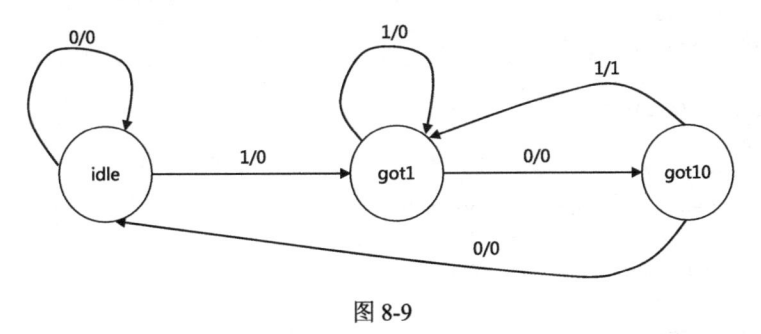

图 8-9

VHDL 代码 8-4

```
1.  --File: mealy_detector_v3.vhd
2.  library ieee;
3.  use ieee.std_logic_1164.all;
4.  use ieee.numeric_std.all;
5.
6.  entity mealy_detector_v3 is
7.    port (
8.      clk  : in std_logic;
9.      rst  : in std_logic;
10.     sin  : in std_logic;
11.     done : out std_logic
12.   );
13. end entity;
14.
15. architecture rtl of mealy_detector_v3 is
16. type state_t is (idle, got1, got10);
17. signal cs, ns : state_t;
18. begin
19.   process(clk)
20.   begin
21.     if rising_edge(clk) then
22.       cs <= idle when rst else ns;
23.     end if;
24.   end process;
25.
26.   process(all)
27.   begin
28.     ns <= idle;
29.     case cs is
30.       when idle =>
31.         ns <= got1 when sin else idle;
32.       when got1 =>
33.         ns <= got1 when sin else got10;
```

```
34.       when got10 =>
35.         ns <= got1 when sin else idle;
36.       when others =>
37.         ns <= idle;
38.     end case;
39.   end process;
40.
41.   process(all)
42.   begin
43.     case cs is
44.       when idle | got1 => done <= '0';
45.       when got10       => done <= sin ?= '1';
46.       when others      => done <= '0';
47.     end case;
48.   end process;
49.
50. end architecture;
```

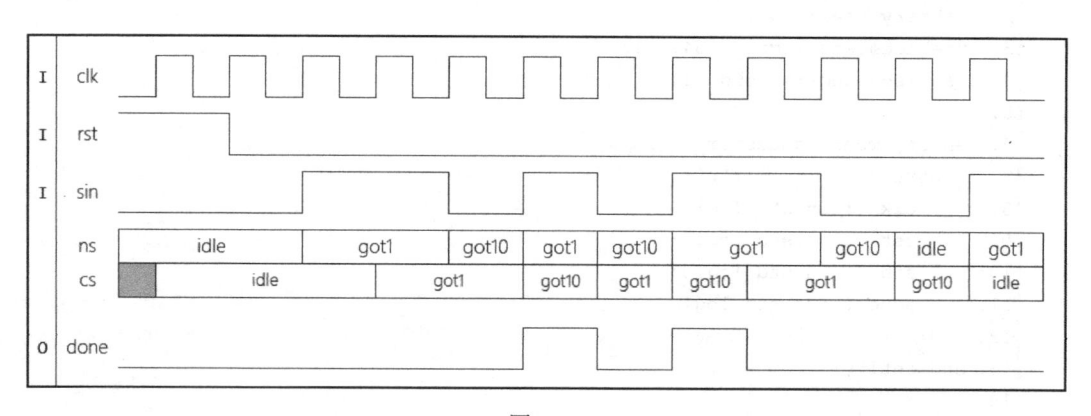

图 8-10

VHDL-2008 新特性

VHDL-2008 引入了匹配相等操作符 "?=",其返回值类型与操作数类型相同,如 VHDL 代码 8-4 第 45 行所示。当 sin 为 1 时返回 1,否则返回 0。实际上,这里可直接描述为 done <= sin。

设计规则 3:使用 case 语句时,当多个条件分支对应的输出完全一致时,可将这几个条件通过操作符 "|" 写在一起,从而达到简化代码的目的,如 VHDL 代码 8-4 第 44 行所示,表明在状态 idle 和 got1 下,输出 done 均为 0。

应用案例 1:获取 FPGA 芯片的 DNA

Xilinx FPGA 在出厂之前都会被 "植入" 一个 DNA,相当于芯片的身份证。每片 FPGA 的 DNA 是唯一的,且是永久的、非易失的。对于 UltraScale/UltraScale+ FPGA,可以通过原语 DNA_PORTE2 获取该芯片的 DNA，共 96 位。DNA_PORTE2 的端口说明如表 8-2 所示，其工作原理如图 8-11 所示：当 read 为高电平时，将 96 位 DNA 数据加载到移位寄存器中，其中，最低位在 read 由高电平变为低电平后会立即出现在 dout 端口；当 shift 为高电平时，在时钟的驱动下，其余 95 位 DNA 数据依次从 dout 端口输出。

表 8-2

名 称	方 向	位 宽	基 本 功 能
clk	输入	1	时钟
din	输入	1	用户数据输入端口
dout	输出	1	DNA 输出端口
read	输入	1	当 read 为高电平时，加载 DNA 信息；当 read 为低电平时，读取用户输入数据
shift	输入	1	移位寄存器使能信号

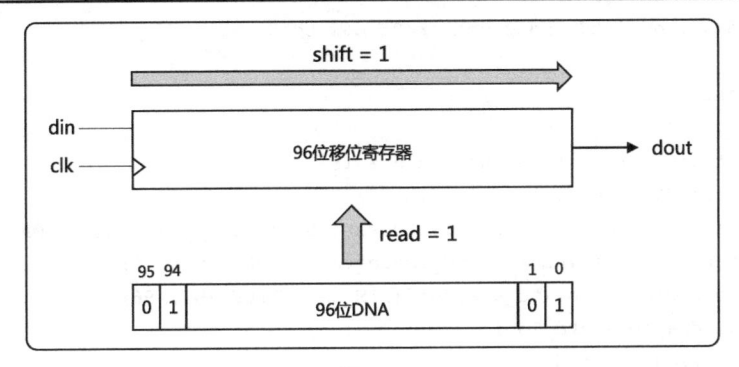

图 8-11

采用状态机实现 DNA_PORTE2 的控制电路，状态转移图如图 8-12 所示。图中，圆圈表示状态，圆圈下的圆角方框对应该状态下的输出。计数器 cnt 用于控制 shift_dna 的持续时间。RTL 代码如 VHDL 代码 8-5 所示。代码第 32 行至第 42 行是对 DNA_PORTE2 原语的实例化，其中，第 34 行的 SIM_DNA_VALUE 仅用于仿真。由于在状态 idle、shift_done 和 write_dna 下输出一致，因此可以使用代码第 68 行的描述方式。输入/输出及内部信号的时序关系如图 8-13 所示。

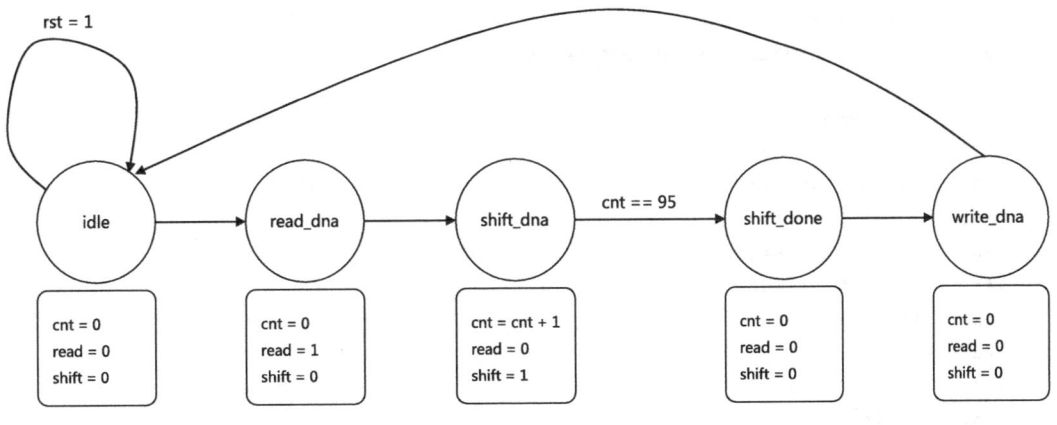

图 8-12

VHDL 代码 8-5

```vhdl
1.  --File: get_device_dna.vhd
2.  library UNISIM;
3.  use UNISIM.vcomponents.all;
4.  library ieee;
5.  use ieee.std_logic_1164.all;
6.  use ieee.numeric_std.all;
7.
8.  entity get_device_dna is
9.    generic ( SIM_DNA_VALUE_INT : integer := 16 );
10.   port (
11.     clk : in std_logic;
12.     rst : in std_logic;
13.     dna : out std_logic_vector(95 downto 0)
14.   );
15. end entity;
16.
17. architecture rtl of get_device_dna is
18.   constant CNT_MAXI : positive := 95;
19.   constant CNT_MAX   : unsigned(6 downto 0) := to_unsigned(CNT_MAXI, 7);
20.   constant SIM_DNA_VALUE : std_logic_vector(95 downto 0)
21.     := std_logic_vector(to_unsigned(SIM_DNA_VALUE_INT, 96));
22.   signal dna_so : std_logic;
23.   signal read   : std_logic;
24.   signal shift  : std_logic;
25.   signal cnt    : unsigned(6 downto 0);
26.   signal cnt_d1 : unsigned(6 downto 0);
27.   signal dna_i  : std_logic_vector(95 downto 0);
28.   type state_t is (idle, read_dna, shift_dna, shift_done, write_dna);
29.   signal cs, ns : state_t;
30. begin
31.
32.   i_DNA_PORTE2 : DNA_PORTE2
33.   generic map (
34.     SIM_DNA_VALUE => SIM_DNA_VALUE
35.   )
36.   port map (
37.     DOUT  => dna_so,
38.     CLK   => clk,
39.     DIN   => '0',
40.     READ  => read,
41.     SHIFT => shift
42.   );
43.
44.   process(clk)
45.   begin
46.     if rising_edge(clk) then
```

```vhdl
47.        cs <= idle when rst else ns;
48.      end if;
49.   end process;
50.
51.   process(all)
52.   begin
53.     ns <= cs;
54.     case cs is
55.       when idle       => ns <= read_dna;
56.       when read_dna   => ns <= shift_dna;
57.       when shift_dna  => ns <= shift_done when cnt ?= CNT_MAX else shift_dna;
58.       when shift_done => ns <= write_dna;
59.       when write_dna  => ns <= idle;
60.       when others     => ns <= idle;
61.     end case;
62.   end process;
63.
64.   process(clk)
65.   begin
66.     if rising_edge(clk) then
67.       case cs is
68.         when idle | shift_done | write_dna =>
69.           cnt   <= (others => '0');
70.           read  <= '0';
71.           shift <= '0';
72.         when read_dna =>
73.           cnt   <= (others => '0');
74.           read  <= '1';
75.           shift <= '0';
76.         when shift_dna =>
77.           cnt   <= cnt + 1;
78.           read  <= '0';
79.           shift <= '1';
80.         when others =>
81.           cnt   <= (others => '0');
82.           read  <= '0';
83.           shift <= '0';
84.       end case;
85.     end if;
86.   end process;
87.
88.   process(clk)
89.   begin
90.     if rising_edge(clk) then
91.       cnt_d1 <= cnt;
92.     end if;
93.   end process;
94.
```

```vhdl
95.   gen0:
96.   for i in 0 to CNT_MAXI generate
97.     process(clk)
98.     begin
99.       if rising_edge(clk) then
100.        if cnt_d1 = to_unsigned(i, 7) then
101.          dna_i(i) <= dna_so;
102.        end if;
103.      end if;
104.    end process;
105.  end generate;
106.
107.  process(clk)
108.  begin
109.    if rising_edge(clk) then
110.      if cs = write_dna then
111.        dna <= dna_i;
112.      end if;
113.    end if;
114.  end process;
115.end architecture;
```

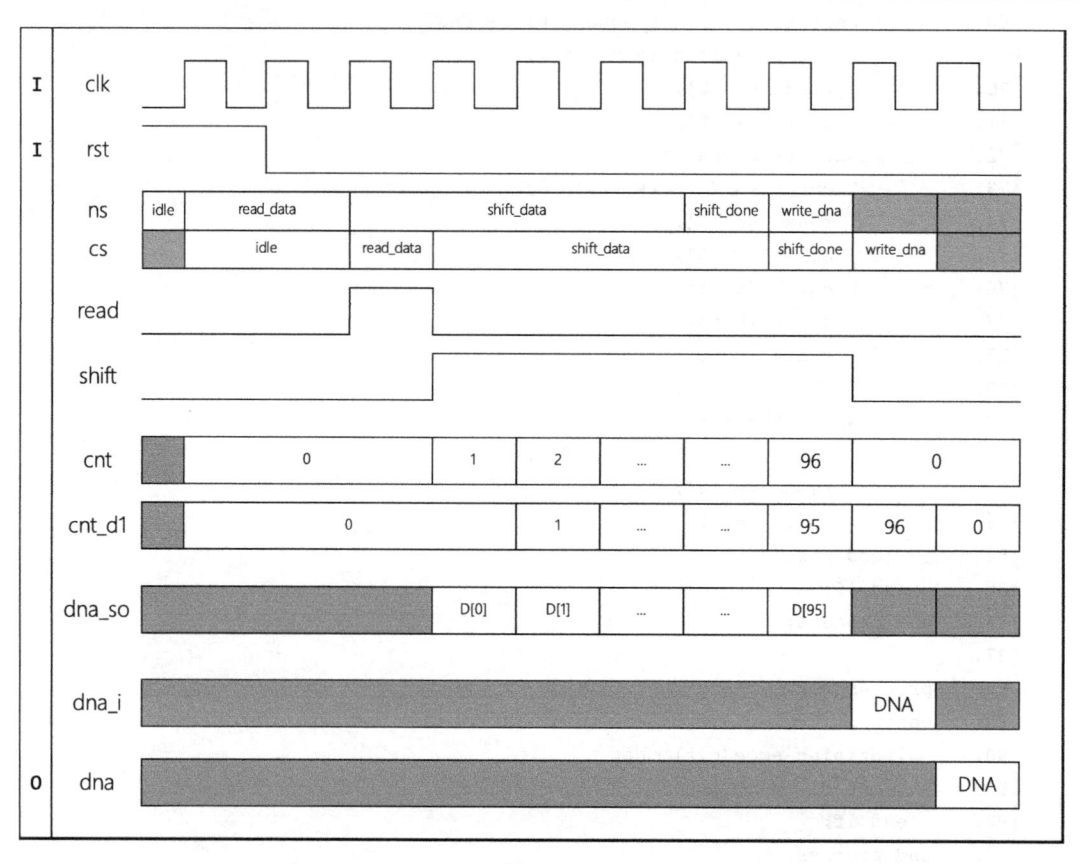

图 8-13

应用案例 2：UART 接口

UART（Universal Asynchronous Receiver Transmitter）为通用异步收发器，是一种常用的接口。它还有其他的名称，如串口、RS-232 接口、COM 端口或 RS-485 接口。常用来在计算机和 FPGA 之间进行数据传输。之所以称为"异步"，是因为该接口在数据传输时仅有数据端口而无时钟端口。UART 有几个重要的参数，如表 8-3 所示。

表 8-3

波 特 率	起 始 位	有效数据位	奇偶校验位	终 止 位
就 UART 而言，调制一个码元需要的比特数为 1，故波特率=比特率。常见的波特率有：4 800、9 600、14 400、19 200、38 400、57 600、115 200 等	通常为 1 位，且为 0	通常为 5~9 位	通常为 0~1 位	通常为 1~2 位，且为 1

注：波特率为每秒传送的码元数，波特率=比特率×调制一个码元所需要的比特数。

这些参数在 UART 的接收端和发送端要保持一致。在信息传输通道中，携带数据信息的信号单元叫作码元，每秒钟通过信道传输的码元数称为码元传输速率，简称波特率。波特率是传输通道频宽的指标。若波特率为 115200，则每秒传送 115200 个码元。这里，每个码元为一比特（一位），因此对应 115200 比特。假定 FPGA 内接收时钟的时钟周期为 PERIOD，单位为纳秒（ns），那么每比特持续的时钟周期数可表示为

$$\text{CYCLES\_PER\_BIT} = \text{ceil}(10^9/\text{Baud\_Rate}/\text{PERIOD})$$

式中，Baud_Rate 为波特率，ceil 表示向上取整。依此公式可知，波特率为 115200，时钟周期为 100ns 时，每比特将持续 87 个时钟周期。

先看 UART 接收器。输入端口包括时钟、串行输入数据（rx_si，由 UART 发送器提供）、并行输出数据（rx_byte，8 位）和输出数据有效标记信号（rx_dv，高电平有效，持续一个时钟周期）。接收器接收的数据的波形如图 8-14 所示。这里以 1 位起始位、8 位数据位和一位终止位为例。起始位固定为 0，终止位固定为 1。假定波特率为 115200，时钟周期为 100ns，则每位数据（包括起始位和终止位）将持续 87 个时钟周期，在设计时要保证采样发生在数据窗的中间位置。显然，采用状态机方式实现比较简单。

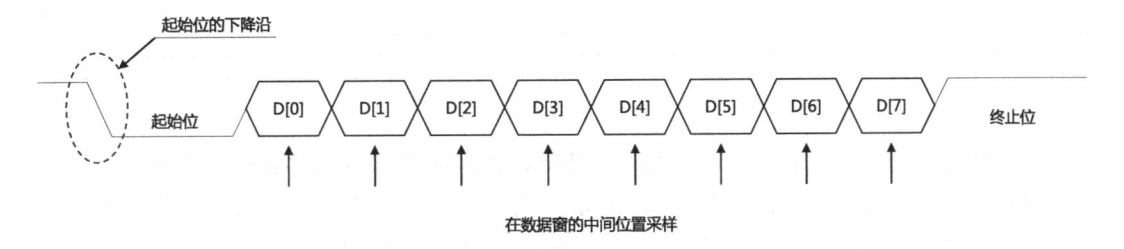

图 8-14

接收器的状态转移图如图 8-15 所示，每个状态的含义如表 8-4 所示。在图 8-15 中，CNT_MAX 为固定常数 86，HALF_CNT_MAX 为 42。rx_cnt 为计数器输出值。每个方框位于相应的状态之下，对应该状态下的输出。基于此，可形成如图 8-16 所示的时序图。当检测到起始位时，计数器开始计数，一旦计数到 42 就清零，此时仍对应起始位。计数器

清零之后继续计数,再次计数到42时,起始位结束,当计数到86时,正好对应D[0]的中间位置。利用此技巧可以保证后续计数到86,即对应数据位的中间位置。

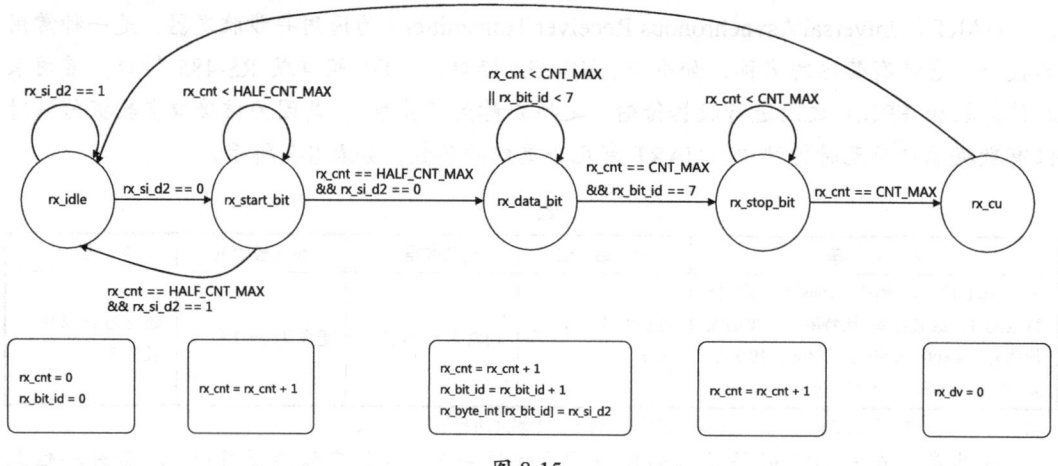

图 8-15

表 8-4

状态	含义
rx_idle	表明接收器处于空闲状态
rx_start_bit	表明接收器检测到起始位
rx_data_bit	表明接收器检测到数据位并对其进行采样
rx_stop_bit	表明接收器检测到终止位
rx_cu(Clean Up)	表明接收器处于复位状态

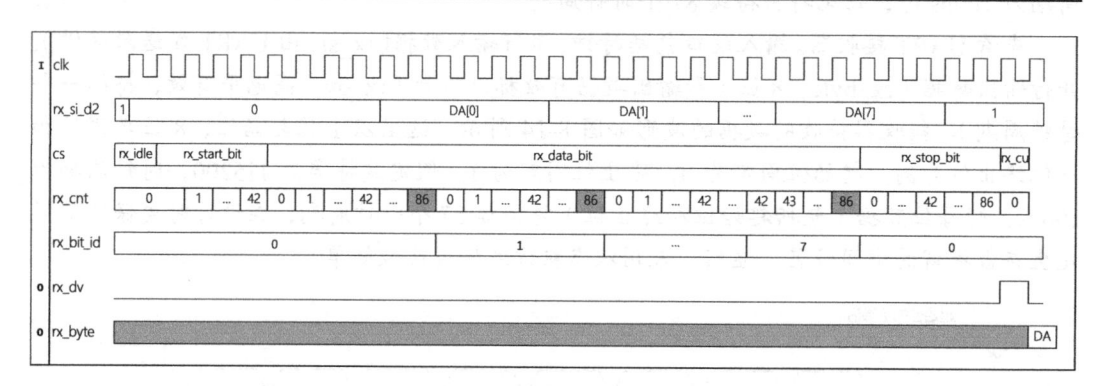

图 8-16

我们将依据如图8-15所示的状态转移图和如图8-16所示的时序图进行代码描述。先看代码的第一部分,如VHDL代码8-6所示。代码第19行利用系统函数ceil完成向上取整的功能,但由于ceil输出为实数,因此用integer的方式将其转换为整数。代码第23行和第24行分别是输入串行数据rx_si的一级延迟信号和二级延迟信号,一定要给其设置初始值'1',代码第36行至第37行正是利用这两个信号,采用双触发器的形式将rx_si同步到时钟clk下的。代码第28行定义了状态寄存器类型,第29行定义了现态和次态两个状态寄存器。代码第35行为状态发生器。

VHDL 代码 8-6

```vhdl
1.  --File: uart_rx.vhd
2.  library ieee;
3.  use ieee.std_logic_1164.all;
4.  use ieee.numeric_std.all;
5.  use ieee.math_real.all;
6.
7.  entity uart_rx is
8.    generic ( CYCLES_PER_BIT : positive := 87 ); --#N of clock cycles/bit
9.    port (
10.     clk    : in std_logic;
11.     rx_si  : in std_logic;
12.     rx_dv  : out std_logic;
13.     rx_byte : out std_logic_vector(7 downto 0)
14.   );
15. end entity;
16.
17. architecture rtl of uart_rx is
18.   constant CNT_MAXI : positive := CYCLES_PER_BIT - 1;
19.   constant CNTW : positive := integer(ceil(log2(real(CNT_MAXI))));
20.   constant HALF_CNT_MAXI : integer := integer(ceil(real(CYCLES_PER_BIT/2-1)));
21.   constant CNT_MAX : unsigned(CNTW - 1 downto 0) := to_unsigned(CNT_MAXI, CNTW);
22.   constant HALF_CNT_MAX : unsigned(CNTW - 1 downto 0) := to_unsigned(HALF_CNT_MAXI, CNTW);
23.   signal rx_si_d1     : std_logic := '1';
24.   signal rx_si_d2     : std_logic := '1';
25.   signal rx_cnt       : unsigned(CNTW - 1 downto 0);
26.   signal rx_bit_id    : unsigned(2 downto 0);
27.   signal rx_byte_int  : unsigned(7 downto 0);
28.   type state_t is (rx_idle, rx_start_bit, rx_data_bit, rx_stop_bit, rx_cu);
29.   signal cs, ns : state_t;
30. begin
31.   --Double-register the incoming data
32.   process(clk)
33.   begin
34.     if rising_edge(clk) then
35.       cs        <= ns;
36.       rx_si_d1 <= rx_si;
37.       rx_si_d2 <= rx_si_d1;
38.     end if;
39.   end process;
```

下面看代码的第二部分，如 VHDL 代码 8-7 所示，描述了状态转移的过程。

VHDL 代码 8-7

```vhdl
1.  process(all)
2.  begin
3.    case cs is
4.      when rx_idle =>
5.        ns <= rx_start_bit when (not rx_si_d2) else rx_idle;
```

```
6.      when rx_start_bit =>
7.        if rx_cnt = HALF_CNT_MAX then
8.          ns <= rx_data_bit when (not rx_si_d2) else rx_idle;
9.        end if;
10.     when rx_data_bit =>
11.       if rx_cnt = CNT_MAX then
12.         ns <= rx_stop_bit when rx_bit_id ?= "111" else rx_data_bit;
13.       end if;
14.     when rx_stop_bit =>
15.       ns <= rx_cu when rx_cnt ?= CNT_MAX else rx_stop_bit;
16.     when rx_cu =>
17.       ns <= rx_idle;
18.     when others =>
19.       ns <= rx_idle;
20.   end case;
21. end process;
```

代码的第三部分如 VHDL 代码 8-8 所示。其中，第 1 行至第 36 行为状态输出部分。这里采用寄存器输出方式，对现态进行译码。不难看出，rx_dv 只持续一个时钟周期，与状态 rx_cu 相关，所以可以将其单独在一个 process 进程中描述，但采用次态 ns 控制，即当 ns 为 rx_idle 时，rx_dv 为 0；当 ns 为 rx_cu 时，rx_dv 为 1。还可以看到，在某些状态下，会有三个输出：rx_cnt、rx_bit_id 和 rx_dv；而在某些状态下，只有一个输出。这样是可以的，最终会形成带使能的触发器，而不会因为 case 语句不完备生成锁存器。代码第 38 行至第 45 行采用 rx_dv 为使能信号捕获 rx_byte_int，将其赋值给 rx_byte。

📄 **VHDL 代码 8-8**

```
1.  process(clk)
2.  begin
3.    if rising_edge(clk) then
4.      case cs is
5.        when rx_idle =>
6.          rx_cnt    <= (others => '0');
7.          rx_bit_id <= (others => '0');
8.          rx_dv     <= '0';
9.        when rx_start_bit =>
10.         rx_cnt <= (others => '0') when rx_cnt ?= HALF_CNT_MAX else
11.                   rx_cnt + 1;
12.       when rx_data_bit =>
13.         if rx_cnt = CNT_MAX then
14.           rx_cnt <= (others => '0');
15.           rx_bit_id <= (others => '0') when rx_bit_id ?= "111" else
16.                        rx_bit_id + 1;
17.           rx_byte_int(to_integer(rx_bit_id)) <= rx_si_d2;
18.         else
19.           rx_cnt <= rx_cnt + 1;
20.         end if;
21.       when rx_stop_bit =>
22.         if rx_cnt = CNT_MAX then
```

```
23.            rx_cnt <= (others => '0');
24.            rx_dv  <= '1';
25.         else
26.            rx_cnt <= rx_cnt + 1;
27.         end if;
28.      when rx_cu =>
29.         rx_dv <= '0';
30.      when others =>
31.         rx_cnt    <= (others => '0');
32.         rx_bit_id <= (others => '0');
33.         rx_dv     <= '0';
34.      end case;
35.   end if;
36. end process;
37.
38. process(clk)
39. begin
40.   if rising_edge(clk) then
41.     rx_byte <= std_logic_vector(rx_byte_int) when rx_dv else
42.                rx_byte;
43.   end if;
44. end process;
45. end architecture;
```

再看 UART 发送端。发送端要与接收端有相同的波特率和数据格式。这里要求发送端按照如图 8-14 所示的格式发送数据。状态转移图如图 8-17 所示，各状态的含义如表 8-5 所示。在此基础上可以形成如图 8-18 所示的时序图。

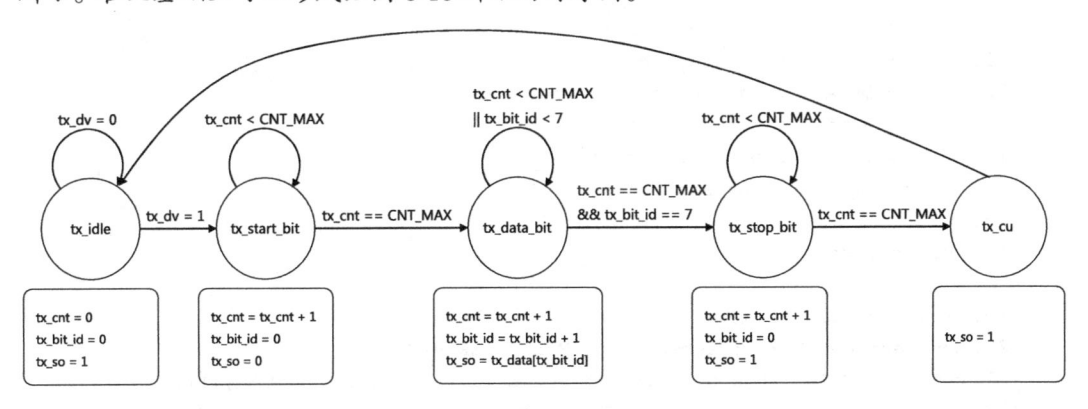

图 8-17

表 8-5

状 态	含 义
tx_idle	表明发送器处于空闲状态
tx_start_bit	表明发送器开始发送起始位
tx_data_bit	表明发生器正在发送数据位
tx_stop_bit	表明发送器开始发送终止位
tx_cu（Clean Up）	表明发送器处于复位状态

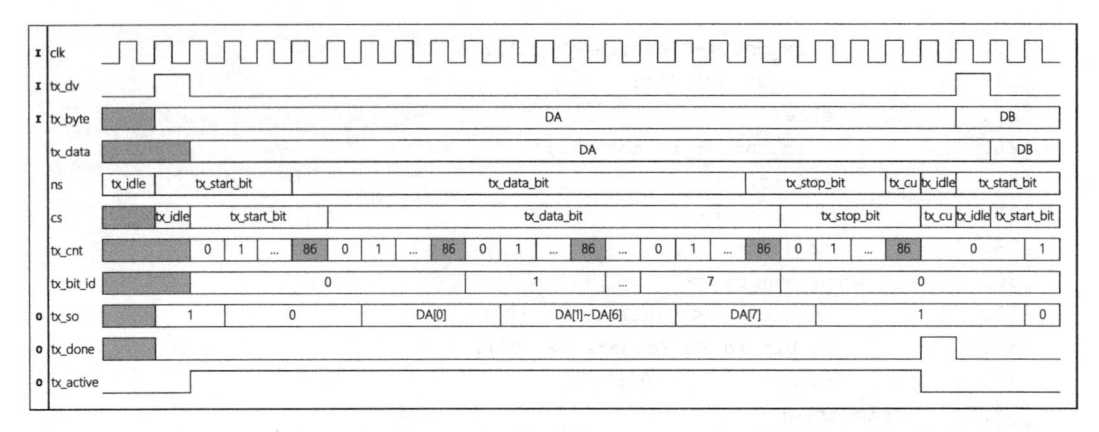

图 8-18

代码第一部分如 VHDL 代码 8-9 所示。代码第 26 行定义了状态寄存器的数据类型，代码第 27 行定义了现态和次态两个状态寄存器，代码第 32 行描述了状态发生器。

VHDL 代码 8-9

```
1.  --File: uart_tx.vhd
2.  library ieee;
3.  use ieee.std_logic_1164.all;
4.  use ieee.numeric_std.all;
5.  use ieee.math_real.all;
6.
7.  entity uart_tx is
8.    generic ( CYCLES_PER_BIT : positive := 87 );
9.    port (
10.     clk        : in std_logic;
11.     tx_dv      : in std_logic;
12.     tx_byte    : in std_logic_vector(7 downto 0);
13.     tx_active  : out std_logic;
14.     tx_so      : out std_logic;
15.     tx_done    : out std_logic
16.   );
17. end entity;
18.
19. architecture rtl of uart_tx is
20. constant CNT_MAXI : integer := CYCLES_PER_BIT - 1;
21. constant CNTW : positive := integer(ceil(log2(real(CNT_MAXI))));
22. constant CNT_MAX : unsigned(CNTW - 1 downto 0) := to_unsigned(CNT_MAXI, CNTW);
23. signal tx_cnt    : unsigned(CNTW - 1 downto 0);
24. signal tx_bit_id : unsigned(2 downto 0);
25. signal tx_data   : std_logic_vector(7 downto 0);
26. type state_t is (tx_idle, tx_start_bit, tx_data_bit, tx_stop_bit, tx_cu);
27. signal cs, ns    : state_t;
28. begin
29.   process(clk)
30.   begin
31.     if rising_edge(clk) then
```

```
32.        cs <= ns;
33.      end if;
34.   end process;
```

代码第二部分如 VHDL 代码 8-10 所示，描述了状态转移的过程。通过计数器 tx_cnt 保证每比特能够持续 87 个时钟周期，这里，CNT_MAX 为 86。

VHDL 代码 8-10

```
1.    process(all)
2.    begin
3.      ns <= tx_idle;
4.      case cs is
5.        when tx_idle =>
6.          ns <= tx_start_bit when tx_dv else tx_idle;
7.        when tx_start_bit =>
8.          ns <= tx_data_bit when tx_cnt ?= CNT_MAX else tx_start_bit;
9.        when tx_data_bit =>
10.         ns <= tx_stop_bit when (tx_cnt ?= CNT_MAX and tx_bit_id ?= "111") else
11.              tx_data_bit;
12.       when tx_stop_bit =>
13.         ns <= tx_cu when tx_cnt ?= CNT_MAX else tx_stop_bit;
14.       when tx_cu =>   ns <= tx_idle;
15.       when others => ns <= tx_idle;
16.     end case;
17.   end process;
```

代码第三部分如 VHDL 代码 8-11 所示，采用寄存器输出方式描述了输出译码逻辑。对于 tx_done，由于仅持续一个时钟周期，因此可将其在次态 ns 下译码，即当 ns 为 idle 时，tx_done 为 0；当 ns 为 tx_cu 时，tx_done 为 1。

VHDL 代码 8-11

```
1.    process(clk)
2.    begin
3.      if rising_edge(clk) then
4.        case cs is
5.          when tx_idle =>
6.            tx_so <= '1';
7.            tx_done <= '0';
8.            tx_cnt <= (others => '0');
9.            tx_bit_id <= (others => '0');
10.           if tx_dv then
11.             tx_data <= tx_byte;
12.             tx_active <= '1';
13.           end if;
14.         when tx_start_bit =>
15.           tx_so <= '0';
16.           tx_cnt <= (others => '0') when tx_cnt ?= CNT_MAX else
17.                     tx_cnt + 1;
18.         when tx_data_bit =>
19.           tx_so <= tx_data(to_integer(tx_bit_id));
```

```
20.            if tx_cnt = CNT_MAX then
21.               tx_cnt <= (others => '0');
22.               tx_bit_id <= (others => '0') when tx_bit_id ?= "111" else
23.                            tx_bit_id + 1;
24.            else
25.               tx_cnt <= tx_cnt + 1;
26.            end if;
27.         when tx_stop_bit =>
28.            tx_so <= '1';
29.            if tx_cnt = CNT_MAX then
30.               tx_cnt <= (others => '0');
31.               tx_done <= '1';
32.               tx_active <= '1';
33.            else
34.               tx_cnt <= tx_cnt + 1;
35.            end if;
36.         when tx_cu => tx_done <= '0';
37.         when others =>
38.            tx_done <= '0';
39.            tx_so   <= '1';
40.         end case;
41.      end if;
42.   end process;
43. end architecture;
```

结合 uart_rx.sv 和 uart_tx.sv 的相关代码,我们可以看到采用多进程描述状态机的好处:简洁直观,便于维护。此外,当某个输出仅与状态中的某个或某两个有关时,可单独将此输出提取出来,采用寄存器输出方式描述。

> 💡 设计规则 4:时序是设计出来的,不是凑出来的,更不是测出来的。

8.3 状态编码方式

常见的状态编码方式包括二进制码、独热码、格雷码和约翰逊码。以 VHDL 代码 8-6 为例,四种常用的编码方式如表 8-6 所示。对于独热码(one-hot)编码方式,可以看到,5 位二进制数只有其中一位为 1。对于格雷码编码方式,相邻两个枚举值只有一位不同。约翰逊编码方式不同于二进制码编码方式,N 位约翰逊计数器可以表示 $2N$ 种状态,二进制码计数器则可以表示 2^N 种状态。

表 8-6

状 态	二进制码	独 热 码	格 雷 码	约翰逊码
rx_idle	000	00001	000	000
rx_start_bit	001	00010	001	100
rx_data_bit	010	00100	011	110
rx_stop_bit	011	01000	010	111
rx_cu	100	10000	110	011

对于如 VHDL 代码 8-6 所示的 uart_rx.sv，分别采用这四种编码方式进行编译，最终布线后的性能对比如表 8-7 所示。其中，目标芯片为 xcvu3p-ffvc1517-1-i，Vivado 版本为 2021.2，时钟频率为 10MHz，可以看到，格雷码方式消耗的 LUT 最多，独热码方式消耗的 FF 最多，但 WNS 最大，这意味着可获得的 F_{max} 最高。

表 8-7

编码方式	LUT	FF	WNS	WHS	最大逻辑级数
二进制码	28	32	98.464	0.042	3
独热码	26	34	98.517	0.040	3
格雷码	29	32	97.750	0.034	3
约翰逊码	27	32	98.178	0.046	3

💡 **设计规则 5**：当状态个数小于或等于 5 时，建议采用二进制码编码方式；当状态个数大于 5 且小于或等于 50 时，建议采用独热码编码方式；当状态个数大于 50 时，建议采用格雷码编码方式。

Vivado 的综合选项设置页面针对状态机提供了一个选项-fsm_extraction，如图 8-19 所示。它有 6 个可选值，分别为 auto、off、one_hot（独热码）、sequential（二进制码）、johnson（约翰逊码）和 gray（格雷码），默认值为 auto，这意味着由工具决定最佳编码方式。

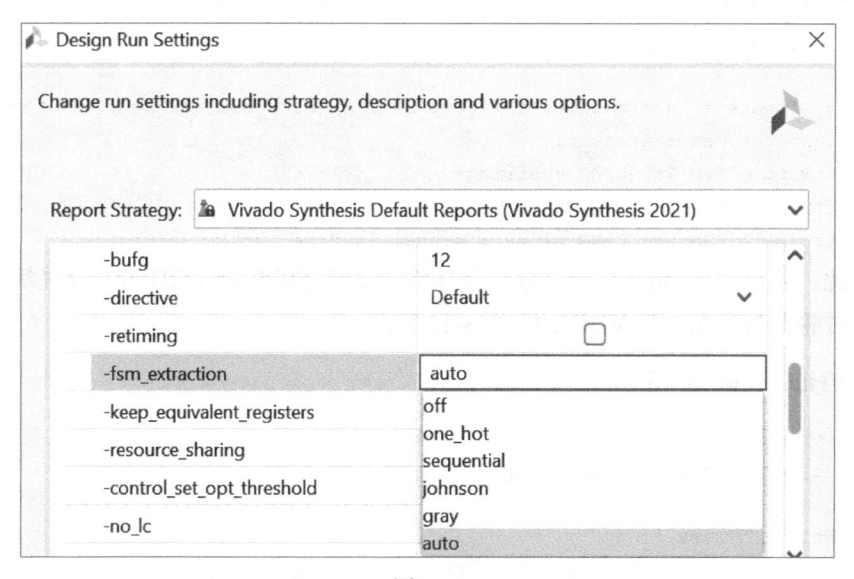

图 8-19

在综合选项设置中通过-fsm_extraction 设定的状态机编码方式的优先级高于 RTL 代码中设定的编码方式。仍以 VHDL 代码 8-6 为例，代码里采用的是二进制码编码方式，-fsm_extraction 采用默认值 auto。在综合日志文件中搜索 FSM 可查看到，实际使用的编码方式为独热码，如图 8-20 所示。

```
INFO: [Synth 8-802] inferred FSM for state register 'cs_reg' in module 'uart_rx'

         State  |      New Encoding  |    Previous Encoding
       rx_idle  |            00001   |            000
   rx_start_bit |            00010   |            001
    rx_data_bit |            00100   |            010
    rx_stop_bit |            01000   |            011
         rx_cu  |            10000   |            100

INFO: [Synth 8-3354] encoded FSM with state register 'cs_reg' using encoding 'one-hot' in module 'uart_rx'
```

图 8-20

Vivado 还提供了综合属性 FSM_ENCODING，共有 7 个可取值，分别为：one_hot（独热码）、sequential（二进制码）、johnson（约翰逊码）、gray（格雷码）、user_encoding（用户定义的编码方式）、none 和 auto。auto 意味着由工具决定最佳编码方式。user_encoding 意味着工具需要使用 RTL 代码中已有的编码方式。FSM_ENCODING 的使用方法如 VHDL 代码 8-12 所示（FSM_ENCODING 可全部小写）。就优先级而言，FSM_ENCODING 大于 -fsm_extraction，-fsm_extraction 大于 RTL 代码中已有的编码方式。在工程实践中，建议直接将-fsm_extraction 设置为 auto，由工具来决定状态机的最佳编码方式。如果设计有特殊需求，可以在 RTL 代码中通过 FSM_ENCODING 指定编码方式。

📄 **VHDL 代码 8-12**

```vhdl
1. type state_t is (rx_idle, rx_start_bit, rx_data_bit, rx_stop_bit, rx_cu);
2. signal cs, ns : state_t;
3. attribute FSM_ENCODING : string;
4. attribute FSM_ENCODING of cs : signal is "one_hot";
5. attribute FSM_ENCODING of ns : signal is "one_hot";
```

基于此，若选用 Vivado 综合，建议采用综合属性 FSM_ENCODING 设置状态机编码方式，同时将其参数化，如 VHDL 代码 8-13 所示。

📄 **VHDL 代码 8-13**

```vhdl
1.  library ieee;
2.  use ieee.std_logic_1164.all;
3.
4.  entity uart_rx is
5.    generic ( FSM_ENCODING_VAL : string := "one_hot" );
6.    port ()
7.  end entity;
8.
9.  architecture rtl of uart_rx is
10. type state_t is (rx_idle, rx_start_bit, rx_data_bit, rx_stop_bit, rx_cu);
11. signal cs, ns : state_t;
```

```
12. attribute FSM_ENCODING : string;
13. attribute FSM_ENCODING of cs : signal is FSM_ENCODING_VAL ;
14. attribute FSM_ENCODING of ns : signal is FSM_ENCODING_VAL ;
```

相比其他编码方案，二进制码编码方案和独热码编码方案更为常用，这里给出二者的优点和缺点，如表 8-8 所示。

表 8-8

优 缺 点	二进制码编码方案	独热码编码方案
优点	需要的触发器个数较少	次态译码电路和输出译码电路需要的组合逻辑较少
		状态信息使用单个触发器表示，因此易于时序收敛
		适用于 ECO（Engineering Change Order，工程设计变更），综合后，所有的状态触发器都会被保留，修改使用某种状态将会非常容易
缺点	次态译码电路和输出译码电路需要较多的组合逻辑	需要的触发器个数较多
	不适用于 ECO	

8.4 基于 ROM 的控制器

除了可以采用 case 语句描述次态译码逻辑和输出译码逻辑，还可以用 ROM 实现这些逻辑。ROM 的读地址由状态机的输入和现态拼接而成，而内部存储的数据由状态输出和次态拼接而成。这样就构成了{输入，现态}到{输出，次态}的映射关系，从 ROM 中读出的数据作为状态机的输出和次态，隐式地完成了状态输出和状态转移功能。

以"101"序列检测器为例，结合如图 8-9 所示的状态转移图，可得 ROM 的地址和数据之间的映射关系，如表 8-9 所示。其中，sin 和 cs 拼接构成 ROM 的地址，done 和 ns 拼接构成存储在 ROM 中的数据，相应的电路结构如图 8-21 所示。

表 8-9

地 址		数 据	
sin	cs	done	ns
0	00	0	00
0	01	0	10
0	10	0	00
0	11	0	00
1	00	0	01
1	01	0	01
1	10	1	01
1	11	0	00

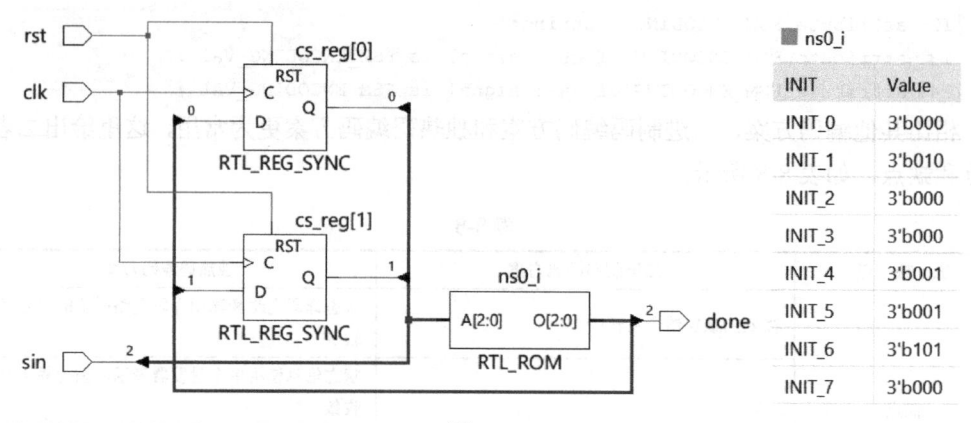

ns0_i	
INIT	Value
INIT_0	3'b000
INIT_1	3'b010
INIT_2	3'b000
INIT_3	3'b000
INIT_4	3'b001
INIT_5	3'b001
INIT_6	3'b101
INIT_7	3'b000

图 8-21

图 8-21 对应的 RTL 代码如 VHDL 代码 8-14 所示。结构体 architecture 中定义了三个函数。其中，函数 init_ram_bin 用于对 RAM 初始化；函数 enum2slv 用于将枚举类型转换为 std_logic_vector 类型；函数 slv2state 用于将 std_logic_vector 类型转换为枚举类型。代码第 18 行定义状态寄存器的数据类型，代码第 19 行定义了两个状态寄存器。代码第 58 行对第 20 行定义的存储单元进行初始化（这部分是可综合的），代码第 63 行至第 68 行描述的是状态发生器。代码第 60 行通过连续赋值语句实现{输入，现态}和{输出，次态}的映射，在这个过程中，需要执行枚举类型和 std_logic_vector 类型之间的相互转换。

VHDL 代码 8-14

```vhdl
1.  --File: mealy_detector_v4.vhd
2.  library ieee;
3.  use ieee.std_logic_1164.all;
4.  use ieee.numeric_std.all;
5.  use std.textio.all;
6.
7.  entity mealy_detector_v4 is
8.    port (
9.      clk  : in std_logic;
10.     rst  : in std_logic;
11.     sin  : in std_logic;
12.     done : out std_logic
13.   );
14. end entity;
15.
16. architecture rtl of mealy_detector_v4 is
17. type mem_t is array(0 to 7) of std_logic_vector(2 downto 0);
18. type state_t is (idle, got1, got10);
19. signal cs, ns : state_t;
20. signal mem    : mem_t;
21. signal ns_slv : std_logic_vector(1 downto 0);
22. signal cs_slv : std_logic_vector(1 downto 0);
23.
24. impure function init_ram_bin(ram_file_name : in string) return mem_t is
25.   file ram_file : text open read_mode is ram_file_name;
```

```
26.    variable ram_file_line : line;
27.    variable ram            : mem_t;
28. begin
29.    for i in mem_t'range loop
30.       readline(ram_file, ram_file_line);
31.       bread(ram_file_line, ram(i));
32.    end loop;
33.    return ram;
34. end function;
35.
36. function enum2slv(state : in state_t) return std_logic_vector is
37.    variable slv : std_logic_vector(1 downto 0);
38. begin
39.    case state is
40.       when idle  => return "00";
41.       when got1  => return "01";
42.       when got10 => return "10";
43.    end case;
44. end function;
45.
46. function slv2state(slv : in std_logic_vector(1 downto 0)) return state_t is
47.    variable state : state_t;
48. begin
49.    case slv is
50.       when "00"   => return idle;
51.       when "01"   => return got1;
52.       when "10"   => return got10;
53.       when others => return idle;
54.    end case;
55. end function;
56.
57. begin
58.    mem <= init_ram_bin("mealy.dat");
59.    cs_slv <= enum2slv(cs);
60.    (done, ns_slv) <= mem(to_integer(unsigned(sin & cs_slv)));
61.    ns <= slv2state(ns_slv);
62.
63.    process(clk)
64.    begin
65.       if rising_edge(clk) then
66.          cs <= idle when rst else ns;
67.       end if;
68.    end process;
69. end architecture;
```

在这种情况下，工具不会检测出状态机，因此也就不存在状态编码问题，换言之，代码第 18 行的枚举值 idle 为 0，got1 为 1，got10 为 2。

对于状态个数较多的场合，使用此方法可以有效减少组合逻辑，尤其是在对时钟频率要求较高的场合，采用此方法很有可能起到立竿见影的效果。

> 🔍 **VHDL-2008 新特性**
>
> VHDL-2008 支持通过聚合操作实现分散赋值，如 VHDL 代码 8-14 第 60 行所示。其结果是将从 mem 中读取的结果的最高位赋值给 done（done 位宽为 1），将其余位赋值给 ns_slv。一个典型的应用场景是进行加法运算时的"和"与"进位输出"可同时赋值，即：
>
>   signal a, b, sum : unsigned (7 downto 0) ;
>   signal cout : std_logic ;
>   ...
>   (cout, sum) <= ('0' & a) + ('0' & b) ;

8.5 思考空间

1. 试用状态机实现一个序列检测器，该序列检测器既可以检测序列"101"，又可以检测序列"110"。要求：

（1）给出状态转移图；（2）分别采用双进程和多进程方式描述。

2. 对于题 1 中采用双进程描述的状态机，试将状态编码方式参数化，并对比不同状态编码（二进制码、独热码和格雷码）下的资源消耗量和 F_{max}（时钟频率为 500MHz，目标芯片为 xcvu3p-ffvc1517-1-i）。

3. VHDL 代码 8-5 是针对 UltraScale/UltraScale+ FPGA 获取其 DNA，尝试将其改为针对 7 系列 FPGA 获取 DNA。

4. 在图 8-14 的基础上，若 UART 发送的数据增加一个奇偶校验位，试给出此时接收器和发送器的时序图、状态转移图，并在此基础上使用 VHDL 描述状态机。

5. 现有一个 4 位计数器，其计数序列依次为 0100,0001,1011,1010,0111,1111,0111,0000,1000。试用 VHDL 代码，采用双进程描述方式实现该计数器，并在此基础上，将其改为用 ROM 方式实现的控制器。

第 9 章

优化扇出

9.1 生成扇出报告

高扇出网线会增加布局布线的压力，很容易导致时序违例。这是因为在布局时过高的扇出使得工具很难将扇出的驱动（源端）与所有的负载（目的端）放置得比较紧凑，从而使有些负载距离驱动比较远，导致线延迟比较大。那么多高的扇出才算高扇出呢？对 7 系列 FPGA 而言，可依据表 9-1 进行判断。这是一个指导性的标准，在实际工程中需要灵活运用。对于 UltraScale/UltraScale+ FPGA，并没有给出这样的标准，而是需要根据时序进行判断。可能会出现扇出为十几或几十的情况，只要其成为时序违例的主要因素，就可尝试从降低扇出的角度优化时序。

表 9-1

时钟频率（MHz）	扇出>5000	扇出>200	扇出>100
Fclk≤125	如果逻辑级数小于 13，那么是可以接受的	—	—
125<Fclk≤250	如果出现时序违例，需要降低扇出或逻辑计数	逻辑级数小于 6 是可以接受的	—
Fclk>250	对绝大多数设计而言不合理	逻辑级数必须更低一些	逻辑级数低一些或复制逻辑，以降低扇出

Vivado 提供了一个非常好用的命令 report_high_fanout_nets，用于生成高扇出网线报告。该命令有诸多选项，表 9-2 给出了主要的几个选项。需要注意的是，-load_types、-clock_regions 和-slr 是两两互斥的，只可以使用其中的一个选项。

表 9-2

选 项	含 义
-timing	在生成报告中显示时序信息（WNS）
-load_types	在生成报告中显示负载类型
-clock_regions	显示负载在每个时钟区域的分布状况（仅适用于布局和布线后的设计）
-slr	显示负载在每个 SLR 的分布状况（仅适用于布局和布线后的设计）
-fanout_greater_than	仅显示扇出大于指定数值的网线
-fanout_lesser_than	仅显示扇出小于指定数值的网线

💡 **设计规则 1**：对综合后的设计就要开始进行扇出分析，以尽早发现高扇出的网线，并评估其可能对设计造成的影响。

report_high_fanout_nets 的具体用法如 Tcl 代码 9-1 所示。代码第 3 行的选项-load_types 生成的报告样例如图 9-1 所示。从此报告中可以看到网线 rectify_reset 的扇出为 10239，是由触发器驱动的（对应 Driver Type 列），其负载由两部分构成：触发器的复位/置位（对应 Set/Reset Slice 列）和 BRAM/DSP/OTHER 的复位/置位（对应 Set/Reset BRAM/DSP/OTHER 列）。如果需要进一步获取网线 rectify_reset 的所有负载以便分析，可先在图 9-1 的报告中选中该网线，再执行 Tcl 代码 9-2 即可。

Tcl 代码 9-1

```
1. #File: report_fanout.tcl
2. report_high_fanout_nets -name fanout1
3. report_high_fanout_nets -load_types -name fanout2
4. report_high_fanout_nets -clock_regions -name fanout3
5. report_high_fanout_nets -timing -name fanout4
```

Net Name	Fanout	Driver Type	Set/Reset Slice	Data & Other Slice	Set/Reset BRAM/DSP/OTHER
rectify_reset	10239	FDRE	10183	0	56
cpuEngine/or1200_cpu/or1200_ctrl/O17	1017	LUT2	0	1017	0
usbEngine0/usb_dma_wb_in/buffer_fifo/O5	912	LUT2	0	912	0
usbEngine1/usb_dma_wb_in/buffer_fifo/O5	912	LUT2	0	912	0

图 9-1

Tcl 代码 9-2

```
1. #File: get_load.tcl
2. set mynet [get_selected_objects]
3. set myload [get_cells -of [get_pins -leaf -of \
4.     [get_nets $mynet] -filter "DIRECTION==IN"]]
5. show_objects $myload -name myload
```

Tcl 代码 9-1 第 5 行选项-timing 生成的报告如图 9-2 所示，其中增加了 Worst Slack（ns）和 Worst Delay（ns）两个信息，这有助于判断该网线是否对时序收敛构成威胁。如果需要查看穿过报告中某个网线的时序路径的时序信息，如这里的 usbEngine1/u1/u3/O3，可先在此报告中选中该网线，再执行 Tcl 代码 9-3，即可生成对应的时序报告，如图 9-3 所示。可以看到，该时序报告中显示 Slack 为 3.998、Net Delay 为 5.562，与图 9-2 中的数据是吻合的。

Net Name	Fanout	Driver Type	Worst Slack(ns)	Worst Delay(ns)
usbEngine1/u1/u3/O3	560	FDRE	3.998	5.562
usbEngine0/u1/u3/O3	560	FDRE	3.134	6.202
n_0_reset_reg_reg_rep	525	FDRE	2.749	5.265
usbEngine0/n_0_csr0_reg[12]_i_2_11	512	LUT2	2.309	2.438

图 9-2

Tcl 代码 9-3

```
1. #File: get_timing_rpt.tcl
2. set mynet [get_selected_objects]
3. report_timing -through $mynet -nworst 10 \
4.     -setup -name timing_rpt
```

Name	Slack ^1	Levels	High Fan...	From	To	Total Delay	Logic Delay	Net Delay
∨ ☐ Constrained Paths (10)								
↳ Path 1	3.998	1	560	usbEngine..._rl_reg/C	usbEngine...eg[12]/D	5.864	0.302	5.562
↳ Path 2	3.998	1	560	usbEngine..._rl_reg/C	usbEngine...eg[12]/D	5.864	0.302	5.562
↳ Path 3	3.998	1	560	usbEngine..._rl_reg/C	usbEngine...eg[12]/D	5.864	0.302	5.562
↳ Path 4	3.998	1	560	usbEngine..._rl_reg/C	usbEngine...eg[12]/D	5.864	0.302	5.562
↳ Path 5	4.161	1	560	usbEngine..._rl_reg/C	usbEngine...eg[18]/D	5.670	0.302	5.368
↳ Path 6	4.161	1	560	usbEngine..._rl_reg/C	usbEngine...eg[18]/D	5.670	0.302	5.368
↳ Path 7	4.161	1	560	usbEngine..._rl_reg/C	usbEngine...eg[18]/D	5.670	0.302	5.368
↳ Path 8	4.161	1	560	usbEngine..._rl_reg/C	usbEngine...eg[18]/D	5.670	0.302	5.368
↳ Path 9	4.165	1	560	usbEngine..._rl_reg/C	usbEngine...eg[12]/D	5.666	0.302	5.364
↳ Path 10	4.165	1	560	usbEngine..._rl_reg/C	usbEngine...eg[12]/D	5.666	0.302	5.364

图 9-3

💡 **设计规则 2**：对综合后的设计进行扇出分析时，在扇出报告里要特别关注由 LUT 驱动的高扇出网线，尽管 Vivado 也会对 LUT 进行复制以降低扇出，但是毕竟 FPGA 芯片中 LUT 的个数远远小于 FF 的个数，且这种复制很难对时序有显著改善。因此，一旦在此阶段发现 LUT 驱动的网线扇出较高，就应从代码层面看是否可以将其修改为由触发器驱动。

9.2 利用设计流程降低扇出

Vivado 提供了不同粒度的扇出优化。在 opt_design 阶段，可以通过添加选项 -hier_fanout_limit 来降低扇出。对于每个层次内的高扇出网线，如果其扇出大于 -hier_fanout_limit 指定值，那么工具就会对该网线的驱动进行复制，并将复制版本的驱动放置在该网线所在的层级内。显然，这是一种模块级的粗粒度的优化方法。同时，在 opt_design 阶段，对于扇出大于 25000 的网线，如果设计中仍然有未用的全局时钟缓冲器（如 BUFG），那么工具会根据时序需求自动将该网线通过 BUFG 引入全局时钟网络。

在布局（place_design）阶段，对于扇出大于 1000 且建立时间裕量小于 2.0 的网线，如果该网线由寄存器驱动，那么工具会自动对寄存器进行复制，以降低扇出。对于扇出大于 10000 的网线，如果设计中仍有未用的全局时钟缓冲器（如 BUFG），那么工具也会自动将其通过 BUFG 引入全局时钟网络。

在布局后的物理优化（phys_opt_design）阶段，工具会根据时序裕量和布局信息自动对高扇出网线的驱动进行复制，通常会对时序有较大改善。这里需要注意的是确保设计中的高扇出网线是由寄存器驱动的，这样便于工具复制和重新布局。在某些情况下，工具并不会对所有关键的高扇出网线的驱动进行复制，此时可以尝试将 phys_opt_design 的 -directive 值修改为 Explore、AggressiveExplore 或 AggressiveFanoutOpt。如果在布线阶段发现某些网线因为扇出较高而成为时序收敛的瓶颈，那么可将该网线标注，在布线之前执行 Tcl 代码 9-4。选项-force_replication_on_nets 后跟的是高扇出网线的名称。在 Vivado Project 模式下，可将该脚本存储在一个.tcl 文件中，并在如图 9-4 所示的界面方框任选一个浏览到该文件即可。在 Vivado Non-Project 模式下，则可以将其放置在 route_design 命令之前，如 Tcl 代码 9-5 所示。

Tcl 代码 9-4

```
1. #File: force_replication.tcl
2. phys_opt_design -force_replication_on_nets [get_nets [list netA netB netC]]
```

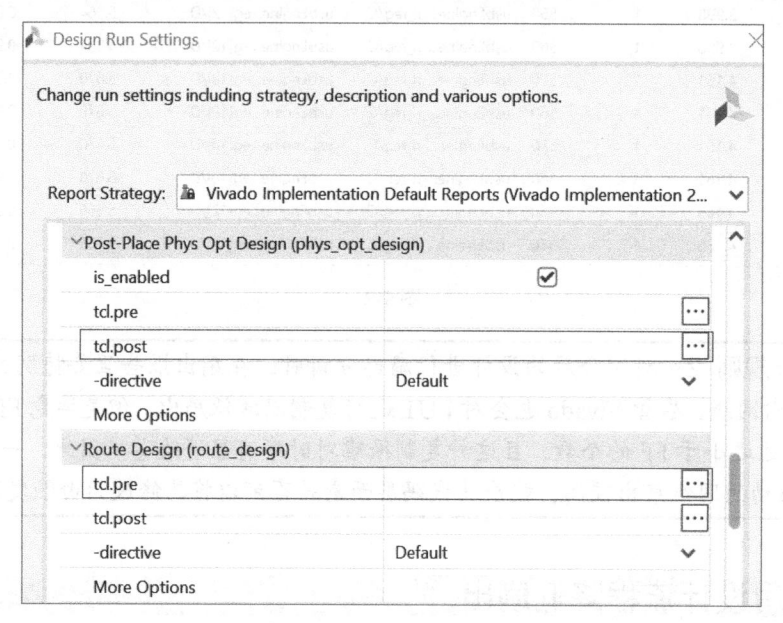

图 9-4

Tcl 代码 9-5

```
1. #File: non_prj_force_replication.tcl
2. opt_design -directive Explore
3. place_design -directive Explore
4. phys_opt_design -directive Explore
5. phys_opt_design -force_replication_on_nets [get_nets [list netA netB netC]]
6. route_design -directive Explore
```

对于 IP 中的高扇出网线,因为 IP 通常源代码不可见及 IP 本身的加密特性,所以采用 phys_opt_design 的选项-force_replication_on_nets 将非常有效。例如,对于 FFT IP 中的 ce_w2c,就可以通过此选项降低扇出,如图 9-5 所示。复制后的寄存器名称带有字符串 "rep"。

Net Name	Fanout	Driver Type	Clock Enable Slice	Clock Enable IO	Clock Enable BRAM/DSP/OTHER
design_1_i/xfft_0/U0/i_synth/axi_wrapper/ce_w2c	1811	FDRE	1299	0	465
design_1_i/xfft_0/U0/i_synth/axi_wrapper/ce_w2c_repN_20	251	FDRE	222	0	28
design_1_i/xfft_0/U0/i_synth/axi_wrapper/ce_w2c_repN_10	249	FDRE	235	0	13
design_1_i/xfft_0/U0/i_synth/axi_wrapper/ce_w2c_repN_9	244	FDRE	220	0	24
design_1_i/xfft_0/U0/i_synth/axi_wrapper/ce_w2c_repN_4	241	FDRE	228	0	12
design_1_i/xfft_0/U0/i_synth/axi_wrapper/ce_w2c_repN_5	236	FDRE	192	0	44
design_1_i/xfft_0/U0/i_synth/axi_wrapper/ce_w2c_repN_21	227	FDRE	224	0	0

图 9-5

9.3 利用约束降低扇出

在综合阶段，Vivado 提供了综合属性 MAX_FANOUT，该属性可指导 Vivado 在综合时对指定网线的驱动进行复制，但要求驱动是触发器或查找表。可以在 RTL 代码中使用该属性，如 VHDL 代码 9-1 所示，也可以在 XDC 约束文件中使用，如 Tcl 代码 9-6 所示，MAX_FANOUT 的施加对象可以是 get_cells 获取的单元，也可以是 get_nets 获取的网线。从工程实践的角度而言，建议采用如 Tcl 代码 9-6 所示的方式。不仅因为这样避免了对 RTL 代码的"污染"，而且这样更为精确。例如，如果只需要对 validForEgressFifo[9]对应的寄存器施加 MAX_FANOUT，可以直接通过 Tcl 命令获取该对象，但在 RTL 代码中，如果采用如 VHDL 代码 9-1 第 3 行所示的方式，那么实际上对 validForEgressFifo 的 14 根网线的驱动都施加了 MAX_FANOUT。

VHDL 代码 9-1

```
1.  --File: set_max_fanout.vhd
2.  signal dv : std_logic;
3.  signal validForEgressFifo: std_logic_vector(7 downto 0);
4.  attribute MAX_FANOUT : integer;
5.  attribute MAX_FANOUT of dv : signal is 5;
6.  attribute MAX_FANOUT of validForEgressFifo : signal is 5;
```

Tcl 代码 9-6

```
1.  #File: set_max_fanout.tcl
2.  set_property MAX_FANOUT 5 [get_cells {validForEgressFifo_reg[9]}]
3.  set_property MAX_FANOUT 5 [get_nets {validForEgressFifo[9]}]
```

> 💡 **设计规则 3**：建议仅对局部扇出相对较低的信号施加 MAX_FANOUT 属性，避免对全局高扇出网线（如全局复位信号）使用 MAX_FANOUT，否则会导致大量的寄存器复制，同时会增加控制集。

在使用 Tcl 代码 9-6 第 2 行之前，validForEgressFifo_reg[9]的扇出为 25，如图 9-6 所示（Elaborated Design）。在使用 Tcl 代码 9-6 第 2 行之后，其扇出降为 5，如图 9-7 所示。为了在 Schematic 视图中查看网线的扇出，只需要点击视图右上角的齿轮标记，在打开的界面中勾选上"Fanout For Scalar Pin"，如图 9-8 所示。

在使用 MAX_FANOUT 时，要注意其施加对象的驱动与负载是否在同一层次内，同时要注意综合设置选项-flatten_hierarchy 的值，有些场合，其值会导致 MAX_FANOUT 无法生效，如表 9-3 所示。

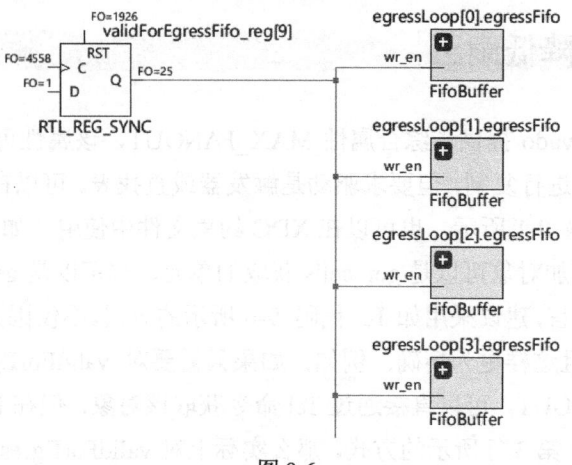

图 9-6

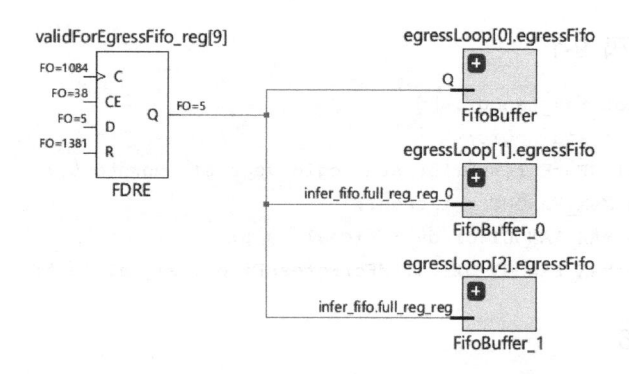

图 9-7

图 9-8

表 9-3

场景	-flatten_hierarchy	MAX_FANOUT 是否生效
MAX_FANOUT 施加对象与其负载在同一层次	none, rebuilt, full	生效
MAX_FANOUT 施加对象与其负载在不同层次	rebuilt, full	生效
MAX_FANOUT 施加对象与其负载在不同层次	none	无效

除了 -flatten_hierarchy 会导致 MAX_FANOUT 无法生效,还有一些 MAX_FANOUT 的无效场景,如表 9-4 所示。

表 9-4

MAX_FANOUT 无效场景	原　因
网线驱动的负载位于 EDIF、NGC 或 DCP 等网表文件内（网线位于网表文件之外）	网表文件被 Vivado 当作黑盒子，因此无法计算网线的负载
网线驱动的负载位于 IP 内（将 IP 以.xci 文件的形式添加到工程中）	IP 都被施加了属性 DONT_TOUCH
网线驱动的负载位于被施加了 KEEP_HIERARCHY 属性的模块	KEEP_HIERARCHY 若为 yes，则会保持层次，这样 Vivado 无法计算该网线在层次内部的扇出

在布局阶段，属性 FORCE_MAX_FANOUT 限制指定网线扇出的最大值，该属性通常结合另一个属性 MAX_FANOUT_MODE 使用。MAX_FANOUT_MODE 用于限定降低扇出的方式，其取值包括 CLOCK_REGION、SLR 和 MACRO（指 BRAM、UltraRAM 和 DSP）。以图 9-9 左侧为例，图中，FF 代表触发器。FF 作为驱动，其输出网线（名为 sig1）的扇出为 4。采用如 Tcl 代码 9-7 所示的方式，代码第 2 行限定网线扇出为 1，代码第 3 行限定该扇出值仅针对 MACRO，因此最终结果如图 9-9 右侧所示。

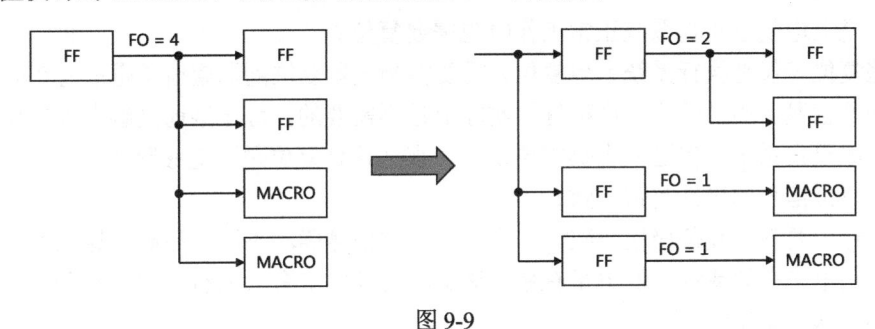

图 9-9

Tcl 代码 9-7

```
1.  #File: force_max_fanout.tcl
2.  set_property FORCE_MAX_FANOUT 1 [get_nets sig1]
3.  set_property MAX_FANOUT_MODE MACRO [get_nets sig1]
```

FORCE_MAX_FANOUT 和 MAX_FANOUT_MODE 是非常好用的降低扇出的方式，如对于 SSI 芯片，如果需要控制指定网线在每个 SLR 内的扇出，就可以将 MAX_FANOUT_MODE 值设置为 SLR。对于任何设计，如果需要确保关键位置的每个 BRAM 的读/写使能是由独立的触发器驱动的，就可以将 MAX_FANOUT_MODE 值设置为 MACRO。

对于高扇出网线，也可以将其引入全局时钟网络，只需要将其连接到全局时钟缓冲器上即可。这可以在 RTL 代码中通过实例化全局时钟缓冲器的方式实现，也可以在 XDC 约束文件中通过 CLOCK_BUFFER_TYPE 实现。从工程实践的角度而言，使用属性 CLOCK_BUFFER_TYPE 更为实用，因为这避免了对 RTL 代码的修改，同时如果不需要插入 BUFG，则只需要将其值设置为 NONE。具体使用方法如 Tcl 代码 9-8 所示。由于 Versal ACAP 提供了针对高扇出网线的全局时钟缓冲器 BUFG_FABRIC，因此相应的 CLOCK_BUFFER_TYPE 值应设置为 BUFG_FABRIC，如代码第 3 行所示。对于其他系列芯片，可将其设置为 BUFG，如代码第 2 行所示。

Tcl 代码 9-8

```
1.  #File: insert_bufg.tcl
2.  set_property CLOCK_BUFFER_TYPE BUFG [get_nets sig1]
3.  set_property CLOCK_BUFFER_TYPE BUFG_FABRIC [get_nets sig1]
```

9.4 从代码层面降低扇出

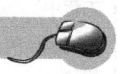

通常在设计中，除了时钟信号的扇出较高，全局复位信号的扇出也较高。时钟由于有全局时钟网络的支持，是完全允许高扇出的。因此，我们需要关注的是非时钟的高扇出信号。

首先，对于全局复位信号，需要从以下 3 个角度审视。

(1) 复位信号驱动的负载是否都是必须复位的？

例如，流水线寄存器是不需要复位的，只需要做好初始化就可以，因为旧数据总会被新数据"冲走"。

(2) 是上电复位还是系统正常工作时也需要复位？

有些复位仅仅是执行系统上电复位，需要将触发器复位到期望值（通常为 0），这实际上完成的是触发器的初始化。这种复位实际上是不需要的，可直接通过触发器初始化操作完成。但如果在系统工作过程中仍需要复位，那么这种复位是不能移除的。

(3) 复位信号的极性是否一致？

在设计中要保持复位极性一致，即当同一复位信号驱动不同负载时，这些负载的复位信号都是高电平有效或都是低电平有效，避免既有高电平有效又有低电平有效，这无形中增加了触发器控制集。

其次，对于局部控制信号，要从电路结构上管理好扇出，这在高速设计中尤为重要。例如，对于 BRAM 和 UltraRAM 的读/写地址端口和读/写使能端口，应尽可能使扇出低一些，且确保其驱动为触发器而非查找表。

最后，我们也可以在代码中通过手工复制寄存器的方式降低扇出。只是需要注意确保复制的寄存器不会被工具优化掉，因为复制的寄存器和原始寄存器被认为是等效寄存器。因此，可对这些寄存器添加综合属性 KEEP 或 DONT_TOUCH（建议直接在 RTL 代码中使用），也可使用模块化综合方式，即对这些触发器所在层次设置 KEEP_EQUIVALENT_REGISTER 属性，将其值设置为 1，如 Tcl 代码 9-9 所示。

Tcl 代码 9-9

```
1.  #File: equivalent_register.tcl
2.  set_property BLOCK_SYNTH.KEEP_EQUIVALENT_REGISTER 1 [get_cells fftEngine]
```

9.5 改善扇出的正确流程

降低扇出有多种方法，可以从设计流程的角度入手，也可以从约束属性的角度入手，还可以从代码层面入手。那么优先选择哪种方式呢？这里建议以设计流程为主，以约束属性为辅，如果两者均未奏效，就需要从代码层面进行人工干预了，形成的流程如图 9-10 所示。

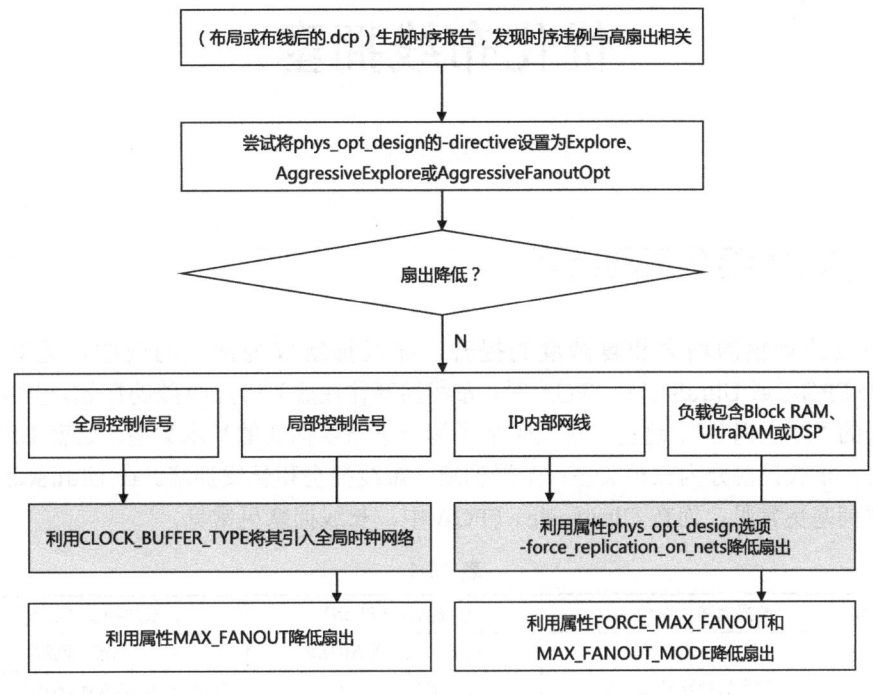

图 9-10

之所以将设计流程作为首选方案，是因为无论是布局还是布局后的物理优化，都是受时序需求驱动的，工具会据此决定是否对高扇出网线的驱动进行复制及复制多少份，这比人工复制更具有针对性。

9.6 思考空间

1. 某时序路径因为扇出过大导致线延迟过大而无法收敛，而该路径位于某 IP 内，如何才能降低扇出？

2. 某触发器输出连接到多个 BRAM 的写使能端口，如何保证这些 BRAM 写使能的驱动是相互独立的？

3. 为什么不能对全局复位信号使用属性 MAX_FANOUT 降低扇出？

4. 如何确保在 RTL 代码中复制的寄存器不会被工具优化掉？

5. MAX_FANOUT 和 FORCE_MAX_FANOUT 有何区别？

第 10 章

优化布线拥塞

10.1 布线拥塞的三种类型

随着设计规模的增大和复杂度的提升，布线拥塞成为常见的问题，尤其是在用 UltraScale FPGA 或 UltraScale+ FPGA 时，布线拥塞往往成为时序收敛的瓶颈，也成为编译时间过长的"罪魁祸首"。这里，我们先来了解一下布线拥塞的基本类型，如表 10-1 所示。总体而言，布线拥塞分为三种类型：全局拥塞、短线拥塞和长线拥塞。在 UltraScale FPGA 中，短线拥塞更常见；而在 UltraScale+ FPGA 中，长线拥塞更常见。

表 10-1

拥塞类型		全局拥塞（Global）	短线拥塞（Short）	长线拥塞（Long）
原因		过多的 LUT 整合	过多且集中的 MUXF	高扇出网线
		过多的触发器控制集	过多且集中的进位链	BRAM/UltraRAM/DSP 利用率过高
		不合理的手工布局		过重的互联
				过多的跨 die 网线
其他			在 UltraScale FPGA 中更常见	在 UltraScale+ FPGA 中更常见

了解拥塞的类型及其根本原因有助于我们找到解决拥塞的办法。那么，现在的问题就成为如何确定当前设计中的拥塞是何种拥塞？这就要利用拥塞报告了。拥塞报告由命令 report_design_analysis 生成，如 Tcl 代码 10-1 所示。代码第 1 行通过选项 -congestion 生成拥塞报告，第 2 行增加了选项 -complexity，这样既可以生成设计复杂度报告，也可以生成拥塞报告。建议在布局后生成拥塞报告。

Tcl 代码 10-1

```
1.  #File: get_cong_rpt.tcl
2.  report_design_analysis -congestion -name cong_rpt1
3.  report_design_analysis -complexity -timing -setup -max_paths 100 \
4.       -congestion -name cong_rpt2
```

拥塞报告由三部分构成。第一部分如图 10-1 所示，显示了拥塞的区域（Window 列）、类型（Type 列）和程度（Level 列）。我们重点关注的是后两个。通过 Type 列可查看拥塞类型，Global 对应全局拥塞，Long 对应长线拥塞，Short 对应短线拥塞。通过 Level 列可查看拥塞程度，该数值越大，拥塞越严重。当拥塞程度为 5 时，表明布线会遇到一些困难；当拥塞程度为 6 时，表明布线会遇到很大困难；当拥塞程度为 7 时，表明几乎无法布线。

Window	Direction	Type	Level
INT_X45Y314->INT_X68Y345 (CLEM_X45Y314->CLEL_R_X68Y345)	South	Global	5
INT_X19Y536->INT_X82Y599 (CLEL_R_X19Y536->CLEL_R_X82Y599)	East	Global	6
INT_X15Y531->INT_X78Y594 (CLEM_X15Y531->CLEM_X78Y594)	West	Global	6
INT_X92Y356->INT_X107Y387 (CLEM_X92Y356->CLEM_R_X107Y387)	North	Long	5
INT_X92Y332->INT_X107Y443 (CLEM_X92Y332->CLEM_R_X107Y443)	South	Long	5

图 10-1

拥塞报告的第二部分显示了每个拥塞区域的资源利用率情况，如图 10-2 所示。结合拥塞类型可进一步判断哪类资源可能会增加。

Combined LUTs	Avg LUT Input	LUT	LUTRAM	Flop	MUXF	RAMB	URAM	DSP	CARRY	SRL
0%	4.574	96%	0%	50%	0%	50%	0%	95%	48%	5%
0%	4.541	86%	0%	44%	0%	59%	0%	87%	39%	5%
0%	4.528	84%	0%	44%	0%	63%	0%	85%	40%	5%
16%	4.418	80%	18%	44%	0%	100%	NA	0%	2%	0%
14%	4.732	70%	10%	42%	0%	72%	NA	0%	7%	0%

图 10-2

拥塞报告的第三部分会显示前三个导致拥塞的设计单元，如图 10-3 所示。

Cell Names		
Top Cell 1	Top Cell 2	Top Cell 3
u_shasha_sys_die1/loop_shasha	u_shasha_sys_die1/loop_sha	u_pcie/pcie_i (0%)
u_shasha_sys_die1/loop_shasha	u_shasha_sys_die1/loop_sha	u_shasha_sys_die1/loop_shasha[4].u_shasha/u_sha256_loop2 (21%)
u_shasha_sys_die1/loop_shasha	u_shasha_sys_die1/loop_sha	u_shasha_sys_die1/loop_shasha[4].u_shasha/u_sha256_loop2 (17%)
u_pcie/pcie_i/pcie_xdma/inst/ud	u_pcie/pcie_i/pcie_xdma/inst	u_pcie/pcie_i (8%)
u_pcie/pcie_i/pcie_xdma/inst/ud	u_pcie/pcie_i/pcie_xdma/inst	u_pcie/pcie_i/pcie_xdma/inst/pcie4_ip_i/inst/gt_top_i/diablo_gt.diablo

图 10-3

有时，我们还需要借助设计复杂度报告查看模块之间的互联程度，如图 10-4 所示。在这个报告中，我们首先要关注的是 Rent 值（对应 Rent 列），当 Rent 值大于或等于 0.65、小于 0.85，且 Total Instances 列的对应值大于 15000 时，该模块与其他模块的互联程度就被认为比较重，而当 Rent 值大于 0.85 时，该模块与其他模块的互联程度就被认为过重，尤其是 Total Instances 列的对应值大于 25000 时。

Instance	Rent	Average Fanout	Total Instances
∨ N top	0.16	2.38	2191006
mb_bsp_top_inst0 (mb_bsp_top)	0.54	3.3	7919
u_kernel_axi_ctrl (kernel_axi_ctrl)	0.46	3.01	2763
u_pcie (pcie_wrapper)	0.41	3.2	109393
u_shasha_sys_die0 (shasha_sys_6kernel_die0)	0.33	2.34	730139
u_shasha_sys_die1 (shasha_sys_5kernel_die1)	0.13	2.34	608965
u_shasha_sys_die2 (shasha_sys_6kernel_die2)	0.3	2.34	730143

图 10-4

10.2 利用设计流程改善布线拥塞

针对布线拥塞，Vivado 在综合和布局布线阶段均提供了相应的策略。在综合阶段，策略 Flow_AlternateRoutability 可用于改善布线拥塞，如图 10-5 所示。

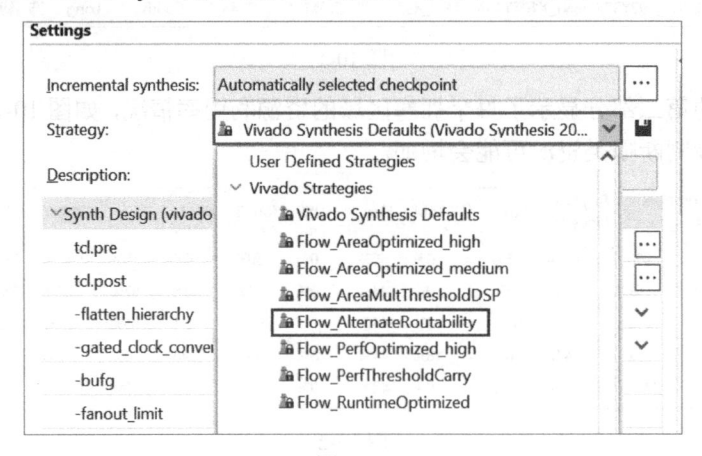

图 10-5

一旦将综合策略选为 Flow_AlternateRoutability，相应的 Settings 里的其他一些选项也会随之变化，如图 10-6 所示。-directive 变为 AlternateRoutability，-no_lc 被勾选，这意味着不再出现 LUT 整合；-shreg_min_size 值更新为 10，意味着深度小于 10 的移位寄存器将被综合为级联触发器，只有当深度大于或等于 10 时，工具才会将其映射为 LUT。

图 10-6

此外，还可以采用更细粒度化的综合策略设置，这是因为 Vivado 支持模块化综合。如果已经探明某个模块是构成布线拥塞的"罪魁祸首"，那么就可以针对该模块设置综合策略，如 Tcl 代码 10-2 所示。这种方法相比全局设置是有好处的：全局设置 Flow_AlternateRoutability 意味着所有模块都无法使用 LUT 整合，这样 LUT 资源利用率就可能上升；这也意味着所有模块中移位寄存器深度小于 10 的都无法映射为 LUT，这样触发器的利用率就可能上升；局部设置更精准地确定了布线拥塞的模块，使得其他模块可以继续使用 LUT 整合和基于 LUT 的移位寄存器。

Tcl 代码 10-2

```
1. #File: set_block_routability.tcl
2. set_property BLOCK_SYNTH.STRATEGY {ALTERNATE_ROUTABILITY} [get_cells usbEngine]
```

从布局布线的角度而言，Vivado 也提供了相应的策略以缓解布线拥塞，如图 10-7 中用方框标记的策略。使用这些策略时需要注意，带有 SSI 字符串的策略仅仅适用于 SSI 器件，不可将其应用于单 die 芯片。

对于 UltraScale FPGA 中出现的布线拥塞，图 10-7 中以 Congestion 打头的几种策略更为适用。而对于 UltraScale+ FPGA 中出现的布线拥塞，图 10-8 中包含 NetDelay 字符串的两种策略更为适用。

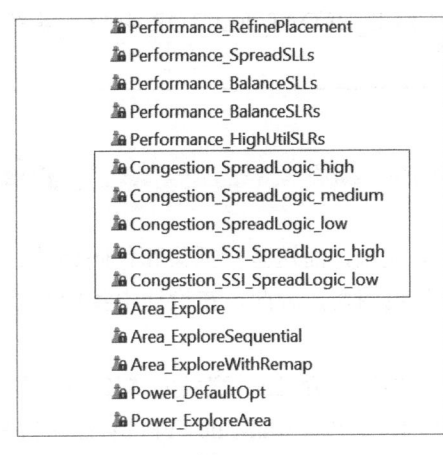

Performance_RefinePlacement
Performance_SpreadSLLs
Performance_BalanceSLLs
Performance_BalanceSLRs
Performance_HighUtilSLRs
Congestion_SpreadLogic_high
Congestion_SpreadLogic_medium
Congestion_SpreadLogic_low
Congestion_SSI_SpreadLogic_high
Congestion_SSI_SpreadLogic_low
Area_Explore
Area_ExploreSequential
Area_ExploreWithRemap
Power_DefaultOpt
Power_ExploreArea

图 10-7

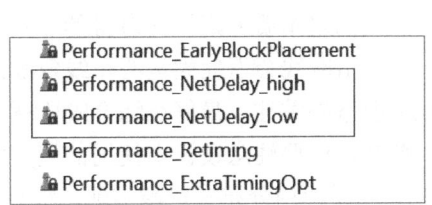

Performance_EarlyBlockPlacement
Performance_NetDelay_high
Performance_NetDelay_low
Performance_Retiming
Performance_ExtraTimingOpt

图 10-8

10.3 利用约束缓解布线拥塞

在 10.1 节，我们介绍了拥塞的类型及其根本原因。一旦明确这两点，结合拥塞报告找到拥塞模块，我们就可以对症下药，有的放矢了。例如，全局拥塞与 LUT 整合和触发器控制集有关，我们可以通过模块化综合技术针对拥塞模块设置这两个约束条件，如 Tcl 代码 10-3 所示。

Tcl 代码 10-3

```
1. #File: set_block_synth.tcl
2. set_property BLOCK_SYNTH.LUT_COMBINING 0 [get_cells usbEngine]
3. set_property BLOCK_SYNTH.CONTROL_SET_THRESHOLD 16 [get_cells fftEngine]
```

短线拥塞与 MUXF 有关，可以采用 Tcl 代码 10-4 阻止工具推断出 MUXF，用 LUT 取而代之。

Tcl 代码 10-4

```
1. #File: set_no_muxf.tcl
2. set_property BLOCK_SYNTH.MUXF_MAPPING 0 [get_cells usbEngine]
```

还可以采用属性 CELL_BLOAT_FACTOR 缓解布线拥塞。该属性有三个可选值：LOW、MEDIUM 和 HIGH，其使用方法如 Tcl 代码 10-5 所示。如果布线拥塞区域较小，就可以将该属性施加在位于拥塞区域内的主要模块上。该属性在布局阶段生效，通过增大模块内单元间的间距来降低拥塞程度。

Tcl 代码 10-5

```
1. #File: set_bloat_factor.tcl
2. set_property CELL_BLOAT_FACTOR high [get_cells cpuEngine]
```

💡 **设计规则**：如果设计已经消耗了大量的布线资源，那么不建议使用属性 CELL_BLOAT_FACTOR。对于较大的模块，使用 CELL_BLOAT_FACTOR 会使工具将子模块放得较远（子模块间距较大）。

10.4 从代码层面降低布线拥塞程度

布线拥塞本质上跟设计自身的一些特征有密切的关系，而这些特征又和代码相关。例如，高扇出网线会消耗大量的布线资源，容易引起长线拥塞，因此，在设计中我们对这类网线就要格外关注，最好在综合阶段就能发现。

此外，还要注意设计中使用的较大的存储单元，如深度为 16、宽度为 512 的单端口 RAM，如果使用分布式资源实现，将会消耗 512 个 LUT。以 UltraScale+ FPGA 芯片为例，每个 SLICE 里有 8 个 LUT，因此这 512 个 LUT 至少要占用 64 个 SLICEM，也意味着 512 个输出对应 512 根网线。从 SLICEM 入口处看，端口密度会比较大，容易引发短线拥塞，而 512 根网线又容易引发长线拥塞。如果使用 BRAM 实现，将会消耗 7 个 36Kb 的 BRAM 和 1 个 18Kb 的 BRAM。由于每个 BRAM 可提供 72 位数据，所以相比分布式 RAM，BRAM 输出的 512 根网线更为集中。因此，从布线拥塞的角度而言，对于大位宽的存储单元，如果运行在较高的时钟频率下（时钟频率大于 300MHz），建议采用 BRAM 实现。

对于基于 SSI 器件的设计，要做好早期设计规划，使得跨 die 路径尽可能少，因为过多的跨 die 路径容易导致长线拥塞。

10.5 缓解布线拥塞的正确流程

我们可以从设计流程、约束和代码层面三个维度缓解布线拥塞，但谁对缓解拥塞的效果最明显，或者说我们应该优选哪种方法呢？好在 Vivado 提供了一个非常好用的命令 report_qor_suggestions，就布线拥塞而言，建议针对布局生成的 .dcp 执行该命令。如果设计出现了布线拥塞，该命令通常会生成一些可缓解拥塞的建议，如图 10-9 所示，需要关注图中方框标记的内容。我们可以直接将这些建议应用于当前设计中，查看是否可以缓解布线拥塞，这正是利用了 Vivado 的智能化，从而减轻了对工程师先验知识的要求。

ID		GENERATED_AT	APPLICABLE_FOR	AUTOMATIC	Incremental Friendly	DESCRIPTION
∨ Utilization	☑					
RQS_UTIL-1-1	☑	route_design	synth_design	No	No	High utilization of certain types of cells
RQS_UTIL-3-1	☑	route_design	place_design	No	No	High utilization of certain types of cells
RQS_UTIL-2-1	☑	route_design	synth_design	No	No	High utilization of certain types of cells
RQS_UTIL-12-1	☑	route_design	opt_design	Yes	No	Remapping small SRL to Registers in th
∨ Congestion	☑					
RQS_CONG-1_1-1	☑	route_design	place_design	Yes	No	Congestion due to high CARRY usage

图 10-9

鉴于此，我们给出的缓解布线拥塞的流程也是从生成拥塞报告开始的，如图 10-10 所示。

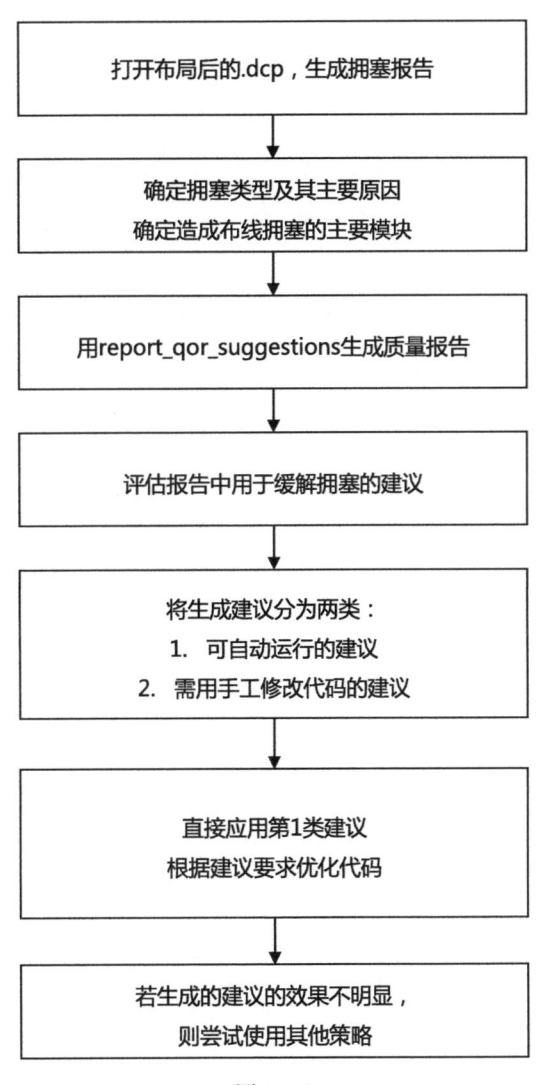

```
┌─────────────────────────────┐
│ 打开布局后的.dcp，生成拥塞报告 │
└─────────────────────────────┘
              │
              ▼
┌─────────────────────────────┐
│ 确定拥塞类型及其主要原因       │
│ 确定造成布线拥塞的主要模块     │
└─────────────────────────────┘
              │
              ▼
┌─────────────────────────────┐
│ 用report_qor_suggestions生成质量报告 │
└─────────────────────────────┘
              │
              ▼
┌─────────────────────────────┐
│ 评估报告中用于缓解拥塞的建议   │
└─────────────────────────────┘
              │
              ▼
┌─────────────────────────────┐
│ 将生成建议分为两类：           │
│ 1. 可自动运行的建议           │
│ 2. 需用手工修改代码的建议     │
└─────────────────────────────┘
              │
              ▼
┌─────────────────────────────┐
│ 直接应用第1类建议             │
│ 根据建议要求优化代码           │
└─────────────────────────────┘
              │
              ▼
┌─────────────────────────────┐
│ 若生成的建议的效果不明显，     │
│ 则尝试使用其他策略             │
└─────────────────────────────┘
```

图 10-10

10.6 思考空间

1. 试阐述布线拥塞的三种类型及其根本原因。
2. 试给出生成拥塞报告的 Tcl 命令。
3. 试给出设计复杂度较高的评判标准。
4. 试给出布线拥塞程度较高的评判标准。
5. 试给出可用于缓解布线拥塞的策略。

反侵权盗版声明

　　电子工业出版社依法对本作品享有专有出版权。任何未经权利人书面许可，复制、销售或通过信息网络传播本作品的行为；歪曲、篡改、剽窃本作品的行为，均违反《中华人民共和国著作权法》，其行为人应承担相应的民事责任和行政责任，构成犯罪的，将被依法追究刑事责任。

　　为了维护市场秩序，保护权利人的合法权益，我社将依法查处和打击侵权盗版的单位和个人。欢迎社会各界人士积极举报侵权盗版行为，本社将奖励举报有功人员，并保证举报人的信息不被泄露。

举报电话：（010）88254396；（010）88258888
传　　真：（010）88254397
E-mail：dbqq@phei.com.cn
通信地址：北京市万寿路173信箱
　　　　　电子工业出版社总编办公室
邮　　编：100036